Phytopathogenic Procaryotes and Plant Diseases

About the Author

Dr B.S. Thind

The author has 35 years' experience in research, teaching and transfer of technology in Plant Pathology. Retired as Professor and Head of the Department of Plant Pathology, Punjab Agricultural University, Ludhiana, India in 2003 and then served there as Emeritus Scientist (ICAR, New Delhi, India) from July 2003-July 2005. Taught various courses in plant pathology and plant bacteriology, including advanced courses in plant bacteriology, and guided 10 Ph. D. and 13 M. Sc. students in the field of plant bacteriology. Published more than 125 research papers in reputed national and international research journals.

Life member of five professional societies and held different positions in the executive committees of these societies. Attended the VII International Conference on Plant Pathogenic Bacteria held in Budapest, Hungary in 1989 and the VIII International Conference on Plant Pathogenic Bacteria held in Versailles, France in 1992. As Member of the National Organizing Committee, organized the IX International Conference on Plant Pathogenic Bacteria held in Chennai, India in 1996 and chaired one session in the conference. Chaired/co-chaired various sessions of different professional societies, the Jeersannidhi Lecture Award and acted as member of panel of judges for Prof. M.J. Narasimhan Academic Award Contest. Attended various national/international symposia, conferences, meetings and workshops of different professional societies

Acted as paper setter and external examiner of four universities and Agricultural Scientists' Recruitment Board, New Delhi. Acted as an expert of selection committees for different posts and as an external examiner of 17 Ph.D. and 10 M.Sc. students of different universities. Acted as reviewer/referee of various National and International Research Journals.

Phytopathogenic Procaryotes and Plant Diseases

B. S. Thind

Published by:
Scientific Publishers
5 A, New Pali Road, P.O. Box 91
Jodhpur 342 001 (India)

E-mail: info@scientificpub.com
Website: www.scientificpub.com

ISBN: 978-81-7233-701-8 (HB)
978-81-7233-780-3 (PB)

Typeset by Rajesh Ojha
Printed in India

Dedicated
to the memory of my beloved
parents, who, in spite of being uneducated,
always inspired me and
helped me in every possible
way to give me the most and
best education possible.

CONTENTS

ABBREVIATIONS OF BACTERIAL GENERA USED IN THE BOOK

Genus	Abbreviation	Genus	Abbreviation
Acetobacter	*Ace.*	*Neisseria*	*Ne.*
Acidovorax	*Aci.*	*Nocardia*	*No.*
Agrobacterium	*Ag.*	*Pantoea*	*Pa.*
Arthrobacter	*Ar.*	*Pectobacterium*	*Pe.*
Bacillus	*Ba.*	*Phytomonas*	*Ph.*
Bacterium	*Bac.*	*Pseudoalteromonas*	*Psa.*
Bradyrhizobium	*Bra.*	*Pseudomonas*	*Psm.*
Brenneria	*Br.*	*Pyrococcus*	*Pyc.*
Burkholderia	*Bu.*	*Pyrodictium*	*Pyd.*
'*Candidatus* Liberibacter'	'*Ca.* Li.'	*Ralstonia*	*Ral.*
'*Candidatus* Phlomobacter'	'*Ca.* Phl.'	*Rathayibacter*	*Rat.*
'*Candidatus* Phytoplasma'	'*Ca.* Phy.'	*Rhizobacter*	*Rhi.*
Clavibacter	*Cla.*	*Rhizobium*	*Rhiz.*
Clostridium	*Clo.*	*Rhodococcus*	*Rho.*
Corynebacterium	*Cor.*	*Saccharomyces*	*Sac.*
Curtobacterium	*Cur.*	*Salmonella*	*Sal.*
Dickeya	*D.*	*Samsonia*	*Sam.*
Enterobacter	*En.*	*Serratia*	*Se.*
Erwinia	*Er.*	*Sphingomonas*	*Sph.*
Escherichia	*Es.*	*Spiroplasma*	*Sp.*
Gluconobacter	*G.*	*Staphylococcus*	*Sta.*
Haemophilus	*Ha.*	*Streptococcus*	*Stc.*
Herbaspirillum	*He.*	*Streptomyces*	*St.*
Janibacter	*Jani.*	*Sulfolobus*	*Sul.*
Janthinobacterium	*Jant.*	*Thermoplasma*	*Thp.*
Leifsonia	*L.*	*Thermus*	*Th.*
Listeria	*Li.*	*Thiobacillus*	*Thb.*
Methanobacterium	*Meb.*	*Xanthomonas*	*Xa.*
Methanopyrus	*Mep.*	*Xylella*	*Xyl.*
Methanospirillium	*Mes.*	*Xylophilus*	*Xylo.*
Methanothermus	*Met.*		

PREFACE

Phytobacteriology or plant bacteriology is a subdiscipline of plant pathology, which deals with plant-associated procaryotes and their interactions with each other and with their hosts. Phytobacteriology is relatively a younger subdiscipline of plant pathology but it is equally important as many plant diseases caused by phytopathogenic bacteria/procaryotes cause enormous losses to the growers. As among the procaryotes, only the bacteria cause plant diseases, the terms 'phytopathogenic bacteria' and 'phytopathogenic procaryotes' are used interchangeably in this book. Some of these diseases such as, bacterial wilt of solanaceous plants, soft rot of fleshy vegetables and fruits, crown gall of plants, citrus canker, citrus huanglongbing, fire blight of pome fruit trees and bacterial blight of rice are of international importance. Plant diseases caused by phytopathogenic bacteria are also disastrous because many of these diseases cannot be effectively controlled due to lack of sound control measures.

In contrast to direct monetary losses and other harmful effects of phytopathogenic bacteria, they have a great deal of commercial value. They are used for the commercial production of many industrial products as well as tools for various scientific studies. Their use in industrial fermentation, production of antibiotics, xanthan gums and polysaccharides like curdlans are the important applications. Production of pectin-degrading enzymes produced by soft-rotting erwinias and restriction nucleases produced by *Agrobacterium* and *Xanthomonas* species are other notable examples of their commercial use. *Gluconobacter* species have various biotechnological applications. Their biotransformed strains are used for the production of rare and special sugars like L-ribose and miglitol, both very promising pharmaceutical lead molecules.

The study of bacteria is also very important because much needed basic data for the benefit of mankind have been obtained with the study of bacteria. *Escherichia coli*, the enteric bacterium, has formed the base for physiological and genetical investigations, while the plant pathogen, *Ag. tumefaciens,* is one of the best known vehicles for gene transfer and most useful genetic engineering tool for developing transgenics. The genome sequencing of many bacteria, including phytopathogens, has also provided the understanding of various life processes and their regulation.

During the last 3 decades, a lot of new information has been published on several aspects of phytopathogenic procaryotes such as, classification, genomics and molecular biology and role of avirulence and hypersensitive response & pathogenicity genes. Rapid advances in DNA sequencing technology have led to a major change in the way the procaryotes are classified. Sequence analysis of highly conserved regions of the bacterial genome such as, small subunit rRNA gene, has provided very valuable information concerning numerous taxonomic changes and realignment of different taxonomic groups. Since the publication of 1st edition of *Bergey's Manual of Systematic Bacteriology* in 1984, the number of published bacterial species has more than tripled, i.e., 390 new genera and 2200 new species have been described. During 1970s, the number of plant pat-hogenic bacterial genera was only six, which now stands at 40, including 3 candidatus categories. This clearly indicates that enormous and elaborate information has been published on the classification of plant pathogenic bacteria. The discovery of many new avirulence (*avr*) genes in plant pathogenic bacteria and their corresponding resistant (*R*) genes in host, and their interaction have greatly facilitated the understanding of mechanisms involved in susceptibility and resistance of host plants and host specificity. A significant progress has also been made in controlling plant diseases caused by phytopathogenic procaryotes using biocontrol agents and bacteriophages. Commercial formulations of many biocontrol agents including bacteriophages are available in the market, which are used at field scale.

There are 13 chapters included in the book. Two chapters (chapters 2 and 3) contain basic information about bacteria. Ten chapters (chapters 1 and 4 to 12) contain general information about phytobacteriology that is necessary for devising effective and long lasting control measures for the control bacterial plant diseases. Chapter 13 contains description of specific bacterial plant diseases. It was not possible to include all the bacterial plant diseases but 52 specific diseases have been described. These diseases include economically important diseases of international importance and representative diseases of different crops and regions. In order to make the phytobacteriology alive, coloured photographs of disease symptoms are given. The latest information published in plant pathological journals and related biological journals have been included in the book. In the beginning of each chapter, a fairly detailed and well-organized account of contents is given, which can serve as an outline of the chapter for the students and the teachers.

The abbreviations of names of bacterial genera, included in the book, are given on page xi. In this case, in some genera the deviation has been made from the tradition of abbreviating the genus name by using only the first letter of the genus, because there is more than one genus included in the book, which start with the same letter. The up-to-date classification of plant pathogenic bacteria is given in chapter 5. The names classified till 2007, has been updated on the basis of information published in International Journal of Systematic Bacteriology (IJSB), International Journal of Systematic and Evolutionary Microbiology (IJSEM) and related journals, and supplemented with the information given in "Names of Plant Pathogenic Bacteria, 1864 – 2004" prepared by Young *et al.* (2004) and "Comprehensive List of Names of Plant Pathogenic Bacteria, 1980 – 2007" prepared by Bull *et al.* (2010). The names classified after 2007 till November 2010 has been updated on the basis of information published in IJSB, IJSEM and related journals. In chapter 13, the names of causal agents of different diseases are given along with author citations. In author citations, the names of all the authors are mentioned when there are only two authors and in case of more than two authors, the name of the first author is mentioned followed by *et al.* One or two and in a few cases up to three recent synonyms, of each causal organism are also given. With a few exceptions, all the references cited in the text are given at the end of a chapter or description of a disease. The tradition of giving all the references in the end of the book must have avoided the repetition of some references but I have preferred this arrangement to the other. In the references cited, the names of all the authors are given when the reference contains only three authors. In case of more than three authors in a reference, the names of first three authors are given followed by *et al.* However, in case of a book, the names of all the authors/editors are given. The references of author citations of the causal organisms are not given. The coloured photographs of symptoms of plant diseases, i.e., Figs. 13.1a to 13.52 are given at the end of the book.

At the undergraduate level, the students have a limited exposure of phytobacteriology. However, at the postgraduate level, the students are taught phytobacteriology in detail. The state agricultural universities and the universities teaching biological sciences have courses dealing exclusively with phytobacteriology. Even the Ph.D. students specializing in phytobacteriology are taught advance courses. This book will serve as a text book for undergraduate and postgraduate students and also as a reference book for research workers engaged in the discipline of plant pathology and phytobacteriology.

I am indebted to numerous colleagues and friends for suggestions in the preparation of the manuscript of this book and for providing me photographs or the permission to use the images of plant disease symptoms. Their names are mentioned along with the photographs supplied by them. In the preparation of this manuscript, I have referred to a number of publications. I would like to acknowledge all the authors of these excellent and valuable studies.

There might be some errors, mistakes and shortcomings in this publication. I will greatly appreciate the healthy criticism and suggestions for the improvement of this publication in future.

B.S. Thind

November 2010

ACKNOWLEDGEMENTS

It is my pleasure to acknowledge the help of all those who contributed in the preparation of this publication. At the outset, I am extremely thankful to the Human Resource Unit, Department of Science and Technology, Ministry of Science and Technology, New Delhi for funding the project entitled, "Phytopathogenic Procaryotes and Plant Diseases" (project No. HR/UR 25/2005).

I am highly thankful to my friend and ex-colleague, Dr R.S. Kahlon, formerly Professor-cum-Head, Department of Microbiology, Punjab Agricultural University (PAU), Ludhiana for going through two chapters entitled, "The procaryotic cell" and "Growth and nutrition of bacteria" and for making highly valuable suggestions. My sincere gratitude is to my ex-colleagues, Dr Kuldip Singh, Senior Molecular Geneticist and Dr R.G. Saini, formerly Sr. Geneticist-cum-Head, Department of Genetics and Biotechnology, PAU, Ludhiana for their critical appraisal of chapters entitled, "Genome evolution of phytopathogenic procaryotes" and "Host-pathogen relationship", respectively. I also feel highly obliged to Dr R. Sridhar, ex-Head, Department of Plant Pathology, Central Rice Research Institute, Cuttack for his highly valuable suggestions concerning the description of bacterial blight and bacterial leaf streak of rice. I am also indebted to Dr A.K. Sood, formerly Professor and Head, Department of Plant Pathology, Chaudhary Sarwan Kumar Himachal Pradesh Krishi Vishvavidyalaya, Palampur and Dr Ram Kishun, ex-Principal Scientist and Head, Crop Protection Division, Central Institute for Subtropical Horticulture, Lucknow for going through the descriptions of bacterial wilt of solanaceous plants and bacterial canker of mango, respectively. The frequent counselling and help rendered by Dr G.S. Miglani, formerly Professor of Genetics and Dr G.S. Dhaliwal, formerly Professor of Entomology, PAU, Ludhiana and Dr Kuldip Singh is gratefully acknowledged. I am also indebted to Dr J.M. Young, Landcare Research, Auckland, New Zealand and Dr N.W. Schaad, ARS-USDA, Foreign Disease-Weed Science Research Unit, Ft. Detrick, MD, USA for their expert advice and frequent counselling. My sincere thanks are also due to Dr N.W. Schaad, Professor Ian K. Toth, The James Hutton Institute Invergowrie, Dundee, DD2 5DA, UK, Dr P.S. Randhawa, President, California Plant and Seed Labs, USA, Dr Indu Jalali, formerly Sr. Scientist (Plant Pathology), CCS Haryana

Agricultural University, Hisar and Dr R. Sridhar for their consents to act as referees/reviewers for this book.

The generous help rendered by my ex-colleagues and friends, namely, Dr T.S. Thind, Sr. Plant Pathologist-cum-Head, Department of Plant Pathology, PAU, Ludhiana, Dr P.S. Sekhon, Sr. Plant Pathologist, Dr Bipen Kumar, Sr. Plant Pathologist and Dr S.K. Thind, Prof. of Plant Protection, members of Plant Pathology Faculty, PAU, Ludhiana; Dr R.P. Maharshi, Ex-Head, Department of Plant Pathology, SKN College of Agriculture, Jobner, Rajasthan, Dr Ram Kishun and Dr Kuldip Singh in lending me the photographs of the symptoms of different plant diseases; Dr Narinder Singh, Sr. Plant Pathologist and Dr S.K. Sharma, Sr. Plant Nematologist, members of Plant Pathology Faculty, PAU, Ludhiana, for taking the photographs of diseased specimens, is highly appreciated. I owe my special thanks to Mr. Parminder Singh, District Extension Specialist (Plant Pathology), Farm Advisory Service Scheme, Jalandhar who helped me immensely in downloading a large number of pictures and photographs from the internet and also for editing and arranging these photographs. I am also thankful to Dr H.S. Randhawa, Sr. Plant Pathologist and Dr Vineet Kumar, Associate Prof. of Plant pathology and Dr Gurvinder Singh Kocher, Microbioligst, PAU Ludhiana for their generous help. I owe a special gratitude to my wife, Mrs. Ravinder Kaur Thind for her continuous and untiring encouragement during this entire period and for her help in reading the proof of the text. I also feel indebted to Mr. Ravinder Singh Gill for typing the manuscript. My sincere thanks are also to those who helped me directly or indirectly in the preparation of this manuscript but their names have not been mentioned here.

B.S. Thind

Chapter 1

INTRODUCTION

1.1. PHYTOBACTERIOLOGY

Phytobacteriology or plant bacteriology is a subdiscipline of plant pathology, which deals with plant-associated procaryotes and their interactions with each other and with their hosts (Dumenyo *et al.*, 2001). The subdiscipline covers all aspects of plant-associated procaryotes including basic as well as of applied nature. The morphology, physiology, taxonomy, genetics and serology of the procaryotes comprise the basic aspects. Host-pathogen interactions including pathogenesis and infection process, epidemiology, effect of environment on disease development and disease management are the applied aspects dealt in this subdiscipline. In the past, emphasis has been mainly on the investigation of plant diseases caused by bacteria due to heavy losses caused by these diseases. However, lately, the field of phytobacteriology has expanded considerably to include beneficial bacterial-plant interactions such as nodulation and nitrogen fixation; promotion of plant growth by phyllosphere, rhizosphere and soil-inhabiting bacteria, and control of ice-nucleation active bacteria by phyllosphere bacteria. Phytobacteriology also includes the management and control of bacterial plant diseases using various methods such as cultural practices, resistant cultivars, chemical and biological control, and molecular genetics. A basic knowledge of various branches of science like microbiology, genetics, chemistry, biochemistry, plant physiology, agronomy, soil science and molecular biology is essential for better understanding of plant diseases caused by micro-organisms. The molecular biology is now becoming increasingly important in every biological science including phytobacteriology.

1.2. PHYTOPATHOGENIC PROCARYOTES

On the basis of cellular organization, the living organisms can be divided into two groups, namely, procaryotes and eucaryotes. The procaryotes have a set of characteristics, which differentiate them from eucaryotes. The main characteristic of the procaryotes is that with a few exceptions, nucleoplasm or genophore (earlier called nucleoid) is not separated from the cytoplasm by a unit-membrane system, i.e., the nuclear membrane is absent. The other characteristic features are the presence of 70S ribosomes (with the exception of archaea, whose ribosomes have a slightly higher S value) dispersed in the cytoplasm, no cytoplasmic streaming and absence of endoplasmic reticulm with attached ribosomes. They take the nutrients in liquid form. Most of them possess a rigid cell wall except one group, i.e., mollicutes. They occur as single cells or simple association of similar cells. The cells may be motile or nonmotile; in case of motile, the motility is mediated by bacterial flagella or gliding motility on solid surface.

The procaryotes are inhabitants of moist environments and predominantly occur as unicellular microorganisms. However, they also occur in other forms such as, filaments, mycelia, or colonies. In some cases, differentiating structures like holdfasts and resting cells are formed. The gene transfer and recombinations occur but in these processes gametogenesis and zygote formation does not ever occur.

The above given characteristics of procaryotes are sufficient to distinguish them from eucaryotic microorganism but some common features are found in some procaryotes and eucaryotes. For example, the hyphae formed in actinomycetes look similar to those of moulds and ability of spirochetes to twist and contort their shape matches the flexibility shown by certain protozoa. Some large bacteria may approach the size of eucaryotic cells while some eucaryotic cells are as small as bacteria. However, procaryotic and eucaryotic cells can be easily differentiated on the basis of fluorescent-labelled genes.

The procaryotes comprise two different groups, namely, the *Bacteria* and the *Archaea*, which have been recognized as different domains for the purpose of classification.

1.2.1. Bacteria

On the basis of Gram reaction and presence or absence of cell wall, the bacteria have been divided into the following three phenotypic subgroups:

(a) Gram-negative bacteria having cell walls

(b) Gram-positive bacteria having cell walls

(c) Bacteria lacing cell walls

1.2.1.1. Gram-negative bacteria having cell walls: These bacteria possess a Gram-negative type of cell wall, which is composed of an outer membrane, and an inner, relatively thin peptidoglycan layer containing muramic acid; the latter may be absent in a few organisms. Reproduction is usually by binary fission, some groups show budding and a very few divide by multiple fission. The cells may occur in the form of spheres, ovals, straight or curved rods, helicals, or filamentous. The cells may be nonmotile or motile showing swimming or gliding motility. In relation to oxygen requirement, the members may be aerobic, anaerobic, facultatively anaerobic, or microaerophlic.

1.2.1.2. Gram-positive bacteria having cell walls: These bacteria have a Gram-positive type of cell wall, in which there is no outer membrane and the peptidoglycan layer is relatively thicker than that of Gram-negative type of cell wall. The cell shapes include rods, spheres, or filaments; the rods and filaments may show true branching. The reproduction is mostly by binary fission but actinomycetes and their relatives produce sporogenous hyphae. Some members produce spores, also called endospores, which help to tide over the adverse conditions but are not a mean of multiplication. The oxygen requirement of the members is the same as that of Gram-negative bacteria.

1.2.1.3. Bacteria lacking cell walls: They are commonly called the **"mycoplasmas"** (also called mollicutes or soft-skin microorganisms). They lack the cell wall but are enclosed by a unit membrane, plasma membrane. As they lack the cell wall (peptidoglycan layer), they are not sensitive to β-lactam antibiotics or other antibiotics, which inhibit cell wall synthesis. They are highly pleomorphic, have different cell forms varying in size and some of these forms are filterable. Cells stain Gram-negative, no resting forms are produced, and usually, they are nonmotile but some members show gliding motility. Different modes of reproduction such as, budding, fragmentation and/or binary fission occur. Most members require complex media and all require cholesterol or related sterols for growth. The typical colony under adequate growth conditions is biphasic having fried-egg appearance, i.e., consisting of an opaque, granular central area which grows down into the medium and a flat, translucent peripheral zone. The mol% G + C contents of r RNA (43 - 48) and of DNA (23 - 46) are lower than that of walled bacteria.

1.2.2. Archaea

Archaea, predominantly inhabitants of terrestrial and aquatic habitats, are found in anaerobic, hypersaline, or hydrothermally and geothermally heated environments. Some members also live symbiotically in the digestive tracts of animals. Archaea grow as mesophiles and thermophiles but some species grow at very high temperatures, i.e., up to 110°C. They grow as aerobes, anaerobes, and facultative anaerobes.

Archaea are not sensitive to β-lactam antibiotics as their cell walls do not contain murein, peptidoglycan containing muramic acid. All archaea contain glycerol isopranyl ether lipids, which is a their unique biochemical property.

Gram reaction varies depending on the composition of cell envelopes. It may be Gram-positive or negative within the same group. The cells are of different shapes, which include spherical, spiral, plate or rod. Unicellular or multicellular forms also occur in filaments or aggregates. The modes of reproduction include binary fission, budding, fragmentation, constriction or mechanisms not yet known. The cell masses exhibit a variety of colours including red, pink, purple, green, yellow, gray, white, etc. Five major groups recognized in archaea are given below.

(a) Methanogenic archaea

(b) Sulphate-reducing archaea

(c) Extremely halophilic archaea

(d) Cell wall-lacking archaea

(e) Extremely thermophilic S°-metabolising archaea

Phytopathogenic procaryotes include cell-walled bacteria, including both Gram-negative and Gram-positive, and cell wall-lacking bacteria (mollicutes). As the procaryotes include both bacteria and archaea and among them only bacteria are pathogenic to plants, the terms "**phytopathogenic procaryotes**" and "**phytopathogenic bacteria**" have been used interchangeably in this book.

Phytopathogenic procaryotes differ from other procaryotes that they have the ability to multiply inside the plants. Phytopathogenic procaryotes cause plant diseases on a wide range of host plants. Some of these diseases are highly devastating and cause serious economic losses.

1.3. ECONOMIC IMPORTANCE OF PHYTOPATHOGENIC PROCARYOTES

There is not enough data available on losses due to plant diseases caused by phytopathogenic procaryotes in spite of the fact that some of these diseases are of international importance. The loss caused by a plant disease varies from region to region due to variation in environmental factors, which influence the disease development, and also the susceptibility of the varieties under cultivation. For example, bacterial blight of rice is severe mainly in Southeast Asia. Moreover, the importance of a crop also varies in a particular region or country. Chemical control of bacterial plant diseases is less effective than those of caused by fungi due to the availability of less number of effective chemicals and low efficacy of the available chemicals. Moreover, many antibiotics, which are effective against many plant pathogenic bacteria, are also used for the control of human and animal bacterial pathogens. Hence, their use in plant disease control in many countries is not permitted due to the risk of transfer of resistance to human and animal pathogens. The crop losses due to plant pathogenic procaryotes in the United States in 1976 are given in Table 1.1 (Kennedy and Alcorn, 1980).

Table 1.1. Estimates of US crop losses caused by phytopathogenic procaryotes in 1976[a]

Pathogen[b]	Disease	Loss (millions of US dollars)
Ralstonia solanacearum	Bacterial wilt of tomato and tobacco	9.4
Agrobacterium tumefaciens	Crown gall of fruit and nut	23.0
Erwinia amylovora	Fire blight of pear	4.7
Er. tracheiphila	Cucumber wilt	0.9
Pantoea stewartii subsp. *stewartii*	Stewart's wilt of corn	0.2
Pectobacterium atrosepticum and/or *Pe. carotovorum* subsp. *carotovorum*	Black leg and/or soft rot of potato	14.0
Clavibacter michiganensis subsp. *insidiosus*	Bacterial wilt of alfalfa	17.0
Pseudomonas savastanoi pv. *glycinea*	Bacterial blight of soybean	65.0
Psm. syringae pv. *syringae*	Bacterial leaf blight of wheat	18.0
Xanthomonas citri subsp. *malvacearum*	Angular leaf spot of cotton	15.0
Xa. phaseoli	Common blight of bean	5.0
Leifsonia xyli subsp. *xyli*	Ratoon stunting disease of sugarcane	10.0
Xylella fastidiosa subsp. *fastidiosa*	Pierce's disease of grapevine	3.0
Spiroplasma citri	Citrus stubborn disease	1.0
'*Candidatus* Phytoplasma spp.'	Pear decline	1.6
	Lethal yellowing of coconut	3.0

[a]Based on data summarized from Kennedy and Alcorn (1980),
[b]Recent names of the pathogens are given.

Bacterial wilt of solanaceous plants is considered the world's single most destructive bacterial plant disease. According to Champoiseau and

Allen (2009), brown rot bacterium is responsible for an estimated loss of one billion US dollars worldwide every year, mostly to poor small-scale farmers in tropical highlands. Bacterial wilt of potato is also considered the second most important disease of potato worldwide after late blight.

Bove (2006) has rightly stated that citrus huanglongbing (greening) is the most important, serious, severe, destructive and devastating disease of citrus in the world. The reduction in yield can range from 30 to 100% depending on proportion of affected canopy. The affected orchards become economically unviable in 7 - 10 years after planting. About 100 million trees have been destroyed in many countries of South and Southeast Asia, Indonesia, Philippines, India, Arabian Peninsula, and South Africa. Since 2004, more than 500 thousand trees have been officially destroyed in Brazil due to this disease and around 300 - 400 thousand trees unofficially destroyed by commercial citrus growers (Gottwald *et al.*, 2007). Catastrophic losses due to the disease from India during 1960s were reported by Fraser *et al.* (1966).

Incomplete eradication of citrus canker from the United States, particularly from Florida state has taken a very heavy toll of citrus trees. After its first introduction in Florida in 1912, it took about 20 years to eradicate the disease from Florida by destroying and burning more than three million nursery and a quarter million fruit bearing trees. The second introduction of disease in Florida in 1986 resulted in the destruction of more than 20 million nursery and orchard trees and the eradication programme continued till 1992. The third introduction of the disease in Florida in 1995 and subsequent years led to the destruction of 1.56 million commercial trees and 6,00,000 dooryard trees leaving the eradication target incomplete. The successful eradication of the disease in Australia, South Africa and New Zealand and partially successful eradication (incidence below 0.22% in Sao Paulo) in Brazil was achieved after destroying citrus trees on a large scale.

Pierce's disease of grapevine caused by *Xyl. fastidiosa* subsp. *fastidiosa* was noticed in California state of the United States in the 1880s but the lack of appropriate vector of casual agent and other non-favouring factors kept the disease under check. The introduction of glassy -winged sharpshooter (*Homalodisca coagulata*), a highly efficient vector of the bacterium, in southern California has changed the status of the disease from minor to severe with enormous increase in losses. It has threatened not only the cultivation of the European grape (*Vitis vinifera*) but also of the many ornamental crops in the southeastern United States.

In 1918, 75% of the bean fields in New York state of the USA were affected with common blight (*Xa. phaseoli*) and the disease caused heavy losses. In the subsequent years, losses ranged from 20 - 50%. In 1953, the disease was wide spread in Nebraska (USA) and it caused an estimated loss of over one million dollars. In 1976 in USA, common blight was economically the most important disease among the bacterial diseases of beans and it caused an estimated loss of 4 million US dollars. According to Gilbertson and Maxwell (1992), in 1967 the disease caused yield losses up to 38% in dry bean in Ontario (Canada) and 10 - 20% in navy bean in Michigan (USA). Yield losses ranging from 17 - 22% were also reported from Colombia. Besides loss in yield, the disease causes reduction in quality of seeds through discolouration of infected seeds.

Although the bacterial blight of rice, caused by *Xa. oryzae* pv. *oryzae*, occurs in many countries, it is most destructive in Southeast Asia. The losses are generally high in an early-infected crop and occurrence of kresek phase may lead to the total failure of the crop. In Japan, prior to introduction of resistant varieties, losses generally ranged from 20 - 30% but in some cases went as high as 50% (Ou, 1972). Reports from Philippines, India and Indonesia estimate that losses due to kresek phase have reached 60 - 75% depending on weather conditions, location and rice variety (Ou, 1985; Reddy *et al.*, 1979). In addition to reduction in yield, the disease may also affect grain quality by interfering with maturation.

Bacterial canker (black spot) of mango causes heavy losses because most of the commercial varieties of mango are highly susceptible to the pathogen and the infection results in drastic yield losses. The fruit infection ranging from 50 - 80% is common on highly susceptible varieties. In 1966 and 1967, severe epiphytotics of the disease caused almost 100 fruit loss on most susceptible cultivars in most of the mango growing areas of South Africa. The fruit loss was mainly due to the fruit drop and non-marketability of the harvested fruit. The estimated loss alone in 1966 was approximately one million US dollar (Gagnevin and Pruvost, 2001). Kishun (1981) observed 20 - 80, 10 - 70, 10 - 40 and 10 - 55% fruit drop in Alphanso, Pairi, Totapuri and local cultivars grown in India, respectively and the maximum fruit drop in all the cultivars occurred 30 - 60 days after fruit set. He further observed that the rot in storage in these varieties was 5 - 80, 11 - 67, 10 - 100 and 5 - 80%, respectively.

Lethal yellowing of coconut palm trees has killed several palm trees in Florida State of the USA and some other countries. First observed in Florida in 1971, it killed 40,000 trees by 1974, and by 1975, 75% of palm trees in Dade county of the state were either dead or dying due to this disease. In Jamaica, 5.4 million trees were killed during a span of 20 years starting from 1961. In Mexico and Tanzania, palm trees in thousands of hectares and in Ghana over a million coconut palm trees were killed within 30 years. The disease also killed more than 60 thousand palm trees in Togo by 1964 (Agrios, 2004).

Common scab of potato is one of the most economically important diseases of the crop. It costs the Tasmanian potato industry in excess of 3.6 million Australian dollars per annum (Wilson, 2004).

An estimated loss of $ 4.7 million due to fire blight occurred in California in 1976 while in Southwest Michigan alone the loss in 1991 was $ 3.8 million. Again in 2000, a severe outbreak of the disease occurred on apples in Michigan, resulting into removal and destruction of 35,000 - 45,000 apple trees from approximately 1550 - 2300 acres and amounting to a loss of $ 42 million. In 1998, apple and pear growers in Washington and northern Oregan suffered a loss of $ 68 million due to fire blight damage.

Stewart's wilt of corn, caused by *Pa. stewartii* subsp. *stewartii,* is a good example of a disease that continues to have enormous implications on trade and commerce. The disease has been a major threat to corn producers in the eastern corn belt of the United States for over 100 years. Severe outbreaks of the disease occurred in Nebraska in 1999 and 2000. In 1999, it was found to occur in 762 out of 1317 fields in Iowa (58% prevalence). The impact of the disease was that the seed produced from 58% of the seed corn fields in Iowa in 1999 could not be exported to the countries with phytosanitary (quarantine) restrictions for this pathogen.

In addition to causing direct loss in crop yield and quality, phytopathogenic bacteria cause other harmful effects to mankind. *Rathayibacter toxicus* causes gumming disease in ryegrass and other fodder grasses and produces toxins, which when consumed often result in fatal poisoning of livestock. Various species of *Anguina* (nematode) act as vector of this bacterium. Glycolipid toxins, known as corynetoxins, are produced by the bacterium as the infected grass matures and becomes senescent. In Australia, *Rat. toxicus* is found most commonly in *Lolium rigidum* (annual ryegrass) with *A. funesta* as a vector and in *Polypogon monspeliensis* (rabbit-foot grass) and *Lachnagrostis filiformis* (syn. *Agrotis avenacea*, annual blowngrass) with an undescribed *Anguina* vector. In Australia more than 100,000 sheep and thousands of cattle have died from this disease in certain years (Edgar, 2004). Besides affecting farmers or growers, bacterial diseases also affect transport and other industries dependent on the affected crops

In contrast to direct monetary losses and other harmful effects caused by phytopathogenic bacteria, they have a great deal of commercial value (Starr, 1984). The studies of these bacteria have yielded very useful products and tools. Several phytopathogenic bacteria are used in industrial fermentation. The most prominent case is the large-scale use of *Xa. campestris* to make xanthan gums, extracellular polysaccharides of considerable economic importance. Xanthan gums are used for various purposes including oil-well drilling muds, tertiary recovery of petroleum from spent wells, ceramic glazes, thickening agents in foods and antidrip agents in paints. Another class of economically important polysaccharides produced by phytopathogenic bacteria includes curdlans, which are produced by various species of *Agrobacterium* and other bacteria. On heating, the curdlans form very elastic and resilient gums. The curdlans have a potential use in various food products as noncaloric gelling, thickening, binding and stabilizing additives. A pseudomonad, isolated from an aquatic plant produces an agar-like polysaccharide that has the potential to be used as a substitute for agar in bacteriological media.

Besides vinegar production, *Gluconobacter* species have many biotechnological applications. In the last decades, many bioconversion routes for rare and special sugars involving *Gluconobacter* spp. have been developed. Among the recent ones, are the biotransformations involved in the production of L-ribose and miglitol, both very promising pharmaceutical lead molecules. Most of these processes make use of *Gluconobacter*'s membrane-bound polyol dehydrogenases.

Various enzymes produced by phytopathogenic bacteria have various applications and are produced commercially. Pectin-degrading enzymes have found an important application in the food industry, where they are used in the clarification of fruit juices. An enzyme produced by an *Agrobacterium* strain is used for converting super coiled, covalently closed, circular DNA into a simple, covalently closed, circular form. Various *Xanthomonas* species and other phytopathogenic bacteria produce restriction endonucleases having much utility in genetic engineering and basic research. An L-asparaginase (protein synthesis inhibitor drug) produced by soft rot bacterium, *Pe. chrysanthemi* has a great potential for use as an antileukemic agent. The potential of Ti-plasmid of *Ag. tumefaciens* to move genetic material from bacteria to plants have made it an important tool in plant genetic engineering and it is widely used for this purpose. This technology is largely responsible for the development of plant biotechnology industry and the latter is revolutionizing agriculture and health care. The production of agrocin by an avirulent bacterium, *Ag. radiobacter* (strains K 84 and K 1026) and its role in controlling crown gall disease on commercial scale are well documented. At present, there are more than 16 commercial biocontrol products based on different bacteria including *Ag. radiobacter*, which are used to control plant diseases caused by different pathogens. The detailed studies of the nodulation process by *Rhizobium and Bradyrhizobium* species have led to the preparations of bacterial inoculum for application to legume crops (Dumenyo *et al.*, 2001).

1.4. HISTORICAL REVIEW OF PHYTOBACTERIOLOGY

Antony van Leeuwenhoek discovered the bacteria for the first time in 1676 and called them animalcules, i.e., small animals. In 1876, Robert Koch showed that anthrax of animals is caused by a bacterium, *Bacillus anthracis*. Robert Koch also devised pour plate method for the purification of bacteria and gave Koch's postulates to prove the pathogenicity of causal agents of pathogenic diseases.

Woronin, a Russian worker, was the first person to show association and multiplication of bacteria in root nodules of leguminous plants in 1866. However, the multiplication of bacteria in root nodules is considered a symbiotic relationship and not a strictly pathogenic action. Professor T.J. Burrill of the University of Illinois was the first to show a bacterium as a causal agent of a plant disease. In a series of experiments starting from 1878 till 1884, he showed that the fire blight of pear was caused by a bacterium, which he identified as *Micrococcus amylovorus.* Burrill, however, did not use the pure culture of the bacterium in his work. J.C. Arthur working at New York Experiment Station confirmed Burrill's work during 1885-1887 using pure culture of the bacterium. In 1883, J H. Wakker from Holland showed that yellow-slime disease of hyacinth was caused by a bacterium. Luigi Savastano, an Italian, demonstrated in 1887 that a bacterium incited the knot disease of olive. Soon after 1890, Erwin F. Smith started work on bacterial diseases and the first disease investigated by him was bacterial wilt of cucurbits. Later on, he worked on many bacterial diseases of plants including bacterial wilt of solanaceous plants and black rot of crucifers. Erwin F. Smith's monumental contributions are included in his monograph entitled, *Bacteria in Relation to Plant Diseases* published in three volumes from 1905 to 1914.

A number of plant pathologists and authors of many books have given credit to T.J. Burrill for providing first credible proof that bacteria could cause plant diseases. However, Kennedy *et al.* (1979), in their critical appraisal of research work on bacterial diseases of plants carried out by T.J. Burrill and others, have opined that the credit for this discovery should be shared by E. Mitscherlich, a German chemist, T.J. Burrill and J.C. Arthur and not given to Burrill alone. They have further stated that Burrill's place in history with respect to these studies is perhaps based on the fact that he was proved by others to have been right, and that he had a strong advocate in Erwin F. Smith.

Most of the work on bacterial diseases of plants from 1878 to 1900 and even later was carried out in the United States because there was a distinct air of disbelief on the part of European botanists regarding the role of bacteria in causing plant diseases. This was mainly due to the influence of De Bary, who considered bacteria as of minor importance as pathogens of plants. De Bary was aware of Wakker's work on the hyacinth disease but not of Burrill's work on fire blight when he published his *Comparative Morphology and Biology of Fungi, Mycetozoa and Bacteria* in 1884. In this publication in the section of bacteria, he devoted just one short paragraph to those parasitic on plants and also pointed out that they had scarcely ever been observed. De Bary in his *Lectures on Bacteria* published in 1887 devoted only 2 out of 146 pages to bacteria parasitic to plants and stated that saprophytic bacteria may also under special conditions, attack the tissues of living plants as facultative parasites, produce disease in them and destroy them.

Alfred Fischer, a professor of botany at Leipzig, Germany, who was a student of De Bary in 1881, challenged the work of E.F. Smith and others concerning the role of bacteria in causing plant diseases in 1897. A. Fischer in his book entitled, *Lectures on Bacteria* refused to accept the findings of Smith and others terming them unreliable. Smith replied to Fischer in 1899 pointing out his complete ignorance of the subject. Fischer wrote back the same year, and Smith replied in two rebuttal papers, which seemed to end the controversy. Smith was the clear winner of the debate. This debate has been duly recorded in one of the Phytopathological Classics published by APS press (Campbell, 1981). Walker (1969) has rightly commented on the controversy created by Alfred Fischer, stating that while this debate is now chiefly of historical importance, it is significant that more than 20 years after the work of Robert Koch and of T.J. Burrill; a rather prominent botanist still challenged the existence of bacterial plant pathogens. By 1900, the concept that bacteria could cause plant diseases was solidly established (Starr, 1984), thereby creating phytobacteriology as a new subdiscipline of plant pathology. Erwin F. Smith's contributions to the field of plant bacteriology and his efforts to put it on sound footing, earned him the title of **"father of plant bacteriology"**.

Although E.F. Smith made highly significant contributions to the field of plant bacteriology, he did not establish a dynasty or a school where young scientist could nurture their enthusiasm and interest in the newly discovered science of phytobacteriology as remarked by Sequeira (2000) in his review article. Therefore, after his death in 1927, phytobacteriology began to wane. Around the middle of 20th century, plant pathology was dominated by mycology and virology.

In the initial stage of plant bacteriology, the emphasis was on reporting new bacterial diseases of plants and description and identification of their causal bacteria. Then the emphasis shifted to the investigations on the physiology of diseased plants and during 1940s to 1960s, the significant contributions were made in this filed along with ecology of plant pathogenic bacteria. The work of R.M. Klein (1954) and of A.C. Braun and Ross Pringle (1959) provided the evidence that the crown gall bacterium, *Ag. tumefaciens* caused permanent transformation of host cells leading to their autonomous and rapid growth in culture. This work along with pioneer work of A.C. Braun and his colleagues on the structure and mode of action of wildfire toxin produced by *Psm. syringae* pv. *tabaci* greatly impacted many plant bacteriologists to determine the pathogenicity determinants in other diseases (Sequeira, 2000).

The discovery of mycoplasma-like organisms by Doi *et al.* (1967) gave the correct etiology of yellow-type diseases, which were earlier thought to be caused by viruses. The discovery of mycoplasma-like organism also resulted in the subsequent discovery of spiroplasmas and fastidious procaryotes. The mycoplasma-like organisms are now called phytoplasmas. The demonstration of conjugative transfer of bacterial genes in *Erwinia* by Chatterjee and Starr (1972) shifted the emphasis of research to molecular genetics of bacterial plant pathogens. Another notable contribution during 1970s was the success achieved in the biological control of crown gall with *Ag. radiobacter* strain K 84 (New and Kerr, 1972). Subsequently, an effective and viable biological control of fire blight has also been achieved. The introduction of pathovar system by Young *et al.* (1978) has greatly streamlined the taxonomy of plant pathogenic bacteria. Rapid advances in DNA sequencing technology and sequence analysis of highly conserved regions of the bacterial genome, such as the small subunit rRNA gene have led to the natural classification of procaryotes that reflects the evolutionary history of bacteria and archaea. The widespread application of new methods of classifying procaryotes has led to an explosive growth in the number of validly published species and higher taxa. Since the completion of first edition of *Bergey's Manual of Systematic Bacteriology* in 1989, the number of published species has more than tripled and accompanied with numerous taxonomic rearrangements and changes in nomenclature. The number of bacterial genera containing phytopathogenic bacteria, which was only six during 1970s, now stands at 40 including three '*Candidatus* genera'.

The first avirulence gene (*avrA*) characterized was cloned from a race 6 strain of the soybean pathogen, *Psm. savastanoi* pv. *glycinea* by Staskawicz *et al.* in 1984. It elicits a resistant reaction on soybean cultivars having resistant gene *Rpg2*. Avirulence genes determine the race specficity of pathogen in a compatible host pathogen interaction. The products of *avr* genes apparently function as recognition factors in resistant hosts (i.e., carrying the *R* gene) and virulence determinants in susceptible hosts (i.e., carrying the *r* allele). In most cases, bacterial avirulence gene function is dependent on interactions with hypersensitive reaction and pathogenicity (*hrp*) genes. The *hrp* genes which control the ability of phytopathogenic bacteria to cause disease on susceptible plants and elicit hypersensitive reaction on resistant or nonhost plants were first discovered for *Psm. syringae* pv. *phaseolicola* by Lindgren and associates in 1986. The role of avirulence genes and of *hrp* genes in host parasite interaction have been reviewed by Leach and White (1996) and Lindgren (1997), respectively.

Investigations carried on *Ag. tumefaceins*, the causal agent of crown gall, during the last 100 years (since its report by Smith and Townsend in 1907) have made this bacterium from merely a plant pathogen to a household name in plant genetic engineering and plant molecular genetics (Nester, 2008). The basic research carried out with this bacterium has provided very useful insights into host-parasite interactions and the mechanism of gene transfer from a procaryote to a eucaryote. The main problem of using Ti-plasmid as a gene vector due to its larger size has been solved by separating it into two parts; the 25 bp borders flanking the T-DNA between which the genes of interest are inserted and the vir region that is necessary for the processing and transfer of the T-DNA. This binary vector system is used in most laboratories. *Agrobacterium* is truly a natural genetic engineer.

A brief account of development of phytobacteriology in India is given below. In India, Cappel (1892) reported bangle blight or bungdi disease of potato from Pune, Maharshtra. Butler in 1903 also reported this disease and stated that it was similar to brown rot of potato caused by *Psm. solanacearum*. Coleman in 1909 proved the bacterial nature of the disease. Later on, Dr M.K. Patel started work on bacterial diseases in 1948. He along with his associates worked for nearly 15 years and reported nearly 40 bacterial diseases, most of which were new records. Based on his contribution to the field of plant bacteriology in India, Dr M.K. Patel is regarded as the **"father of plant bacteriology"** in India. A review of bacterial plant disease investigation published by Patel and Kulkarni (1953) included 27 species of bacterial pathogens reported from India. The first book on bacterial plant diseases in India was written by G. Rangaswami in 1962, in which he listed 73 bacterial plant pathogens. In 1965, Mathur and associates published a list of bacterial plant pathogens found in India. Phytopathogenic bacteria of India and bibliography published by Chakravarti *et al.* (1973) contains a list of 483 references on plant pathogenic bacteria and 148 bacterial plant pathogens belonging to six genera, namely, *Agrobacterium, Bacillus, Corynebacterium, Erwinia, Pseudomonas* and *Xanthomonas*. A summer institute on Plant Bacteriology under the directorship of Dr P.N. Patel was held in the Division of Mycology and Plant Pathology, Indian Agricultural Research Institute, New Delhi in 1972 and its proceedings were published in 3 volumes edited by Dr P.N. Patel. First National Symposium on Plant Bacterial Diseases was held at Sri Venkateswara University, Triputi in 1980 under the directorship of Dr M.V. Nayudu and several important recommendations were made. Ninth International Conference on Plant Pathogenic Bacteria was held in 1996 at University of Madras, Chennai under the chairmanship of Dr A. Mahadevan.

At present, phytobacteriology is one of the leading areas in plant pathology and has attracted a large number of scientists from areas other than plant pathology. It is evident from the large number of papers on phytobacteriology published in *Molecular Plant-Microbe Interactions, The Plant Cell* and *Applied and Environmental Microbiology*.

SELECTED REFERENCES

Agrios, G.N. (2004). *Plant Pathology.* 5th Ed. Elsevier Academic Press, San Diego, 922 pp.

Bove, J.M. (2006). Huanglongbing: A destructive, newly emerging century-old disease of citrus. *J. Plant Pathol.* 88: 7 - 37.

Braun, A.C. and Pringle, R.B. (1959). Pathogen factors in the physiology of disease-toxins and other metabolites. In: S. Holton, G.W. Fischer, R.W. Fulton, H. Hart and S.E.A. McCallan (eds). *Plant Pathology- Problems and Progress.* Univ. Wisconsin, Madison, pp. 88 - 99.

Campbell, C.L. (ed.) (1981). *The Fischer-Smith Controversy: Are There Bacterial Diseases of Plants?* Phytopathol. Classics No.13, APS Press, St. Paul Minn. XVIII + 65 pp.

Cappel, E.L. (1892). A note on a potato disease prevalent in Poona district and elsewhere. Bombay Dir. Dept. Land Records and Agri. (Abstr. Butler, 1903).

Chakravarti, B.P., Hegde, S.V., Ahmed, S.R. *et al.* (1973). Phytopathogenic bacteria of India and bibliography (1892 - 1972). Bulletin, Department of Plant Pathology, University of Udaipur, Udaipur.

Champoiseau, P.G. and Allen, C. (2009). *Ralstonia solanacearum* race 3 biovar 2: Tropical losses, temperate anxieties. APSnet Feature Story, [http://www.apsnet.org/online/feature/ralstonia/].

Doi, Y.M., Teranaka, M., Yora, K. *et al.* (1967). Mycoplasma- or PLT group-like microorganisms found in the phloem elements of plants infected with mulberry dwarf, potato witches' broom, aster yellows, or paulownia witches' broom. *Ann. Phytopathol. Soc. Jpn.* 33: 259 - 66.

Dumenyo, C.K., Chatterjee, A. and Chatterjee, A.K. (2001) Phytobacteriology. In: O.C. Maloy and T.D. Murray (eds). *Encylopedia of Plant Pathology.* Vol. II, John Wiley and Sons, Inc., New York, pp. 762 - 69.

Edgar, J. (2004). Future impact of food safety issues on animal production and trade: Implication for research. *Aust. J. Exp. Agri.* 44: 1073 - 78.

Fraser, L.R., Singh, D., Capoor, S.P. *et al.* (1966). Greening virus, the likely cause of citrus dieback in India. *FAO Plant Prot. Bull.* 14: 127 - 30.

Gagnevin, L. and Pruvost, O. (2001). Epidemiology and control of mango bacterial black spot. *Plant Dis.* 85: 928 - 35.

Gilbertson, R.L. and Maxwell, D.P. (1992). Common blight of beans. In: H.S. Chaube, U.S. Singh, A.N. Mukhopadhyay and J. Kumar (eds). *Plant Diseases of International Importance.* Vol. II, *Diseases of Vegetables and Oilseed Crops.* Prentice Hall, Englewood Cliffs, New Jersey, pp. 18 - 39.

Goto, M. (1992). *Fundamentals of Bacterial Plant Pathology.* Academic Press, Inc., San Diego, 342 pp.

Gottwald, T.R., da Graca, J.V. and Bassanezi, R.B. (2007). Citrus huanglongbing: The pathogen and its impact. APSnet Feature Story, [http://www.apsent.org/online/feature/huanglongbing/].

Kennedy, B.W. and Alcorn, S.M. (1980). Estimates of U.S. crop losses to procaryote plant pathogens. *Plant Dis.* 64: 674 - 76.

Kennedy, R.W. and Lacy, G.H. (1982). Phytopahthogenic procaryotes: An overview. In: M.S. Mount and G.H. Lacy (eds). *Phytopathogenic Procaryotes.* Vol. 1, Academic Press, New York, pp. 3 - 17.

Kennedy, B.W., Widin, K.D. and Baker, I.S.F. (1979). Bacteria as the cause of disease in plants: A historical perspective. *ASM News* 45: 1 - 5.

Kishun, R. (1981). Loss in mango fruit due to bacterial canker *Xanthomonas mangiferaeindicae. Proc. 5th Int. Conf. Plant Path. Bact.* Cali, Colombia, pp. 181 - 84.

Krieg, N.R. (2001). Procaryotic domains. In: D.R. Boone and R.W. Castenholz (eds). *Bergey's Manual of Systematic Bacteriology.* Vol. I, *The Archaea and the Deeply Branching and Phototrophic Bacteria.* 2nd Ed. Springer Verlag, Inc. New York, 21-25 pp.

Leach, J.E. and White, F.F. (1996). Bacterial avirulence genes. *Annu. Rev. Phytopathol.* 34: 153 - 79.

Lindgren, P.B. (1997). The role of *hrp* genes during plant bacterial interactions. *Annu. Rev. Phytopathol.* 35: 129 - 52.

Nester, E. (2008). *Agrobacterium:* The natural genetic engineer: 100 years later. APSnet Feature Story, [http://www.apsnet.org/online/feature/Agrobacterium/].

Ou, S.H. (1972). *Rice Diseases.* 1st Ed. Commonwealth Mycological Institute, Kew, Surrey England, 368 pp.

Ou, S.H. (1985). *Rice Diseases.* 2nd Ed. C. A. B. International Slough, UK, 380 pp.

Patel, M.K. and Kulkarni, Y.S. (1953). A review of bacterial plant disease investigation in India. *Indian Phytopathol.* 5: 131 - 40.

Reddy, A.P.K., MacKenzie, D.R., Rouse, D.I. *et al.* (1979). Relationship of bacterial leaf blight severity to grain yield of rice. *Phytopathology* 69: 967 - 69.

Sequeira, L. (2000). Legacy for the millennium: A century of progress in plant pathology. *Annu. Rev. Phytopathol.* 38: 1 - 17.

Starr, M.P. (1984). Landmarks in the development of phytobacteriology. *Annu. Rev. Phytopathol.* 22: 169 - 88.

Walker, J.C. (1969). *Plant Pathology.* 3rd Ed. McGraw-Hill Book Co., New York, 819 pp.

Wilson, C.R. (2004). A summary of common scab disease of potato research from Australia. In: S. Naito, N. Kondo, S. Akino, *et al.* (eds). *Proc. Int. Potato Scab Symp. 2004,* Hokkaido Univ., Sapporo, Japan, pp. 198 - 214.

Young, J.M., Dye, D.W., Bradbury, J.F. *et al.* (1978). A proposed nomenclature and classification for plant pathogenic bacteria. *N. Z. J. Agric. Res.* 21: 153 – 77.

Chapter 2

THE PROCARYOTIC CELL

2.1. Size
2.2. Shape
2.3. Arrangement
2.4. Cell structure
    2.4.1. Flagella and pili
    2.4.2. Surface layers (capsule and slime layer)
    2.4.3. Cell wall
        2.4.3.1. Cell wall of Gram-positive bacteria
        2.4.3.2. Cell wall of Gram-negative bacteria
        2.4.3.3. Gram staining
    2.4.4. Cytoplasmic (plasma) membrane
    2.4.5. Cytoplasm
    2.4.6. Ribosomes
    2.4.7. Inclusion and storage products
    2.4.8. Gas vesicles
    2.4.9. Nucleoplasm (genophore)
    2.4.10. Plasmids
    2.4.11. Dormant forms
        2.4.11.1. Spores
        2.4.11.2. Cysts

On the basis of cellular organization, the living organisms are divided into two groups, namely, procaryotes and eucaryotes. The members of the procaryotic world comprise a vast heterogeneous group of very small, mostly unicellular organisms. Procaryotes include two types of organisms, namely, bacteria and archaea. The majority of procaryotes, including the photosynthetic cyanobacteria are included in the bacteria. The below given description of a procaryotic cell conforms mainly to bacteria because they have been studied more intensively than archaea. The bacteria look quite simple when we look at size, shape and arrangement of their cells with an ordinary light microscope. However, the modern electron microscopy has revealed an amazing complexity and details of external and internal cell structures, which were not possible with an ordinary light microscope.

2.1. SIZE

Being microscopic, the bacteria are measured in micrometers (μm), which are equivalent to 1/1000 mm (10^{-3} mm). The cell size of bacteria varies with the species but most cells are approximately 0.5 to 1.0 μm in diameter or width and 2.0 - 5.0 μm in length. *Staphylococcus* and *Streptococcus* cells show a slightly larger diameter measuring 0.75 to 1.25 μm and some filamentous forms may be as long as 100 μm.

A few cases of bacteria having extremely large size have come to the knowledge of bacteriologists. *Epulopiscium fishelsoni* found in the gut of surgeon-fish measures 50 - 100 μm in width and 0.5 mm in length. Another bacterium, *Thiomargarita nomibiensis* measures 100 - 300 μm in diameter. This appears to be the largest bacteria discovered so far. The eucaryotic cells on an average measure from 2 - 200 μm in diameters.

The size of the bacterial cells is influenced by age and cultural conditions. Actively growing cells are usually larger than the aged cells. In stained preparations, the cell size is also reduced due to shrinkage of cells.

The extent of small size of a bacterium is evident from the fact that approximately 1 trillion (1,000,000,000,000 or 10^{12}) bacterial cells weigh a mere 1 gram.

The importance of extremely minute size of a bacterial cell is reflected due to the very high ratio of its surface area to the volume. This highlights the facts that there is a large surface through which nutrients can enter for a relatively small volume of cell substance to be nourished. This is partly responsible for the high rate of metabolism and growth of bacteria. Due to their rapid growth and multiplication, the bacteria are used more frequently in biological research.

2.2. SHAPE

Every bacterial species is associated with a constant shape. There are three basic shapes of bacteria, namely, spherical, cylindrical or spiral. Most commonly found bacteria are either cocci or bacilli.

2.2.1. Spherical

Spherical cells are called **cocci** (singular, **coccus**). They are usually round but they can be ovoid or ellipsoidal (Fig. 2.1a). Their size is measured in diameter.

2.2.2. Cylindrical

They are rod-shaped and have length and breadth. They are called **bacilli** (singular, **bacillus**). The ends of the cells may be flat, rounded, pointed or tapered (Fig. 2.1b). Their size is measured in length and breadth. There is not much difference in the width of different species but differences in length are considerable.

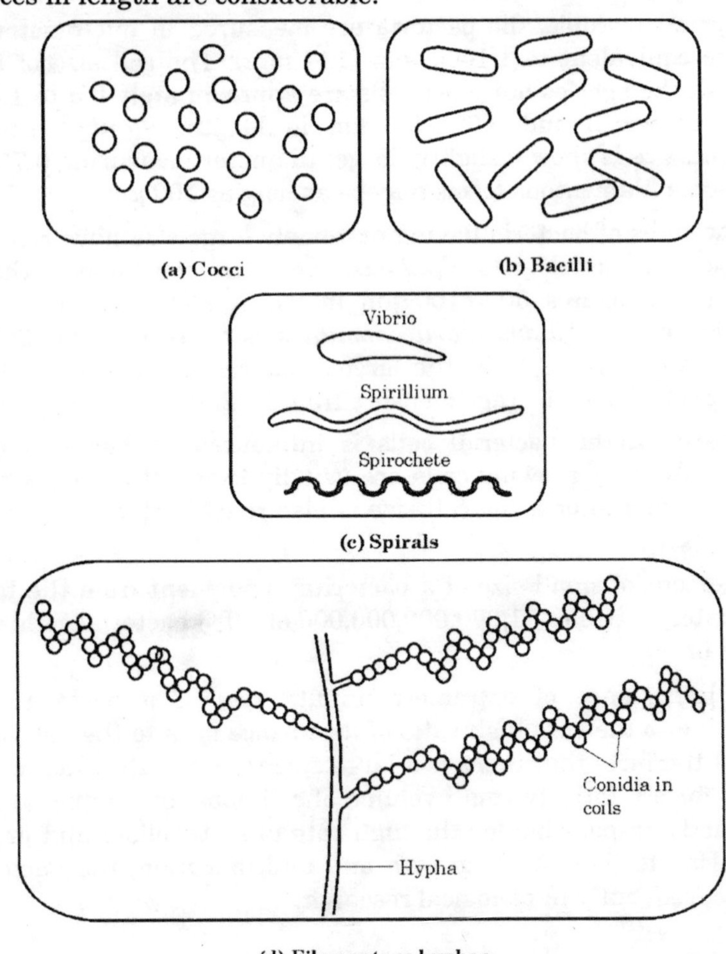

Figure 2.1 Shapes of bacterial cells:
(a) Cocci (b) Bacilli (c) Spirals (d) Filamentous hyphae

2.2.3. Spiral

The cells are helical in shape and are called **spirilla** (singular, **spirillum**). Spiral bacteria have one or more twists and they are never straight (Fig. 2.1c). Certain spiral bacteria are curved short rods and resemble distinctive commas. These are called **vibrios** and *Vibrio cholerae*, the causal agent of cholera, is a typical example of vibrios. The other spiral bacteria are long rods twisted into spirals; if the rods are rigid having thick cell walls, they are called **spirilla,** and **spirochetes** when cells have thin flexible cell walls. *Treponema pallidum*, the causal agent of syphilis, is a typical example of a spirochete

There are many modifications of these three basic shapes. The cells of *Pasteuria* are pear-shaped while those of *Caryophanon* are disc-shaped. Square and star-shaped forms also occur. The genus *Caulobacter* contains appendaged bacteria. The members of Actinomycetes produce long filamentous hyphae that may branch to produce a mass of hyphae called mycelium (Fig. 2.1d).

Although the shape of cells of most bacterial species is constant under a definite set of environmental conditions, a few species have a variety of cell shapes and are called pleomorphic. Mycoplasmas are pleomorphic.

2.3. ARRANGEMENT

The majority of bacteria exist in unicellular forms; however, some species show arrangements or attachment of cells. This arrangement is usually found in spherical and cylindrical forms. In spherical cells, the following five types of arrangement of cells are found.

2.3.1. Diplococci

A coccus cell divides in one plane forming a pair of two cells e.g. some spp. of *Neisseria* (Fig. 2.2a).

2.3.2. Streptococci

The cells divide in one plane forming a chain of cells e.g. *Streptococcus* and *Lactococcus* spp. (Fig. 2.2b).

2.3.3. Tetrads

The cocci divide at a right angle to first plane of division forming a group of four cells in the shape of a square e.g. *Pediococcus* and *Micrococcus* spp. (Fig. 2.2c).

2.3.4. Sarcinae

In this case, the third division occurs in a plane different to the first two divisions resulting in cuboidal or packet arrangement of eight cells e.g. *Sarcina* spp. (Fig. 2.2d).

2.3.5. Staphylococci

When the divisions in the three planes occur in an irregular pattern, the arrangement of cells look like a bunch of grapes e.g. *Staphylococcus* spp. (Fig. 2.2e).

In cylindrical bacteria, only the following two types of cell arrangement are found.

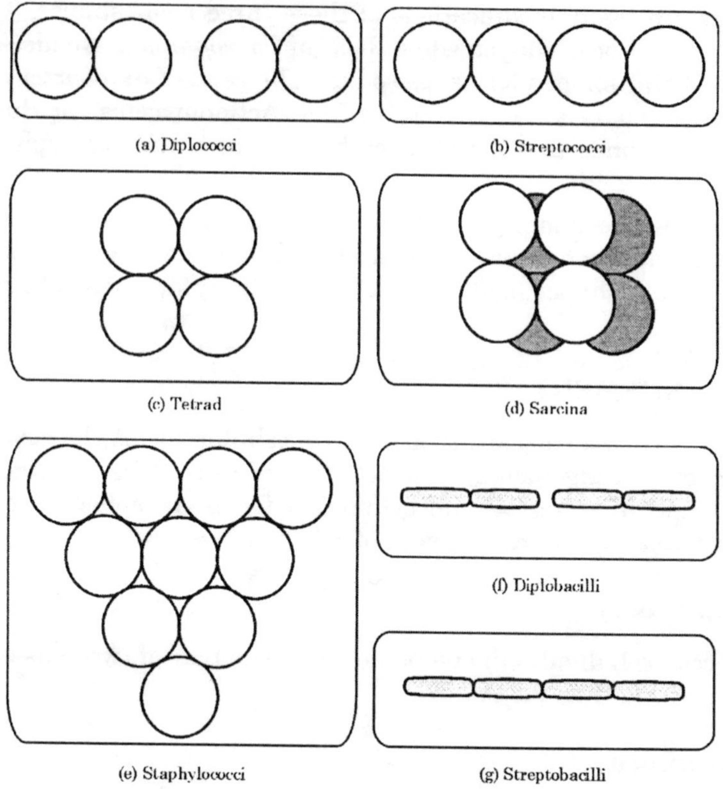

Figure 2.2 Arrangement of cells: (a) Diplococci (b) Streptococci (c) Tetrads (d) Sarcina (e) Staphylococci (f) Diplobacilli (g) Streptobacilli

2.3.6. Diplobacilli

Two cells are arranged from end to end (Fig. 2.2f).

2.3.7. Streptobacilli

Three or more cells are arranged from end to end forming a chain e.g. *Bacillus* spp. (Fig. 2.2g).

Each of these arrangements is typical of a particular species and is helpful in its identification. However, rarely all the cells of a given species are arranged exactly in the same pattern. It is the predominant arrangement that is taken into account while studying bacteria. The size, shape and arrangement of bacteria are important parameters of their gross morphology and are used in their identification.

2.4. CELL STRUCTURE

The cell structures described below are not common to all the bacteria. Some of these structures such as cytoplasmic membrane, cytoplasm, ribosomes and nucleoplasm are found in all the bacteria while the others are associated with certain particular species. Some of these structures are present inside the cell while others are present outside the cell wall (Fig. 2.3).

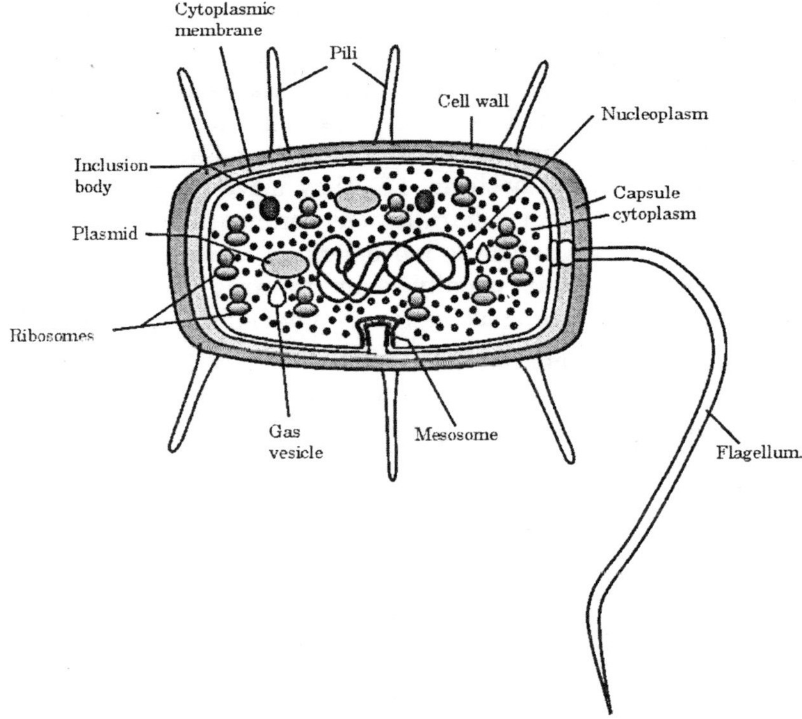

Figure 2.3 A bacterial cell

2.4.1. Flagella

Many bacteria are motile and this motility is due to the presence of flagella (singular, flagellum). Many species of bacilli and spirilla possess flagella while cocci rarely have these organelles. Flagella are thin hair like filaments with a helical shape. These originate from the cytoplasmic membrane and extend to the surface after penetrating the cell wall. Flagella are many times longer than the cells, measuring 15 - 20 μm in length. However, the diameter of a flagellum is only a fraction of the cells width, i.e., 12 - 20 nanometers (nm). Being too thin in width, unstained flagella cannot be seen with an ordinary light microscope. The staining procedures using a mordant make flagella thicker due to the deposition of dye and thus making them visible by light microscopy.

A flagellum consists of three parts, the basal body, hook and a long filament. The basal body is embedded in the cell and consists of a central rod surrounded by rings. Gram-positive bacteria have only one pair of rings surrounding the central rod, one ring present in each of cytoplasmic membrane and cell wall. In Gram-negative bacteria, two pairs of rings are present, outer rings attached to the cell wall and inner rings lying in the cytoplasmic membrane. The helical filament is made of protein molecules called flagellin. These molecules are made within the cell and then transported along the hollow core of the flagellum for adding to the distal end, thereby confining the growth to the distal end of the flagellum.

Flagella function by rotating in a corkscrew like fashion and propel the bacteria through liquid. The rings of the basal body rotate the flagella through chemical reaction. The bacteria having polar flagella show rapid movement than those having peritrichous flagella. The environment affects the motility of the bacteria, i.e., they move towards a favourable environment or away from a harmful environment. Swimming bacteria also exhibit chemotaxis, which is movement in response to chemicals in the environment. Bacteria usually swim towards increasing levels of nutrients and away from increasing levels of harmful substances.

The number of flagella and the pattern of flagellar attachment on a bacterial cell are used to classify bacteria. Four types of flagellar attachment found in bacteria are given below:

2.4.1.1. Monotrichous: One flagellum at one pole of the cell e.g. *Xanthomonas* spp. (Fig. 2.4a).

2.4.1.2. Lophotrichous: A cluster of flagella at one pole of the cell e.g. *Pseudomonas* spp. (Fig. 2.4b).

2.4.1.3. Amphitrichous: One flagellum or cluster of flagella at both poles of the cell e.g. *Spirillum* spp. (Fig. 2.4c).

2.4.1.4. Peritrichous: The flagella scattered over the entire surface of the cell e.g. *Erwinia* and *Escherichia* spp. (Fig. 2.4d).

Some bacteria e.g. spirochetes have specialized flagella called **periplasmic flagella** (also called **axial filaments**). The helical filaments of these flagella are not free and loose like that of normal flagella. These arise at the poles and twine around the protoplasmic cylinder beneath the outer membrane of the cell wall. The spirochetes move in corkscrew like manner with the help of these flagella.

The flagella, being the organs of locomotion, play a role in the virulence of some plant pathogenic bacteria. The bacteria swim towards their host plants often after sensing a chemical signal from the host. In *Agrobacterium tumefaciens,* aflagellated mutants show about 38% reduction in virulence and produce small and less number of tumours on the host in comparison to their wild-type parent. Also, Rvi$^-$ (reduced in virulence) mutants of *Pectobacterium atrosepticum* that are deficient in flagella biosynthesis and motility but produce extracellular enzymes levels similar to the wild-type, are less virulent than the parents.

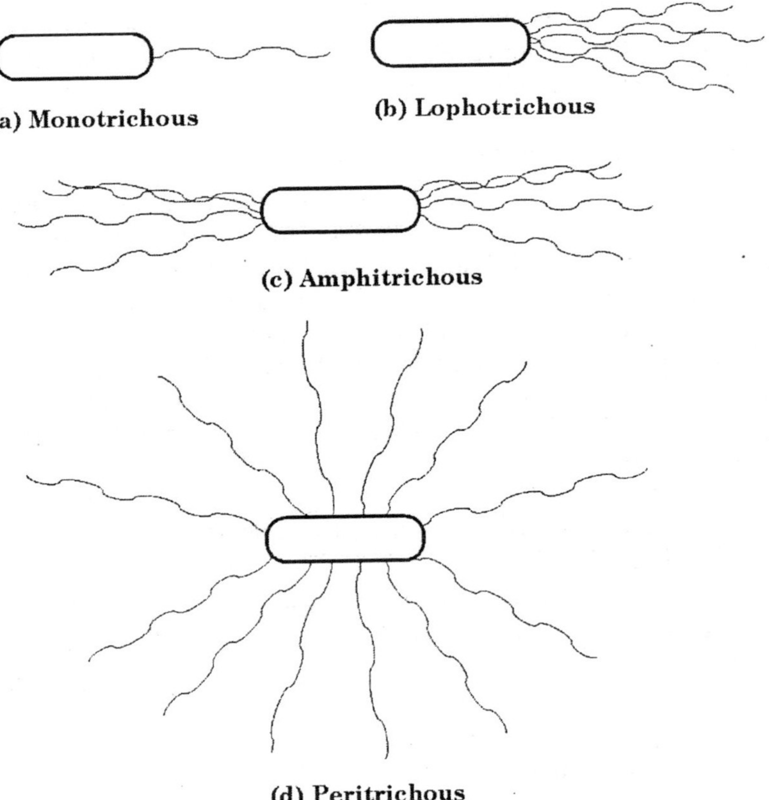

Figure 2.4 Arrangement of flagella:
(a) Monotrichous (b) Lophotrichous (c) Amphitrichous (d) Peritrichous

Pili: The pili (singular, pilus), also called fimbriac, are short, straight, non-helical structures present on the surface of bacterial cells. These are thinner (3 to 10 nm in diam) than flagella and can be seen only with electron microscope. The pili are more common in Gram-negative bacteria and are no way associated with the motility of the cells. These are hollow like flagella and are made of protein sub-units called pilin, which are arranged in a spiral fashion around the central core. These originate beneath the cell wall but no anchoring structures analogous to the basal bodies of flagella are found. They are present all along the cell and their number can range from one to several hundred per cell. In *Es. coli,* there are 100 to 300 pili arranged peritrichously around a cell, each pilus measuring 2 to 20 nm in length and 7 nm in diameter. Different morphological types of pili occur, each associated with a different function but attachment to the substrate or host cell is the primary function. A special type of pilus called **F** or **sex pilus** is associated with the bacterial mating. Bacteria that have this pilus are called donor cells and those without it are called recipient cells. In animal pathogenic bacteria, pili help bacteria to attach to host cells lining the respiratory and intestinal tracts. *Ne. gonorrhoeae,* the causal agent of gonorrhea, bears pili that recognize and adhere to receptor on human cells. Pili also provide the site of attachment to some bacteriophages.

2.4.2. Surface layers (capsule and slime layer)

Mainly two types of surface layers, namely, capsule and slime layer are found in bacteria. If the layer is definite in outline and shape and is attached firmly to the cell wall, it is called a **capsule**. Capsule may be found in many bacteria but is not universal in occurrence. Capsule is generally made of polysaccharides but in some cases it may consist of polypeptides. It can be best seen by negative staining or through electron microscopy. The capsule is a protective layer and often plays a major role in pathogenicity of pathogenic bacteria. The capsules of *Stc. pneumoniae* protect the bacterial cells from being engulfed by the white blood cells (phagocytosis) of eucaryotic cells, thereby increasing the chances of infection. In many pathogenic species, the loss of capsule may result in the loss of pathogenicity but not the viability.

When the surface of external layer is not of definite shape, disorganized and is attached loosely to the cell wall, it is called a **slime layer** (e.g. *Rhizobium trifolii*). The slime layer being soluble in water creates many problems in the industry. The slime producing bacteria clog filters in sugar industry by increasing the viscosity of the fluids and coat pipes and other equipment.

The slime layer and capsule (also termed as **glycocalyx**) perform other functions also depending on the bacterial species. Among these, adherence of cells to rocks in fast-moving water, plant roots, teeth etc. is a major function. Capsule may prevent attachment and lysis of cells by bacteriophages. They may also act as a reservoir of stored food and protect against temporary drying of cells by binding water molecules.

2.4.3. Cell wall

Barring a few groups of bacteria e.g. mycoplasmas, cell wall is the universal outer most layer of bacterial cells. It is a rigid structure responsible for the shape of cell. It accounts for 20 - 30% dry weight of the cell. Besides giving a characteristic shape to the cell, the cell wall serves as a selective barrier, allowing the influx of desired nutrients and liquids and preventing the escape of certain enzymes. It also prevents the movement of dyes, bile salts, antibiotics, heavy metals and degradative enzymes into the cell.

In bacteria, peptidoglycan (also called murein or mucopeptide) is mainly responsible for determining the shape of cell wall. The cell wall of bacteria is also unique in the sense that peptidoglycan is neither found in archaea nor in eucaryotes. The basic structure of peptidoglycan in all the bacterial species consists of the following three building blocks: 1) N-acetyl glucosamine (NAG), 2) N-acetylmuramic acid (NAM) and 3) a peptide made of four amino acids or tetrapeptide.

The alternating units of NAG and NAM linked together by glycosidic bonds form the backbone and give strength to the cell wall. The small peptides attached to NAM give an additional rigidity to the cell wall. In Gram-positive bacteria and actinomycetes, these peptides consist of 4 amino acids: L-alanine, D-glutamic acid, L-lysine and D-alanine. Diaminopimelic acid (DAP) replaces L-lysine in Gram-negative bacteria, myxobacteria, cyanobacteria and some Gram-positive bacteria. The D-configuration of glumatic acid and alanine makes the peptide bond resistant to pepsin and trypsin enzymes. The peptide chains are cross-linked with each other either through a direct bond (Gram-negative bacteria) or through a pentaglycine bridge that is formed between terminal D-alanine of one peptide chain and the third amino acid (lysine or DAP) of the other.

The enzyme lysozyme can completely destroy the cell wall of Gram-positive bacteria leaving a spherical cell called protoplast. However, the cell walls of Gram-negative bacteria are relatively resistant to this enzyme and are not completely destroyed; the spheroplast produced in this manner is equally susceptible to osmotic lysis as the protoplast.

Penicillin is a potent inhibitor of cell wall synthesis of Gram-positive bacteria. It acts by preventing the formation of cross-linkages. Penicillin does not affect the pre-existing cross-linkages but does not allow the formation of fresh cross-linkages. Therefore, the non-growing population of bacteria is not inhibited. Mycoplasmas are also not affected by penicillin because they lack the cell wall.

The cell walls of bacteria and archaea differ in both chemical function and structure. The cell walls of archaea as already stated, lack peptidoglycan but contain proteins, glycoproteins or complex polysaccharides. Some species contain pseudomurein, which consists of N-acetyltalosaminuronic acid in place of N-acetylmuramic acid and lacks D-amino acid.

In spite of the fact that peptidoglycan is common to the cell walls of Gram-positive and Gram-negative bacteria, the cell walls of both the groups differ in thickness and chemical composition. These differences help in the classification of some bacteria and also explain the basis of Gram's staining reaction. The cell walls of Gram-positive species are generally thicker (20 - 25 nm) than those of Gram-negative species (10 - 15 nm). The similar situation occurs with respect to Gram-positive and Gram-negative archaea.

2.4.3.1. Cell walls of Gram-positive bacteria: Their cell walls have much greater amount of peptidoglycan and hence much thicker than those of Gram-negative bacteria. In some species, the weight of this polymer may go up to 50% or more of the dry weight of the cell wall compared to 10% in Gram-negative species. Teichoic acids (acidic polysaccharides) are also found in walls of many species. Teichoic acids, which are polymers of a carbohydrate like glucose, an alcohol (glycerol or ribitol) and phosphate, are attached to the peptidoglycan and to the cytoplasmic membrane. These may help in the storage of phosphorus and being negatively charged also help in the transport of positive ions into and out of the cell. Gram-positive bacteria have neither outer membrane nor periplasmic space as found in Gram-negative species.

2.4.3.2. Cell walls of Gram-negative bacteria: These are more complex but thinner than the cell walls of Gram-positive bacteria. The cell walls of these bacteria are bilayered structures consisting of an outer membrane and a peptidoglycan layer; the latter is present in the periplasmic space (space between outer membrane and cytoplasmic membrane). Peptidoglycan layer of Gram-negative bacteria constitutes only 5-10% of the dry weight of the cell wall. Outer membrane is responsible for the distinguishing character of these bacteria.

The outer membrane also acts as a selective barrier for the movement of some substances into and out of the cell. This selective permeability is attributed to the presence of proteins called porins that form diffusion channels for the passage of molecules. The permeability is also related to the electric charge and size of molecules. Both specific and non-specific porins are present. Outer membrane is attached to the underlying peptidoglycan layer by a lipoprotein.

Outer membrane, a bilayered membrane contains phospholipids, lipopolysaccharides (LPSs), integral membrane proteins and lipoproteins. These components are synthesized in the cytoplasm or at the inner leaflet of the inner membrane and have to be transported across the inner membrane and through the periplasm to assemble eventually in the correct membrane (Bos *et al.,* 2007). In the outer membrane, lipopolysaccharides are present exclusively in the upper layer and phospholipids almost entirely in the lower layer.

Lipopolysaccharides found only in Gram-negative bacteria consist of lipid A, core polysaccharide and O antigen. The lipid part of a lipopolysaccharide called an endotoxin is poisonous in nature. It can cause diarrhoea, fever and destruction of red blood cells in the host.

The O antigens present in the outer membrane are responsible for many serological reactions of these bacteria. They also act as sites for the attachment of bacteriophages. Some outer member proteins also act as receptor sites for the attachment of bacteriocins and bacteriophages.

2.4.3.3. Gram staining: The differences in chemical composition and structures of cell walls of Gram-positive and Gram-negative bacteria are the bases for differential Gram reaction. The treatment of bacterial cells with crystal violet followed by Lugol's iodine solution results in the formation of crystal violet-iodine (CVI) complex within the cells. The subsequent washing of Gram-negative bacterial cells with ethanol dissolves the lipids and removes them. This disrupts the outer membrane, increases its permeability and results in washing away of the dye complex. Hence, decolourized cells are stained pink with the counter stain, safranin. On the other hand, in the Gram-positive bacteria, ethanol makes the pores in peptidoglycan layer to shrink, thus trapping the CVI dye complex inside the cells.

2.4.4. Cytoplasmic (plasma) membrane

Cytoplasmic membrane present below the cell wall is the seat of many enzymatic activities. It also acts as a selective barrier for the transport of water soluble molecules into and out of the cell and is much more selective than the cell wall. It contains various enzymes, some of which are involved in energy production and cell wall synthesis. Permeases proteins present in the membrane transport small molecules into the cell.

Cytoplasmic membrane is 7.5 nm thick and when seen by electron microscopy, it looks as a bilayered membrane. It is mainly composed of phospholipids and proteins. The phospholipids form a bilayer in which most of the proteins are partially or totally buried (intrinsic proteins). Some of the proteins called extrinisic proteins are also present on the surface of the phospholipid bilayer. The fluidity nature of the membrane allowing both proteins and lipids to move within phospholipids matrix is called the **fluid mosaic model**.

Cytoplasmic membranes of most procaryotes do not contain sterols such as cholesterol and hence they are less rigid than those of eucaryotes. However, among the procaryotes, the cytoplasmic membranes of the mycoplasmas contain the sterols, which impart rigidity to these membranes. In the mycoplasmas, cytoplasmic membrane is the outer most membrane of the cells.

The electron micrographs of thin sections of bacterial cells reveal extensive invaginations of the cytoplasmic membranes, called **mesosomes**. They are more prominent in Gram-positive bacteria. Mesosomes may be present near the cytoplasmic membrane or deep inside the cytoplasm. Earlier, these were thought to the associated with various functions like cell division, metabolic process, enzyme secretion etc. But now, these are considered to be more of an artifact than the real structure.

In the absence of membrane-bounded organelles in bacterial cells, the cytoplasmic membranes act as multiple structures. The bacteria in which the metabolism is based on the use of light energy or change of gases, their cytoplasmic membranes show extensive and elaborate intracellular extensions. Such type of extensions increases the surface area available for these activities. In phototrophic bacteria, the photosynthetic pigments are attached to these membrane systems. The nature of these membrane systems varies with different groups of phototrophic bacteria. In purple bacteria, these modifications are quite elaborate and often occupy a sizable internal cell area. In *Ectothiorhodospira,* the systems consist of flat sheets called **lamellae** while in *Chromatium* these occur as round tubes or vesicles. In *Pelodictyon*, a green bacterium, chlorosomes constitute the photosynthetic apparatus which are cylindrical structures attached to the lower side of the cytoplasmic membrane. The lamellar membrane systems of oxygenic cyanobacteria are often complex and consist of multilayered structures called **thylakoids**.

Cytoplasmic membrane plays an important role in regulating the transport/movement of water soluble molecules. This involves simple diffusion or facilitated diffusion down a concentration gradient or energy-based active transport.

2.4.5. Cytoplasm

Below the cytoplasmic membrane is the cytoplasmic area. The cytoplasmic area contains the cytoplasm, ribosomes, nucleoplasm (genophore, nucleoid), granules, vacuoles etc. The cytoplasm is a matrix of gel-like consistency and consists of water (about 80%), nucleic acids, carbohydrates, proteins, lipids, inorganic ions etc. Many chemical reactions, involving the synthesis of cell components from nutrients, take place in cytoplasm. Contrary to eucaryotic cells, there is no cytoplasmic streaming in procaryotic cells, i.e., cytoplasm does not flow around within the cell. There is also no evidence of presence of cytoskeleton, a network of fibrils responsible for maintaining the shape of the cell, in the procaryotic cytoplasm.

2.4.6. Ribosomes

The ribosomes, the site of protein synthesis, are densely packed throughout the cytoplasm. During protein synthesis, many ribosomes may be held together to form polyribosomes or polymers. In procaryotic cells, the ribosomes are 70S (Svedberg unit) type (compared to 80S found in eucaryotic cells) each having two subunits, the smaller 30S, squash-shaped and the larger 50S, somewhat spherical. The Svedberg unit is a coefficient of sedimentation in the ultra centrifugation. High levels of Mg^{2+} ions are required to keep both the subunits of ribosomes associated. The antibiotics like streptomycin, neomycin, tetracycline etc. that inhibit protein synthesis cause malfunctioning of the ribosomes. In *Es. coli*, the ribosomes are made up of 66% RNA, known as ribosomal RNA (r RNA) and 34% proteins.

2.4.7. Inclusions and storage products

Different kinds of chemical substances often accumulate forming insoluble deposits in the cytoplasm. Although they differ in their chemical composition and other aspects, they function as the reservoir of energy and other structural building blocks. Most of them are bounded by non-unit, thin, lipid-based membrane and are called inclusion bodies. They can be easily seen with light microscope in suitably stained preparations.

Some phototrophic (purple and green) bacteria store large amounts of elemental sulphur in the form of granules by oxidizing sulphur compounds such as hydrogen sulphide, thiosulphate etc. These granules may be present inside the cells (e.g. *Chromatium* spp.) or on the surface of cells (e.g. *Chlorobium* spp.) These granules may serve as stored energy for the bacteria. Another commonly stored substance in bacteria is a lipid material called poly-β-hydroxybutyrate (PHB). Poly-β-hydroxyalkanoate (PHA), a derivative of PHB, also accumulates as granules in bacteria. The plastic-like consistencies of many of these PHAs and biodegradable nature of their plastics have generated a lot of interest in the recent years. These granules can be stained with fat soluble dyes like Sudan black and Nile blue. Glycogen, a polymer of glucose subunits, is another stored product found in bacterial cells. Being polysaccharide, glycogen granules stain brown with iodine solution. Both PHB and glycogen granules serve as reservoir for carbon and energy.

Volutin granules or metachromatic granules composed of polyphosphates are also found in some bacteria. In electron microscopy, they appear as round, dark areas and stain intense reddish purple with dilute methylene blue dye.

Intracellular crystals of iron oxide, magnetite and Fe_3O_4 are found in some aquatic bacteria. These crystals are bounded by a membrane and are called **magnetosomes**. They vary in shape from square to rectangular or look like a spike. As such bacteria possess ability of orienting and migrating along geomagnetic or exhibit magnetotaxis, they are called magnetotactic bacteria.

2.4.8. Gas vesicles

The bacteria living in water bodies possess gas vesicles, with the help of which they can move up and down in response to different environmental conditions. The large blooms produced by some cyanobacteria are endowed with these gas vesicles. They are hollow but rigid, spindle shaped structures and proteinaceous in nature. They vary in length and width and also in number from a few to several hundred per cell. The vesicle membrane is permeable to gasses but does not allow the water and solutes to enter the vesicle.

2.4.9. Nucleoplasm (genophore)

The nuclear material in bacteria is called the nucleoplasm or genophore. The earlier belief that the nuclear material of bacteria is dispersed throughout the cell was due to the inadequacy of the staining techniques. With advancement of science, the exact nature of bacterial hereditary material can be easily demonstrated. It is a fairly compact clump, organized into a definite body and occupies roughly one-third volume of the cell. The total nuclear material in bacteria is called nucleoplasm or genophore. The most prominent feature of bacterial nucleoplasm is that it is **not surrounded by a nuclear membrane**. In contrast, the nuclear material in eucaryotes or higher organisms is surrounded by a well-defined nuclear membrane. Fluorescent microscopy has been found very helpful for delineating the gross boundaries of the nucleoplasm. The fluorescent probes like 4, 6-diamidino-2-phenylindole dihydrochloride (DAPI) capable of entering the cell and binding to DNA, at a non-toxic concentration, greatly help in visualization of nucleoplasm boundaries. The electron micrograph of *Es. coli* revealed that its nucleoplasm has a highly clefted appearance and is confined to a ribosomes-free zone in the cell.

Chromosome: The bacterial nucleoplasm (genophore) or genome basically means bacterial chromosome as the plasmids, extra chromosomal elements, may or may not be present. Most bacteria have a single chromosome. However, some bacteria such as *V. cholerae* and *Rohodobacter sphaeroides* have two chromosomes. The chromosome of *Es. coli* is a closed circular DNA molecule and is approximately 1 mm long. Taking into account 1 - 2 μm length of this bacterial cell, the chromosome must have condensed about 500 - 1000 times to fit into the cell. The condensation of the chromosome occurs by organizing the DNA into a series of loops and super coiling of these loops in an under wound fashion. The transcription, recombination and other functions are not hindered by the folding pattern. In most bacteria, the chromosome is circularly closed but in a few cases linear chromosomes have also been reported e.g. *Streptomyces lividans* and *Rhodococcus fascians.* Among the procaryotes, *Mycoplasma genitalium* possesses the smallest genome containing only 470 genes (~10 - 15% of *Es. coli*)

2.4.10. Plasmids

Plasmids are extra chromosomal genetic material found in most bacteria and they play a significant role in their adaptation and evolution. Most plasmids are covalently closed circular structures consisting of double stranded DNA and the DNA is negatively supercoiled. However, some bacteria such as *Borrelia burgderferi* and some species of genus *Streptomyces* possess linear plasmids. The plasmids have the ability to replicate independently. The number of copies of a plasmid per cell varies with different plasmids, F-plasmid or P1 prophage-plasmid exists as a single copy while 40 - 300 copies of plasmid PIJ 101 are found in *St. coelicolor*. Many bacteria may contain more than one type of plasmids. Initially the plasmids were named on the basis of functions which they performed e.g. F-plasmid as it mediates fertility and Ti-plasmid of *Ag. tumefaciens* induces tumour formation in some vascular plants. Now a more standardized method to name plasmids has been worked out. In this method, 'p' letter is placed before the capital letters that describe either the plasmid or denote the initials of person(s) who isolated or constructed the plasmid e.g. pPATH$_{Pag}$, pathogenicity plasmid of *Pantoea agglomerans* pv. *gypsophilae*. The methods for the detection and isolation of plasmids have also been devised.

Plasmids are of common occurrence in phytopathogenic bacteria and contribute significantly to host evolution in a multi-faceted manner. Plasmids tend to encode determinants of virulence and ecological fitness that can enhance adaptation to a specific niche or can influence niche expansion. Many of these determinants appear to have been acquired from other bacteria via horizontal transfer, showing an important function of plasmids in the acquisition of sequences that enable rapid evolution. Ultimately, these genes are transferred to the host chromosomes through plasmid integration events; thereby resulting in stabilization of important acquired determinants within the genome. Most plasmids characterized in phytopathogenic bacteria are self-transmissible and possess suites of genes encoding type IV secretion systems. Recently, Sundin (2007) has given an excellent account of genomic insights into the contribution of phytopathogenic bacterial plasmids to the evolution of their hosts. The role of plasmids in the evolution of phytopathogenic bacteria was also reviewed earlier by Coplin (1989).

Most genera of plant pathogenic bacteria harbour one or more plasmids containing genes that encode traits of importance to ecological fitness and/or virulence. Till 2007, complete genome sequence information for 22 medium- to large-sized plasmids belonging to six plant pathogenic species was obtained (Sundin, 2007).

Bacterial plasmids are well known carriers of antibiotic and heavy metal resistance genes in all ecosystems. This fact is also true for plasmids of plant pathogenic bacteria. Plasmids-encoded resistance determinants to copper and streptomycin in *Psm. syringae* pv. *tomato* and *Xa. vesicatoria* were some of the first plasmid-encoded determinants to be cloned and characterized in these bacteria. With few exceptions, all strains of *Er. amylovora* isolated from diseased tissues contain a nonconjugative plasmid called the ubiquitous plasmid pEA29. This plasmid can be cured in the laboratory, and the cured strains are reduced in virulence and show thiamine auxotrophy. The presence of pEA29 influences the expression of the chromosomal amylovoran biosynthetic operon during infection. The exopolysaccharide amylovoran is a pathogenicity determinant of the fire blight pathogen and thiamine biosynthesis is required for the high level expression of the amylovoran biosynthetic operon during infection. Of the other plasmids found in *Er. amylovora*, the plasmid pEL60 is of interest because it is an IncL/M plasmid with a high degree of similarity to the backbone of pCTX-M3, a multiple antibiotic resistance plasmid identified in the opportunistic human pathogen, *Citrobactor freundii.* pEL60 is a cryptic plasmid, encoding only genes involved in replication, maintenance and conjugative transfer, while pCTX-M3 contains two additional insertions totalling 29.6 kb that encode a number of antibiotic resistance genes.

The acquisition and expression of novel sequences may confer pathogenicity on a saprophytic strain. A classical example is the transformation of *Pa. agglomerans*, a widespread epiphytic and commensal bacterium into a host specific tumorigenic pathogen by acquiring a plasmid called pPATH containing pathogenicity island (PAI). The PAI was evolved on an interon plasmid of IncN family, which is distributed among genetically diverse populations of *Pa. agglomerans*. The horizontal gene transfer was the major contributing force in the creation of PAI, although a pathoadaptive mechanism might also be involved. More details on this aspect are given in chapter 4 under transformation.

Vivian *et al.* (2001) found that most strains of *Psm. syringae* analyzed, regardless of pathovar, contained at least one indigenous plasmid. Genetic studies of plasmids from different pathovars reveal that the large majority of plasmids in *Psm. syringae* share a major replication gene *repA*. Sequencing and functional characterization of *repA* from 100 kb plasmid pPT23A of *Psm. syringae* pv. *tomato* PT 23 and analysis of the distribution of *repA* in diverse pathovars of *Psm. syringae* led to the recognition of the pPT23A plasmid family as a group, called pPT23A family plasmids (PFPs). Most plasmids of this family range from approximately 35 – 100 kb and are believed to have originated from a common ancestor because they share *repA* gene that encodes an essential replication protein. The pPT23A plasmid family is considered to be universally distributed among *Psm. syringae* pathovars. The *repA* replication gene is the only gene conserved in all PFPs studied by Zhao *et al.* (2005). Genes encoding virulence factors such as type III effectors, coronatine toxin biosynthesis, plant hormone production, and determinants important for epiphytic fitness including UV radiation tolerance are commonly found on PFPs (Vivian *et al.*, 2001). Genes of potential ecological importance or virulence genes of general importance to *Psm. syringae* are more widely distributed among pathovars. The loss of virulence in *Psm. syringae* pv. *phaseolicola* to bean is associated with the loss of plasmid pAV511.

The plasmids act as carriers of virulence/avirulence genes in *Xanthomonas* species. Additional avirulence/effector genes such as those of *avrBs3/pthA* family and *avrBs4* are plasmid encoded in *Xanthomonas* spp. and in some cases, these genes are located on conjugative plasmids. The *avrBs3/pthA* factors play a role in pathogen aggressiveness and virulence. Fujikawa *et al.* (2006) have shown that *avrBs3/pthA* effectors function as suppressors of host defence responses.

The integration of plasmids into the host chromosomes occurs and it represents one delivery method for horizontally acquired sequences. Integration of plasmid pMC7105 from *Psm. syringae* pv. *phaseolicola* was first demonstrated in mid-1980s. It was shown to replicate autonomously or integrate into the host chromosome. Integration was achieved via recombination between copies of the insertion sequence IS*801* on the plasmid and chromosome. An extensive sequence homology found between pMC7105 and plasmid DNA from *Psm. syringae* pv. *glycinea* indicates that integration of plasmids may occur in other *Psm. syringae* pathovars. Plasmid pFKN, a PFP from *Psm. syringae* pv. *maculicola*, has also been shown to integrate into the chromosome through recombination between alleles of the effector gene *avrPphE*. This integration was considered to comprise an evolutionary event resulting in the formation of a new pathogenicity island in the *Psm. syringae* pv. *maculicola* chromosome. The integration of plasmids into chromosomes can be viewed as the first step in the generation of new island sequence.

Insertion sequence (IS) elements play a role in the evolution of bacteria in various ways. These sequences move within plasmids, between plasmids and between plasmid and chromosome. IS elements can facilitate recombination between plasmids and chromosomes. Plasmid -encoded IS elements effect pathogen virulence and evolution. The element IS*801* appears prominently in the evolution of certain *Psm. syringae* pathovars while absent in other pathovars. IS*801*-like sequences occur close to effector genes in *Psm. syringae* pvs. *glycinea* and *phaseolicola*, and downstream of plant hormone biosynthetic genes in *Psm. savastanoi* pv. *savastanoi*. On account of this close proximity, IS*801* has been considered to be involved in the mobility of virulence genes. IS*3* family is one of the largest families of IS element. IS*476* encoded on plasmids of *Xa. campestris* pv. *campestris* and *Xa. vesicatoria*, IS*1389* of *Xa. campestris* pv. *amaranthicola*, IS *Xac3* of *Xa. citri* subsp. *citri*, IS *Psy26* of *Psm. syringae* pv. *tomato* and an unnamed element of pEA29 of *Er. amylovora* are members of the IS*3* family. IS elements like IS*801* serve as sites for recombination facilitating integration of plasmids into chromosomes, and these elements can also mobilize adjacent sequences during the transposition process.

The plasmids also play a role in other plant pathogenic bacteria. *Psm. syringae* pv. *syringae*, incitant of bacterial canker of stone fruits contains plasmid-mediated genes for copper resistance. Patro and Jindal (2000) have shown the existence of plasmids in *Xa. axonopodis* pv. *vignaeradiatae*, incitant of bacterial leaf spot of green gram, and their correlation with virulence. The highly virulent isolates contained the maximum number of plasmids. The plasmid-borne genes were shown to govern virulence, exopolysaccharide production, antibiotic resistance and pigmentation.

The plasmids have played a very important role as vectors in genetic engineering. They have been used to transfer required genes into the strains of the same species, into different species or even into organisms of different kingdoms. Their small size, easy isolation and introduction, more number of copies, possession of an easily selectable marker, presence of sites for restriction endonucleases and amenability to desired modifications have greatly enhanced their utility in genetic engineering. Tumour inducing (Ti) plasmid of *Ag. tumefaciens* is commonly used for transferring genetic material into different bacterial species or even into plants. *Ag. tumefaciens* is an excellent genetic engineering tool. The DNA transformation capabilities of the bacterium have been extensively exploited in biotechnology as means of inserting foreign genes into plants. Therefore, Nester (2008) has appropriately called it a natural genetic engineer and also a household name in plant genetic engineering and plant molecular genetics.

2.4.11. Dormant forms

Some bacteria produce dormant forms or resting stages called spores and cysts. These dormant forms help the bacteria to tide over or withstand the adverse conditions like heat or drying. They are metabolically inactive, i.e., not growing but under favourable conditions, they can germinate and grow into metabolically active vegetative cells.

2.4.11.1. Spores: Spores are thick-walled, dormant and highly refractile structures. As they are formed inside the cells, they are also called **endospores**. The spores are highly resistant to heat, drying and chemicals toxic to the vegetative cells. The endospores of *Clostridium botulinum,* the cause of botulism (food poisoning), can resist boiling for several hours while vegetative cells of most bacteria are killed above 70°C. As the spores are resistant to adverse environmental conditions, they help the bacteria to tide over the adverse environmental conditions.

Among the rod-shaped bacteria, the genera *Bacillus* and *Clostridium* are exclusively spore formers. Among the spherical bacteria, the members of genus *Sporosarcina* form the spores. Normally one spore is produced per cell and they are usually produced in old cultures. The spores vary in shape (spherical or ovoid) and location within the cell. The location in the cell can be central, sub-terminal or terminal (Fig. 2.5a, b and c, respectively). In some cases, terminal spores are formed in swollen sporangia

(Fig. 2.5d). Spore position in the mother cell or sporangium frequently differs among species and is helpful in the identification of bacteria. It is essential to apply heat to the endospores when staining for light microscopy. In bacteria, endospores are not the means of multiplication but means to tide over the adverse conditions because one vegetative cell produces one spore and on germination one spore produces one cell.

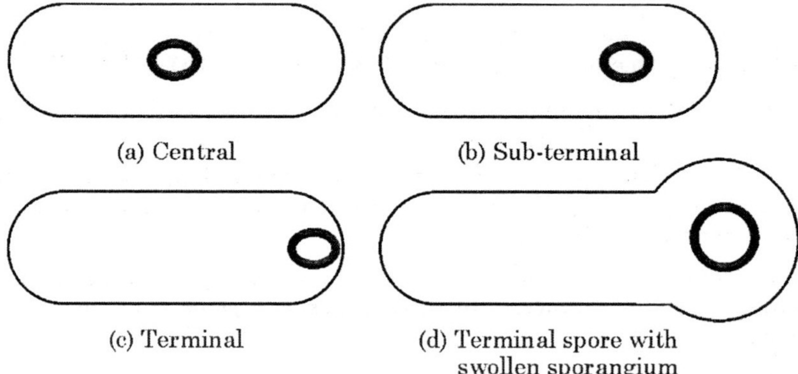

Figure 2.5 Location of spores in the cell:
(a) Central (b) Sub-terminal (c) Terminal (d) Terminal spore with swollen sporangium

During sporulation, dehydration process occurs that expels most of the water from the spore. All endospores contain large amounts of dipicolinic acid (DPA), a compound not found in vegetative cells. DPA accounts for 5 - 10% of the endospore's dry weight and it occurs in combination with large amounts of calcium, both accounting for 15% of the spore's dry weight. Earlier, it was thought that DPA was directly responsible for resistance of spores to heat. However, the occurrence of heat-resistant mutants lacking DPA has seriously weakened this viewpoint. Calcium does aid in resistance to wet heat, oxidizing agents and sometimes dry heat. It may be that calcium-dipicolinate often stabilizes spore nucleic acids. The calcium-dipicolinate complex in combination with acid-soluble protein stabilization of DNA, protoplast dehydration, DNA repair, the spore coat and the greater stability of cell proteins in bacteria adapted to growth at high temperatures are considered responsible for imparting resistance to spores against heat and other lethal agents.

Another type of spores produced by bacteria is **conidia** (singular, conidium), produced by actinomycetes. These spores are not much resistant to heat but are resistant to drying. As one thallus of actinomycetes produces large number of conidia, these are the means of reproduction, but not the means to tide over the adverse conditions.

2.4.11.2. Cysts: The cysts are also dormant, thick-walled structures that resist drying. However, they are not highly resistant to heat like endospores. The vegetative cells give rise to cysts and under favourable conditions cysts germinate to give rise to vegetative cells. The vegetative cells of several bacterial genera have been reported to produce cysts but those produced by the members of *Azotobacter* are considered the typical bacterial cysts.

REFERENCES

Bos, M.P., Robert, V. and Tommassen, J. (2007). Biogenesis of the Gram-negative bacterial outer membrane. *Ann. Rev. Phytopathol.* 61:191 - 214.

Coplin, D.L. (1989). Plasmids and their role in the evolution of phytopathogenic bacteria. *Ann. Rev. Phytopathol.* 27: 187 - 212.

Fujikawa, T., Ishihara, H., Leach, J.E. *et al.* (2006). Suppression of defence response in plants by the *avrBs3/pthA* gene family of *Xanthomonas* spp. *Mol. Plant-Microbe Interact.* 19: 342 - 47.

Nester, E. (2008). *Agrobacterium*: The Natural genetic engineer: 100 years later. APSnet Feature Story, [http://www.apsnet.org/online/feature/*Agrobacterium/*].

Patro, T.S.S.K. and Jindal, J.K. (2000). Expression of plasmid-borne characters in *Xanthomonas axonopodis* pv. *vignaeradiatae*. *Trop. Agri. Res.* 12: 86 - 94.

Pelczar, M.J. Jr., Chan, E.C.S. and Krieg, N.R. (2004). *Microbiology: Concepts and Applications*. 5th Ed. Tata McGraw-Hill, New Delhi, 918 pp.

Prescott, L.M., Harley, J.P. and Klein, D.A. (2005). *Microbiology*. 6th Ed. McGraw-Hill Companies, Inc., New York, 1130 pp.

Srivastava, S. and Srivastava, P.S. (2003). *Understanding Bacteria*. Kluwer Academic Publishers, Dordrecht, 469 pp.

Sundin, G.W. (2007). Genomic insights into the contribution of phytopathogenic bacterial plasmids to the evolutionary history of their hosts. *Ann. Rev. Phytopathol.* 45: 129 - 51.

Tortora, G.J., Funke, B.R. and Case, C.L. (2007). *Microbiology: An Introduction*. 9th Ed. Benjamin Cummings, San Francisco, 958 pp.

Vivian, A., Murillo, J. and Jackson, R.W. (2001). The roles of plasmids in phytopathogenic bacteria: Mobile arsenals? *Microbiology* 147: 763 - 80.

Zhao, Y., Ma, Z. and Sundin, G.W. (2005). Comparative genomic analysis of the pPT23A plasmid family of *Pseudomonas syringae*. *J. Bacteriol.* 187: 2113 - 26.

Chapter 3

GROWTH AND NUTRITION OF BACTERIA

3.1. GROWTH

In bacteria growth or multiplication means increase in cell numbers. Bacteria multiply by asexual reproduction as the cells arise from just one parent cell. One cell divides into two cells; two cells divide into four cells and so on until the conditions are favourable for the growth of organisms. This increase where the cell number is doubled every time is called **geometric progression** and is expressed as given below:

$1 \rightarrow 2^1 \rightarrow 2^2 \rightarrow 2^3 \rightarrow 2^4 \rightarrow \ldots\ldots\ldots\ldots \rightarrow 2^n$, where the exponent n refers to the number of generations.

In most unicellular procaryotes, asexual reproduction is through transverse binary fission, where a single cell divides into two daughter cells. Before the cell division, the cell contents roughly double in amount and the nucleoplasm or genophore (nuclear material) is replicated. Along with the enlargement of the parent cell, the cytoplasmic membrane extends inwards transversely and the nuclear material gets separated into two equal halves. The cytoplasmic membrane further moves inward dividing the cell in the zone between the two nucleoplasms. Simultaneously, the cell wall material also grows inward to form a cross wall (septum) between the two daughter cells. In unicellular species, the daughter cells separate from each other while in others they remain attached to form different arrangement of cells like pairs, chains or clusters.

Some procaryotes reproduce asexually by methods other than binary fission. *Rhodopseudomonas acidophila* multiplies by budding. In this method, a small bud is produced at one end of the cell, which grows eventually becoming a new full-fledged cell and then separating from the parent cell. The members of actinomycetes such as *Streptomyces* spp. produce condia (spores) in chains at the hyphal tips. These conidia germinate and produce new organisms. The species of *Nocardia* that produce filamentous growth, multiply by fragmentation of filaments into small rod-shaped or coccoid cells, which give rise to new growth.

3.1.1. Growth rate

The growth rate is a parameter by which the growth of a microorganism can be assessed. The growth rate can be determined by assessing the change in number of cells or mass per unit of time. Another parameter of growth rate is to know the generation time.

3.1.2. Generation time

It is the time interval required for each microbial cell to divide or for the population in a culture to double. Hence, it is also known as **doubling time**. The generation time is not constant for all the microbial species and they exhibit a wide range. It is 20 minutes for *Escherichia coli,* 6 hr for *Mycobacterium tuberculosis* and 33 hr for *Treponema pallidum.* Moreover, the generation time is also not constant for a particular species under all conditions. The generation time is strongly influenced by the nutrition of the medium and physical conditions of incubation.

3.1.3. Growth curve of unicellular microorganisms

The routine microbial cultures growing in a flask or a tube of liquid medium are considered to be in a closed system, because no new nutrients are added and no metabolic waste products are removed from the medium. On the addition of inoculum to the medium, the cells start growing after some time and the cell population continues to increase at a constant rate for some period. With the passage of time, the nutrients get exhausted and metabolic waste products accumulate which result in decreasing the growth rate, then decline in cell number and finally the death of the culture. When a microbial culture is inoculated into a fresh medium, it will start multiplying after some time. If the number of viable cells is plated at regular time intervals and logarithmic values of these cells are plotted versus time, a growth curve is obtained (Fig. 3.1). This growth curve consists of various phases given below:

1. Lag phase
2. Log phase
3. Stationary phase
4. Decline phase

3.1.3.1. Lag phase: Immediately after inoculation, the cells do not divide and there is no increase in cell number. The cells are metabolically active, repairing cellular damage and synthesizing enzymes but they do not divide. The occurrence of a lag phase or its durations is influenced by many factors. If the inoculum is taken from a log phase, the lag phase may not occur but if the same is derived from a stationary phase, the lag phase will usually appear. This is due to the reason that the old cells get depleted of some cellular constituents and essential enzymes, and thus require some time to re-establish their synthesis.

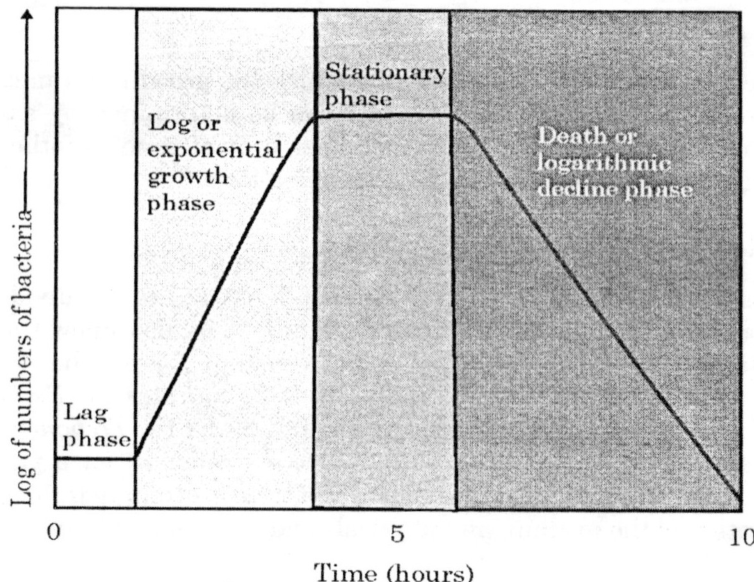

Figure 3.1 Bacterial growth curve showing four typical phases of growth

3.1.3.2. Log phase: Log phase, also called **exponential, logarithmic** or **growth phase,** is a period of rapid balanced growth. In this phase, the number of cells doubles (multiplies in geometric progression) with each generation time of the microorganisms. As the culture grows, nutrients of the medium go on depleting and waste metabolic products go on accumulating. Hence, the log phase or the constant rate of increase in cell number stops after sometime. The repercussions of an unlimited exponential phase can be gauzed from the following example. A bacterium of the size of *Es. coli* dividing every hour, in six days it would produce bacterial mass having a total volume 10,000 times that of the earth.

3.1.3.3. Stationary phase: After the log phase, there is stationary phase in which there is no net increase in cell number. The multiplication of cells does occur but it compensates for the death of cells.

3.1.3.4. Decline phase: Decline phase, also called the **death phase**, sets in due to further accumulation of inhibitory metabolic products and depletion of essential nutrients. The death rate increases and the number of viable cells decreases exponentially. Finally all the cells die ranging from days to months.

3.1.4. Measurement of growth

The growth of a bacterial population can be measured by the following two methods:

3.1.4.1. Cell count

3.1.4.2. Cell weight/mass

3.1.4.1. Cell count: The cell count can be obtained by microscopic or by colony count.

3.1.4.1.1. Microscopic count: It is the simplest and most direct method to determine the bacterial population. The haemocytometers containing counting chambers are used for this purpose. In this, a grid consisting of 25 large squares is marked on a slide. The bacterial cell suspension is placed in the chamber and covered with a cover slip. The total area of 25 squares is 1 sq. mm and a total volume of 0.02 mm^3. The numbers of cells in different squares are counted and the average thus obtained, is multiplied by the required factor to calculate the number of cells per milliliter.

In spite of being least cumbersome and quickest, this method suffers from some major drawbacks. Firstly, it does not differentiate between living and dead cells. Secondly due to small size of bacteria, the viewing and counting of the cells becomes difficult and may cause inaccuracy in the result.

3.1.4.1.2. Colony count: As in most studies, only the viable cells are taken into account, a method giving count of viable cells is preferred over the microscopic count. This method is called colony count or **plate count** and is based on the assumption that every viable cell will produce a colony when grown on a suitable medium. It can be performed by **spread plate method** or **pour plate method** and in both the methods an appropriate dilute cell suspension is used. In spread plate method, 0.1 ml of cell suspension is spread evenly on an agar plate by a glass spreader. In pour plate method, 0.1 ml cell suspension is poured in the Petri plate and mixed with cooled molten agar medium. In both the methods, the plates are incubated at the desired temperature and the number of colonies counted. The number of viable cells in the suspension is determined by multiplying the number of colonies per plate by 10 and is expressed as colony forming units (cfu) per milliliter. Pour plate method is considered better as it gives more accurate results due to uniform and thorough spreading of the cells.

3.1.4.2. Cell weight/mass: The cell weight is obtained by centrifuging the cell suspension and weighing the pellet of cells. The dry weight of such cells can be obtained by drying the cells in an oven at ~ 100°C, which in most of the cases is approximately 10 - 20% of the wet mass. The cell mass may also be determined by measuring the amount of various cellular constituents like DNA, RNA, or proteins present in the cells or by measuring the amount of various metabolic products, namely, organic acids.

The turbiditimetric method is also used to determine the cell mass, which is based on the principle that a cell suspension looks turbid because every cell placed in the path of light scatters it. The presence of more number of cells will scatter more light and hence making the cell suspension more turbid. The turbidity is measured with spectrophotometer or colorimeter in terms of absorbance at a specific wavelength. The turbiditimetric method is used in those cases where a cell suspension having a known number of viable cells is to be used immediately but by pour plate method, this count will be available after some time, say 24 or 48 hr. In such cases, a standard curve between turbidity (optical density) and the number of viable cells over the growth period is made beforehand. The cell suspension is standardized to obtain the desired number of cells by matching with the corresponding optical density. This method is quick and popular but does not differentiate between living and dead cells.

3.2. FACTORS AFFECTING GROWTH OF BACTERIAL CULTURES

The microbial cultures are generally grown in the laboratory for various experimental purposes. In some bacterial species, the maximum population is obtained in 24 - 48 hr while others require a long incubation period to produce the maximum growth. Every organism must find in its environment all the substances required for energy generation and cellular biosynthesis. In fact, the growth is the direct outcome of an interaction among the microorganism's inherent capacity to grow and the conditions/environments under which it grows. The conditions, which affect the growth are broadly classified as nutritional and physical factors.

3.2.1. Physical factors

3.2.1.1. Temperature: Among the physical factors, temperature is probably the most important one to influence the growth of bacteria. It is due to the fact that all the processes of growth are dependent on chemical reactions that are influenced by temperature. The bacteria have been found in almost all the habitats of the earth suggesting a very broad range of temperature for growth. This temperature range may be wider for some species e.g. *Bacillus subtilis* grows from 8 - 53°C, a range of 45 degrees. In contrast, *Neisseria gonorrhoeae* has a range of only 10°, i.e., from 30 to 40°C.

All the microbial species have three important temperatures, namely, minimum, optimum and maximum and these are known as cardinal temperatures. The cardinal temperatures of a species may vary with the availability of the nutrients or with the stage of the life cycle of the microorganisms. Besides affecting the growth rate, temperature may also affect the cellular activities like metabolic processes, kind of reproduction, morphology etc. It is not ironical to assume that the optimum temperature lies in the middle of the temperature range. Rather, it is nearer the upper limit of the range. On the basis of the temperature range, the microorganisms are divided into following five groups (Fig. 3.2).

1. Psychrophiles
2. Psychrotrophs
3. Mesophiles
4. Thermophiles
5. Hyperthermophiles

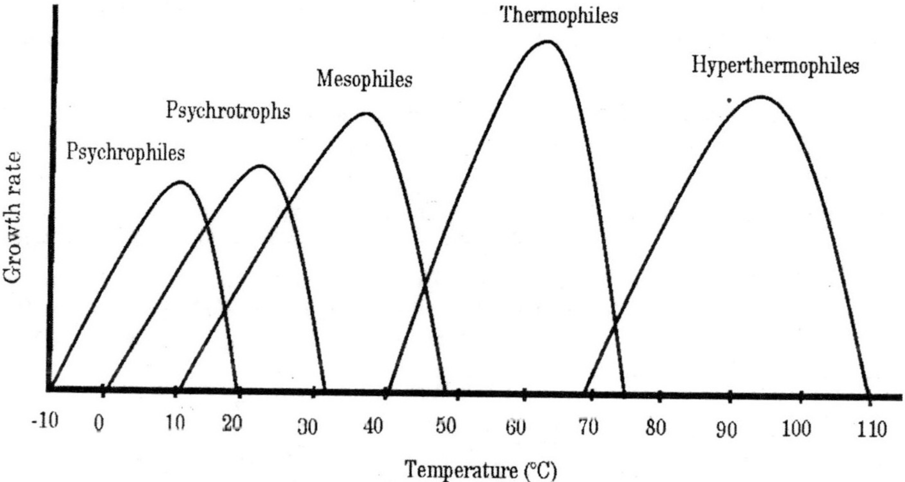

Figure 3.2 Growth rate of five environmental classes of procaryotes in relation to temperature. The curves exhibit three cardinal points, i.e., minimum, optimum and maximum temperatures for growth.

3.2.1.1.1. Psychrophiles: They are cold-loving microbes and grow best at temperature around 15°C but they can grow in the temperature range of 0 to 20°C. They are found in cold waters and in soils of Polar Regions and oceans. The majority of the marine bacteria belong to this group. Many forms of life flourish in the icy waters of Arctic and Antarctic regions. Some of the psychrophiles are the species of *Pseudomonas, Alcaligenes, Flavobacterium, Vibrio, Bacillus* and *Arthrobacter.*

The adaptation of psychrophilic bacteria to cool environment is due to the occurrence of mainly unsaturated fatty acids in their plasma membranes. Some psychrophiles, especially from the Antarctic region have been found to contain polyunsaturated fatty acids, generally not found in other procaryotes. The degree of unsaturation of a fatty acid is correlated with its solidification temperature and unsaturated fatty acids remain liquid at low temperatures. The better efficiency of psychrophilic enzymes in catalyzing reactions at lower temperatures (at or near 0°C) is another reason for their adaptation to low temperatures.

3.2.1.1.2. Psychrotrophs: Psychrotrophs, also called **facultative psychrophiles** or **psychrotolerants,** are the variants of psychrophiles.

They can grow at 0 to 7°C but have the optimum temperature range between 20 and 30°C and maximum at about 35°C. The psychrotrophs are the main cause of food spoilage in the refrigerators because they can grow fairly well at temperatures maintained in the refrigerators.

3.2.1.1.3. Mesophiles: Most bacteria are mesophiles and they produce maximum growth in the temperature range of 25 to 40°C. They often have a minimum temperature range of 15 to 20°C and maximum temperature around 45°C. Most of the saprophytic bacteria, pathogens of humans and warm-blooded animals belong to this group. Human pathogens grow best at about body temperature, which is 37°C. Plant pathogens also belong to this group.

3.2.1.1.4. Thermophiles: The thermophiles are heat-loving microorganisms. Most of the thermophiles grow in the temperature range of 40 to 85°C, but most of them have optimum temperature between 55 and 65°C. They are mostly found in hot springs, volcanic regions, compost heaps etc. No eucaryotic cells are known to grow at temperature higher than 60°C, hence thermophiles are procaryotes. *Ba. flavothermus, Thermus aquaticus* and *Thermoplasma acidophilum* are the example of thermophiles. *Th. aquaticus* is the source of taq polymerase that is largely used in polymerase chain reaction in these days.

3.2.1.1.5. Hyperthermophiles: Some procaryotes have an optimum temperature of 80°C or even much higher. Such procaryotes are called hyperthermophiles or **extreme thermophiles**. In 1982, German microbiologists discovered a new genus, called *Pyrodictium,* which had an optimal temperature of 105°C and could even grow at 110°C. Most of the hyperthermophiles belong to archaea (archaeobacteria) group. *Pyrococcus furiosus, Pyc. woesei,* and species of *Methanothermus* and *Methanopyrus* belong to this category.

Thermophiles and hyperthermophiles have various provisions, which enable them to withstand such high temperatures. The membranes of thermophiles contain highly saturated fatty acids, which remain stable at high temperature. The high G + C content found in many thermophiles keeps their DNA stable at high temperatures. The membranes of hyperthermophiles contain repeating subunits of C5 compound, phytane, a branched, saturated isoprenoid. This compound imparts considerable strength to these bacteria to withstand the superheated temperatures. The better stability of structural proteins and enzymes of thermophiles in comparison to those of mesophiles also contributes to the survival of thermophiles at such higher temperatures.

3.2.1.2. Gaseous environment: Different gases present in the environment such as oxygen, carbon dioxide, nitrogen and methane are required in varying amounts for the growth of microorganisms. Among these oxygen and carbon dioxide are the main gases, which affect the growth of microorganisms. On the basis of their requirement for oxygen, the procaryotes are divided into following four groups:

1. Aerobes
2. Facultative anaerobes
3. Anaerobes
4. Microaerophiles

3.2.1.2.1. Aerobes: The bacteria that require oxygen for growth and can grow in a standard air atmosphere of 21% oxygen belong to this group. The low levels of oxygen can affect the growth of some aerobes. There is a no problem when the growth occurs on the surface of solid media but in liquid cultures, the growth is retarded after some time due to lack of oxygen. To overcome this problem, the liquid cultures are grown on shakers to increase the supply of dissolved oxygen. Most plant pathogens are aerobic in nature.

3.2.1.2.2. Facultative anaerobes: These microorganisms can grow in the presence as well as in the absence of oxygen. They do not need oxygen for growth but may use it for energy-yielding chemical reactions. In the presence of oxygen (aerobic conditions), they grow by aerobic respiration while in the absence of oxygen; they switch over to fermentation or anaerobic respiration. The enteric bacteria such as *Es. coli* and *Saccharomyces cerevisiae* belong to this group.

3.2.1.2.3. Anaerobes: They do not need or use oxygen for growth and grow in the absence of oxygen. They grow by fermentation, anaerobic respiration, bacterial photosynthesis and methanogenesis. For strict anaerobes, oxygen is a toxic substance and a brief exposure to it will kill them. The members of *Methanobacterium* and *Methanospirillium* genera and many spp. of *Clostridium* are strict anaerobes.

A sub-group called **aerotolerant anaerobes** has been recognized among the anaerobic microorganisms. They live exclusively by anaerobic (fermentative) type of metabolism but are not affected by low levels of oxygen. Some species of *Clostridium* like *Clo. tetani* and *Clo. perfringens* and some streptococci belong to this group.

3.2.1.2.4. Microaerophiles: These are the aerobic microorganisms, which grow only at the reduced concentrations of oxygen and usually grow best at oxygen levels ranging from 5 - 15%. They cannot grow at atmospheric level of oxygen (21%). Their high susceptibility to superoxide radicals and hydrogen peroxide inhibits their growth at saturated levels of oxygen (21%) because these toxic compounds are formed in cultures grown under aerobic conditions.

The response of an organism to oxygen depends upon the chemical molecules produced during its growth and the enzymes produced by it. The addition of a single electron to an oxygen molecule results in the formation of a superoxide radical (O_2^-) toxic to the cells. Superoxide radicals are converted into hydrogen peroxide (H_2O_2) and these are further converted to hydroxyl radicals (OH^-). Both hydrogen peroxide and hydroxyl radicals can destroy various vital cell components.

Aerobic, facultative anaerobic and microaerophilic bacteria produce various enzymes to protect the cells from these toxic radicals. The bacteria produce enzyme superoxide dismutase that rapidly converts superoxide radicals to peroxide. Two other enzymes, i.e., catalase and peroxidase produced by bacteria eliminate hydrogen peroxide, the former converts hydrogen peroxide into molecular oxygen and water, and the latter converts it into water. The elimination of both superoxide radicals and hydrogen peroxide ensures the non-production of hydroxyl radicals. As obligate anaerobes do not produce superoxide dismutase, catalase and/or peroxidase, they are killed by different oxygen radicals when exposed to oxygen.

The bacteria containing chlorophyll and other pigments can reduce oxygen to form singlet oxygen, which is toxic to the cells, in the presence of light. All photosynthetic and some non-photosynthetic bacteria that contain carotenoid pigments are protected from lethal oxidation of singlet oxygen. The carotenoid pigments react with the singlet oxygen and convert it to nontoxic triplet form of oxygen.

3.2.1.3. Hydrogen ion concentration (pH): Hydrogen ion concentration is another important component of environment media, which affects the growth of procaryotes. Like temperature, pH also has three cardinal values, i.e., minimum, optimum and maximum pH. However, unlike the optimum temperature, the optimum pH of a microorganism falls approximately in the middle of the pH range over which the growth occurs. Every microorganism has a pH range within which it can grow and within this range there is optimum pH at which maximum growth takes place. Most bacteria prefer near neutral pH for their growth and their pH range lies between 6.5 and 7.5. The microorganisms that have an optimum pH well below neutrality (pH 7.0) are called **acidophilies** while those that grow best under alkaline conditions are called **alkalophiles**. The microorganisms that grow best at neutral pH are called **neutrophiles** (Fig. 3.3). Some species of *Thiobacillus* and certain genera of Archaea like *Sulfolobus* and *Thermoplasma* are obligate acidophiles. Some *Bacillus* and *Nitrobacter* species are alkalophiles. During the growth of the bacteria, the pH of the medium usually changes due to the production of acidic or alkaline metabolites and may make the media unfit for further growth. This problem is overcome by adding buffers to the medium as these buffers stabilize the pH of the medium.

The optimum pH defined for the growth of a microorganism represents the pH of the external environment and not the intracellular pH of the cell. The intracellular pH of the cell is maintained around neutrality irrespective of the pH of the external environment. In extreme acidophiles or alkalophiles, the intracellular pH may vary by 1.0 - 1.5 pH units from neutrality.

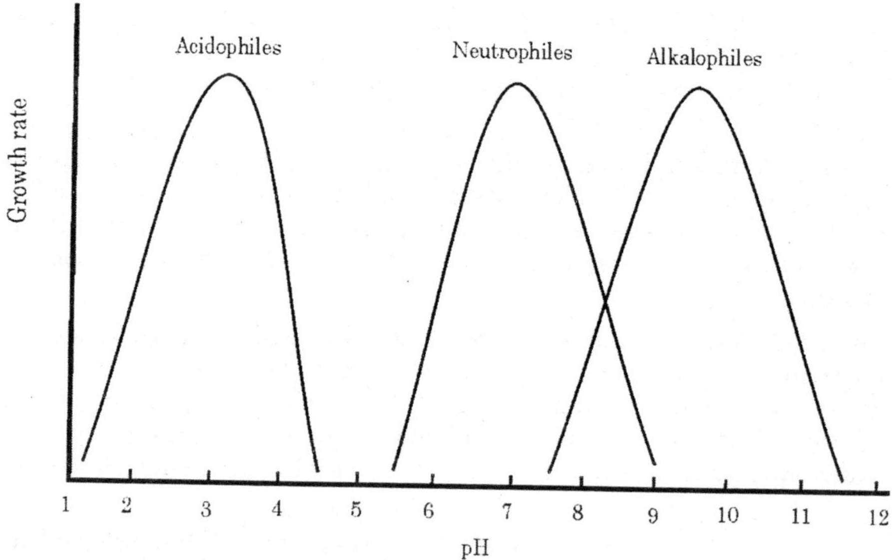

Figure 3.3 Growth of acidophilic, neutrophilic and alkalophilic procaryotes in relation to pH

3.2.1.4. Water availability: Water is an essential component for all living cells. It is also a universal solvent and hence required for all biochemical reactions. The water activity (A_w) is measured in terms of amount of free water available for reaction. It is equivalent to amount of water present in a solution or substance or measure of atmospheric water known as relative humidity. The water activity of pure distilled water is 1.0 (100% water) and of seawater is 0.98. The water activity of agricultural soils varies between 0.9 and 1.0. The addition of solutes such as salts and sugars in the water decreases the water activity.

The water activity requirement of microorganisms ranges from 0.7 to 1.0 while most bacteria grow optimally at 0.9 water activity. The bacteria do not grow in air on account of relatively low availability of water in air. Many bacteria have developed mechanisms to withstand desiccation e.g. production of endospores, formation of thick walls and slime layers and production of extracellular polysaccharide layers including trehalose.

3.2.1.5. Light: Light is not required for growth by bacteria but all photosynthetic bacteria need light for photoautotrophic generation of ATP.

3.2.2. Nutritional factors (requirements)

The bacteria have simple nutritional requirements. Some of them can grow in a solution containing a few chemical compounds. The process of growing bacteria in synthetic media is known as **culturing**. The composition of media will vary with the species of bacteria to be grown but all the culture media are designed to provide all the essential nutrients required for energy generation and cellular biosynthesis. As the nutritional requirements of most bacteria are not fully known, only a very small fraction (~<1%) of all the bacteria existing in nature can be cultured.

The nutritional requirements of bacteria can be broadly categorized as major elements, trace elements and growth factors. These is no hard and fast rule for categorization of major and trace elements because the trace elements of one organism may act as major elements of another and vice-versa.

3.2.2.1. Major elements: These are required in large amount in comparison to trace elements for bacterial growth. Although these may vary with different bacterial species, C, H, O, N, S, P, K, Mg, Fe, Ca etc. are usually classified as major elements. They perform either structural or functional roles in the cells. The brief role of some major elements is given below.

3.2.2.1.1. Carbon: It is a main constituent of cellular material and accounts for 50% of dry weight of the cells. Carbon forms the backbone of three major classes of organic nutrients, namely, carbohydrates, lipids and proteins. The photoautotrophs or chemoautotrophs obtain carbon from CO_2; the others obtain it from organic compounds. Although bacteria use a variety of carbon compounds including higher hydrocarbon like petroleum, glucose is the most preferred one.

3.2.2.1.2. Oxygen: It is a second major constituent of bacterial cell and accounts for 20% of dry weight of the cells. In aerobic respiration, it acts as electron acceptor. It is obtained from organic compounds, water, carbon dioxide and atmospheric oxygen.

3.2.2.1.3. Nitrogen: It is an important constituent of nucleic acids, proteins/amino acids, nucleotides and coenzymes. It accounts for 14% of dry weight of the cells. It is generally obtained from nitrates, ammonia and organic compounds while some bacteria, including many of the photosynthetic cyanobacteria use gaseous nitrogen directly from the atmosphere. This process is called **nitrogen fixation.** Some bacteria that fix atmospheric nitrogen are free-living, mostly in the soil, but some do it in association with plants. The latter group includes *Rhizobium* and *Bradyrhizobium* species in association with legume plants.

3.2.2.1.4. Hydrogen: It is the main component of organic compounds and cell water. It is obtained from water, organic compounds and H_2.

3.2.2.1.5. Sulphur: It is a component of some amino acids like cysteine, methionine and glutathione, and several coenzymes. The sulphates are the major source of sulphur but it can also be obtained from organic sulphur compounds and hydrogen sulphide. Some bacteria are capable of utilizing elemental sulphur also.

3.2.2.1.6. Phosphorus: It is an important constituent of phospholipids, nucleic acids, nucleotides, lipopolysaccharides and coenzymes. It is also found in the energy bonds of ATP. It is usually derived from inorganic phosphates.

3.2.2.1.7. Metal ions: The metal ions like potassium, magnesium and calcium are the important inorganic cations required for bacterial growth and are supplied by their respective salts. They function as cofactor in several enzymatic reactions and as structural components. Calcium is also a major component of endospores. Iron derived from iron salts, acts as a component of cytochromes and a cofactor for some enzymatic reactions. Chloride is an important anion that must be added in the medium for bacterial growth.

3.2.2.2. Trace elements: These are the metal ions, required by certain species, in very small amounts that is difficult to detect or measure. As these are required in such small amounts, it is not necessary to add them in the media because they are present as contaminants in water or other media ingredients. The trace elements generally act as cofactors for essential enzymatic reactions. The cations that usually serve as trace elements in bacterial nutrition are Mn, Co, Mo, Zn, and Cu.

3.2.2.3. Growth factors: There are certain organic compounds, which are required in small amounts by some bacteria because these are essential substances for their growth and the organisms are unable to synthesize them from the available nutrients. These compounds are called growth factors. Such compounds must be added to the media. The inability of the bacteria to synthesize the growth factors results either from the missed/blocked metabolic pathway in the cells. The vitamins, amino acids, purines and pyrimidines are the examples of growth factors.

Not all the bacteria require growth factors e.g. *Es. coli* can synthesize all the purines, pyrimidines, animo acids and vitamins required for growth.

Auxotrophs: Auxotrophs are the mutant strains of bacteria that show some specific nutritional requirements not needed by their wild-type counter parts. A strain of *Es. coli* requiring tryptophan for growth would be called a tryptophan auxotroph and would be referred as *Es. coli* trp$^-$, while the wild-type is referred as prototroph (*Es. coli* trp$^+$)

3.2.2.4. Carbon and energy sources: The procaryotes are divided into following four nutritional groups on the basis of carbon and energy sources utilized by them.

3.2.2.4.1. Photoautotrophs: These use light as source of energy and CO_2 as sole source of carbon. Cyanobacteria, some purple and green bacteria belong to this group.

3.2.2.4.2. Photoheterotrophs: These procaryotes also use light as source of energy but organic compounds as source of carbon e.g. some green and purple bacteria.

3.2.2.4.3. Chemoautotrophs: They use inorganic compounds like H_2, NH_3, NO_2 and H_2S as source of energy and CO_2 as source of carbon e.g. a few bacteria and many archaea. These are also called **lithoautotrophs.**

3.2.2.4.4. Chemoheterotrophs: They use organic compounds for both energy and carbon sources e.g. most bacteria and some archaea. They are also called **heterotrophs.**

REFERENCES

Pelczar, M.J. Jr., Chan, E.C.S. and Krieg, N.R. (2004). *Microbiology: Concepts and Applications*. 5th Ed. Tata McGraw-Hill, New Delhi, 918 pp.

Prescott, L.M., Harley, J.P. and Klein, D.A. (2005). *Microbiology*. 6th Ed. McGraw-Hill Companies, Inc., New York, 1130 pp.

Srivastava, S. and Srivastava, P.S. (2003). *Understanding Bacteria*. Kluwer Academic Publishers, Dordrecht, 469 pp.

Tortora, G.J., Funke, B.R. and Case, C.L. (2007). *Microbiology: An Introduction*. 9th Ed. Benjamin Cummings, San Francisco, 958 pp.

Chapter 4

VARIATION IN BACTERIA

- 4.1. Mutation
  - 4.1.1. Selection
  - 4.1.2. Screening
- 4.2. Recombination
  - 4.2.1. Transformation
  - 4.2.2. Conjugation
    - 4.2.2.1. $F^+ \times F^-$ mating
    - 4.2.2.2. Hfr conjugation
    - 4.2.2.3. $F' \times F^-$ mating
  - 4.2.3. Transduction
    - 4.2.3.1. Generalized transduction
    - 4.2.3.2. Specialized or restricted transduction
- 4.3. Plasmids

The bacteria are haploid microorganisms. They have simple growth requirements and a very rapid rate of multiplication. All these features make them excellent research tools for studying variability. The bacteria does not have well-defined sexual reproduction but have evolved different parasexual methods, which cause genetic variations. A brief account of these methods is given below.

4.1. MUTATION

A sudden heritable change in the genetic material of an organism is called mutation. The term refers both to the change in the genetic material and to the process by which the change occurs. An organism in which this change has occurred is called a **mutant**. A mutant results from the change in DNA base sequence but it may or may not give rise to a changed phenotype. Sometimes the phenotypic changes are so minor that they can be detected only by special genetic or biochemical techniques.

The mutation may be spontaneous or induced; the latter is caused by treatment with mutagens. As the spontaneous mutations are very rare, mutagens are used to create mutants needed in various types of studies. The commonly used mutagens include physical agents (UV light, X-ray etc.) and chemicals like nitrous oxide, acridine dyes etc. The mutations are also classified on the basis of nature of change, i.e., addition/deletion/substitution occurred in the DNA. The mutations that result from the change of one base in DNA are called point mutations while those resulting from the removal of a part or whole of gene are called deletion mutations. Insertion mutations result from the addition of a chunk of DNA into the base sequence of a gene.

The mutants have played an important role in understanding variation in bacteria. Being the main source of allelic variation, they have contributed enormously toward biological variations. Various biochemical processes and pathways have been elucidated employing different mutants. They have also contributed in understanding the functioning of various genes. The haploid nature of bacteria greatly facilitates the isolation and characterization of mutants. The mutants can be detected either through selection or screening.

4.1.1. Selection

The selection is based on the principle that the mutant is able to grow in the presence of a specific chemical while the parent cells are unable to grow. An antibiotic resistant mutant can be isolated by growing it on a medium containing that particular antibiotic, because only mutant cells will grow.

4.1.2. Screening

In this method, both mutants and wild-type strains are able to grow on the routine media but they are differentiated by growing on some specific media or minimal media. MacConkey agar is used to isolate mutants unable to use lactose (*Lac*$^-$). Both *Lac*$^+$ and *Lac*$^-$ bacteria can grow on this medium but the utilization of lactose by *Lac*$^+$ cells leads to acid production and consequently a drop in local pH, making the colonies to turn pink. On the other hand *Lac*$^-$ mutant cells, unable to utilize lactose, use amino acids as a carbon source and produce ammonia. The production of ammonia increases the local pH resulting in decolourization of the dye due to which colonies appear white.

The detection of the mutants unable to grow on a minimal medium is generally carried out by two methods: (1) **patch replication** and (2) **replica-plating**. In patch replication, the mutagenized colonies are first grown on the master plate on the complete medium. After the growth, the colonies on the master plate are numbered on the backside of the plate and then transferred individually to a minimal medium. The mutants do not grow on a minimal medium. The cells of a parent colony, which do not grow on the minimal media, are the mutant cells. Replica-plating method developed by Lederberg and Lederberg in 1952 is more accurate and efficient than patch replication in detecting the mutants. It is also more helpful in screening different types of mutants like auxotrophic and temperature-sensitive conditional mutants. The bacterial culture exposed to the mutagen is first grown on the master plate containing the complete medium. After the appearance of colonies, the plate is inverted and pressed against sterile velvet mounted on a wooden block. Some of the cells from each colony stick to the velvet. A sterile plate of a minimal medium is pressed onto the velvet and incubated. The failure of the imprinted colony to grow on the minimal medium indicates the presence of mutant cells in the corresponding colony in the master plate (Fig. 4.1a). To select an antibiotic resistant mutant, the medium containing that particular antibiotic is used in place of minimal medium (Fig. 4.1b).

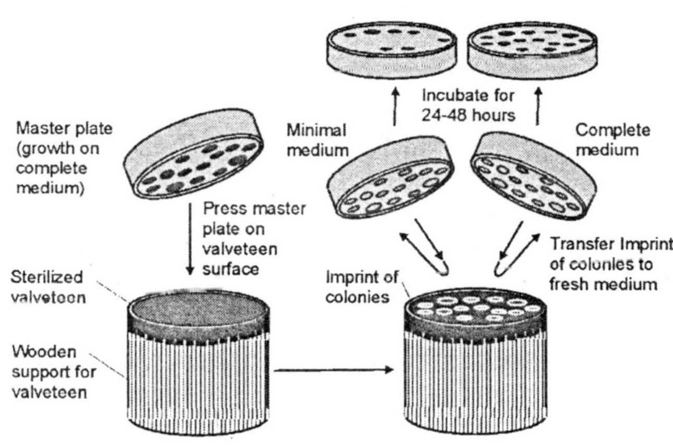

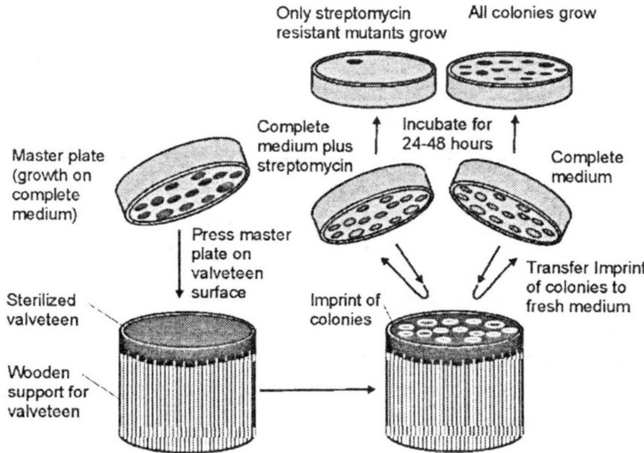

Figure 4.1 Replica-plating: (a) Selection of auxotrophic mutants (b) Selection of drug resistant mutants

Most point mutations revert, restoring the lost function and giving rise to a wild-type allele. This process of restoring the original phenotype is called **reversion** and the restored organism is called **revertant**. Reversion can be spontaneous or induced by mutagens. The reversion rate for single point mutations is comparatively higher than those for double point mutations. A deletion mutation does not revert at all.

4.2. RECOMBINATION

In recombination, the genetic material from two related organisms is physically exchanged. As the recombination generally occurs between two related or homologous DNAs, it is also called a **homologous recombination**. This mechanism plays an important role in generating variation in bacteria. Recombination along with mutation has contributed immensely to the evolution. Non-homologous recombinations also occur in bacteria in which the combining molecules are not similar in their DNA sequences. Integration of phage genome into a bacterial genome, both sharing very limited homology, is a typical example of this kind of recombination. Recombination in bacteria occurs through three mechanisms, namely, **transformation, transduction** and **conjugation**. The major difference among these three processes is the mode of transfer of DNA from one cell to another cell.

Transformation involves the uptake of DNA molecules from one bacterium by another bacterium. In transduction, the transfer of bacterial genes from a donor cell to a recipient cell is mediated by a phage. In conjugation, the genetic material from the donor cell is transferred to the recipient female cell through a conjugation tube or sex pilus. These three mechanisms can be distinguished by two criteria, i.e., (1) sensitivity to the presence of deoxyribonuclease (DNase) and (2) occurrence of cell contact. The occurrence of the recombination in the absence of DNase but not in its presence indicates that the recombination process is transformation. Transduction and conjugation processes are differentiated by growing the bacteria in Davis U-tube (Fig. 4.2). Donor and recipient cells are put in separate arms of this tube. Two arms of the tube are separated by a sintered glass filter. The bacterial cells cannot pass through this filter but DNA molecules and viruses including bacteriophages can pass through it. The liquid in one arm of the tube is sucked in and pushed back with force to mix contents of both the arms. If the recombination occurs in Davis U-tube, i.e., in the absence of cell contact, it is transduction.

4.2.1. Transformation

It can be defined as an uptake of a naked DNA molecule or fragment from the environment by a cell and the incorporation of this molecule into the chromosome of the recipient cell in a heritable form. This phenomenon was first discovered by Frederick Griffith in 1928 while working with *Streptococcus pneumoniae*. After years of research, Oswald Avery *et al.* in 1944 showed that the component responsible for transformation was DNA. Their work also provided one of the conclusive indications that DNA was indeed the carrier of genetic information.

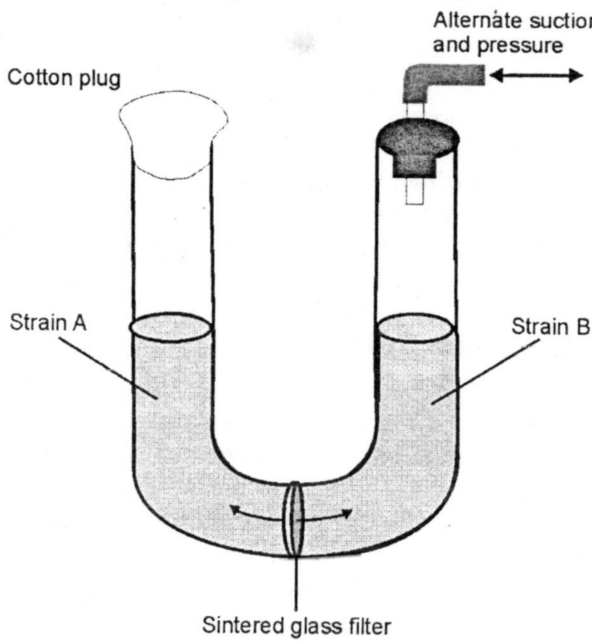

Figure 4.2 Davis U-tube

After lysis, bacteria release considerable DNA into the surrounding environment. Gene transfer by this process occurs in soil and marine environments and may be an important way of genetic exchange in nature. In nature, transformation does not occur in all species of bacteria. It occurs only in those species having competency factor. The transformation frequency of very competent cells is around 10^{-3} for most genera when sufficient DNA is available. Natural transformation has been reported only in a few genera, namely, *Bacillus, Pseudomonas, Azotobacter, Helicobacter, Acinetobacter, Haemophilus, Thermoactinomyces, Neisseria* etc. Most of the work on transformation has been carried out with *Stc. pneumoniae, Ba. subtilis* and *Ha. influenzae,* but the mechanism of transformation has been intensively studied in *Stc. pneumoniae.*

The number of bacterial cells in a culture that are in competent state varies with the growth conditions and the stage of growth curve. The cells of *Stc. pneumoniae* become competent during the exponential phase and secrete a small protein called the competence factor, which stimulates the production of 8 - 10 new proteins needed for transformation. A component cell binds a double-stranded DNA fragment. The process of binding is random and DNA fragments compete with each other for binding. After binding, endonucleases cleave the DNA into double-stranded fragments ranging from approximately 5 - 15 kilo bases in size. During uptake of the DNA, envelope-associated exonuclease hydrolyses one strand while the other strand associates with small proteins and moves through the plasma membrane. The single-stranded fragment then aligns with a homologous region of the chromosome of the recipient cell and gets integrated with the chromosome after replacing a part of it. The replaced segment of the host DNA degrades. The integration of segment of heterologous DNA also occurs but at frequencies very much lower than the frequencies observed using homologous DNA. The mechanism of transformation in *Ha. influenzae* differs from that of *Stc. pneumoniae* in several respects. *Ha. influenzae* does not produce any competency factor and secondly the absorption of DNA occurs only from the closely related species. The basis for specificity of absorbed DNA is that *Ha. influenzae* has a special 11 base pair sequence (5'AAGTGCGGTCA3') that is repeated over 1400 times in its DNA (Prescott *et al.*, 2005).

The cell walls of Gram-positive and Gram-negative bacteria are thick and complex. Hence an efficient and elaborate machinery or enzyme system is required to transport the exogenous DNA bound to protein complex across the cell walls and membranes. In competent cells, the correct genes are expressed for proper functioning of the transformation machinery. The expression of these genes is controlled by growth conditions. Generally very small fragments of DNA are taken up by competent cells but a minimum length of about 500 nucleotide-pairs seems to be needed for the integration to occur. In most cases, donor DNA fragments are about 20,000 nucleotide-pairs (or about 1/200 of the total chromosome length).

Besides natural transformation, artificial or induced transformation can also be achieved in the laboratory using different techniques. Chilling cells in the presence of divalent cations such as Ca^{2+} makes the cell walls permeable to plasmid DNA. Cells suspended in the calcium chloride solution containing DNA are incubated on ice and then briefly heat shocked at 42°C for 30 - 120 seconds, which causes the DNA to enter the cells. This method works very well for circular plasmid DNAs. This approach also succeeds with species like *Escherichia coli,* which are not naturally competent. To increase the transformation frequency, relatively higher concentrations of DNA, than present in the nature, are used. An excellent preparation of competent cells gives $\sim 10^8$ transformants per microgram of plasmid. A good non-commercial preparation gives 10^5 - 10^6 transfomants per microgram of plasmid. This method usually does not work well with linear DNA, such as fragments of chromosomes, because exonuclease in the cell degrades the linear DNA. In contrast, naturally competent cells are more efficiently transformed with linear DNA than with plasmid. When linear DNA fragments are used for transformation of *Es. coli,* the cells are rendered deficient in one or more exonuclease activities to protect the DNA fragments. As plasmids are not easily degraded as linear fragments, it is more appropriate to transform bacteria with plasmid DNA. The discovery of calcium chloride treatment has enabled researchers to use *Es. coli* as an efficient tool for genetic engineering. Moreover, DNA from any source can be incorporated into bacteria by splicing it into a plasmid before transformation (Prescott *et al.,* 2005).

Another method employed for the entry of plasmid DNA into the bacterial cells is electroporation in which cells are briefly shocked with an electric field of 10 - 20 kV/cm. It makes holes in the cells through which the plasmid DNA, even of large plasmids, can enter. Natural membrane-repair mechanisms rapidly close these holes after the shock.

Transformation usually produces a mixture of very few transformed cells and abundant non-transformed cells. The methods are available to ascertain whether the transformation has occurred or not. The plasmids used in transformation usually contain genes imparting resistance to antibiotics. The incorporation of the antibiotic into the medium and the ability of the cells to grow on this medium confirms the occurrence of the transformation. Another marker used for identifying *Es. coli* cells that have acquired recombinant plasmids, is the *lacZ* gene, which codes for β-galactosidase.

As β-galactosidase is a homo-tetramar, with each monomer made up of one *lacZ*-∝ and one *lacZ*-ω protein, no functional enzyme will be formed if only one of the two requisite proteins is expressed in the resulting cell. Thus, if a strain of *Es. coli* without *lacZ*-∝ in its genome is transformed using a plasmid containing thc missing gene fragment, transformed cells will produce β-galactosidase, whereas untransformed cells will not, as they are only able to produce the omega half of the monomer. In this kind of transformation, the polylinker region of the plasmid lies in the *lacZ*-∝ gene fragment, meaning that successfully produced recombinant plasmids will have the desired genes inserted somewhere within *lacZ*-∝. The expression of this disrupted gene fragment by *Es. coli* does not result into the production of any useable *lacZ*-∝ protein and hence no usable β-galactosidase is formed. When grown on media containing the colourless modified galactose sugar X-gal, colonies that are able to metabolise the substrate and have been transformed but not by recombinant plasmids will appear blue, while the colonies that are not able to metabolise the substrate but have been transformed by recombinant plasmids will appear white (Wikipedia, 2008).

Introduction of DNA into animal cells is usually called **transfection**.

Evolution or emergence of new bacterial plant pathogens has long been a subject of great interest to plant pathologists or plant bacteriologists. Barash and Manulis-Sasson (2009), in their most recent review article, have described that how *Pantoea agglomerans*, a widespread epiphyte and commensal bacterium, has evolved or transformed into an hrp-dependent and host-specific tumorigenic pathogen by acquiring a plasmid containing pathogenicity island (PAI). The PAI was evolved on an interon plasmid of the IncN family, which is distributed among genetically diverse populations of *Pa. agglomerans*. The horizontal gene transfer was the major contributing force in the creation of PAI, although a pathoadaptive mechanism might also be involved. Plant pathogenic strains of *Pa. agglomerans* isolated from gypsophila contain a pathogenicity plasmid designated as pPATH. The pathogenicity plasmid of *Pa. agglomerans* pv. *gypsophilae* (*Pag*), designated as $pPATH_{Pag}$, is approximately 135 kb and it contains a pathogenicity island of ~75 kb. The $pPATH_{Pag}$ harbours an *hrp/hrc* gene cluster, type III effectors, a gene cluster encoding for biosynthesis of indole acetic acid (IAA) and cytokinins, and many diverse insertion sequence elements and partial sequences from known genes of other bacteria. Mutations in genes encoding components of the type III secretion system or the *hrp/hrc* regulatory gene cascade completely abolished gall formation while mutations in genes encoding type III effectors or biosynthesis of IAA and cytokinins reduced gall development substantially or almost totally. These findings suggest that pathogenicity of *Pa. agglomerans* pv. *gypsophilae* and *Pa. agglomerans* pv. *betae* is hrp-dependent, and that acquisition of the plasmid-borne PAI was a crucial step in the transformation of *Pa. agglomerans* into a gall-forming and host-specific plant pathogen.

The pathogenicity plasmid of *Pa. agglomerans* pv. *gypsophilae* is nonconjugative. The nonconjugative nature of this plasmid was also supported by the lack of putative genes that showed significant homology to *tra* genes. Plant synthesized IAA and cytokinins are absolutely required for gall initiation, whereas the pathogen-secreted phytohormones are not needed. However, pathogen-secreted IAA and cytokinins do have a significant effect on gall size and might be important for expression of the *hrp* regulatory genes as signalling molecules.

4.2.2. Conjugation

Conjugation, also called sexual reproduction, is the transfer of genetic material from one bacterium to another bacterium by direct cell to cell contact. The transfer of genetic material is unidirectional, i.e., only from donor to a recipient cell. The donor cells possess F factor, also called fertility or sex factor and are designated as F^+. The recipient cells do not contain F factor and are designated as F^-.

The initial evidence for conjugation was provided by Lederberg and Tatum (1946) while working with K-12 strain of *Es. coli.* They used two auxotrophic strains of *Es. coli* K-12, one strain required biotin (Bio^-), phenylalanine (Phe⁻) and cysteine (Cys⁻) for growth while the other needed threonine (Thr⁻), leucine (Leu⁻) and thiamine (Thi⁻). Both the strains were inoculated together, incubated for several hours and then plated on the minimal medium. Recombinant phototrophic colonies appeared on the minimal medium after incubation indicating that the chromosomes of both the auxotrophs came in contact and underwent recombination. They purposely used double or triple auxotrophs assuming that two or three reversions would not often occur simultaneously. However, Lederberg and Tatum (1946) did not directly prove that the bacterial cells came in direct contact with each other for the transfer of genetic material. This evidence was provided by Davis (1950) who used U-tube (Fig. 4.2) fitted with a sintered glass filter in the centre. The filter allows the passage of medium and very small particles but not bacterial cells. Each side of the tube filled with nutrient medium was inoculated with a different auxotroph of *Es. coli*. During the incubation, the medium was sucked in and pushed back with force to ensure exchange of medium between two halves. After incubation, the cultures from both the sides were plated on the minimal medium. Non-appearance of colonies indicated the lack of gene transfer because the bacterial cells did not come in contact with each other. The conjugation is mainly of the following three kinds.

4.2.2.1. $F^+ \times F^-$ mating: It is a basic type of conjugation. The F^+ strains contain an extra chromosomal F factor carrying the genes for pili formation and plasmid transfer. It is a circular DNA molecule, contains 19 genes and replicates independently of bacterial chromosome. The donor (F^+) and recipient (F^-) cells of *Es. coli* do not differ morphologically except that the donor or male cells contain F pilus or sex pilus. Each male cell may carry 1 - 5 pili, which are 1 - 2 μm long and formed at right angle to the cell surface. They are hollow, made up of a protein called pilin, with an external diam of 85 A° and an axial hole of 20 - 25 A° in diam. The sex pili unite the donor and recipient cells together and also act as a conduit for the transfer of the genetic material. At the time of transfer, one strand of F factor is nicked; it starts unwinding and separates from the other strand. The nicked strand starts moving to the F^- cell through the pilus and during this process, the entering strand is replicated to produce double-stranded DNA. The circular strand left behind in the F^+ cell is also replicated. Eventually, the nicked strand completely passes through to the recipient cell and is completely replicated. The pilus gets broken separating the cells. As a result of this process, F factor is introduced into a recipient cell making it F^+ (Fig. 4.3a). This new F^+ cell can act as a donor and can initiate conjugation with another F^- cell. As both the cells contain an F factor, both cells are F^+. As the bacterial chromosome genes are rarely transferred with the independent F factor, the recombination frequency is low. This is due to the fact that the F factor is not homologous to the DNA of bacterial chromosome.

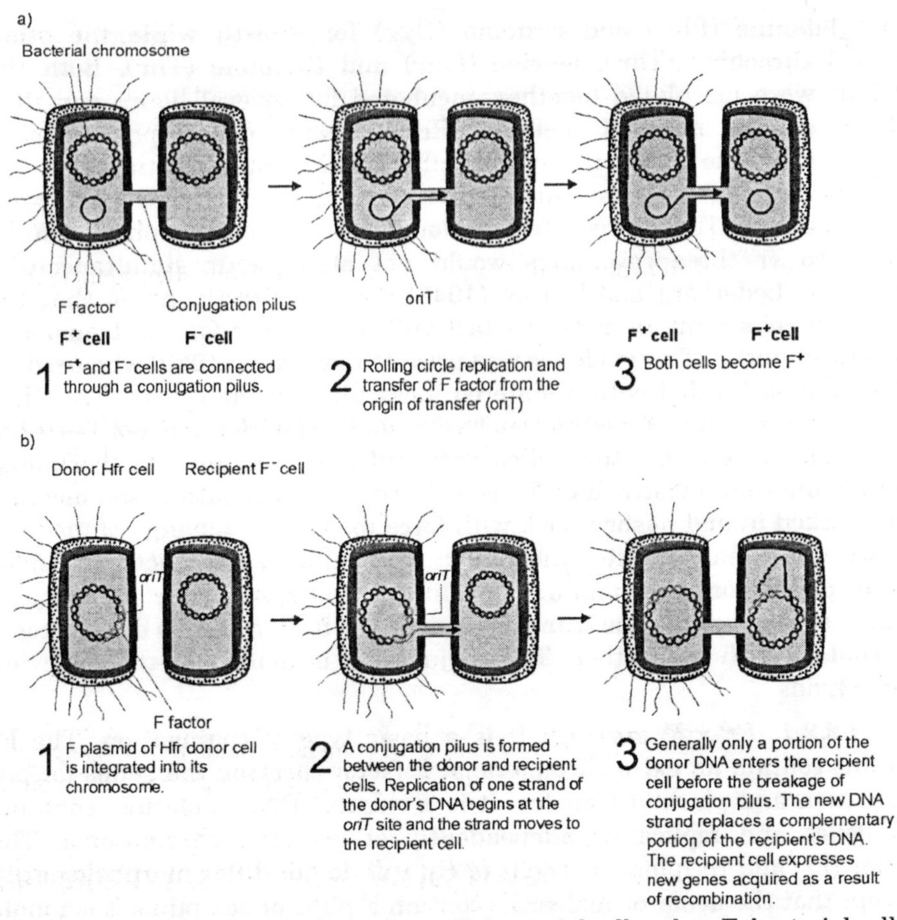

Figure 4.3 (a) Conjugation between an F^+ bacterial cell and an F^- bacterial cell (b) Conjugation between an Hfr bacterial cell and an F^- bacterial cell

Most of the work on plasmids and conjugation has been carried out with *Es. coli* and other Gram-negative bacteria. However, some Gram-positive bacterial genera, namely, *Enterococcus, Streptococcus, Bacillus, Staphylococcus* and *Streptomyces* also contain self-transmissible plasmids. Very little is known about the transfer of genetic material in such cases. The recipient cells of *Es. faecalis* release short peptide chemical signals, which activate transfer genes in donor cells having the required plasmid. Activated donor cells release special plasmid-encoded proteins which make donor and recipient cells to adhere with each other. Thereafter, the plasmid transfer occurs.

4.2.2.2. Hfr conjugation: It is also known as Hfr × F^- mating. Hfr donor strains give 10^3 times higher frequency recombination than F^+ strains and hence are called high frequency of recombination (Hfr). In F^+ cells, the F factor/plasmid occurs in an autonomous state, independent of bacterial chromosome. Occasionally, the F factor gets integrated into bacterial chromosome at a random location by recombination between homologous insertion sequences (IS) present on the plasmid and host chromosome. In *Es. coli,* the sites of integration are IS with F plasmid carrying two copies of IS 3 and one copy each of IS 2 and $\gamma\ \delta$ (TN 1000). Bacterial chromosome carries several copies of these homologous elements, which can be used by F plasmid for integration. The nature and orientation of chromosome transfer of an Hfr is governed by the location and orientation of these elements. The transfer of chromosome by Hfr initiates with nicking the chromosome at origin of transfer *(OriT)* of the integrated F plasmid. As it is replicated, the chromosome moves into the recipient cell, the leading part being origin of transfer along with a part of F plasmid followed by the chromosome. As the main body of the F plasmid is at the tail end of the chromosome being transferred, it rarely gets transferred. Hence, the recipient cell is rarely converted into a F^+ strain (Fig. 4.3b). The DNA received from the Hfr donor can undergo recombination with the host chromosome at a high frequency, as the DNA sequences are homologous. This recombination can also result in gene conversion, if the transferred DNA and the corresponding region of the host DNA contain different alleles of the same gene.

The finer details of chromosome transfer by Hfr donor were revealed by Jacob and Wollman (1955) using **"interrupted-mating"** procedure. They disrupted the mating at different time intervals by vortexing the mating population. The cells were then plated to determine their genotypes. In general, the transfer starts from a specified point, i.e., origin of transfer or *OriT* and proceeds linearly till the two cells remain in contact. There is a clear relationship between the time at which a given marker is transferred from Hfr to F^- and its map location on the chromosome. After ascertaining the time of entry of several other donor markers in the subsequent studies, a temporal map of *Es. coli* was constructed indicating the location of genes on a time scale (in minutes). These findings also revealed that chromosomal transfer by Hfr is never complete leading to only a part of the chromosome transferred. The markers which are close to the origin of transfer are transferred efficiently and those which are far-off do not get a chance to be transferred and hence do not appear in the recombinants. Therefore, it leads to the formation of a **merozygote**. The map of *Es. coli* has been standardized for 100 temporal map units. Time taken for complete transfer of Hfr chromosome to F^- is 100 minutes (Srivastava and Srivastava, 2003).

4.2.2.3. F' ×F- mating: It is also called F' conjugation or F-mediated sexduction. As F plasmid is an episome, it can leave the bacterial chromosome. During this process, if only plasmid portion is excised from the chromosome, the Hfr cell becomes an F^+ cell again. However, sometimes the excision is not proper and the F factor takes a segment of the bacterial chromosome along with it. When a part of the chromosomal

DNA becomes associated with the episome, it is called an F' or F-prime. The F' factors range in size from those carrying a single bacterial marker to those carrying up to half of the bacterial chromosome. F' plasmid is transferred to F^- cell and it converts the recipient cell into an F' cell. Adelberg and Burns (1959) were the first to identify the F' factor in *Es. coli*.

The mechanism of F' factor transfer to the recipient F^- cell is the same as that of F factor transfer in $F^+ \times F^-$ mating and chromosome transfer in Hfr $\times$ F^- mating. The recipient cell having its own complete chromosome becomes diploid for the segment of the chromosomal DNA on the episome. Such a partial diploid cell is called a **merozygote**. The chromosomal DNA on the episome can undergo recombination at high frequency with its homologous sequence on the chromosome. F' conjugation is very important to microbial geneticists because F' factors play important roles in genetic analysis of bacteria. Partial diploidy provides an important method for determining dominance relationship between alleles and defining genes by complementation tests in bacteria. As F' factors are available carrying almost any part of *Es. coli* chromosome, sexduction analysis can be made for almost any mutation of interest. The creation of F' plasmid is also useful in mapping the chromosome.

4.2.3. Transduction

Zinder and Lederberg (1952) discovered another mechanism of gene transfer in bacteria while working with *Salmonella typhimurium*. They mixed two multiple auxotrophic strains, one requiring phenylalanine, tyrosine and tryptophan (phe^- tyr^- trp^-) and the other requiring methionine and histidine (met^- his^-). After incubation, when this mixture was plated on an amino acid-free minimal medium, prototrophic colonies appeared at a frequency of $\sim 10^{-5}$. The appearance of the prototrophic strain even after placing the auxotrophic strains in two arms of Davis U-tube (Fig. 4.2) showed that the cell to cell contact was not required and hence this process is different from conjugation. The transfer of the genetic material was due to a filterable agent and this agent was identified as phage P 22 of *Sal. typhimurium*. This process of gene transfer mediated by a phage was referred to as **transduction.**

Later on in 1956, Lederberg's associates discovered the transfer of *gal* marker in *Es. coli* through Lambda (λ) phage. However, the gene transfer by these two phages, namely, P 22 and Lambda differs remarkably. The former can transfer virtually any gene and hence the process is called **generalized transduction**. Phage Lambda transfers only *gal* or *bio* or very rarely both the genes resulting in **restricted** or **specialized transduction**. These two transduction pathways are related in a way that both employ a phage for the transfer of genetic material but differ in many other respects. The recombinant cells formed through either of these pathways are called **transductants**.

4.2.3.1. Generalized transduction: Phage P 22 of *Sal. typhimurium* and phage P 1 of *Es. coli* are well known examples which bring about generalized transduction although many other phages are also capable of doing it. During the lytic cycle of the phage, viral DNA and proteins are synthesized for the assembly of phage particles. At the same time, bacterial genome is degraded into random DNA fragments. By mistake in some of the protein capsids, DNA fragments of bacterial genome are packed instead of viral DNA and these particles act as transducing phages (Fig. 4.4). The evidence for the presence of bacterial DNA, but not a mixture of bacterial and viral DNA, was provided by Ikeda and Tomizawa in 1965 based on density labelling of the DNA of donor cells with 5-bromouracil before infection and replacing 5-bromouracil with thymidine at the time of phage infection. The quantity of the bacterial DNA carried by transducing phages is governed mainly by the size of the capsid. Phage P 22 carries about 1% of the bacterial genome while in phage P 1 of *Es. coli* and other Gram-negative bacteria, it ranges from 2.0 - 2.5% of the genome.

Transducing phages are capable of injecting the DNA into other bacterial cells but does not cause lysis of the cells, as they do not contain viral DNA. They simply act as carrier of genetic information from the donor bacterium to the recipient bacterium. As the DNA fragment enclosed in the capsid can be any segment of the bacterial genome, it will transfer any gene(s), hence it is called generalized transduction. After the transducted DNA is delivered to the recipient cell, three possible fates await this DNA: (a) degradation by nucleases, (b) recombination with homologous recipient DNA and (c) occurrence in the non-replicating stable form. About 70 - 90% of the transferred DNA is not integrated but often it is able to survive and express. The bacteria which contain this non-integrated DNA are called **abortive transductants** and are **partial diploids**. The process of occurrence of donor DNA in a recipient cell without recombination and replication is known as **abortive transduction**. The non-integrated DNA exists in the cells in a circular super coiled form and the circularity is brought about by association of ends with a protein. Its absence in infectious phage particles is explained as being cellular rather than a phage-encoded protein. Circularized DNA is protected from degradation by the enzymes. Once destined to abort, this DNA does not form a complete transductant. Complete transductants are formed when transducing DNA integrates with the genome of the recipient cell (Srivastva and Srivastava, 2003)

4.2.3.2. Specialized or restricted transduction: It involves the transfer of one or a few but specified genes from the donor to the recipient cell. It is caused by the temperate phages, which attach their genomes at one or a few specified sites of the bacterial chromosome. Phage λ is the best-studied example for this purpose. As mentioned earlier, this

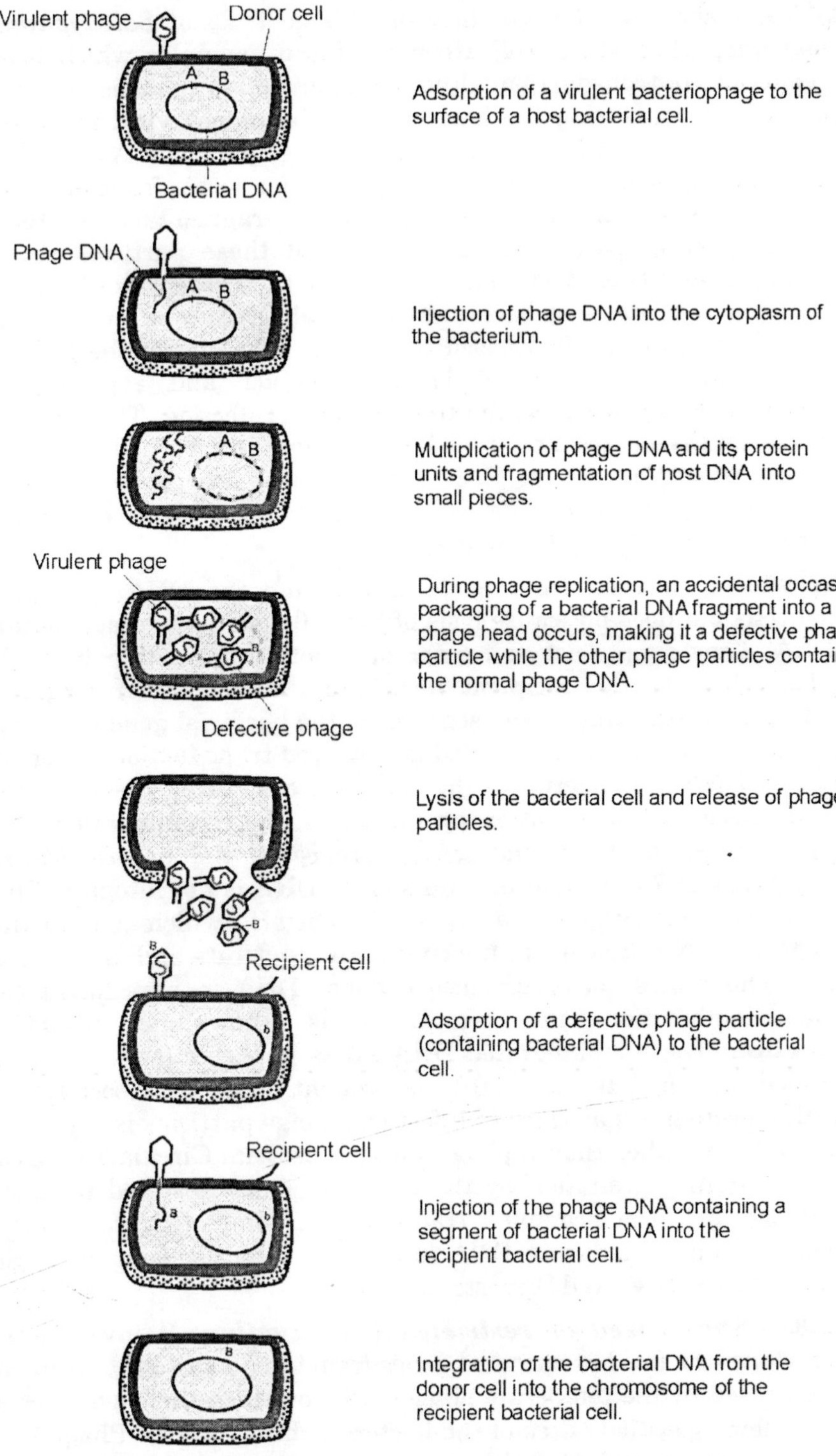

Figure 4.4 Generalized transduction

phenomenon was discovered by Lederberg and his associates in 1956 while they were working to know whether phage λ could transfer genes of *Es. coli* from a donor to a recipient cell. They lysogenized prototrophic wild-type *Es. coli* K-12 strain with λ Phage and treated this bacterium with UV irradiation to produce phage progeny. Different kinds of non-lysogenic mutants (deficient in different characters) of *Es. coli* were infected with these phages and then grown on different selective media to know the gene transfer. The results were negative in all, except in one case, where about 10^{-6} *gal*$^{-}$ (deficient for galactose utilization) cells were converted into *gal*$^{+}$ after phage infection. Thus it became evident that phage λ can also transfer the genes from one cell to another but the transfer is limited to only a few gene(s). Later on, it was found that phage λ can also transfer *bio* (for biotin utilization) gene and rarely both *gal* and *bio* genes (Fig. 4.5).

The life cycle of λ phage and generation of transducing phage in *Es. coli* is given below. After adsorption to the host cells, λ Phage injects its DNA into the host cell. The DNA, which is originally double stranded, circularizes and finally gets converted into a covalently closed circular DNA. Then the phage genome is attached to the bacterial chromosome at a site called *att* λ situated between gal and bio genes. After integration, it multiplies as prophage with the multiplication of bacterial cells and is passed on to all the progeny cells. It can also be transferred through conjugation to the recipient cells. Temperate phages undergo spontaneous transition from the lysogenic state to lytic state but it is very rare, i.e., about one in 10^5 cell divisions. The change from the lysogenic state to lytic state can also be induced artificially by irradiation with ultraviolet light. During this change, prophage is freed from the host chromosome. The excision of prophage from the host cell is site specific like the integration process. Usually the excision process is very precise and phage genome is cut out exactly in the form in which it existed prior to integration. However, rarely the error occurs in the excision of phage genome and in such cases a segment of the bacterial chromosome is excised with the phage genome leaving behind a segment of the phage genome. Only the host genes situated close to attachment sites of the prophage are excised with the phage genome and packaged in the phage particles. As the genome of λ phage is attached between *bio* and *gal* genes of *Es. coli* chromosome, it can take along with it, either *gal* gene or *bio* gene or very rarely both. This event is similar to the excision of F plasmid and creation of F′s. Another specialized transducing phage ϕ 80 integrates near *trp* (for the synthesis of tryptophan) genes of *Es. coli* chromosome and transduces *trp* marker.

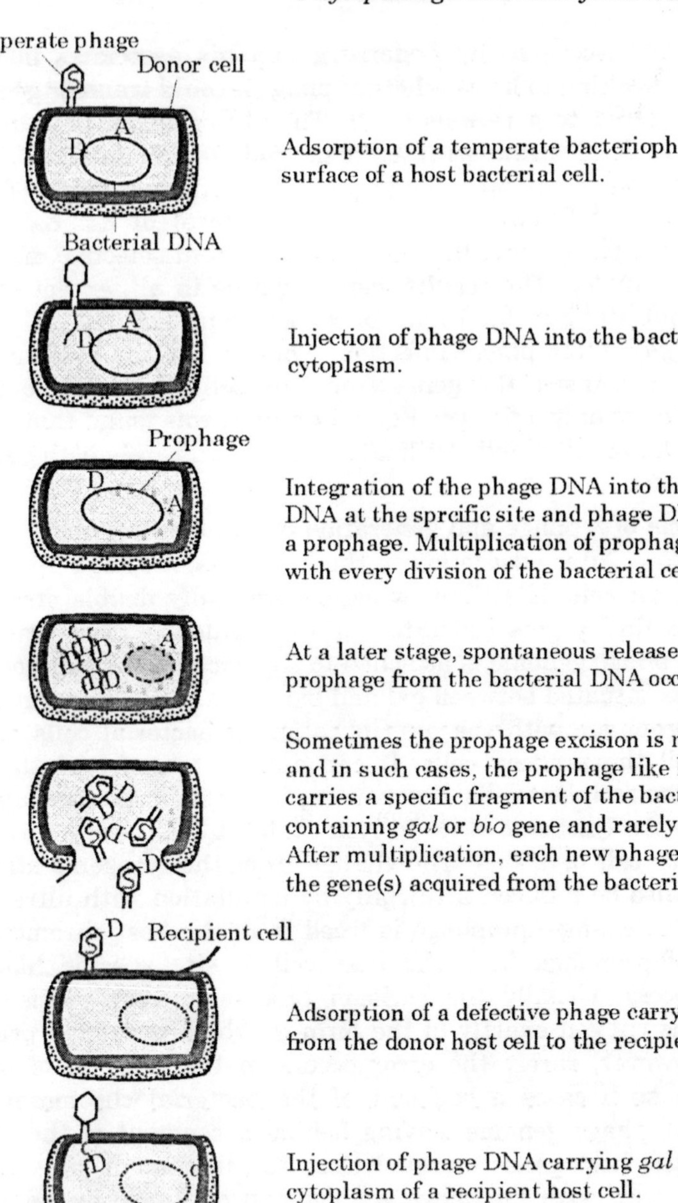

Figure 4.5 Specialized transduction

The induction of *Es. coli* cells lysogenized with λ phage produces a mixture of abundant normal phage particles and very few, i.e., about 10^{-6}

defective transducing particles. As the lysate contains only very few transducing particles, it is often called low-frequency transduction lysate (LFT lysate). The transducing λ *gal* phage, formed due to the integration of *gal* gene into λ genome is defective and is called λ *dgal* (lambda defective *gal*). It has been rendered defective due to the replacement of phage genes, required for growth and maturation under lytic conditions, with bacterial DNA. Being defective, it cannot multiply without the help of another phage called helper phage. A wild-type λ phage or any other appropriate phage that can provide the missing function of the defective phage can act as a helper phage. In contrast, the creation of λ *bio* phage does not result in loss of essential genes, generating a normal phage called λ *pbio*. Out of these two phages, only λ *pbio* can multiply at low multiplicity of infection, i.e., only one phage particle infects a single host cell. λ *dgal* reproduces at high multiplicity of infection meaning that a host cell is infected with one particle of λ *dgal* and one or more particles of wild-type λ phage.

Lambda phage can transfer bacterial genes in two ways. In the first method, two crossovers replace the gal^- gene of the recipient's chromosome with donor gal^+ gene. The transductants produced by this method are haploid, stable and non-lysogenic. In the second method, there is a single crossover between the recipient and phage-carried alleles. This results in addition of phage genome and donor genes to the genome of the recipient cell without any replacement of donor DNA. The transductants produced as a result of single crossover are partially diploids gal^+/gal^- heterogenotes, unstable and lysogenic. As the gal^+/gal^- heterozenotes are unstable, the segregation of gal^- cells occurs at a frequency of about one in 1000 cell divisions. Since λ *dgal* phage particles cannot reproduce without the help of a helper phage, λ *dgal* chromosome is excised and may simply be lost or diluted out during cell divisions. Recombination may also occur between gal^+ exogenote (donor DNA fragment) and a gal^- endogenote (recipient chromosome) resulting in the transfer of gal^+ marker to the endogenote and production of stable gal^+ transductants (Gardner *et al.*, 1991). The host cells infected by both the phages, i.e., λ *dgal* and a normal λ (λ^+) phage are called double lysogens. This results from the insertion of λ^+ genome at attachment site *attB* and that of λ *dgal* on the left attachment site. The inducement of these heterozygous lysogens with UV irradiation produces 50% λ *dgal* particles and 50% λ^+ particles in the lysate. Such lysates are called high-frequency transduction (Hft) lysates because in these lysates 50% of the phage particles contain transducing *gal* genes.

Transduction is an important tool for transferring genes in bacteria. It is also used in genetic mapping although it is not an efficient and reliable tool for this purpose. Hft lysates greatly facilitate the genetic analysis by increasing the frequency of transduction events. As the amount of bacterial DNA carried by a phage is limited, the transduction is suitable for transferring a small but specified portion of a bacterial genome. Many specialized phages or their regions have been used to develop cloning vectors.

4.3. PLASMIDS

Plasmid is an extra chromosomal unit of genetic material that replicates independently of chromosome and is stably inherited to the offsprings. Most, but not all, plasmids are dispensable, because they are not essential for the survival of cells in which they reside. However, in some cases, they are required under certain environmental conditions e.g. the presence of R-plasmid is essential in a bacterium to enable it to grow on a medium containing that particular antibiotic to which it imparts resistance. Based on their ability to mediate conjugative self-transfer, they are divided into two groups, namely, conjugative and non-conjugative plasmids. Conjugative or transmissible plasmids mediate transfer of DNA by conjugation. All F and F' plasmids, many R plasmids and some col plasmids belong to this group. Non-conjugative or non-transmissible plasmids do not mediate transfer of DNA by conjugation. Many R and col plasmids are non-conjugative.

The ability of conjugative plasmids to transfer the genetic material has played a major role in bacterial evolution. This ability has also been used to affect horizontal gene transfer not only between the strains of the same species but also across the species and even in organisms belonging to different kingdoms. The most important achievement made through the use of plasmids is their role as vectors in genetic engineering experiments. Due to their some desirable properties, they have been used extensively, as such or after required modifications, as cloning vectors in molecular biology.

The spread of multiple antibiotic and drug resistance in bacteria including pathogens and instability of industrially important bacteria used in fermentation have resulted in serious consequences. The evolution of compound R plasmids, i.e., carrying two or more R- determinants (the segment carrying the resistance gene or genes) has aggravated the situation further. The occurrence of bacteria containing compound R plasmids has resulted in their ability to acquire resistance to commonly used antibiotics like penicillin, streptomycin, tetracycline etc. A study in Japan revealed that frequency of cells carrying R plasmid-mediated antibiotic resistance in natural population of sewers and in polluted lakes and streams, increased from less than 1% to 50 - 80% in less than 10 years. Moreover, R plasmids are not rapidly dispersed only within a bacterial species but also transmitted among different species and even genera. The dispersal of *Es. coli* R plasmids to different genera, namely, *Salmonella, Pasteurella, Shigella, Proteus* and *Haemophilus* that too including pathogenic species, is well known (Gardner *et al.,* 1991). The dispersal of R plasmids is not as rapid as those of F plasmids because only about 0.02% cells containing R plasmids are competent donors. However, if after the transfer of R plasmid, a large number of R^- recipient cells are added, R plasmid quickly spreads in the recipient cells. The recipient cell, which receives R plasmid, becomes a donor capable of subsequent transfer

Stc. lactis and *Stc. cremoris* are commonly used as starter cultures in dairy fermentation process for making cheese. Both the bacteria contain multiple plasmids and these plasmids carry genes that encode enzymes, which play essential roles in the fermentation processes. Two important key factors for proper cheese ripening are lactose metabolism and proteinase activity. The genes carried by the plasmids encode for three enzymes involved in lactose metabolism and proteinase activity. As most plasmids are unstable by nature, the presence of key genes on these plasmids indicates the unpredictable functioning of these fermentation starter cultures.

REFERENCES

Adelberg, E.A. and Burns, S.N. (1959). A variant sex factor in *Escherichia coli. Genetics* 44: 497.

Barash, I. and Manulis-Sasson, S. (2009). Recent evolution of bacterial pathogens: The gall-forming *Pantoea agglomerans* case. *Ann. Rev. Phytopathol.* 47: 133 – 52.

Davis, B.D. (1950). Nonfiltrability of the agents of genetic recombination in *Escherichia coli. J. Bacteriol.* 60: 507.

Gardner, E.J., Simmons, M.J. and Snustad, D.P. (1991) *Principles of Genetics.* 8th Ed. John Wiley and Sons Inc., New York, 811 pp.

Jacob, F. and Wollman, E.L. (1955). Etude génétique d'un bactériophage tempéré *d'Escherichia coli.* III. Effet du rayonnement ultraviolet sur la recombinaison génétique. *Ann. Inst. Pasteur Paris,* 88: 724 – 49.

Lederberg, J. and Tatum, E.L. (1946). Novel genotypes in mixed cultures of biochemical mutants of bacteria. *Cold Spring Harbor Symposia Quant. Biol.* 11: 113.

Prescott, L.M., Harley, J.P. and Klein, D.A. (2005). *Microbiology.* 6th Ed. McGraw-Hill Companies, Inc., New York, 1130 pp.

Srivastava, S. and Srivastava, P.S. (2003). *Understanding Bacteria.* Kluwer Academic Publishers, Dordrecht, 469 pp.

Tortora, G.J., Funke, B.R. and Case, C.L. (2007). *Microbiology: An Introduction.* 9th Ed. Benjamin Cummings, San Francisco, 958 pp.

Wikipedia (2008). Transformation (genetics). [http://en.wikipedia.org/wiki/ Transformation_(genetics)].

Zinder, N.D. and Lederberg, J. (1952). Genetic exchange in *Salmonella. J. Bacteriol.* 64: 679 – 99.

Chapter 5

CLASSIFICATION OF BACTERIA

5.3.1.3.28. *Ralstonia*
5.3.1.3.29. *Rathayibacter*
5.3.1.3.30. *Rhizobacter*
5.3.1.3.31. *Rhizobium*
5.3.1.3.32. *Rhodococcus*
5.3.1.3.33. *Samsonia*
5.3.1.3.34. *Serratia*
5.3.1.3.35. *Sphingomonas*
5.3.1.3.36. *Spiroplasma*
5.3.1.3.37. *Streptomyces*
5.3.1.3.38. *Xanthomonas*
5.3.1.3.39. *Xylella*
5.3.1.3.40. *Xylophilus*

Classification is an arrangement of organisms into groups, called taxonomic ranks, on the basis of similarities or relationships. Two other areas closely related with the classification are nomenclature and identification. Nomenclature is the assignment of names to the taxonomic groups according to the nomenclature code. Identification is the practical use of classification scheme to determine the identity of an organism as a member of an established taxon or as member of a previously unidentified species.

5.1. NOMENCLATURE CODE

For many years, the Botanical Code of Nomenclature was the basis for bacterial nomenclature. In the International Congress of Microbiology held in London in 1936, it was decided to have a separate code for bacterial nomenclature. Therefore, the first separate code for bacteria, i.e., International Bacteriological Code of Nomenclature was published in 1948. The code was emended many times but a major revision occurred in 1976 (Lapage *et al.*, 1975). The name of the Code has also been changed several times but since 2000, it is known as the **International Code of Nomenclature of Procaryotes** and is governed by the **International Committee on Systematics of Procaryotes**. Before the adoption of above-mentioned last name of the code, it was called **International Code of Nomenclature of Bacteria** and was governed by the **International Committee on Systematic Bacteriology**.

The 1976 revision of the International Code of Nomenclature of Bacteria (Lapage *et al.*, 1975) provides that after 1st January 1980, priority of publication shall date from January 1, 1980 and on that date all names published prior to January 1, 1980 and included in the **Approved Lists of Bacterial Names** of ICSB (International Committee on Systemic Bacteriology) shall be treated validly published for the first time on that date. Those names validly published prior to January 1, 1980 but not included in the Approved Lists will have no further standing in nomenclature. After January 1, 1980, the priority of publication shall date from the date of their publication in the **International Journal of Systematic Bacteriology** (from 2000 in **International Journal of Systematic and Evolutionary Microbiology**). Since 1980, the names effectively published outside the above-mentioned two journals can only make a claim to priority, if they are subsequently published in validation lists published in these two journals. The priority shall date from the date of their publication in the validation lists and not from the date of their original publication.

The Bacteriological Code (1990 Revision) does not cover taxa above the rank of a class. Therefore, in the second edition of *Bergey's Manual of Systematic Bacteriology*, other codes of nomenclature have been used for naming the higher ranks. As there is no recognition of a rank above kingdom, Bergey's Trust, after considerable deliberations, concluded to incorporate the rank of **domain** into the hierarchy. Further, the rank of kingdom has not been used in this edition to avoid possible conflicts with other codes of nomenclature.

5.2. SPECIES

The bacterial species is the basic and most important taxonomic rank in the bacterial systematics. The concept of a bacterial species is poorly understood in comparison to higher organisms. This is mainly due to the reason that bacteria being procaryotic organisms differ markedly from higher organisms. Unlike higher organisms, sexuality is not used in bacterial species classification because relatively few bacteria undergo conjugation. Moreover, most of the procaryotic microorganisms are relatively simple and similar in morphological characters. Therefore, the morphological features alone are usually of little classificatory significance in comparison to higher organisms.

Brenner *et al*. (2001) defined the bacterial species as "a distinct group of strains that have certain distinguishing features and that generally bear a close resemblance to one another in the more essential features of organization". Each species differs considerably and can be distinguished from all other species. The above given definition of the species was generally loosely followed. It was mainly due to the reasons that one cannot accurately determine 'a close resemblance', 'essential features' or how many 'distinguishing features' are sufficient to create a species.

5.3. CLASSIFICATION

Von Naegeli (1857) placed the bacteria in class **Schizomycetes**. In the 7th edition of *Bergey's Manual of Determinative Bacteriology*, class

Schizomycetes was placed in one the five divisions, namely, Protophyta, recongnised under the kingdom Plantae. Kingdom Plantae was proposed by Carolus Linnaeus in 1753. E. Chatton (1937), on the basis of cellular organization, divided the living organisms into two groups, **procaryotes** and **eucaryotes**. Later on, R.G.E. Murray (1968) proposed two kingdoms, namely, Procaryotae and Eucaryotae to include these two groups. During 1970s, the comparison of oligonucleotide catalogs of the 16S rRNA of procaryotes and of 18S rRNA of eucaryotes of a broad range of living organisms revealed that there are two different kinds of procaryotes: the ***Archaea*** (also called archaebacteria or archaeobacteria) and the ***Bacteria***. On the basis of these findings, Woese, Kandler and Wheelis (1990) opined that there are three fundamentally different kinds of living organisms: the *Archaea,* the *Bacteria* and the eucaryotes (now called the ***Eucarya***).

Different criteria have been used to classify bacteria. First of all, it was the morphological approach in which the major emphasis was placed on morphological characters. Later on, the emphasis was also placed on physiological and biochemical characters along with the morphological characters. As a result of this, the classification of bacteria became tedious and time consuming because a large number of tests were to be carried out and some of these tests required long time for completion. With the rapid advances in DNA sequencing technology, sequence analysis of highly conserved regions of the bacterial genome, most notably genes coding for the RNA of the small ribosomal subunit has now led to a natural classification that reflects the evolutionary history of bacteria.

Bergey's Manuals of Determinative Bacteriology and Bergey's Manuals of Systematic Bacteriology are the most comprehensive and authoritative publications in the field bacterial taxonomy. The first edition of *Bergey's Manual of Determinative Bacteriology* was published in 1923 and the subsequent editions up to 7th were published in 1925, 1930, 1934, 1939, 1948 and 1957, respectively. As the new information became available it was added in the subsequent editions. The eighth edition of *Bergey's Manual of Determinative Bacteriology* (Buchanan and Gibbons, 1974) is different in format, presentation of information and approach to classification from the 7th edition. This manual is meant mainly for the identification of bacteria and these are divided into 19 groups. The ninth edition of *Bergey's Manual of Determinative Bacteriology* (Holt *et al.,* 1994) published in 1994 is a departure from past editions that it is intended solely for the identification of those bacteria that have been described and cultured. In this edition, the bacteria have been divided into 35 groups including 5 groups of Archaeobacteria (archaea). These groups are not formal taxonomic ranks but easily recognized phenotypic groups useful for identification purposes.

The second edition of *Bergey's Manual of Systematic Bacteriology* is the latest edition of the manual. It represents a major departure from the first edition, as well as from the eighth and ninth editions of *Bergey's Manual of Determinative Bacteriology*, in that the organization of the content follows a phylogenetic framework based on analyses of the nucleotide sequence of the ribosomal small subunit RNA, rather than one based on phenotypic characters. The classification in eighth and ninth editions of *Bergey's Manual of Determinative Bacteriology* and in the first edition of *Bergey's Manual of Systematic Bacteriology* was organized in a non-hierarchical scheme because information about higher taxa was insufficient for construction of a formal hierarchical classification such as those used in all the previous editions. Instead, the genera were organized into phenotypic groupings, and these groupings were called parts, sections or groups.

It is generally agreed that procaryotes (also called ***Procarya***) fall into two major lines of evolutionary decents, the *Archaea* and the *Bacteria*. These have been recognized as two different domains in the second edition of *Bergey's Manual of Systematic Bacteriology*. The domains are subdivided into phyla that represent the main organizational units in this edition. With the exception of two phyla, i.e., Cyanobacteria and Actinobacteria, phyla are further subdivided into classes, classes into orders, orders into families, families into genera and genera into species. In case of Cyanobacteria, sub-sections replace the families. In case of Actinobacteria, subclasses and suborders are included. Different taxonomic ranks and their numbers recognized till 2003 are given in Table 5.1 (Garrity *et al.*, 2005).

Table 5.1. Different taxonomic ranks used in classification of bacteria

Taxonomic rank	Archaea	Bacteria	Total
Domain	1	1	2
Phylum	2	24	26
Class	9	32	41
Subclass	0	5	5
Order/subsection	13	75	88
Suborder	0	17	17
Family	23	217	240
Genus	79	1115	1194
Species	281	6185	6466

The domain *Archaea* is divided into 2 phyla and 9 classes. The domain *Bacteria* is divided into 24 phyla and 32 classes. The domain *Bacteria* is much larger and more complex than the other procaryotic domain, *Archaea*. The size of various phyla of domain *Bacteria* differs considerably. The phylum Proteobacteria is the largest and at the end of

October 2003, it contained 2279 validly named species belonging to 521 genera. The phylum Dictyoglomi is the smallest and contains only a single recognized species. The phyla of the domain *Bacteria* along with their phyla numbers are listed in Table 5.2.

Table 5.2. Phyla of the domain *Bacteria*

Phylum number	Phylum name	Phylum number	Phylum name
B1	*Aquificae*	B13	*"Firmicutes"*
B2	*Thermotogae*	B14	*"Actinobacteria"*
B3	*Thermodesulfobacteria*	B15	*"Planctomycetes"*
B4	*"Deinococcus-Thermus"*	B16	*"Chlamydiae"*
B5	*Chrysiogenetes*	B17	*"Spirochaetes"*
B6	*Chloroflexi*	B18	*"Fibrobacteres"*
B7	*Thermomicrobia*	B19	*"Acidobacteria"*
B8	*"Nitrospirae"*	B20	*"Bacteroidetes"*
B9	*Deferribacteres*	B21	*"Fusobacteria"*
B10	*Cyanobacteria*	B22	*"Verrucomicrobia"*
B11	*"Chlorobi"*	B23	*"Dictyoglomi"*
B12	*Proteobacteria*	B24	*Gemmatimonadetes*

" "Used for provisional arrangement, the rest are established

Since the publication of Garrity *et al.* (2005), three new phyla have been recognized in the domain *Bacteria*. These phyla are candidate phylum '*Synergistetes*' (Jumas-Bilak *et al.*, 2007), phylum *Caldiserica*, originally called the candidate phylum OP5 (Mori *et al.*, 2009) and phylum *Tenericutes*. The phylum '*Synergistetes*' was first named as candidate phylum '*Synergistes*' but this name is illegitimate due to its prior use for a genus.

5.3.1. Classification of plant pathogenic bacteria

The classification of plant pathogenic bacteria has also seen many revisions and rearrangements. Before 1967, i.e., prior to the discovery of role of mycoplasma-like organisms in causing plant diseases, there were only 6 genera of bacteria, which included plant pathogens. These genera were *Agrobacterium, Corynebacterium, Erwinia, Pseudomonas, Streptomyces and Xanthomonas*. The 1976 revision of International Code of Nomenclature of Bacteria (Lapage *et al.*, 1975) had a major effect on the taxonomy of bacterial plant pathogens. As a result of revision of the code, the names of many plant pathogenic bacteria were not accepted and hence, not included in the approved lists. Some of these were not accepted because of their poor descriptions or absence of suitable type cultures. The others were not accepted because some of the plant pathogenic species reported were similar in morphological, cultural and biochemical characters and could not be separated from each other on the basis of these characters. However, they could be separated from each other on the basis of their pathogenicity on different host plants. Most of the names in the latter category were nomenspecies of *Xa. campestris* and *Psm. syringae* groups. More than 40 species included in the genus *Pseudomonas* were identical to *Psm. syringae* on the basis of cultural, morphological and biochemical characters. According to the revised code, the specific epithets of these species (not included in the approved lists) would thereafter be available for reuse to name new taxa. The loss of species designations for recognised plant pathogens and subsequent potential reuse of these names for new taxa was bound to result in unnecessary confusion and loss of the knowledge previously published about these pathogens.

5.3.1.1. Pathovar concept: The executive committee of **International Society for Plant Pathology** (ISPP) set up a **Committee on Taxonomy of Plant Pathogenic Bacteria** (CTPPB) to see the consequences of 1976 revision of code for plant pathologists and to make necessary recommendations for retaining the names of plant pathogenic bacteria not included in the approved lists. The infrasubspecific taxon "pathovar" records most accurately the differences exhibited by plant pathogenic bacteria in terms of host range or in producing different symptoms (diseases) on the same host. The standards define a pathovar as "a strain or a group of strains with the same or similar characteristics, differentiated at the infrasubspecific level from other strains of the same species or subspecies on the basis of distinctive pathogenicity to one or more plant hosts." These differences in pathogenicity can be in symptoms or host range. Pathovars may also differ in other characteristics as well as pathogenicity but such differences must be of minor importance, otherwise there would be grounds for separation at subspecies or even species level (Bradbury, 1982). CTPPB of ISPP requested the Judicial Commission of International Committee on Systematic Bacteriology to alter the rules of the code to include the taxon pathovar. However, the Judicial Commission denied the request because it could set a precedent for the inclusion of other infrasubspecific subdivisions.

The consensus of opinion of ISPP members and its CTPPB was that the pathovar designation most accurately records the known status of nomenspecies of plant pathogenic bacteria not included in the approved lists and that formal recognition is essential to avoid confusion in the nomenclature. Young *et al.* (1978) proposed that bacterial plant pathogens which do not have a claim to species rank be classified as pathovars and also listed all validly published names of plant pathogens having authentic cultures as species or pathovars. In order to formalize the use of pathovar names, the ISPP through its executive committee and its CTPPB, decided to publish a list of pathovar names and standards, to be known as the International Standards for Naming Pathovars of Phytopathogenic Bacteria.

5.3.1.2. International Standards for Naming Pathovars of Phytopathogenic Bacteria: The rules for the Standards were adopted from the Code and were so written that if the Code ever does expand to assume authority over the infrasubspecific sub-divisions, the pathovar standards would be transferred easily to the Code. The Standards were adapted to the Code in a manner that allows both the Standards and the Code to be applied without conflict. The principal requirements of the Code for legitimate names are essentially the same for pathovars.

Unlike the valid publication of species names, no journal is specified for the valid publication of pathovar names. Although there is requirement of submission of pathotype culture to one international culture collection, the ISPP-CTPPB suggests that researchers should follow the Code and submit the new pathotype strains at least to two international culture collections located in two different countries to ensure stability and access. Priority for pathovars included in the list of pathovars names published with the standards (Dye *et al.*, 1980) and also those published before the publication of this list is assigned in a manner similar to those of species. These are treated as though they had been validly published for the first time on January 1, 1980. Priority for pathovar names published after January 1, 1980 is determined according to the date of effective publication of legitimate pathovar names. For more details on nomenclature of bacterial plant pathogens, the readers may consult the review articles by Young (2008) and Bull *et al.* (2008).

Plant pathologists or plant bacteriologists have to follow two separate but, complimentary set of rules: the rules set forth in International Code of Nomenclature of Procaryotes for naming down to subspecies level, and the standards in the International Standards for Naming Pathovars of Plant Pathogenic Bacteria to name pathovars. A list of pathovar names and pathotype strains is maintained by the CTPPB of ISPP. The first such list was an appendix to the Standards and only contained the names of organisms for which pathovar names and pathotype or neopathotype strains needed to be assigned (Dye *et al.,* 1980). Names of plant pathogenic species in the approved lists were not included in the ISPP-CTPPB list. These two documents together contained the names of all bacterial plant pathogens validly published as of 1980. Since then, the ISPP-CTPPB periodically publishes lists of recently published legitimate names and updates an annotated comprehensive list of names of plant pathogenic bacteria. The subsequent lists published by ISPP-CTPPB are given below.

1. Names of plant pathogenic bacteria 1864 - 1995 (Young *et al.*, 1996) is the first comprehensive list of names published by the ISPP-CTPPB in which all the names proposed for plant pathogenic bacteria were included in a single list. This was originally published in the Review of Plant Pathology (Vol. 75, pages 721 - 863) and in 2000 was placed on the ISPP website, http://www.isppweb.org/names_bacterial_revised.asp.
2. Names of plant pathogenic bacteria published since 1995 (Young *et al.*, 2004; http://www.isppweb.org/names_bacterial.asp) is a list of names published from 1995 to March 2004. It is published only online and includes all names that were published during this period.
3. Names of plant pathogenic bacteria, 1864 - 2004 (Young *et al.*, 2004; http://www.isppweb.org).
4. Names of plant pathogenic bacteria, 1864 - 2006 (Bull *et al.*, 2008; http://www.isppweb.org/about_tppb.asp).
5. Comprehensive list of names of plant pathogenic bacteria, 1980 - 2007 (Bull *et al.*, 2010).

In the second edition of *Bergey's Manual of Systemic Bacteriology*, phytopathogenic bacteria have been placed in three phyla, namely, Proteobacteria, "Firmicutes", and "Actinobacteria". Among these phyla, Proteobacteria is the largest and is divided into five classes, namely, "Alphaproteobacteria", "Betaproteobacteria", "Gammaproteobacteria", "Deltaproteobacteria" and "Epsilonproteobacteria". Out of these five classes, first three classes contain the plant pathogenic bacteria. The phyla, "Firmicutes" and "Actinobacteria" contain 3 and 1 class, respectively and all these classes contain plant pathogenic bacteria. The important characteristics of the phyla containing plant pathogenic bacteria are given in Table 5.3.

5.3.1.3. Genera of phytopathogenic bacteria: Till now 40 genera of phytopathogenic bacteria have been reported in the literature (Young *et al.*, 2004; Pal and Jalali, personal communication). These include 37 established genera and three candidatus categories, whose relatedness has been determined and verified by *in-situ* probing. A brief description of these genera along with their systematic position is given below. The authenticated species and pathovars of these genera are also listed alphabetically.

5.3.1.3.1. Acetobacter Beijerinck 1898: The genus was established by Beijerinck in 1898. The important characteristics of the genus are given below.

Cells are Gram-negative, nonsporing, straight or slightly curved rods measuring 0.6 - 0.8 × 1.0 - 4.0 μm. Cells may occur singly, in pairs, or in chains and may be motile or nonmotile; if motile, the motility is by means of peritrichous flagella. Colonies of most strains are nonpigmented, while some produce pale colonies and a small percentage of strains produces brown water soluble pigments. Obligately aerobic; metabolism is strictly respiratory, never fermentative, with oxygen as the terminal electron acceptor. Tests for oxidase, gelatin liquefaction, indole and H_2S production are negative. Catalase test is usually positive and some strains reduce nitrates to nitrites. Chemoorganotrophs and some strains need growth factors such as thiamine, niacin, pantothenic acid, or *p*-aminobenzoic acid. Optimal temperature for growth is 30°C and optimal pH is in the acidic range, i.e., 4.0 - 6.0. Acetate and lactate are oxidised to CO_2 and H_2O and ethanol to acetic acid. Major quinone is Q-9. *Acetobacter* strains are found in fermented foods, flowers, fruits, vinegar and palm wine, and cause spoilage of cocoa and grape wine, beer, cider and fermented foods. The genus does not contain any human and animal pathogen. The G + C content of the DNA varies from 50.5 - 60.3 mol%.

Typc species – *Acetobacter aceti*

Beijerinck and Folpmers (1916) described two species, namely, *Ace. aceti* and *Ace. pasteurianus* isolated mainly from flowers, fruits, palm wine, vinegar, kefir and fermented food. Later on, these species were also found to cause brown rot of apple and pear fruits and pink disease of pine apple (Kontaxis and Hayward, 1978; Dhanwantri *et al*., 1978).

Table 5.3. Phyla of the domain *Bacteria* encompassing phytopathogenic bacteria

Phylum number	Phylum name/properties	Class	Genera containing phytopathogenic bacteria
B12	*Proteobacteria* Largest phylum of the domain *Bacteria,* contains 2279 species assigned to 521 genera of which 26 genera contain plant pathogens. It is divided into five classes.	*"Alphaproteobacteria"*	*Acetobacter, Agrobacterium,'Candidatus* Liberibacter', *Gluconobacter, Rhizobium, Sphingomonas*
		"Betaproteobacteria"	*Acidovorax, Burkholderia, Herbaspirillum, Janthinobacterium, Ralstonia, Xylophilus*
		"Gammaproteobacteria"	*Brenneria, 'Candidatus Phlomobacter', Dickeya, Enterobacter, Erwinia, Pantoea, Pectobacterium, Pseudoalteromonas, Pseudomonas, Rhizobacter, Samsonia, Serratia, Xanthomonas,*

			Xylella
		"Deltaproteobacteria"	None
		"Epsilonproteobacteria"	None
B13	*"Firmicutes"* Contains 1503 species belonging to 223 genera. Contains phenotypically diverse groups of bacteria such as thermophilic and hyperthermophilic bacteria; anaerobic, straight, curved and helical Gram-negative rods; phototrophic and nonphototrophic bacteria; Gram-negative and Gram-positive cocci; Gram-positive and Gram-negative rods; mycoplasmas and thermoactinomycetes.	*"Bacilli"* *"Clostridia"* *Mollicutes*	*Bacillus* *Clostridium* *'Candidatus* Phytoplasma', *Spiroplasma*
B14	*"Actinobacteria"* A very high level of morphological, physiological and genomic diversity is found among its members. The members are divided into two major phenotypic groups: unicellular nonsporulating actinobacteria and filamentous sporulating sporoactinomycetes. The unicellular nonsporulating group contains Gram-negative aerobic rods and cocci; budding and/or appendaged bacteria; Gram-positive cocci; regular and irregular nonsporulating Gram-positive rods and mycobacteria. The sporoactinomycetes include nocardioform actinomycetes, actinomycetes having multilocular sporangia, *Streptomyces* and other related genera.	*Actinobacteria*	*Arthrobacter, Clavibacter, Curtobacterium, Corynebacterium, Janibacter, Leifsonia, Nocardia, Rathayibacter, Rhodococcus, Streptomyces*

Systematic position: Domain-*Bacteria*; Phylum-*Proteobacteria*; Class-*"Alphaproteobacteria"*; Order-*Rhodospirillales*; Family-*Acetobacteraceae*; Genus-*Acetobacter*; Species-*Ace. aceti, Ace. pasteurianus*.

5.3.1.3.2. Acidovorax Willems *et al.* 1990 emend. Willems *et al.* 1992: The genus was created by Willems *et al.* (1990) to include bacteria isolated from soil, water, activated sludge and clinical samples. Two years later, Willems *et al.* (1992) transferred four plant pathogenic species and subspecies of the genus *Pseudomonas* to this genus. Two more pathogens, i.e., causal agents of bacterial leaf spot of anthurium (*Anthurium andraeanum)* and bacterial leaf spot of lamb's lettuce (*Valerianella locusta*) have been included in this genus (Gardan *et al.*, 2000, 2003a). On the basis of genetic and phenotypic data, Schaad *et al.* (2008) proposed an emendation of the species *Aci. avenae* and elevated *Aci. avenae* subsp. *citrulli* and *Aci. avenae* subsp. *cattleyae* to species rank as *Aci. citrulli* and *Aci. cattleyae*, respectively. They also proposed a new taxon, *Aci. oryzae* sp. nov. with FC-143^T=ICPB 30003^T=ICMP 3960^T=ATCC 19882^T as the type strain

The below given description of the genus is as proposed by Willems *et al.* (1990) and emended by Willems *et al.* (1992). Cells are Gram-negative, straight to slightly curved rods and measure 0.2 – 1.2 × 0.8 - 5.0 μm. Cells occur singly, in pairs, or in short chains and are motile by means of one or rarely two or three polar flagella. Most strains produce non-pigmented growth on nutrient agar but some phytopathogenic strains produce a yellow to slightly brown diffusible pigment. Optimal temperature for growth is 30 - 35°C. Oxidase test is positive while urease activity is variable among strains. Aerobic, having a strictly oxidative type of metabolism with oxygen as the terminal electron acceptor. However, some strains of *Aci. temperans* and *Aci. delafieldii* are capable of heterotrophic denitrification of nitrate. Most strains are chemoorganotrophs, although strains of *Aci. facilis* and *Aci. delafieldii* can grow lithoautotrophically, using the oxidation of H_2 as an energy source. Organic acids, amino acids and peptone support good growth of the bacterium but only a limited number of sugars support the growth. Two fatty acids, namely, 3-hydroxyoctanoic acid and 3-hydroxydecanoic acid are always present while 2-hydroxy-substituted fatty acids are absent. The G + C content of the DNA varies from 62.0 - 70.0 mol% (T_m). *Acidovorax* strains can be obtained from infected plants, soil, water, and clinical samples and activated sludge.

Type species- *Acidovorax facilis*

Systematic position: Domain-*Bacteria*; Phylum-*Proteobacteria*; Class-"*Betaproteobacteria*"; Order-"*Burkholderiales*"; Family-*Comamonadaceae*; Genus-*Acidovorax*; Species-*Aci. anthurii, Aci. avenae, Aci. avenae* subsp. *avenae, Aci. cattleyae, Aci. citrulli, Aci. konjaci, Aci. oryzae, Aci. valerianellae.*

5.3.1.3.3. Agrobacterium Conn 1942: The genus was established by Conn in 1942 to include *Ag. tumefaciens* and *Ag. rhizogenes* which cause hypertrophies of the affected plant parts. The pathogenicity character is encoded by genes present on the transmissible plasmids. The ease with which the capacity to induce hypertrophies is transferred between species of *Agrobacterium* suggests that a stable taxonomy based on pathogenicity may not sound well. The use of names of *Ag. tumefaciens* and *Ag. rhizogenes* with natural classification recognizing pathogenicity of strains according to their 'tumorigenic' or 'rhizogenic' and 'non-pathogenic state' has been supported earlier. According to Young *et al.* (2001), this also posed a problem as the epithet *'tumefaciens'* was too closely associated with tumorigenicity to stand independently as a species name. In 1993, it was proposed to replace the name of *Ag. tumefaciens* with *Ag. radiobacter* to resolve this problem but it could not be adopted because *Ag. tumefaciens* was conserved as the type species of this proposed combination (Sawada *et al.*, 1993). Young *et al.* (2001) found that three genera of family *Rhizobiaceae*, namely, *Rhizobium*, *Agrobacterium* and *Allorhizobium* do not differ phenotypically and also do not have unique phenotypic and genetic circumscription. Moreover, their phylogenetic differentiation on the basis of algorithm and sequences of 16S rRNA is also not strong. Therefore, they amalgamated these three genera in a single genus *Rhizobium* as the senior subjective synonym. However, Garrity *et al.* (2005) have recognised both the genera, *Rhizobium* and *Agrobacterium* in the family *Rhizobiaceae* for the time being. *Agrobacterium* is treated as a separate genus here also. Bauzar and Jones (2001) have included the causal agent of aerial tumours of *Ficus benjamina* in this genus.

The genus has the following important characteristics. Cells are Gram-negative, nonsporing rods, measuring 0.6 - 1.0 × 1.5 - 3.0 μm. Cells are found singly or in pairs and are motile by means of 1 - 6 peritrichous flagella. Aerobic and chemoorganotrophs. Metabolism is strictly respiratory type in which oxygen is used as terminal electron acceptor. Colonies are generally circular, smooth, convex and colourless to light beige. Growth on media containing carbohydrates is usually copious and slimy. Optimal temperature for growth is 25 - 28°C. Catalase and urease are always positive while oxidase is generally positive. Growth in mineral salts media containing mannitol and other carbohydrates is accompanied by an acidic reaction. Common inhabitants of soil. Tumour forming strains occur in soils where diseased plants grew earlier or, are growing. The G + C content of the DNA ranges from 57.0 - 63.0 mol%.

Type species- *Agrobacterium tumefaciens*

Ag. tumefaciens causes crown gall of plants. Virulent strains of the bacterium contain a tumour-inducing plasmid, called Ti-plasmid. For successful infection of the host, T-DNA, called transfer DNA is cut from the Ti-plasmid and is transferred to the host cell genome. The penetration of *Ag. tumefaciens* into the host is through wounds. Although *Agrobacterium* species generally infect plants, they have been shown to cause opportunistic infections in humans with weakened immune systems (Hulse *et al.*, 1993), but have not been shown to be a primary pathogen in otherwise healthy persons. The DNA transmission capabilities of *Agrobacterium* have been extensively exploited in biotechnology, as a means of inserting foreign genes into plants. The plasmid T-DNA that is transferred to the plants is an ideal vehicle for genetic engineering. Tumour inducing genes are deleted from the plasmid and the desired gene sequence is cloned into the T-DNA.

Systematic position: Domain-*Bacteria*; Phylum-*Proteobacteria*; Class-"*Alphaproteobacteria*"; Order-"*Rhizobiales*"; Family-*Rhizobiaceae*; Genus-*Agrobacterium*; Species-*Ag. larrymoorei, Ag. rhizogenes, Ag. rubi, Ag. tumefaciens, Ag. vitis.*

5.3.1.3.4. Arthrobacter Conn and Dimmick 1947: The genus was created by Conn and Dimmick in 1947. The bacterium causing bacterial blight of leaves and twigs of American hollyhock, and earlier classified as *Corynebacterium ilicis*, was transferred to this genus. However, according to Young *et al.* (2004), *Ar. ilicis* is a validly named species, but probably is not a plant pathogen. The Judicial Commission ruled (opinion 87) that the name *Ar. ilicis* is represented by the type strain DSM 20138 =ATCC 14264 =NCPPB 1228 and is not a homotypic synonym of *Cor. ilicis* Mandel *et al.* 1961, and is reported not to be a plant pathogen (*Int. J. Syst. Evol. Microbiol.* 58: 1976 - 78).

Arthrobacter (from the Greek, "**jointed small stick**") are coryneform bacteria characterized by pleomorphism and Gram variable reaction. In young cultures, slender rods are Gram-negative and in about 30-hour-old cultures, very short rods and coccoids are Gram-positive. Coccoid cells measure from ~ 0.6 - 1.0 μm in diameter. Colonies on blood agar are small, measuring 2 mm in diam and range in colour from yellow to white. The bacterium can be readily isolated on sheep blood agar. Moderate to good growth occurs on yeast extract-peptone medium. Optimal temperature for growth is between 25 and 30°C. They have a complex life cycle composed of two distinct stages when grown in complex media. It is also called a **rod-coccus growth cycle** and is a most distinctive feature of the genus. In exponential phase, the bacteria are irregular, branched rods and may reproduce by a process called snapping division. As they enter stationary phase, the cells change to coccoid form. Upon transfer to fresh complex medium, the coccoid cells swell and produce one or more overgrowths and again form actively growing irregular rods. When viewed under the microscope, some cells are seen forming V-shaped arrangements and also more complex angular arrangements.

Most species of arthrobacteria are obligate aerobes, but all exhibit a pure respiratory, never fermentative, metabolism. However, two species, namely, *Ar. globiformis* and *Ar. nicotianae* exhibit anaerobic metabolism. Both the species utilize nitrate as an electron acceptor at the end of their respiratory chain, reducing it to ammonia via nitrite. Little or no acid is produced from glucose and other sugars in peptone media. Nonsporulating. Many species exhibit a week motility that often goes unnoticed. Are nutritionally versatile and use a variety of substrates in their oxidative metabolism including nicotine, nucleic acids and various herbicides and pesticides. Pyridone is used as a sole source of carbon. Chemoorganotrophs. Catalase positive.

Although arthrobacters are often isolated from sewage, fish and plant surfaces, soil is their most important habitat, where they form a significant component of the microbial flora. They are well adapted to this habitat because the cocci are resistant to desiccation and starvation. Due to their metabolic diversity, they degrade a variety of chemicals including agricultural pesticides. Hexavalent chromium is 100 times more toxic than trivalent chromium due to its oxidation state and is also much more soluble in water, thus allowing it to seep into ground water very easily. Arthrobacteria cannot only grow in its presence but can also reduce it to trivalent chromium, a much less toxic form. *Arthrobacter* species in association with several strains of *Streptomyces* can completely degrade an organophosphate insecticide, diazinon. They can use this as the only source of carbon and energy but individually none can grow on this compound. The genome of *Arthrobacter* sp. FB 24 has been sequenced and it contains 5011598 base pairs. Currently TIGR is sequencing another species, *Ar. aurescens* TC 1 (http://microbewicki.kenyon.edu/index.php/Arthrobacter). The G + C content of the DNA ranges from ~ 59.0 - 70.0 mol% (T_m).

Type species- *Arthrobacter globiformis*

Systematic position: Domain-*Bacteria*; Phylum-"*Actinobacteria*"; Class-*Actinobacteria*; Subclass-*Actinobacteridae*; Order-*Actinomycetales*; Suborder-*Micrococcineae*; Family-*Micrococcaceae*; Genus-*Arthrobacter*; Species-*Ar. ilicis.*

5.3.1.3.5. Bacillus Cohn 1872: It is a very old genus and was created by Cohn in 1872. The important characteristics of the genus are given below. Cells are Gram-positive, endospore forming rods that are usually motile and peritrichously flagellate. The presence of spores makes the cells (rods) to bulge at one end. The spores may be thin- or thick-walled. The bacterium is heterotrophic, aerobic or sometimes facultative, and catalase positive. Ubiquitous in nature, the genus includes both free-living and pathogenic species. The type species, *Ba. subtilis* is one of the best understood procaryotes, in terms of molecular biology and cell biology. The cell wall is composed of teichoic and teichuronic acids. *Ba. subtilis* and *Ba. megaterium* are strict aerobes and do not require growth factors. The former is a powerful producer of extracellular amylases and proteases and serves as a source of commercial production of these enzymes.

Many species of *Bacillus* are of considerable importance. Some species produce antibiotics, namely, bacitracin, polymyxin, gramicidin etc. *Ba. cereus* causes some forms of food poisoning and can infect humans. *Ba. anthracis* is the causal agent of anthrax disease that afflicts both farm animals and humans. *Ba. polymyxa* fixes atmospheric nitrogen. *Ba. thuringiensis* and *Ba. sphaericus* have insecticidal values. Both the species form a solid protein crystal, the parasporal body, next to their spores during endospore formation. The parasporal body of *Ba. thuringiensis* contains protein toxins that can kill larvae of more than 100 lepidopterous insects. Most of these *Ba. thuringiensis* toxin genes are carried on large plasmids. The genes of the bacterium which encode toxin production (BT genes) have been incorporated into cotton, corn and brinjal. These genetically modified crops are called BT cotton, BT corn and BT brinjal and possess the insecticidal properties. The parasporal body of *Ba. sphaericus* contains proteins toxic for mosquito larvae and may prove useful in controlling mosquitoes that carry the malaria parasite, *Plasmodium*. The 4.2 million base pairs genome of *Ba. subtilis* has been sequenced. The genome contains genes for the catabolism of many diverse carbon sources, and for protein secretion and antibiotic synthesis. The G+C content of the DNA ranges from 32.0 - 69.0 mol% (T_m).

Type species- *Bacillus subtilis*

Generally the members of this genus are not active plant pathogens. However, some of its members are found associated with plant material and cause its breakdown particularly at or after the crop harvest. This breakdown also occurs at relatively higher temperatures in comparison to the temperature required by other plant pathogenic bacteria for their pathogenicity. In 1982, a pathogen of cereals, namely, *Ba. megaterium* pv. *cerealis* was transferred to this genus.

Systematic position: Domain-*Bacteria*; Phylum-"*Firmicutes*"; Class- "*Bacilli*"; Order- *Bacillales*; Family-*Bacillaceae*; Genus-*Bacillus*; Species-*Ba. megaterium, Ba. megaterium* pv. *cerealis.*

5.3.1.3.6. Brenneria Hauben *et al.* 1999: Hauben *et al.* (1999) created this genus by transferring the species of cluster III of the then genus *Erwinia*. These species were transferred on the basis of complete 16S rDNA sequence analysis. One newly described species, *Er. alni* was also transferred. However, Samson *et al.* (2005) transferred one species, *Br. paradisiaca* to a newly created genus *Dickeya*.

The genus was named after Don J. Brenner, an American bacteriologist. The below given description of the genus is based on the data of Hauben *et al.* (1998) and of Wilson *et al.* (1957), Dye (1968), Lelliott and Dickey (1984), Verdonck (1992), Rijckaert (1994) and Surico *et al.* (1996). Cells are Gram-negative, nonsporing rods having rounded ends and measure 0.5 - 1.0 × 1.3 - 3.0 μm. Cells occur singly, rarely in pairs and are motile by means of peritrichous flagella. Facultatively anaerobic, catalase positive and oxidase negative. Optimal temperature for growth is 27 - 30°C. The bacterium does not produce arginine dicarboxylases, arginine dihydrolase, lysine dicarboxylases and ornithine dicarboxylases. Members do not hydrolyze starch. Acid is produced from D-glucose, D-mannose, D-galactose, fructose, sucrose and salicin but not from adonitol, inositol and dulcitol. Metabolism is fermentative type.

The major fatty acids include 12:0, 14:0, unknown 14.503, 16:0 and 17:0 cyclopropane. The species of the genus comprise a distinct phylogenetic group, as determined by 16S rRNA gene sequence comparisons and are characterized of having 12 signature nucleotides. The G + C contents of the members range from 50.1 - 56.1 mol%. The species of the genus cause tree diseases like wilts, necrosis, rots, blights and cankers. The entry of the bacteria into the host tissue generally occurs through wounds and natural openings.

Type species- *Brenneria salicis*

Systematic position: Domain-*Bacteria*; Phylum-*Proteobacteria*; Class-"*Gammaproteobacteria*"; Order-"*Enterobacteriales*"; Family-*Enterobacteriaceae*; Genus-*Brenneria*; Species-*Br. alni, Br. nigrifluens, Br. quercina, Br. rubrifaciens, Br. salicis.*

5.3.1.3.7. Burkholderia Yabuuchi *et al.* 1993 emend. Gillis *et al.* 1995: The genus was created by Yabuuchi *et al.* in 1993 to include some species of the then genus *Pseudomonas*. Gillis *et al.* (1995) further emended this genus on the basis of polyphasic taxonomy and also proposed a new species for this genus. All the 19 species of this genus belong to the ribosomal RNA similarity group II, which can be differentiated from the other groups of aerobic pseudomonads by rRNA/DNA hybridization or by rDNA sequencing. As one plant pathogenic species of this genus has been transferred to a newly created genus *Ralstonia*, it contains only six plant pathogenic species.

The genus was named after W.H. Burkholder, an American bacteriologist, who first discovered the etiological agent of sour skin of onions. The following description of the genus is based on its proposal by Yabuuchi *et al.* (1993) and emendment by Gillis *et al.* (1995). Cells are Gram-negative, nonsporing, straight or curved rods but not helical and measure generally 0.5 - 1.0 × 1.5 - 4.0 μm. Cells may occur singly or in pairs and are motile by one or generally several polar flagella. One species, namely, *Bur. mallei* lacks flagella and is nonmotile. Most strains accumulate poly-β-hydroxybutyrate as carbon reserve material and are capable of *ortho* cleavage of protocatechuate. Catalase is produced and oxidase activity varies among species. Chemoorganotrophs. Metabolism is strictly respiratory type with oxygen as the terminal electron acceptor. Some species are capable of anaerobic respiration using nitrate. Strains of some species such as, *Bur. vietnamiensis* and *Bur. cepacia* are capable of fixing N_2. Most species grow at 40°C. The species are characterized by the presence of hydroxy fatty acids having 14, 16 and 18 carbon atoms and most characteristic of these acids is the $C_{16:0\ 3OH}$. The type strains of several species are characterized by the presence of two types of ornithine lipids. All species can utilize glucose, galactose, glycerol, inositol, sorbitol and mannitol as sole sources of carbon. More than half of the species are pathogenic on plants, animals and humans. Isolated from plant materials, soil and clinical samples. The G + C content of the DNA ranges from 59.0 - 69.6 mol%.

Type species- *Burkholderia cepacia*

Systematic position: Domain-*Bacteria*; Phylum-*Proteobacteria*; Class-"*Betaproteobacteria*"; Order-"*Burkholderiales*"; Family-"*Burkholderiaceae*"; Genus-*Burkholderia*; Species-*Bur. andropogonis, Bur. caryophylli, Bur. cepacia, Bur. gladioli, Bur. gladioli-* pv. *agaricicola,* pv. *alliicola*, pv. *gladioli*; *Bur. glumae, Bur. plantarii.*

***Candidatus* category**: There are many bacteria that cannot be cultured by currently available methods. Therefore, these cannot be investigated in ways that provide the description for proposal as legitimate taxa. Keeping this in view, Murray and Schleifer (1994) proposed the rank or category of *Candidatus* for classification of such procaryotes. The International Committee on Systematic Bacteriology accepted the proposal of Murray and Schleifer (1994) and recommended the category '*Candidatus*' as a possible taxonomic rank for the unculturable procaryotic organisms (Murray and Stackebrandt, 1995). In addition to genomic information, such as sequences apt to determine the phylogenetic position of the organism, all information, including structural, metabolic and reproductive features, should be included in the description of a provisional taxon, together with the natural environment in which the organism can be identified by in situ-hybridization or other similar techniques for cell identification. *Candidatus* species are differentiated primarily on the basis of the comparative analyses of 16S rRNA sequences, other sequence data, ecological information, and metabolic data where available. To make clear the identification of organisms as *Candidatus*, names are preceded by the word *Candidatus* (italicized) but the name itself is not italicized and the entire designation is written between single quotation marks e.g. '*Candidatus* Liberibacter asiaticus'. *Candidatus* nomenclature is neither governed by the Code nor by the Standards for Naming Pathovars. However, to prevent any nomenclature confusion, the Phytoplasma/Spiroplasma Working Team of the International Research Project for Comparative Mycoplasmology (IRPCM) has suggested rules for the description of organisms as novel taxa within '*Candidatus* Phytoplasma' (IRPCM, 2004) and by extension to, other plant pathogenic '*Candidatus* genera' (Young, 2008). As the *Candidatus* microorganisms cannot be cultured, their type strains cannot be isolated or maintained for reference. The reference material for *Candid-*

atus is either a DNA sample or a specific rDNA sequence. Till now, three *Candidatus* genera have been recognised which are described below.

5.3.1.3.8. 'Candidatus Liberibacter' Jagoueix *et al.* 1994: Citrus huanglongbing (earlier called citrus greening) disease was initially considered to be caused by **mycoplasma-like organisms** (MLOs) and **rickettsia-like organisms** (RLOs). Later studies revealed that its causal agent is enclosed by 25 nm thick cell envelop and was called bacterium-like organism (BLO). Jagoueix-Eveillard *et al.* (1994) characterized two uncultured phloem-restricted bacteria causing citrus huanglongbing in Asia and Africa and named them as *'Candidatus* Liberobacter asiaticum' and *'Candidatus* Liberobacter africanum', respectively. Following the rules of the International Code of Nomenclature of Bacteria, these two pathogens have been renamed as *'Candidatus* Liberibacter asiaticus' and *'Candidatus* Liberibacter africanus'. Teixeira *et al.* (2005) have reported *'Candidatus'* Liberibacter americanus' associated with citrus huanglongbing in Sao Paulo state of Brazil.

The below given description of the 'Candidatus Liberibacter' is based mainly on Jagoueix *et al.* (1994). Bacteria are Gram-negative, filamentous in shape and occur in phloem sieve tubes of plants. Round forms are also observed but they are considered to be degenerating bacteria. They are transmitted by two insect vectors (Psyllidae), namely, *Trioza erytriae* and *Diaphornia citri*. They also multiply in the haemolymph and cells of salivary glands of both the vectors. The above given description of the *'Candidatus* genus' is based on the bacteria present in the phloem sieve tubes of sweet orange trees affected by huanglongbing (ex. greening disease) in Pune, India and in Nelspruit, South Africa. *'Ca.* Li. asiaticus' requires higher temperature than *'Ca.* Li. africanus' for optimum expression of the disease symptoms.

Most recently, Sechler *et al.* (2009) designed a medium called Liber A to culture all the three *'Candidatus* Liberibacter spp.' which cause citrus huanglongbing. The medium containing citrus vein extract and a growth factor supported the growth of *'Ca.* Liberibacter spp.' for four or five single colony transfers before the decline of viability. The colonies of *'Ca.* Li. asiaticus' were irregularly-shaped, convex and 0.1 – 0.3 mm after 3 – 4 days of incubation. Two strains of *'Ca.* Li. asiaticus' and one of *'Ca.* Li. americanus' grown on this medium proved pathogenic to citrus and were isolated from uninoculated tissues of inoculated trees and seedlings 9 and 2 months later, respectively. The identity of the bacteria was confirmed by RT-PCR and 16S rDNA sequencing. This is the first report of the culturing and pathogenicity of *'Ca.* Li. asiaticus' and *'Ca.* Li. americanus' associated with citrus huanglongbing.

Type species- *'Candidatus* Liberibacter asiaticus'

Systematic position: Domain-*Bacteria*; Phylum-*Proteobacteria*; Class-"*Alphaproteobacteria*"; Order-"*Rhizobiales*"; Family-"*Phyllobacteri-*

aceae"; Genus-'*Candidatus* Liberibacter'; Species-'*Ca.* Li. africanus', '*Ca.* Li. africanus subsp. capensis', '*Ca.* Li. americanus', '*Ca.* Li. asiaticus' '*Ca.* Li. solanacearum'.

5.3.1.3.9. '*Candidatus* Phlomobacter' Zreik *et al.* (1998): Zreik *et al.* (1998) created '*Candidatus* Phlomobacter' to include bacterium-like organisms (BLOs) responsible for causing marginal chlorosis of strawberry. At present, it contains only one plant pathogenic bacterium.

The below given description of the '*Candidatus* Phlomobacter' is taken from Zreik *et al.* (1998). Gram-negative, walled bacterium, restricted to the sieve tubes of the phloem tissue of strawberry plants. The bacterium is 0.20 - 0.27 μm in diameter and 4.0 μm long. The bacterium is unable to grow in culture media, could not be transmitted to periwinkle plants through dodder, and insect vectors are not known. Sequence analysis of the cloned rDNA revealed that it is a new bacterium within group 3 of class "Gammaproteobacteria" while citrus huanglongbing agents belong to class "Alphaproteobacteria". '*Candidatus* Phlomobacter fragariae' name was proposed for strawberry marginal chlorosis agent.

Type species- '*Candidatus* Phlomobacter fragariae'

Systematic position: Domain-*Bacteria*; Phylum-*Proteobacteria*; Class-"*Gammaproteobacteria*"; Order-"*Enterobacteriales*"; Family-*Enterobacteriaceae*; Genus-'*Candidatus* Phlomobacter'; Species -'*Ca.* Phl. fragariae'.

5.3.1.3.10. '*Candidatus* Phytoplasma': Several wall-less procaryotes are associated with plants and cause yellow-type diseases. These were, earlier called mycoplasma-like organisms due to their morphological similarities to mycoplasmas. Some of these microorganisms, which are spiral-shaped and can grow on artificial culture media; have been assigned to the genus *Spiroplasma*. The remaining wall-less, non-helical and not yet cultured procaryotes that colonize plant phloem and sap feeding insects are called phytoplasmas. In 1994, the trivial name "Phytoplasma" was adopted by the Phytoplasma Working Team at the 10th Congress of International Organization of Mycoplasmology, replacing the term "mycoplasma-like organisms" or MLOs. Zreik *et al.* (1995) proposed the genus '*Candidatus* Phytoplasma' to include '*Candidatus* Phytoplasma aurantifolia', the causal agent of witches' broom disease of lime. Since then, many other '*Candidatus* phytoplasmas' causing different plant diseases have been added to this genus (Jung *et al.*, 2002, 2003, 2003a; Verdin *et al.*, 2003; Hiruki and Wang, 2004; Lee *et al.*, 2004, 2004a, 2006; Schneider *et al.*, 2005; Marcone *et al.*, 2004, 2004a; Seemuller and Schneider, 2004; Arocha *et al.*, 2005, 2007; Valiunas *et al.*, 2006; Al-Saady *et al.*, 2008; Zhao *et al.*, 2009). Most plant pathogenic *Candidatus* species are members of this genus.

Classification and taxonomic assignment of phytoplasmas have been based primarily on molecular analysis of 16S rRNA gene sequences because of the inaccessibility of measurable phenotypic characters suitable for conventional microbial characterization. In the recent past, many molecular techniques have been developed for the identification and classification of phytoplasmas. Wei *et al.* (2007), on the basis of distinctive virtual RFLP patterns and calculated similarity coefficients (obtained by using computer-simulated RFLP analysis of 16S rRNA genes), classified phytoplasma strains into 28 groups. They were able to classify hundreds of previously unclassified phytoplasmas and delineate 10 new phytoplasma groups representing three recently described and seven novel putative *'Candidatus* Phytoplasma' taxa. Wei *et al.* (2008), extending their above-mentioned efforts in the exploitation of computer-simulated 16S rRNA gene RFLP analysis and virtual gel plotting for rapid classification of phytoplasmas, developed a Perl program for automated RFLP pattern comparison and similarity coefficient calculation. By streamlining virtual RFLP pattern analysis, the program has led to the establishment of a criterion for phytoplasma 16Sr subgroup classification and to the delineation of new and distinct subgroup lineages in the clover proliferation phytoplasma group (16SrVI).

Martini *et al.* (2007) showed that ribosomal protein (rp) genes, rplV (rpl22) and rpsC (rps3) based phylogeny allowed for finer resolution of distinct lineages within the phytoplasma 16Sr groups. RFLP analysis of rp gene sequences permitted finer differentiation of phytoplasma strains in a given 16Sr group. Hodgetts *et al.* (2008) have highlighted the utility of virtual RFLP analysis of secA gene sequences to differentiate 16Sr groups and subgroups suggesting that this gene may provide an informative alternative molecular marker for pathogen identification and diagnosis of phytoplasma diseases. Zhao *et al.* (2009a) developed an interactive online tool, iPhyClassifier, to expand the efficacy and capacity of the current 16S rRNA gene sequence-based phytoplasma classification system. iPhyClassifier performs sequence similarity analysis, simulates laboratory restriction enzyme digestions and subsequent gel electrophoresis and generates virtual restriction fragment length polymorphism (RFLP) profiles. iPhyClassifier makes instant suggestions on tentative phytoplasma 16Sr group/subgroup classification status and *'Candidatus* Phytoplasma' species assignment based on calculated RFLP pattern similarity coefficients and overall sequence similarity scores. Using iPhyClassifier, the authors have revised and updated the classification of strains affiliated with the peach X-disease phytoplasma group. The online tool is available at http://www.ba.ars.usda.gov/data/mppl/iPhyClassifier.html.

'Ca. Phytoplasma' cells lack rigid cell walls, are surrounded by a single unit membrane and are pleomorphic in shape. When observed by transmission electron microscopy, they appear as round to filamentous

pleomorphic bodies with a mean diameter of 200 - 800 nm. Some of the filamentous forms are branched. Phytoplasmas are sensitive to tetracyclines but nonsensitive to penicillin and other cell wall inhibiting antibiotics. They are resistant to digitonin and sensitive to hypotonic salt solutions, and are therefore, similar to those mollicutes that do not require sterols. Members use UGA as a stop codon and not as a tryptophan codon, as in several other mycoplasmas. Phytoplasmas are closest relatives of the members of the genus *Acholeplasma*. The genome size of different phytoplasmas varies considerably, ranging from 530 to 1350 kilo base pairs The Bermuda grass white leaf phytoplasma has the smallest genome size of 530 kbp found in phytoplasma todate and may represent the smallest chromosome known for any living cell. Like other members of the mollicutes, phytoplasmas contain one circular double-stranded chromosomal DNA molecule. Short, circular extra chromosomal DNAs or plasmids were found in all the members of the aster yellows group and stolbur group and in some members of the X-disease and clover proliferation groups. Large plasmids may also be present in some phytoplasmas. The G + C contents of phytoplasma chromosomal DNA are estimated to be between 23 and 29 mol% (Bd) (Lee *et al.*, 2000). Members inhabit the phloem sieve elements of vascular plants, and the gut, haemolymph, salivary glands and other organs of sap-sucking insects (leafhoppers and planthoppers). In insect hosts, they may cause premature mortality. In plant hosts, they cause a wide range of symptoms, such as virescence, phyllody, sterility of flowers, proliferation of auxiliary shoots, abnormal elongation of internodes and many others, often less specific symptoms.

Type species- '*Candidatus* Phytoplasma aurantifolia'

Systematic position: Domain-*Bacteria*; Phylum-"*Firmicutes*"; Class-*Mollicutes*; Order-Not yet established; Family-Not yet established; Genus-'*Candidatus* Phytoplasma'; Species–'*Ca*. Phy. allocasuarinae', '*Ca*. Phy. americanum', '*Ca*. Phy. asteris', '*Ca*. Phy. aurantifolia', '*Ca*. Phy. australasia', '*Ca*. Phy. australiense', '*Ca*. Phy. brasiliense', '*Ca*. Phy. caricae', '*Ca*. Phy. castaneae', '*Ca*. Phy. cynodontis', '*Ca*. Phy. fragariae', '*Ca*. Phy. fraxini', '*Ca*. Phy. graminis', '*Ca*. Phy. japonicum', '*Ca*. Phy. lycopersici', '*Ca*. Phy. mali', '*Ca*. Phy. omanense' '*Ca*. Phy. oryzae', '*Ca*. Phy. phoenicium', '*Ca*. Phy. pini', '*Ca*. Phy. prunorum', '*Ca*. Phy. pyri', '*Ca*. Phy. rhamni', '*Ca*. Phy. spartii', '*Ca*. Phy. tamaricis', '*Ca*. Phy. trifolii', '*Ca*. Phy. ulmi', '*Ca*. Phy. ziziphi'.

5.3.1.3.11. Clavibacter Davis *et al.* 1984: The genus was created by Davis *et al.* (1984) to accommodate some phytopathogenic coryneform bacteria whose cell walls contain diaminobutyric acid. Two newly described plant pathogens, *Cla. xyli* subsp. *xyli* and *Cla. xyli* subsp. *cynodontis*, causal agents of ratoon stunting disease of sugarcane and Bermuda grass stunting disease, respectively were classified in this genus. The creation of this genus was considered necessary, because no other genus as circumscribed for coryneform bacteria was suitable for these phytopathogens. Later on, both *Cla. xyli* subsp. *xyli* and *Cla. xyli* subsp. *cynodontis* were transferred to *Leifsonia.* Three other species of this genus, namely, *Cla. toxicus, Cla. rathayi* and *Cla. tritici* have been transferred to genus *Rathayibacter*.

The below given description of the genus is based on Davis *et al.* (1984). Cells are Gram-positive, not acid-fast, pleomorphic rods, which are often arranged at an angle to give V-shaped arrangements; a marked rod-coccus cycle is not observed. Cells are nonmotile and endospores are not formed. Obligately aerobic. Optimal temperature for growth is ca. 20 to 29°C and maximum is ca. 29 to 35°C. The bacterium has exact nutritional requirements and growth occurs on suitable solid media. Slow and weak oxidative production of acids occurs from some carbohydrates. Some organic acids are utilized. Neither nitrate is reduced nor utilized. Oxidase, urease, lipase and tyrosinase are not generally produced.

Cell wall peptidoglycan contains 2, 4-diamino butyric acid. Rhamnose is the major component of the cell wall while arabinose is absent. Mycolic acids are also absent. Non-hydroxylated fatty acids consist predominantly of *anteiso*- and *iso*-methyl branched chains; small amounts of straight-chain saturated acids are found. Respiratory quinones are menaquinones. Polar lipids comprise diphosphatidylglycerol, phosphatidylglycerol and lipids. The G + C content of the DNA is about 70 ± 5 mol%.

Type species- *Clavibacter michiganensis*

Systematic position: Domain-*Bacteria*; Phylum-*"Actinobacteria"*; Class-*Actinobacteria*; Subclass-*Actinobacteridae*; Order-*Actinomycetales*; Suborder-*Micrococcineae*; Family- *Microbacteriaceae*; Genus-*Clavibacter*; Species-*Cla. michiganensis, Cla.* michiganensis- subsp. *insidiosus,* subsp. *michiganensis,* subsp. *nebraskensis,* subsp. *sepedonicus*, subsp. *tessellarius.*

5.3.1.3.12. Clostridium Prazmowski 1880: It is a very old genus established way back in 1880. At present, it contains only one plant pathogenic species associated with spoilage of potatoes and carrots.

The genus has the following salient characteristics. Cells are Gram-positive, spore forming rods. Obligate anaerobes. Most species are obligate fermenters and they ferment by pathways that generate organic solvents such as butantol. No doubt, clostridia are rod-shaped, but when producing spores, they appear more like drum sticks with a bulge at one end. Do not carry out dissimilatory sulphate reduction. An exposure of vegetative cells to oxygen for 5 - 10 minutes may cause massive death of the population. However, spores are not killed by the oxygen. In many cases, the death results from the production of hydrogen peroxide by the bacteria, which is highly toxic. Most species completely lack heme proteins, including catalase, and are incapable of decomposing hydrogen peroxide.

Clostridium species are commonly found in the environment. They inhabit soil, sewage, marine sediments and the intestines of both animals and humans. Some species produce toxins and cause diseases. *Clo. tetani* causes tetanus. The bacterium releases an exotoxin called tetanus toxin, which blocks the release of neurotransmitters from the presynaptic membrane of inhibitory interneurons of spinal cord and brainstem of mammals that regulate muscle contraction. *Clo. botulinum* is an etiological agent of botulism and it is contracted usually through the ingestion of botulinum toxin. *Clo. perfringens* is one of the several species of the genus that causes gas gangrene and is the causative agent in 95% gas gangrene cases. *Clo. sordellii* has been linked to the death of a few women after child birth.

Apart from causing diseases and harmful effects, *Clostriduim* species have commercial uses and industrial value. *Clo. thermocellum* can utilize lignocellulosic waste and generate ethanol. *Clo. ljungdahlii* can produce ethanol from single-carbon sources including synthesis gas, a mixture of carbon monoxide and hydrogen that can be generated from the partial combustion of either fossil fuels or biomass. *Clo. butyricum* and *Clo. pasteurianum* fix nitrogen. The genomes of *Clo. tetani, Clo. perfringens* and *Clo. acetobutylicum* have been sequenced. *Clo. tetani* also contains a plasmid that measures 74082 bp long with 61 open reading frames and encodes the tetanus toxin. The genome of *Clo. perfringens* encodes the typical anaerobic fermentation enzymes leading to gas production but no enzymes for the tricarboxylic acid cycle of respiratory chain (Clostridium – Microbe Wiki). The G + C content of the DNA ranges from 22.0 - 55.0 mol% (T_m).

Type species- *Clostridium butyricum*

Systematic position: Domain-*Bacteria*; Phylum - *"Frimicutes"*; Class-*"Clostridia"*; Order – *Clostridiales*; Family - *Clostridiaceae*; Genus-*Clostridium*, Species-*Clo. puniceum.*

5.3.1.3.13. Corynebacterium Lehmann and Neumann 1896: The genus was created by Lehmann and Neumann in 1896 with *Cor. diphtheriae* as type species. It contained human, animal and plant pathogens. All the plant pathogenic species of this genus except *Cor. ilicis* have been transferred to different genera, namely, *Clavibacter, Curtobacterium, Rathayibacter and Rhodococcus*. The Judicial Commission ruled (opinion 87) that the name *Cor. ilicis* Mandel *et al.* 1961 is represented by the type strain ICMP 2608 =ICPB CI144 and is reported to be a plant pathogenic species (*Int. J. Syst. Evol. Microbiol.* 58: 1976 - 78).

Type species - *Corynebacterium diphtheriae*

Systematic position: Domain-*Bacteria*; Phylum-"*Actinobacteria*"; Class-*Actinobacteria*; Subclass-*Actinobacteridae*; Order-*Actinomycetales*; Suborder-*Corynebacterineae*; Family-*Corynebacteriaceae*; Genus-*Corynebacterium*, Species-*Cor. ilicis.*

5.3.1.3.14. Curtobacterium Yamada and Komagata 1972: The genus was proposed by Yamada and Komagata in 1972. In 1983, Collins and Jones proposed the transfer of *Cor. flaccumfaciens* to *Curtobacterium* as *Cur. flaccumfaciens* and transfer of *Cor. betae, Cor. oortii* and *Cor. poinsettiae* as pathovars of *Cur. flaccumfaciens*.

The characteristics of the genus as described by Yamada and Komagata (1972) are given below. Cells are weakly Gram-positive, small, short rods. Coccoid cells are found in old cultures and old cells frequently loose the character of Gram-positive reaction. Generally motile by means of lateral flagella. Cells multiply by bending type of cell division. Pleomorphism is slightly recognized. Metachromatic granules are not recognized. Ornithine is a principal amino acid found in the cell walls. Acid production is slow and weak from several sugars. Chemoheterotroph. Various kinds of organic acids are assimilated in addition to acetate, pyruvate and lactate. Gelatin is slowly hydrolyzed. The G + C content of the DNA is 66.0 - 71.0 mol%. The bacterium is widely distributed in soil and plant material.

Type species - *Curtobacterium citreum*

Systematic position: Domain-*Bacteria*; Phylum-"*Actinobacteria*"; Class-*Actinobacteria*; Subclass-*Actinobacteridae*; Order-*Actinomycetales*; Suborder-*Micrococcineae*'; Family- *Microbacteriaceae*; Genus-*Curtobacterium;* Species-*Cur. flaccumfaciens, Cur. flaccumfaciens-* pv. *betae,* pv. *flaccumfaciens,* pv. *ilicis,* pv. *oortii,* pv. *poinsettiae.*

5.3.1.3.15. Dickeya Samson *et al.* 2005: The genus was created by Samson *et al.* (2005). The bacterium causing hollow stalk and blight of *Chrysanthemum morifolium* was described and identified as *Erwinia chrysanthemi*. Subsequently, similar bacteria were isolated from several plants causing typical soft rot and wilt symptoms. After extensive biochemical testes, all these pathogens were finally grouped into a single species, *Er. chrysanthemi* and designated as *Er. chrysanthemi* complex. However, plant bacteriologists for their convenience, divided the members of this complex into six pathovars, namely, pv. *chrysanthemi* pv. *dianthicola*, pv. *dieffenbachiae*, pv. *parthenii*, pv. *zeae* and pv. *paradisiaca*. Later on, Hauben *et al.* (1998) renamed *Er. chrysanthemi* pv. *paradisiaca* as *Brenneria paradisiaca* and transferred the other members to *Pectobacterium chrysanthemi.* Although these pathogens were grouped together as pathovars of *Pe. chrysanthemi*, lot of diversity existed among these pathogens. Samson *et al.* (2005) have resolved this problem by using DNA-DNA hybridization and analysis of 16S rRNA gene sequences. They found that on the basis of analysis of 16S rRNA gene sequences, *Pe.*

chrysanthemi strains and *Br. paradisiaca* formed a clade distinct from the genera *Pectobactrium* and *Brenneria*, therefore; they transferred all these pathogens to a newly created genus *Dickeya*. On the basis of DNA-DNA hybridization, all the strains of *Pe. chrysanthemi* formed six genomic species and hence were divided into five species, one species having two biovars. Most recently, Parkinson *et al.* (2009) developed a phylogenetic method, using sequences from *recA* locus, to differentiate species of *Dickeya* genus. The method provides a simple means for identification to species and intraspecies level.

The below given description of the genus is taken from Samson *et al.* (2005), which is based on their data and those of Waldee (1945) and Hauben *et al.* (1998). The genus was named in honour of an American plant pathologist, Robert S. Dickey, for his contributions to the research on the *Er. chrysanthemi* complex. Cells are Gram-negative rods, measuring 0.5 - 1.0 × 1.0 - 3.0 μm having rounded ends. They occur mostly singly or in pairs but sometimes in chains, and are usually motile by means of peritrichous flagella. Facultatively aerobic/anaerobic bacteria that catabolize glucose by a fermentative pathway and reduce nitrates to nitrites. The bacterium produces pectolytic enzymes and indole, and grows at 36°C. Catabolizes (+)-L-arabinose, myo-inositol, (+)-D-malate, malonate, D-mannose, mucate, saccharate and meso-tartrate but do not catabolize (+)-D-trehalose, methyl α-glucoside, (+)-D-arabitol or sorbitol. Members of the genus form a distinct clad based on the sequence analyses of 16S rRNA gene. The G + C contents of its members range from 56.4 to 59.5 mol%. Cause vascular wilts or soft rots on a range of host plants.

Type species- *Dickeya chrysanthemi*

Systematic position: Domain-*Bacteria*; Phylum-*Proteobacteria*; Class-"*Gammaproteobacteria*"; Order-"*Enterobacteriales*"; Family-*Enterobacteriaceae*; Genus-*Dickeya*; Species-*D. chrysanthemi, D. chrysanthemi* pv. *chrysanthemi, D. chrysanthemi* pv. *parthenii, D. dadantii, D. dianthicola, D. dieffenbachiae, D. paradisiaca, D. zeae.*

5.3.1.3.16. Enterobacter Hormaeche and Edwards 1960: The genus was established in 1960 by Hormaeche and Edwards. Earlier, the genus did not include any plant pathogen. Now it contains four plant pathogenic species transferred from *Erwinia*. In order to infer an encompassing phylogeny of this genus, 16S rDNA sequences of additional members of *Enterobacter* are required.

The genus has the following important characteristics. Cells are Gram-negative, nonsporing, straight rods, which measure 0.6 - 1.0 × 1.2 - 3.0 μm. Some strains are encapsulated. Motile by peritrichous flagella. Facultative anaerobes, and some Enterobacter bacteria as facultative anaerobes ferment both glucose and lactose as a carbon source. Both acid and gas are produced from fermentation of glucose. *Enterobacter* species can be distinguished from other Gram-negative rods (*Escherichia coli* and *Klebsiella* spp.) by virtue of being fast fermenter of lactose. Gas is produced from metabolic processes. Nitrate is reduced to nitrite. H_2S is not produced from thiosulphate. Methyl red and Voges-Proskauer tests are positive for many strains. Tween 80 and gelatin are either very slowly hydrolysed or not hydrolyzed. Optimal temperature for growth is 30°C.

Enterobcter are found on plants and human skin, in soil, water, sewage and intestinal tracts of humans and animals. Some species of *Enterobacter* such as, *En. sakajakii* are opportunistic human pathogens. They cause infections in immunocompromised (usually hospitalized) hosts and rarely in otherwise healthy individuals. *En. sakajakii* is well known for infections in infants who were fed milk-based powdered infant formulae. The bacterium survives when the contaminated powdered formula is heated and prepared. Since it has caused several outbreaks of infection in the past, baby formulae are effectively screened for *En. sakajakii* before they are sold. *En. cloacae* A-11 and similar bacteria are found on cucumber, radish, peas, soybean, sunflower and sweet corn seeds. *En. cloacae* is also used for biological control of plant diseases. The G + C content of the DNA is 52.0 - 60.0 mol%.

Type species- *Enterobacter cloacae*

Systematic position: Domain-*Bacteria*; Phylum-*Proteobacteria*; Class-"*Gammaproteobacteria*"; Order-"*Enterobacteriales*"; Family-*Enterobacteriaceae*; Genus-*Enterobacter;* Species-*En. cancerogenus, En. cloacae, En. cloacae* subsp. *dissolvens, En. nimipressuralis, En. pyrinus.*

5.3.1.3.17. Erwinia Winslow *et al.* 1920 emend. Hauben *et al.* 1999: The genus was coined to contain all plant pathogenic bacteria that were Gram-negative, peritrichously flagellate and facultatively anaerobic rods. The genus was named in honour of Erwin F. Smith, a pioneer plant bacteriologist. The heterogenicity of this genus was generally acknowledged right from its inception. Several proposals were put forth to emend or revise this genus. Waldee (1945) suggested that *Erwinia* should be emended to contain pathogens that cause only necrosis or wilt diseases (*Er. amylovora, Er. salicis* and *Er. tracheiphila*), utilize a restricted range of carbon compounds for growth; and the biochemically more active soft rotting pathogens (*Er. carotovora* and *Er. chrysanthemi*) should be placed in a new genus, *Pectobacterium*. Several workers supported Waldee's suggestion. However, in 8th edition of *Bergey's Manual of Determinative Bacteriology*, three clusters of organisms, namely, *amylovora, herbicola* and *carotovora* groups were recognized under *Erwinia*. According to Kim *et al.* (1999), the phylogenetic group of seven species of the then genus *Erwinia* cluster *amylovora* (cluster-I), based on the complete 16S rDNA sequences, represents the true authentic erwinias. Some of the species and pathovars of *Erwinia* have been transferred to other genera, namely *Brenneria, Dickeya, Enterobacter, Pantoea* and *Pectobacterium*. However

Gardan *et al.* (2004) added one species, the causal agent of pink canker of papaya, to this genus.

The below given description of the genus is based on the emended description of the genus published by Hauben *et al.* (1999). Cells are Gram-negative, nonsporing, straight rods which measure 0.5 - 1.0 × 1.0 - 3.0 μm. Cells occur mostly singly or in pairs and are motile by peritrichous flagella. They are facultative anaerobes but anaerobic growth of some species is poor. Optimal temperature for growth is between 27 and 30°C. Strains do not produce indole and do not oxidize gluconate. Catalase is positive but oxidase and pectinase are negative. Acid is produced from fructose, sucrose, D-glucose, D-galactose, mannitol and β - methylglucoside but not from adonitol, arabitol, dulcitol, dextrin, inulin, starch, maltose and tagatose. Utilize fumarate, gluconate, malate, succinate, D-glucose, D-galactose, glycerol and β-methylglucoside but not benzoate, oxalate, propionate, L-arabitol, methanol, butanol and sorbose as carbon- and energy-yielding sources. Strains utilize L-alanine, L-serine, glycylglycine and L-glutaminic acid but not kynureninic acid and trigonelline as nitrogen sources. The members are sensitive to chloramphenicol, furazolidone, oxytetracycline, tetracycline and nalidixinic acid. Strains do not possess arginine dihydrolase, caseinase, urease and phenylalanine dreaminase.

The species of the genus comprise a distinct phylogenetic group, as determined by 16S rRNA gene sequence comparisons and are characterized by 14 generic signature nucleotides. The major fatty acids include 12:0, 14:0 and 16:0. The G + C content of the DNA ranges from 49.8 - 54.1 mol%. Members of the genus occur as saprophytes, epiphytes and plant pathogens. Plant pathogens cause plant diseases that include mainly wilts and blights. The entry of the pathogens into host is mainly through wounds and natural openings.

Type species- *Erwinia amylovora*

Systematic position: Domain-*Bacteria*; Phylum-*Proteobacteria*; Class-"*Gammaproteobacteria*"; Order-"*Enterobacteriales*"; Family-*Enterobacteriaceae*; Genus-*Erwinia*; Species-*Er. amylovora, Er. mallotivora, Er. papayae, Er. persicina, Er. psidii, Er. pyrifoliae, Er. rhapontici, Er. tracheiphila.*

5.3.1.3.18. Gluconobacter Asai 1935: The genus was established by Asai in 1935. At present it contains only one plant pathogenic species, the causal agent of bacterial rot of mature fruits (pear, apple and grape) and pink disease of pineapple.

The genus has the following characteristic features. Cells are Gram-negative, ellipsoidal to rod-shaped and measure 0.5 - 1.0 × 2.6 - 4.2 μm. Cells may be motile or nonmotile and motile cells have 3 - 8 polar flagella. The cells may occur singly and/or in pairs but rarely in chains.

Nonsporing. Obligately aerobic. Metabolism is strictly respiratory type in which the oxygen acts as a terminal acceptor. The best growth occurs at 25 - 30°C. Optimum pH for growth is 5.0 - 6.0 though most strains can even grow at pH 3.5. Oxidase test is negative but catalase test is positive. Indole and hydrogen sulphide are not produced, gelatin is not liquefied and nitrates are not reduced. Ethanol is oxidized to acetic acid but acetic acid is not further oxidized to carbon dioxide and water, as the bacterium does not have a full set of Kreb's cycle enzymes. Acid is generally produced from maltose, glucose, fructose and xylose and growth can occur even up to concentration of 10.0% glucose. Major quinone possessed is ubiquinone of Q-10 type. *Gluconobacter* strains occur in sugar rich environments like fruits, flowers, soft drinks, grapes, wine, palm sap, coca wine, cider, beer etc. The G + C content of the DNA varies from 52.0 - 64.0 mol%.

In addition to vinegar production, *Gluconobacter* species have many biotechnological applications. In the last decades, many bioconversion routes for rare and special sugars involving *Gluconobacter* spp. have been developed. Among the recent ones, are the biotransformations involved in the production of L-ribose and miglitol, both very promising pharmaceutical lead molecules. Most of these processes make use of *Gluconobacter*'s membrane-bound polyol dehydrogenases. The other enzymes of recent interest in industrial biotechnology include dextran-dextrinase, capable of transglucosylating substrate molecules and intracellular NAD-dependent polyol dehydogenases involved in co-enzyme regeneration.

Type species- *Gluconobacter oxydans*

Systematic position: Domain-*Bacteria*; Phylum-*Proteobacteria*; Class-"*Alphaproteobacteria*"; Order-*Rhodospirillales*; Family-*Acetobacteraceae*; Genus-*Gluconobacter;* Species- *G. oxydans.*

5.3.1.3.19. Herbaspirillum Baldani *et al.* 1986 emend. Baldani *et al.* 1996: The genus was created in 1986 to include a root-associated nitrogen fixing bacterium. Later, a plant pathogen, earlier described as *Psm. rubrisubalbicans* (causal agent of mottled stripe disease of sugarcane) was transferred to this genus.

The below given description of the genus is based on the proposal of Baldani *et al.* (1986) and emended by Baldani *et al.* (1996). Cells are Gram-negative, generally vibroid and sometimes helical, measuring 0.6 - 0.7 μm in width while the cell length varies from 1.5 to 5.0 μm depending on the culture media. Cells are motile, having one to three or more flagella on one or both the poles. The bacterium has a typical respiratory metabolism, and sugars may be oxidized but are not fermented. Among the species included in the genus, three species are capable of fixing atmospheric N_2 under microaerobic conditions and grow well with N_2 as the sole nitrogen source, even in the presence of 10% sucrose. Oxidase and urease tests are positive while catalase test is weak or variable. The salts of certain organic acids such as malate, fumarate, succinate, pyruvate, citrate and trans-aconitate are favoured carbon substrates for both NH_4^+- and N_2-dependent growth. Other carbon sources, such as glycerol, mannitol, sorbitol, D-glucose, galactose, L-arabinose, ribose, D-xylose, D-fructose, D-mannose, but not sucrose, are also utilized. Starch and gelatin are not hydrolyzed. Optimal temperature for growth is between 30 and 34°C and optimal pH between 5.3 and 8.0. The genus constitutes a separate rRNA cluster within the β-subclass of the *Proteobacteria*. The G + C content of the DNA ranges from 60.0 - 65.0 mol% (T_m). The bacterium occurs mainly in association with graminaceous plants, endophytically colonizing roots, stems and leaves, and also in soil.

Type species-*Herbaspirillum seropadicae*

Systematic position: Domain-*Bacteria*; Phylum-*Proteobacteria*; Class-"*Betaproteobacteria*"; Order-"*Burkholderiales*"; Family-"*Oxalobacteraceae*"; Genus-*Herbaspirillum*; Species-*He. rubrisubalbicans*

5.3.1.3.20. Janibacter Martin *et al.* 1997: Originally the genus was created to accommodate a Gram-positive actinomycete, isolated from environmentally polluted samples and whose cell walls contained meso-diaminopimelic acid. Yoon *et al.* (2004) described a bacterium isolated from abnormally spoiled oriental melons and transferred it to this genus on the basis of 16S rRNA gene sequence analyses and genomic relatedness.

The below given description of the genus is as proposed by Martin *et al.* (1997). The name of the genus is derived from a Latin word, *Janus*, a God in Roman mythology, who is said to have had two faces. Cells are Gram-positive, coccoid to rod-shaped and nonmotile. Cells occur singly, in pairs or occasionally in irregular clumps. Nonsporing and not acid-fast. Aerobic, chemoorganotrophs. Catalase is produced while oxidase test is negative. The cell wall type is A1γ (directly cross-linked *meso*-A_2 pm). Mycolic acids are absent; the menaquinone is MK-8 (H_4). The complex fatty acid profile contains straight-chain saturated and monounsaturated fatty acids as the major components and the polar lipids are diphosphatidylglycerol, phosphatidylglycerol and phosphatidylinositol. The strains of the genus are inhabitants of sludge and sewage waste. The G + C content of the DNA is 70.0 mol%.

Type species- *Janibacter limosus*

Systematic position: Domain-*Bacteria*; Phylum-"*Actinobacteria*"; Class-*Actinobacteria*; Subclass-*Actinobacteridae*; Order-*Actinomycetales*; Suborder-*Micrococcineae*; Family-*Intrasporangiaceae*; Genus-*Janibacter*; Species-*Jani. melonis.*

5.3.1.3.21. Janthinobacterium De Ley *et al.* 1978 emend. Lincoln *et al.* 1999: The genus was created in 1978 to include some Gram-negative psychrotrophic and violacein producing bacterial species associated with food spoilage. In 1999, Lincoln and associates described a new species causing soft rot of mushrooms (*Agaricus bisporus*) and assigned it to this genus.

The description of the genus given below is as proposed by De Ley *et al.* (1978) and emended by Lincoln *et al.* (1999). Cells are Gram-negative, straight and sometimes slightly curved rods having rounded ends and measure 0.8 - 1.5 × 1.8 - 6.0 μm. Cells occur mostly singly but sometimes in pairs or short chains. Nonsporing. Motile. Strictly aerobic, having a respiratory type of metabolism with oxygen as the terminal electron acceptor. Temperature range for growth is 4 - 30°C but best growth occurs at 25°C. Optimal pH lies between 7.0 and 8.0. Many strains produce the violet pigment, violacein, hence the genus name *Janthinobacterium,* which is derived from janthinus meaning violet-coloured. Chemoorganotrophs. Citrate is used as a sole source of carbon and ammonium as a sole source of nitrogen. The major phospholipids are phosphatidyl ethanolamine, phosphatidylglycerol and diphosphatidylglycerol as polar lipids. The major respiratory lipoquinone is Q-8. In Hugh and Leifson medium, acid is always produced from galactose, xylose, L-arabinose and *m*-inositol, and oftenly from lactose, mannitol, cellobiose, sucrose, inulin and glycerol. Growth occurs on L-leucine but not on ornithine and phenylalanine. The G + C content of the DNA varies from 61.0 - 67.0 mol%. The bacterium is commonly found in cold and temperate climates in soil and water but can also be isolated from diseased mushrooms.

Type species- *Janthinobacterium lividum*

Systematic position: Domain-*Bacteria*; Phylum-*Proteobacteria*; Class-"*Betaproteobacteria*"; Order-"*Burkholderiales*"; Family-*Oxalobacteraceae*; Genus-*Janthinobacterium*; Species-*Jant. agaricidamnosum.*

5.3.1.3.22. Leifsonia Evtushenko *et al.* 2000 emend. Dastager *et* al. 2009: The genus was created by Evtushenko and associates in 2000 to accommodate two new species, namely, *L. poae* and *L. aquatica* found in *Poa annua* root galls induced by the nematode *Subanguina radicicola* and *Cor. aquaticum. Cor. aquaticum* was reclassified as *L. aquatica*. *Cla. xyli* was reclassified with two subspecies, namely, *L. xyli* subsp. x*yli* and *L. xyli* subsp. c*ynodontis* in this genus. Members of the genus are characterized by coryneform morphology. The species clearly differ from each other phylogenetically and phenotypically. The species can be distinguished on the bases of their morphologies, composition of cell wall sugars, the requirement of complex media for growth and many physiological characteristics, including the oxidase reaction. Currently, the genus contains only one phytopathogenic species having two subspecies.

The below given description of the genus is taken from Evtushenko *et al.* (2000). The colonies are yellow or white, circular, somewhat convex, glistening, opaque and butyrous. The cells are Gram-positive, nonsporing, irregular rods or filaments, which usually fragment into shorter rods or coccoid elements. Primary branching occurs in young cultures of some species. Cells are usually motile, not acid-fast, obligately aerobic and mesophilic. Catalase test is negative while oxidase test is variable among species. Cell wall peptidoglycan of B-type contains glycine, glutamic acid, diaminobutyric acid (DAB) and alanine in a molar ratio close to 1:1:2:1; both isomers, L-DAB and D-DAB, are usually present in almost equal proportions. The major menaquinone is MK-11; MK-10 is also present. Rhamnose is a major constituent of cell wall sugars with small amounts of glucose, galactose and mannose while some species also contain fucose. Cell wall teichoic acids and mycolic acids are not present. The principal phospholipids are phosphatidylglycerol and diphosphatidylglycerol. Among fatty acids, *anteiso*-15:0, *anteiso*-17:0 and *iso*-16:0 predominate. The concentrations of polyamines are low while putrescine is the predominant compound. The G + C content of the DNA ranges from 66.0 - 73.0 mol%. Some species can be distinguished by their tolerance to antibiotics. The genus forms a coherent phylogenetic cluster attached to the branch of *Agromyces* spp.

Type species- *Leifsonia aquatica*

Systematic position: Domain-*Bacteria*; Phylum-"*Actinobacteria*"; Class-*Actinobacteria*; Subclass-*Actinobacteridae*; Order-*Actinomycetales*; Suborder-*Micrococcineae*; Family-*Microbacteriaceae*; Genus-*Leifsonia*; Species-*L. cynodontis, L. xyli, L. xyli* subsp. *cynodontis, L. xyli* subsp. *xyli.*

5.3.1.3.23. Nocardia Trevisan 1889: The genus was created by Trevisan in 1889. In 1952, one phytopathogenic bacterium causing galls, abundant proliferation of buds and whiches' broom effect on blue berry was included in this genus.

The genus was named after Edmond Nocard, a French veterinarian, who discovered the first member of the genus in 1889 as the cause of bovine infection. *Nocardia* are strict aerobes and catalase positive bacteria, which stain from Gram-positive to Gram-variable. Nonsporing, nonmotile, chemoorgantrophs, having an oxidative type of metabolism. Rudimentary to extensively branched vegetative hyphae measuring 0.5 - 1.2 μm in diameter are produced. These hyphae often fragment into rod-shaped and coccoid forms. Conidia are occasionally produced in chains on the aerial hyphae. Colonies have a variable shape and may be pigmented or white. Colonies are also of variable appearance and most species appear to have aerial hyphae when viewed with a dissecting microscope. Some strains are partially acid-fast, i.e., they require less concentration of sulphuric or hydrochloric acid during the staining procedure due to the presence of mycolic acids of intermediate length in their cell walls. *Nocardia* species have a wide range of temperature for growth but grow slowly on non-selective culture media. Colonies become evident in 3 - 5 days on most commonly used media. However, sometimes a prolonged incubation period of 2 - 3 weeks may be needed.

Members are distributed world widely in soils that are rich in organic matter and are also found in aquatic habitats. They are involved in the degradation of hydrocarbons and waxes and can contribute to the biodeterioration of rubber joints in water and sewage pipes. Although most are free-living saprophytes, some species such as *No. asteroides,* are opportunistic pathogens, which cause nocardiosis in humans and animals. Persons with a weak immune system like children, the elderly and the immunocompromised are most at risk. The most commonly infected organs are lungs but the central nervous system and other organs may also be involved. Some *Nocardia* species also cause a variety of cutaneous infections like actinomycetoma. The virulence factors are the catalase and superoxide dismutase, which inactivate reactive oxygen species that would otherwise prove toxic to the bacteria, and a cord factor (trehalose 6-6′ dimycolate). The cord factor interferes with fagocytosis by macrophages by preventing the fusion of the phagosome with the lysosome. The G + C content of the DNA varies from 64.0 - 72.0 mol%.

Type species- *Nocardia asteroides*

Systematic position: Domain-*Bacteria*; Phylum-"*Actinobacteria*"; Class-*Actinobacteria,* Subclass-*Actinobacteridae;* Order-*Actinomycetales;* Suborder-*Corynebacterineae;* Family-*Nocardiaceae*; Genus-*Nocardia*; Species-*No. vaccinii*.

5.3.1.3.24. Pantoea Gavini *et al.* 1989 emend. Mergaert *et al.* 1993: Gavini and associates established this genus in 1989 to include some species isolated from diverse geographical and ecological sources. Kageyama *et al.* (1992) described a new species from citrus fruits and transferred it to this genus. Later, some species of the then genus *Erwinia*, based mainly on the differences in DNA-DNA hybridization were also transferred to this genus. On the basis of multilocus sequence analyses and amplified fragment length polymorphism analysis, Brady *et al.* (2009) identified bacteria isolated from eucalyptus leaves and shoots, showing symptoms of blight and die-back, collected in Uganda, Uruguay and Argentina and from maize displaying brown stalk rot symptoms in South Africa as *Pantoea vagans* sp. nov., *Pantoea eucalypti* sp. nov. and *Pantoea deleyi* sp. nov. The below given description of the genus is taken from Gavini et al. (1989) and emended by Mergaert *et al.* (1993).

The cells are Gram-negative, nonencapsulated, nonsporing straight rods and measure 0.5 - 1.3 × 1.0 - 3.0 μm. Cells are peritrichously motile or nonmotile. Cultures grow well on nutrient agar at 30°C. Colonies on nutrient agar are smooth, translucent and more or less convex having entire margins. The colonies are yellow, pale beige to pale reddish-yellow, or non-pigmented. Facultatively anaerobic and oxidase test is negative. No lysine decarboxylase, tryptophan dreaminase, or urease activity. Pectate is not degraded, and hydrogen sulphide is not produced from thiosulphate. Glucose dehydrogenase and gluconate dehydrogenase are produced. No gas is produced from glucose. Acid is produced from D-xylose, D-ribose, D-galactose, D-fructose, trehalose, maltose, D-mannose, D-mannitol and *N*-acetyl-D-glucosamine but not from D-arabinose, L-xylose, L-fucose, or D-turanose. The bacterium has been isolated from plant surfaces, seeds, fruits, soil and water as well as from humans (wounds, blood, urine and internal organs) and animals in several parts of the world. The G + C content of the DNA ranges from 49.7 to 60.6 mol% (T_m).

Type species-*Pantoea agglomerans*

Systematic position: Domain-*Bacteria*; Phylum-*Proteobacteria*; Class-"*Gammaproteobacteria*"; Order-"*Enterobacteriales*"; Family-*Enterobacteriaceae;* Genus-*Pantoea;* Species-*Pa. agglomerans, Pa. agglomerans* pv. *betae, Pa. agglomerans* pv. *gypsophilae, Pa. agglomerans* pv. *millettiae*; *Pa. ananatis, Pa. ananatis* pv. *ananatis, Pa. ananatis* pv. *uredovora*; *Pa. citrea, Pa. deleyi, Pa. dispersa, Pa. eucalypti, Pa. stewartii, Pa. stewartii* subsp. *indologenes, Pa. stewartii* subsp. *stewartii, Pa. vagans.*

5.3.1.3.25. Pectobacterium Waldee 1945 emend. Hauben *et al.* 1999: Waldee (1945) proposed the emendment of the genus *Erwinia* and creation of a new genus *Pectobacterium* to include biochemically more active and soft rotting bacteria like *Er. carotovora* and *Er. chrysanthemi.* Graham (1964), while reviewing the taxonomy of soft rot coliform bacteria supported Waldee's suggestion, followed by many other workers. However, the genus *Pectobacterium* was not recognized in the 8th edition of *Bergey's Manual of Determinative Bacteriology* but it appeared in the 'Approved Lists of Bacterial Names' (Skerman *et al.*, 1980). Later in 1998, the species of genus *Erwinia* were further divided into different phylogenetic groups based on complete 16S rDNA sequences. Hauben *et al.* (1998) united the species constituting the cluster II group and transferred them to this genus. Gardan *et al.* (2003) elevated three subspecies of *Pe. carotovorum* to the species level.

The below given description of the genus is taken from the emended description of the genus given by Hauben *et al.* (1999). Cells are Gram-negative, straight rods with rounded ends and measure 0.5 - 1.0 × 1.0 - 3.0 μm. Occur singly or in pairs and motile by peritrichous flagella. Optimum temperature for growth is between 27 and 30°C. Facultatively anaerobic. Metabolism fermentative. Catalase positive and oxidase negative. Esculin is hydrolyzed but not starch. No decarboxylases are formed for arginine, lysine, or ornithine. Strains do not possess tryptophan dreaminase or urease. Acid is produced from fructose, D-glucose, D-mannose, D-galactose, *N*-acetylglucosamine, L-rhamnose, D-ribose, sucrose and salicin but not from adonitol, L-arabitol, L-sorbose, starch, D-lyxose and β-methylmannoside. Strains utilize fructose, D-galactose, D-glucose, ribose, sucrose, mannose, acetate, fumarate, succinate, salicin, arbutin, glycerol, D-mannitol and β-methylglucoside but not betaine, benzoate, gallate, oxalate, propionate, butanol, methanol, or sorbose as carbon- and energy-yielding sources. Utilise L-alanine, arginine, asparagine, citrulline, glucosamine, glutamine, glycine, histidine, leucine, L-serine, tyrosine, L-tryptophan, ammonium chloride, asparaginic acid, urea, L-methionine and phenylalanine but not betaine, choline, quinolinic acid, spermine, spermidine, trigonelline, trimethylammonium, anthranilic acid, hydroxyproline and sarcosine as nitrogen sources.

The major fatty acids of the type strains include 12:0, 14:0, 15:0, 16:0, 17:1ω8C and 17:0. The species of the genus comprise a distinct phylogenetic group, as determined by 16S rRNA gene sequence comparisons, and are characterized by 17 signature nucleotides. The G + C content of the DNA ranges from 50.5 - 56.1 mol%. *Pectobacterium* species produce pectolytic enzymes and cause plant diseases, mainly the soft rots and also blights, cankers, necroses and wilts. The entry of the bacteria into the host plant is mainly through wounds and natural openings.

Type species- *Pectobacterium carotovorum*

Systematic position: Domain-*Bacteria*; Phylum-*Proteobacteria*; Class-*"Gammaproteobacteria"*; Order-*"Enterobacteriales"*; Family-*Enterobacteriaceae*; Genus-*Pectobacterium;* Species-*Pe. atrosepticum*, *Pe. betavasculorum, Pe. cacticida, Pe. carotovorum*, *Pe. carotovorum*- subsp. *carotovorum,* subsp. *odoriferum*; *Pe. cypripedii, Pe. wasabiae.*

5.3.1.3.26. Pseudoalteromonas Gauthier *et al.* 1995: The genus was created by Gauthier *et al.* (1995). They determined the sequences of small-subunit rDNA of 17 strains of genera, *Alteromonas, Shewanella, Vibrio* and *Pseudomonas*, and these sequences were analyzed by phylogenetic methods. On the basis of data obtained, they emended the existing genus *Altermonas* and created a new genus *Pseudoalteromonas* to accommodate 11 species that were previously included in *Alteromonas* and transferred one species to genus *Pseudomonas.* Later on, an aerobic, Gram-negative, marine plant pathogenic bacterium, the causal agent of red spot disease of 'makonbu' (*Laminaria japonica*), identified as *Pseudoalteromonas bacteriolytica* sp. nov. was transferred to this genus (Sawabe *et al.*, 1998).

The following description of the genus is taken from Gauthier *et al.* (1995). Cells are Gram-negative, nonsporing, straight rods, which measure 0.2 - 1.5 × 1.8 - 3.0 μm and occur singly or in pairs. Motile by a single polar flagellum, either unsheathed or sheathed. Strains do not accumulate poly-β-hydroxybutyrate. Nonbioluminescent. Several species produce pigments. Chemoorganotrophs. Aerobic, possessing a strictly respiratory type of metabolism but not fermentative, with oxygen as the terminal electron acceptor. Oxidase is positive but catalase is generally weak and irregular. None of the strains has a constitutive arginine dihydrolase system. One species, namely, *Psa. denitrificans* is capable of denitrification. Mostly psychrotolerant, with growth occurring between 0 and 40°C but all species grow at 20°C. Many strains require amino acids for growth. All species require seawater base for growth. The major quinone is ubiquinone-8 (Q-8). All the above mentioned 12 species of the genus are positive for gelatinase, lipase, lecithinase and DNase activities and utilization of D-glucose as a sole source of carbon; and negative for utilization of D-ribose, L-rhamnose, salicin, erythritol, sorbitol, *meso*-inositol and adonitol. The mol% G + C of the DNA is 38.0 - 48.0 (T_m).

Type species- *Pseudoalteromonas haloplanktis*

Systematic position: Domain-*Bacteria*; Phylum-*Proteobacteria*; Class-"*Gammaproteobacteria*"; Order-"*Alteromonadales*"; Family *Alteromonadaceae;* Genus-*Pseudoalteromonas;* Species-*Psa. bacteriolytica.*

5.3.1.3.27. Pseudomonas Migula 1894: The genus was established by Migula in 1894. It contains different types of bacteria, i.e., epiphytic and saprophytic species, animal, human and plant pathogens. The genus is characterized by straight or slightly curved rods but not helical, measuring 0.5 - 1.0 × 1.5 - 5.0 μm. The cells are motile by one to several polar flagella. Xanthomonadins are not produced. No resting stages are known, Gram-negative. Aerobic, having a strictly respiratory type of metabolism with oxygen as the terminal electron acceptor. However, in some cases, nitrate can be used as an alternate electron acceptor, allowing anaerobic growth. Most species do not accumulate polyhydroxybutyrate granules. Catalase positive, oxidase positive or negative. Chemoorganotrophs. Organic growth factors are not required for most species. The mol% G + C of the DNA is 58.0 - 69.0.

Type species- *Pseudomonas aeruginosa*

In first edition of *Bergey's Manual of Systematic Bacteriology,* Palleroni (1984) enlisted 92 species in this genus, out of which 23 were plant pathogens. Taxonomy of this genus has been revised many times. Several phytopathogenic species of this genus have been transferred to other genera, namely, *Acidovorax* (Willems *et al.,* 1992), *Herbaspirillum* (Baldani *et al.,* 1996) *Burkholderia* and *Ralstonia* (Urakami *et al.*, 1994; Yabuuchi *et al.*, 1993, 1995) and *Xanthomonas*. Many phytopathogenic species of the genus have been designated as pathovars of *Psm. savastanoi* and *Psm. syringae*. Some new species, namely, *Psm. salomonii* and *Psm. palleroniana* (Gardan *et al.*, 2002), *Psm. syringae* pv. *broussonetiae* (Takahashi *et al.*, 1996) and *Psm. costantinii* (Munsch *et al.*, 2002) have been added to this genus.

Systematic position: Domain-*Bacteria*; Phylum-*Proteobacteria*; Class-"*Gammaproteobacteria*", Order-*Pseudomonadales*; Family-*Pseudomondaceae*; Genus-*Pseudomonas*; Species-*Psm. agarici, Psm. amygdali, Psm. asplenii, Psm. avellanae, Psm. beteli, Psm. cannabina, Psm.*

caricapapayae, Psm. cichorii, Psm. cissicola, Psm. corrugata, Psm. costantinii, Psm. ficuserectae, Psm. flectens, Psm. fuscovaginae, Psm. hibiscicola, Psm. marginalis, Psm. marginalis- pv. *alfalfae*, pv. *marginalis*, pv. *pastinacae*; *Psm. mediterranea, Psm. meliae, Psm. palleroniana, Psm. salomonii, Psm. savastanoi, Psm. savastanoi-* pv. *fraxini*, pv. *glycinea*, pv. *nerii*, pv. *retacarpa*, pv. *savastanoi*; *Psm. syringae, Psm. syringae-* pv. *aceris*, pv. *actinidiae*, pv. *aesculi*, pv. *alisalensis*, pv. *antirrhini*, pv. *apii*, pv. *aptata*, pv. *atrofaciens*, pv. *atropurpurea*, pv. *avii*, pv. *berberidis*, pv. *broussonetiae*, pv. *castaneae*, pv. *cerasicola*, pv. *ciccaronei*, pv. *coriandricola*, pv. *coronafaciens*, pv. *coryli*, pv. *cunninghamiae*, pv. *daphniphylli*, pv. *delphinii*, pv. *dendropanacis*, pv. *dysoxyli*, pv. *eriobotryae*, pv. *garcae*, pv. *helianthi*, pv. *hibisci*, pv. *japonica*, pv. *lachrymans*, pv. *lapsa*, pv. *maculicola*, pv. *mellea*, pv. *mori*, pv. *morsprunorum*, pv. *myricae*, pv. *oryzae*, pv. *panici*, pv. *papulans*, pv. *passiflorae*, pv. *persicae*, pv. *phaseolicola*, pv. *philadelphi*, pv. *photiniae*, pv. *pisi*, pv. *porri*, pv. *primulae*, pv. *rhaphiolepidis*, pv. *ribicola*, pv. *sesami*, pv. *solidagae*, pv. *spinaceae*, pv. *striafaciens*, pv. *syringae*, pv. *tabaci*, pv. *tagetis*, pv. *theae*, pv. *tomato*, pv. *ulmi*, pv. *viburni*, pv. *zizaniae*; *Psm. tolaasii, Psm. tremae, Psm. viridiflava.*

5.3.1.3.28. Ralstonia Yabuuchi *et al.* 1996. The genus was created in 1996 by Yabuuchi *et al.* (1995) to accommodate bacteria from ecological diverse niches that were classified previously in *Burkholderia* and *Alcaligenes*. The genus was established on the basis of phenotypic characterization, cellular lipid and fatty acids analysis, phylogenetic analysis of 16S rDNA sequences and rRNA-DNA hybridization within the homology group II of pseudomonads. The genus contains plant pathogens, human pathogens and metal resistant bacteria. Yabuuchi *et al.* (1995) transferred *Bur. solanacearum* (syn. *Psm. solanacearum*) at the time of creation of the genus but *Psm. syzygii*, causal agent of Sumatra disease of clove, was transferred later on by Vaneechoutte *et al.* (2004).

The below given description of the genus is based on Yabuuchi *et al.* (1995). The genus was named after E. Ralston, an American bacteriologist who first described *Psm. pickettii* and suggested a taxonomic relationship of this species to *Psm. solanacearum* based on DNA homology between these species. Cells are Gram-negative, motile or nonmotile rods. When motile, the motility is by means of a single polar flagellum or peritrichous flagella. Oxidase activity varies with species. Lysine and ornithine decarboxylases are not produced. Major quinone is ubiquinone Q-8. Cellular lipids of the members contain two kinds of phosphatidylethanolamines, PE-1 and PE-2. PE-2 contains 2-hydroxy fatty acid at *sn*-2 position of the glycerol moiety. Monosaccharides, disaccharides and polyalcohols are oxidized and assimilated as a sole source of carbon and energy. The inability of *Ral. pickettii, Ral. eutropha* and *Ral. solanacearum* to assimilate galactose, mannose, sorbitol and mannitol distinguishes these species from *Burkholderia* species. Two-dimensional thin layer chromatography failed to detect two kinds of ornithine-lipids and C19 cyclopropanoic acid in cellular fatty acids in the type strains of the above-mentioned three *Ralstonia* species while the presence of these compounds was clearly demonstrated in six species of *Burkholderia* including *Bur. cepacia*. Species have different flagellar morphology. The members of the genus are plant and human pathogens, and knallgas bacteria and metal-resistant bacteria. Strains of *Ral. pickettii* are found in various clinical specimens and hospital environments. *Ral. solanacearum* is taxonomically heterogeneous and infects a large number of host plants. The G + C content of DNA ranges from 64.0 – 68.0 mol%.

Type species- *Ralstonia pickettii*

Systematic position: Domain-*Bacteria*; Phylum-*Proteobacteria*; Class-"*Betaproteobacteria*"; Order-*Burkholderiales,* Family-*Burkholderiaceae;* Genus-*Ralstonia;* Species-*Ral. solanacearum, Ral. syzygii.*

5.3.1.3.29. Rathayibacter Zgurskaya *et al.* 1993. The genus was created by Zgurskaya *et al.* (1993) to accommodate three species of Gram-positive, aerobic, coryneform bacteria previously classified in the genus *Clavibacter*, namely, *Cla. rathayi, Cla. tritici* and *Cla. iranicus* as well as six strains that were isolated from annual cereal grasses, responsible for ryegrass toxicity and are very similar to *Cla. toxicus*. Later on, Sasakai *et al.* (1998) transferred *Cla. toxicus* to this genus.

The below given description of the genus is taken from Zgurskaya *et al.* (1993). The genus was named after E. Rathay, an Australian plant pathologist who first isolated members of this genus. Cells are irregular, pleomorphic, short rods and measure 0.4 - 0.8 × 0.5 - 1.8 μm. V-forms and sometimes short cell chains are formed but a clear rod-coccus cycle is not observed. Gram-positive, nonsporing and not acid-fast. Obligately aerobic. Chemoorganotrophic, having an oxidative type of catabolism. Nutritionally exacting. Optimal growth occurs between 24 and 28°C; no growth occurs at 10 or 37°C. Cell wall peptidoglycan contains alanine, glutamic acid, glycine and diaminobutyric acid (1:1:1:2). Cell wall sugars include glucose, mannose and rhamnose. Cell walls of some strains also contain xylose and galactose. Menaquinones are of the MK-10 type. Mycolic acids are absent. Branching saturated fatty acids of the *anteiso* type (*anteiso*-15:0 and *anteiso*-17:0) are predominant. Diagnostic phospholipids are represented by phosphatidylglycerol, diphosphatidylglycerol and unknown glycosyldiacylglycerols; several strains contain lyso-phosphatidylglycerol.

Most strains utilize cellobiose, galactose, fructose, glucose, maltose, mannose, glycerol, sucrose, raffinose, trehalose and salicin as sole carbon sources but do not utilize adonitol, dulcitol, erythritol, inositol, mannitol, sorbitol, dextrin, melibiose, or L-rhamnose. Acid is not produced from dulcitol, inositol, mannitol, sorbitol, lactose, melibiose, or L-rhamnose. Phenylalanine and methionine are assimilated but most amino acids are not used as sole sources of nitrogen. Casein, guanine, Tween 60, Tween 80, starch and urea are not hydrolysed or decomposed. Nitrate is not reduced to nitrite. Methyl red and Voges-Proskauer tests are variable. The G + C content of the DNA ranges from 63.0 - 72.0 mol%. The bacterium is isolated mainly from cereal crops and annual cereal grasses.

Type species- *Rathayibacter rathayi*

Systematic position: Domain-*Bacteria*; Phylum-"*Actinobacteria*"; Class-*Actinobacteria;* Subclass-*Actinobacteridae;* Order-*Actinomycetales;* Suborder-*Micrococcineae;* Family-*Microbacteriaceae*; Genus-*Rathayibacter;* Species-*Rat. iranicus, Rat. rathayi, Rat. toxicus, Rat. tritici.*

5.3.1.3.30. Rhizobacter Goto and Kuwata 1988: The genus was created by Goto and Kuwata (1988) to classify a bacterium causing galls on carrots. The following description of the genus is based on the publication of Goto and Kuwata (1988). Cells are Gram-negative, nonsporing, encapsulated, straight to slightly curved rods, measuring 0.9 - 1.3 × 2.1 - 2.5 μm. Cells are motile by polar or lateral flagella or by both but motile cells are rare in the populations. Poly-β-hydroxybutyrate granules are formed. Chemoorganotrophic. Aerobic, having respiratory type of metabalism of glucose, and D-glucose is utilized as a sole source of carbon and energy. On agar plates, the colonies are white or yellowish-white, plicated, tough or viscid; if the colour is yellow, the pigment differs from xanthomonadins. In liquid media, floccular growth consisting of globular units occurs without finger-like projections. Oxidase and catalase tests are positive. Nitrates are reduced to nitrites and H_2S is produced from cysteine and auxin. Growth is inhibited by KCN and *o*-nitrophenyl-β-D-galactopyranoside test is positive. Arginine dihydrolase, nitrogenase, fluorescent pigment, indole and acetoin are not produced and the tests for methyl red and denitrification are also negative. Starch, dextrin and glycogen are hydrolyzed. The bacterium is susceptible to 10 μg vibriostatic agent O/129 phosphate. Does not require NaCl or growth factors. Contains ubiquinone Q-8. Benzene derivatives are not utilized as carbon sources. The G + C content of the DNA is 67 - 71 mol%. The type species is a plant pathogen that causes galls on carrot roots and is found in soil.

Type species- *Rhizobacter dauci*

Systematic position: Domain-*Bacteria*; Phylum-*Proteobacteria*; Class-"*Gammaproteobacteria*"; Order-*Pseudomonadales;* Family-*Pseudomonadaceae*; Genus-*Rhizobacter*; Species-*Rhi. dauci.*

5.3.1.3.31. Rhizobium Frank 1889: The salient characteristics of the genus are given below. Cells are Gram-negative, nonsporing, small to medium sized rods measuring 0.5 - 1.0 × 1.2 - 3.0 μm. Motile by 1 - 6 flagella, which may be subpolar, bipolar, or peritrichous. Cells contain characteristic granules of polymerized β-hybroxybutyrate, which stain with Sudan Black and appear as highly refractive bodies under phase contrast illumination. Optimal temperature for growth lies between 25 and 30°C. Optimum pH lies between 6.0 and 7.0. Aerobic, having a respiratory type of metabolism with oxygen as the terminal electron accepter. Colonies are circular, convex, opaque or semi-translucent, raised and mucilaginous on yeast mannitol/mineral salt agar. Colony colour is generally white or beige. Growth on carbohydrate media is usually accompanied by abundant amounts of extracellular polysaccharide slime. *Rhizobium* species can grow on relatively simple synthetic media. Chemoorganotrophs and utilize a wide range of carbohydrates and salts of organic acids as sole carbon sources, without gas formation. Produce an acidic reaction in mineral salts medium containing mannitol or other carbohydrates. Glutamate is much superior to nitrate or the ammonium ons as nitrogen source. Vitamin requirements differ for different species and strains. Thiamine, biotin and calcium pantothenate are required for growth while nicotinic acid, folic acid, *p*-aminobenzoic acid, inositol, vitamin B12 and riboflavin do not promote growth.

Casein, starch, chitin and agar are not hydrolysed. Starch and ellulose are not utilized. An acid reaction is produced in mineral-salts nedium having mannitol or other carbohydrates. Strains of all species ause hypertrophies in plants resulting in formation of root nodules with r without symbiotic nitrogen fixation. In root nodules, bacteria live ndophytically and reduce or fix atmospheric nitrogen into a form tilizable by the host plant. Plant-host specificity is usually for a few egume genera but, in some strains, may extend to many legume genera nd is largely determined by the chemical structure of lipochito-ligosaccharide Nod factors produced. The genes, which code for odulation (*nod*) and nitrogen fixation (*nif*), are clustered on large plasm-ls or megaplasmids (pSyms) present in the bacteria. Plasmid transfer ccurs between species and results in expression and stable inheritance of haracters possessed by the plasmid-donor species. The G + C content of e DNA varies from 57.0 - 66.0 mol%. Members are common inhabitants f arable land.

Type species- *Rhizobium leguminosarum*

Systematic position: Domain-*Bacteria*; Phylum-*Proteobacteria*; lass-"*Alphaproteobacteria*"; Order-"*Rhizobiales*"; Family-*Rhizobiaceae*; enus-*Rhizobium*. Both *Agrobacterium* and *Rhizobium* have been ecognized as separate genera in this book, as done by Garrity *et al.* 2005). The plant pathogenic species included under *Rhizobium* by Young *al.* (2001) are included under the genus *Agrobacterium* in this book.

5.3.1.3.32. Rhodococcus Zopf 1891: The genus was described in 1891 include a saprophytic Gram-positive bacterium producing red pigment and forming irregular masses of cells. Later on, Goodfellow (1984) transferred *Cor. fascians*, the causal agent of fasciation of sweet pea, and leafy galls and shoot proliferations on many plants, to this genus.

The genus is characterized by Gram-positive, nonsporeforming, nonmotile, aerobic bacteria. The cells are nocardioform, meaning they may have mycelial growth with fragmentation into rod-shaped or coccoid forms. The cocci may germinate into short rods, form filaments with side projections, show elementary branching, or in some cases produce extensively branched substrate hyphae. The fragmentation of rods, filaments and hyphae produces next generation of short rods or cocci. In some strains, feeble, aerial hyphae are produced. No conidia are produced. Colony appearance may be rough, smooth or mucoid, and pigmented but rarely colourless. Catalase positive. Chemoorganotrophs, having an oxidative type of metabolism. A wide range of organic compounds is utilized as sole sources of carbon for growth and energy. The growth is generally good at 30°C on laboratory media. The cell walls contain mycolic acids.

The genus *Rhodococcus* is genetically diverse and well adapted to the soil environment. As a group, the rhodococci have tolerance for starvation, which may aid in their persistence in soil. The other habitats are aquatic habitat and herbivore dung, and a few reside in eucaryotic cells. The members of the genus are important due to their ability to metabolize/degrade harmful environmental pollutants such as toluene naphthalene, polychlorinated biphenyls, petroleum hydrocarbons, detergents, benzene and various herbicides/pesticides. Strains of *Rhodococcus* are also important due to their ability to catabolize a wide range of compounds and produce bioactive steroids, acrylamide and acrylic acid and their involvement in fossil fuel biodesulfurization. *Rho. qingshengii* is capable of degrading carbendazim, a fungicide widely used for plant disease control. The genome of *Rhodococcus* sp. strain RHA 1 was fully sequenced in October 2006 (McLeod *et. al.*, 2006). The genome is 9.7 mega base pairs long and is one of the largest bacterial genomes completel sequenced todate. The genetic and catabolic diversity of the genus is no only due to the large bacterial chromosome, but the presence of thre large linear plasmids. *Rho. equi* contains a circular plasmid while *Rho fascians* harbours a linear plasmid (about 200 kb). Bacterial isolate identified as *Rho. fascians* have been found in diverse habitats, includin polar seas and ice sediments in Greenland. Two isolates have bee reported as fixing nitrogen. The G + C content of the DNA ranges fro 63.0 - 72.0 mol% (T_m).

Type species- *Rhodococcus rhodochrous*

Systematic position: Domain-*Bacteria*; Phylum-"*Actinobacteria* Class-*Actinobacteria;* Subclass-*Actinobacteridae*; Order-*Actinomycetale* Suborder-*Corynebacterineae;* Family-*Nocardiaceae*; Genus-*Rhodococcu* Species-*Rho. fascians*.

5.3.1.3.33. Samsonia Sutra *et al.* 2001. The genus was established in 2001 and named in honour of R. Samson, a French phytobacteriologist, who did a pioneering work on pectolytic species of genus *Erwinia*. At present, the genus contains only one plant pathogenic species, *Sam. erythrinae*, which causes bark necrotic lesions on *Erythrina* species. The strains of this species isolated from *Erythrina* species formed a homogenous phenon among plant pathogenic species of *Enterobacteriaceae* on the basis of phenotypic characters, 16S rDNA sequence analyses and DNA-DNA hybridization.

The below given description of the genus is based on the data of Sutra *et al.* (2001) and that of Sutra *et al.* (1999). Cells are Gram-negative, nonsporing, catalase positive and oxidase negative rods having rounded ends. Motile by means of peritrichous flagella. The strains are β-galactosidase positive, produce acetoin but not hydrogen sulphide, and do not possess arginine dihydrolase, lysine decarboxylase, ornithine decarboxylase, urease or tryptophan dreaminase. Glucose is fermented without gas production. Nitrate is reduced. Strains do not utilize citrate and malonate as carbon sources and do not produce indole from tryptophan. They hydrolyse esculin but not gelatin, casein, lecithin, or Tween 80.

Type species- *Samsonia erythrinae*

Systematic position: Domain-*Bacteria*; Phylum-*Proteobacteria*; Class-"*Gammaproteobacteria*; Order-*Enterobacteriales;* Family-*Enterobacteriaceae*; Genus-*Samsonia*; Species-*Sa. erythrinae.*

5.3.1.3.34. Serratia Bizio 1823: The genus was established by Bizio in 1823. The genus is characterized by Gram-negative, straight rods (0.5 - 0.8 × 0.9 - 2.0 μm) with rounded ends and generally motile by peritrichous flagella. Facultatively anaerobic. Growth factors are generally not required. Temperature range for growth lies between 10 and 36°C. Colonies on nutrient agar are generally opaque, somewhat iridescent and appear white, pink or red in colour. Red colour of the colonies is due to the production of a characteristic red colour, prodigiosin. Members of the genus can be distinguished from other members of the family Enterobacteriaceae by their unique production of three enzymes, namely, DNase, lipase and gelatinase. Glucose is fermented and gluconate is oxidized to 2-ketogluconate. The bacterium ferments and utilizes maltose, fructose, D-galactose, D-mannose, ribose, D-mannitol and trehalose as sole carbon sources. Esculin is hydrolysed by most strains. Tween 80 is hydrolysed by all species except one. Acetoin is produced from pyruvate by all species except *Se. fonticola*. Nitrate and chlorate are reduced anaerobically. Generally, the members of this genus occur on plant surfaces, in soil and water, and as opportunistic human pathogens. *Se. marcescens* is mainly the human pathogen and causes nosocomical infections. The genus did not contain any plant pathogen till 1919, when Paine and Stansfield (1919) suggested the role of *Se. proteamaculans* in causing leaf spot disease of king protea (*Protea cynaroides*), which was later on confirmed by Grimont *et al.* (1978). The mol% G + C of the DNA is 52.0 – 60.0.

Type species- *Serratia marcescens*

Systematic position: Domain-*Bacteria*; Phylum-*Proteobacteria*; Class-"*Gammaproteobacteria*"; Order-*Enterobacteriales;* Family- *Enterobacteriaceae*; Genus-*Serratia*; Species-*Se. marcescens*, *Se. proteamaculans.*

5.3.1.3.35. Sphingomonas Yabuuchi *et al.* 1990 emend. Yabuuchi *et al.* 1999 emend. Takeuchi *et al.* 2001 emend. Yabuuchi *et al.* 2002: Yabuuchi *et al.* (1990) created the genus *Sphingomonas* to include strictly aerobic, chemoheterotrophic, yellow pigmented and Gram-negative rod shaped bacteria that contain glycosphingolipids as cell envelop components. Later on, Yabbuchi *et al.* (1999) emended the genus and transferred *Rhizomonas suberifaciens*, the causal agent of corky root disease of lettuce, to this genus along with two other species. On the basis of morphological, physiological and chemo-taxonomic characteristics along with DNA-DNA hybridization and 16S rDNA sequences comparison data, Takeuchi *et al.* (1995) included four new species in this genus. These species were *Sph. rosa* (isolated from rose plants and formerly named as *Ag. rhizogenes*), *Sph. pruni* (isolated from *Prunus persica*), *Sph. asaccharolytica* and *Sph. mali* (isolated from apple trees). They also identified two strains, isolated from *Psychotria nairobiensis* and formerly named as *Chromobacterium lividum*, as *Sph. yanoikuyae*. Buonaurio *et al.* (2002) described a novel pathogen causing brown spot on yellow Spanish melon (*Cucumis melo* var. *indodorus*) fruits and identified it as *Sph. melonis*.

The below given description of the genus is taken mainly from the emended description of the genus proposed by Yabuuchi *et al.* (2002). Cells are Gram-negative, nonsporing, straight or slightly curved rods or ovoid-shaped cells, which measure 0.2 - 1.4 × 0.5 - 4.0 μm. Reproduction is generally by binary fission but budding or asymmetric division as visualized by electron microscopy has been found in two species, namely *Sph. natatoria* and *Sph. ursincola.* Motility is usually by a single polar flagellum but *Sph. trueperi* is peritrichously flagellate. Many species are nonmotile, since they do not show spreading growth on 0.3% semi-solid agar plates. Bipolar staining in Gram stain (*Sph. ursincola)* and a rosette like aggregation caused by polar–fimbriae occurs in some species. Aerobic having a strictly respiratory type of metabolism with oxygen as the terminal electron accepter. Colony colour is variable, for instance, deep yellow, orange, lemon-yellow or non-pigmented. In certain cases, the pigmentation varies with the medium used. The colonies of some non pigmented strains such as *Sph. yanoikuyae* become lemon-yellow after incubation of more than 3 days at room temperature. Esculin is hydro lysed. Two species, namely, *Sph. ursincola* and *Sph. natatoria* containing bacteriochlorophyl a are facultative photoorganotrophs. Catalase positive. Oxidase positive or negative. Cell walls contain sphingolipids instead of lipopolysaccharides. Respiratory quinone is ubiquinone 10. Glucuronosyl-(1→1)-ceramide (SGL-1), galacturonosyl-β(1→1)-ceramide (in several species), and 2-hydroxymyristic acid, but not 3-hydroxy fatty acids are found. Major long-chain bases of SGL-1 are C-18:0, C-20:1 and C-21 cyclopropane. The G + C content of the DNA ranges from 59.0 - 68.0 mol%. Inhabitans of natural and man-made environments. Some species are opportunistic pathogens of humans and a few are plant pathogens.

Type species- *Sphingomonas paucimobilis*

Systematic position: Domain-*Bacteria*; Phylum-*Proteobacteria*; Class-"*Alphaproteobacteria*"; Order-"*Sphingomonadales*"; Family-*Sphingomonadaceae*; Genus-*Sphingomonas*; Species-*Sph. asaccharolytica, Sph. mali, Sph. melonis, Sph. pruni, Sph. rosa, Sph. suberifaciens, Sph. yanoikuyae.*

5.3.1.3.36. Spiroplasma Saglio *et al.* 1973: The genus was first described by Saglio *et al.* (1973) to classify a mycoplasma-like organism associated with citrus stubborn disease. Later on, two more species, namely, *Sp. kunkelii* and *Sp. phoeniceum*, causal agents of corn stunt and stubborn of periwinkles, respectively were added to this genus.

The salient characteristics of the genus are given below. Cells are pleomorphic and vary in shape and size. Different shapes found are helical and branched nonhelical filaments, and spherical and ovoid forms. The helical forms are generally 100 - 200 nm in width and 3 - 5 μm long. Spherical cells are approximately 300 nm in diameter. The colonies have a typical **fried-egg appearance**. The colony margin is frequently diffuse due to the movement of growing cells. Colony size varies from 0.1 - 4.0 mm. Cells divide by binary fission. Facultatively anaerobic. Flagella, periplasmic fibrils or other organs of locomotion are not found. However, helical filaments show motility due to flexional and twitching movements. Chemoorganotrophs. Acid is produced from glucose but fermentation of other carbohydrates is variable. Arginine is generally hydrolysed but urea hydrolysis, esculin fermentation and serum digestion are negative. Cholesterol or possibly other sterols are required for growth. Sensitive to tetracycline, erythromycin and amphotericin B but resistant to penicillin. The genome size is very small and G + C content of the DNA varies from 25.0 - 31.0 mol%. Members can be isolated from phloem fluid of vascular plants and insects that feed on this fluid and from surfaces of flowers and other plant parts. *Sp. citri* causes stubborn disease of citrus and corn stunt.

Type species- *Spiroplasma citri*

Systematic position: Domain-*Bacteria*; Phylum-"*Firmicutes*"; Class - *Mollicutes*; Order-*Entomoplasmatales;* Family-*Spiroplasmataceae*; Genus - *Spiroplasma*; Species-*Sp. citri, Sp. kunkelii, Sp. phoeniceum.*

5.3.1.3.37. Streptomyces Waksman & Henrici 1943: The genus was described in 1943 by Waksman and Henrici. The members of the genus are characterized by the production of **coenocytic mycelium** and Gram-positive reaction. The aerial mycelium produces chains of three to many spores, which may be round, oval or cylindrical in shape. More than 500 species of *Streptomyces* have been described but most of them are saprophytes. However, some species attack tuber and root crops such as potato, sweet potato, beet, radish, turnip and carrot and cause scabbing. Till now, 17 phytopathogenic species of this genus have been validly published (Bouchek-Mechiche *et al.*, 2000, 2006; Goyer *et al.*, 1996; Miyajima *et al.*, 1998; Park *et al.*, 2003). A few species are also pathogens of animals including humans. Recently, Guo *et al.* (2008) reported the results of a multilocus sequence analysis scheme developed to address the phylogeny of *St. griseus* 16S rRNA gene clade. Sequence fragments of six housekeeping genes, *atpD*, *gyrB*, *recA*, *rpoB*, *trpB* and 16S rRNA, were obtained for 53 reference strains representing 45 valid species and subspecies. Analysis of each individual locus confirmed the suitability of loci and the congruence of single-gene trees for concatenation. Concatenated trees of three, four, five and all six genes were constructed, and the stability of the topology and discriminatory power of each tree were analysed. It was concluded from the results that phylogenetic analysis based on multilocus sequences is more accurate and robust for species delineation within *Streptomyces*. A multilocus phylogeny of six genes proved to be optimal for elucidating the interspecies relationships within the *St. griseus* 16S rRNA gene clade. Therefore, this multilocus sequence analysis scheme provides a valuable tool that can be applied to other *Streptomyces* clades for refining the systematic framework of this genus.

The characteristic feature of the genus is the production of an extensively branched mycelium resembling fungi. The diameter of vegetative hyphae ranges from 0.5 - 2.0 μm. The chains of three to many spores, called **conidia**, are formed on mature aerial hyphae. In a few species, substrate mycelium produces conidia in short chains. The conidia are often pigmented having smooth, hairy, or spiny outer surface. Conidia are nonmotile and responsible for the reproduction and dispersal of *Streptomyces*. The colonies are discrete and may be leathery or butyrous. Initially the colonies have a smooth surface but later on they appear powdery, granular, or velvet-like due to the formation of a weft of aerial mycelium. Produce a wide variety of pigments responsible for the colour of colonies. Gram-positive, strict aerobes and not acid-fast. Chemoorganotrophs having an oxidative type of metabolism. Use many organic compounds as sole source of carbon for energy and growth. Catalase positive. Nitrates are generally reduced to nitrites and gelatin, casein and starch are hydrolysed. Members of the genus are widely distributed and found predominantly in soil and decaying vegetation.

They play an important role in the degradation of organic matter, most commonly noted in compost piles. In fact, the odour of the moist earth is largely the result of production of volatile substances such as **geosmin** by *Streptomyces* spp. Members are a great source for the production of many antibacterial pharmaceutical agents. Some of these are erythromycin, neomycin, streptomycin, vancomycin, tetracycline, rifamycin, chloramphenicol and lincomycin. The complete genome sequence of *St. coelicolor* strain A 3 was published in 2002. The genome sequences of *St. scabiei,* the cause of potato common scab, and of *St. avermetilis* have also been completed. In all these species, the chromosome is a linear structure, in contrast to most bacterial chromosomes, which are circular in shape. The G + C content of the DNA is 69.0 - 78.0 mol% (T_m).

Type species- *Streptomyces albus*

Systematic position: Domain-*Bacteria*; Phylum-*"Actinobacteria"*; Class-*Actinobacteria*; Subclass-*Actinobacteridae;* Order-*Actinomycetales;* Suborder-*Streptomycineae;* Family-*Streptomycetaceae*; Genus-*Streptomyces*; Species-*St. acidiscabies, St. albidoflavus, St. aureofaciens, St. candidus, St. caviscabies, St. collinus, St. europaeiscabiei, St. intermedius, St. ipomoeae, St. luridiscabiei, St. niveiscabiei, St. puniciscabiei, St. reticuliscabiei, St. scabiei, St. setonii, St. steliiscabiei, St. turgidiscabies, St. wedmorensis.*

5.3.1.3.38. Xanthomonas Dowson 1939 emend. Vauterin *et al.* 1995: The genus was established by Dowson in 1939 to include plant pathogens which cause leaf spots, cankers, blights etc. on various crop plants. The cells are Gram-negative rods, which are 0.4 - 0.6 × 0.8 - 2.0 μm, occur mostly singly or in pairs, occasionally in short chains and rarely as filaments. Cells are motile by means of a single polar flagellum. Usually produce yellow colonies, which are smooth and butyrous, mucoid or viscid. The exopolysaccharide xanthan, which is responsible for the mucoid or viscous growth, is characteristic of the genus. No growth occurs at pH 4.5 or 4.0, and at 37°C. Growth is inhibited by 6% NaCl, 30% glucose, 0.01% methyl green, 0.01% lead acetate, or 0.1% triphenyl tetrazolium chloride. Chemoorganotrophs and may require one or more growth factors. Glycine, L-glutamine and L-asparagine are not utilized as sole sources of both carbon and nitrogen. Obligately aerobic and metabolism is strictly respiratory type with oxygen as the terminal electron acceptor. Catalase is produced, oxidase production negative or weak but urease and indole or acetoin are not produced. Nitrate not reduced and litmus milk is not acidified.

Small amounts of acid are produced from many carbohydrates but not from L-rhamnose, adonitol, D-sorbitol, *meso*-erythritol, *meso*-inositol, or sorbose. In Biolog GN microplate tests, D-fructose, ∝-D-glucose, D-mannose, methylpyruvate and ∝-ketoglutaric acid are oxidized and these reactions are the characteristic of the genus, while ∝-cyclodextrin, adonitol, D-arabitol, *meso*-erythritol, *meso*-inositol, xylitol, Dglucosaminic acid, γ-hydroxybutyric acid, itaconic acid, sebacic acid, L-ornithine, L-pyroglutamic acid, D-serine, DL-carnitine, γ-aminobutyric acid, phenylethylamine, putrescine, 2-aminoethanol, and 2, 3 butanediol are not oxidized. Among the nine fatty acids that predominate in whole cell preparations, three acids, namely, 9-methyl decanoic acid, 3-hydroxy-9-methyl decanoic acid and 3-hydroxy-11-methyl dodecanoic acid are highly characteristic of the genus and serve as useful criteria for differentiating *Xanthomonas* strains from other bacteria. The ubiquinone that is present has eight isoprene units. Spermidine is the major polyamine present in the *Xa.* strains. Spermine is usually present in detectable amounts while 2-hydroxyputrescine and 1,3- diaminopropane are never present. Species described are either plant pathogens or associated with plants.

In eighth edition of *Bergey's Manual of Determinative Bacteriology*, only five species, namely, *Xa. albilineans, Xa. ampelina, Xa. axonopodis, Xa. campestris* and *Xa. fragariae* were included in this genus. The number of species increased with the passage of time but with the introduction of pathovar system (Young *et al.*, 1978), many species were downgraded and designated as pathovars of *Xa. campestris* and *Xa. axonopodis*.

Vauterin *et al.* (1995), on the basis of DNA-DNA hybridization study of 183 strains of the genus, divided it into 20 DNA homology groups and proposed a comprehensive reclassification of the genus. They found that the previously described species *Xa. campestris* was heterogeneous and it comprised of 16 DNA homology groups and one of these groups exhibited a high level of DNA homology with *Xa. axonopodis.* They emended *Xa. campestris* to include only the pathovars infecting crucifers and transferred its 34 pathovars to *Xa. axonopodis*. They recognized a total of 16 species in the genus leaving 66 former pathovars of *Xa. campestris* unassigned to any species of the genus. The other changes in the taxonomy of the genus include reclassification of *Psm. cissicola* as *Xa. cissicola*, transfer of the bacterium isolated from passion fruit plants (*Passiflora* spp.) and originally classified as *Xa. campestris* pv. *passiflorae* to *Xa. axonopodis* pv. *passiflorae* on the basis of DNA-DNA hybridization data (Goncalves and Rosato, 2000) and inclusion of *Xa. cynarae*, the causal agent of bacterial bract spot of artichoke (*Cynara scolymus*), as a new species (Trebaol *et al.*, 2000). Roumagnac *et al.* (2004), on the basis of DNA-DNA hybridization, thermal stability of DNA reassociation and fluorescent amplified fragment length polymorphism analysis, identified the causal agent of bacterial blight of onion, garlic and Welsh onion (*Allium spp.*) as *Xa. axonopodis* pv. *allii*. The pathogens causing bacterial spot of tomato, pepper or both have been divided into four species, namely, *Xa. vesicatoria, Xa. euvesicatoria, Xa. perforans* and *Xa. gardneri* (Jones *et al.*, 2004). Parkinson *et al.* (2007) showed that phylogenetic analysis of *Xanthomonas* species based on comparison of partial gyrase B gene sequences was a useful method to determine inter-species relatedness and for rapid and accurate species-level identification. Most recently, Ah-You *et al.* (2009) used amplified fragment length polymorphism, multilocus sequence analysis and DNA–DNA hybridization for genotypic classification of *Xanthomonas* pathovars associated with the plant family Anacardiaceae and concluded that the causal agents of mango bacterial canker and cashew bacterial spot should be classified as pathovars of *Xa. citri*, namely, *Xa. citri* pv. *mangiferaeindicae* and *Xa. citri* pv. *anacardii*, respectively. J.P. Euzeby has listed 30 species under this genus (List of Prokaryotes with Standing in Nomenclatur: http://www.bacterio.cict.fr/). The most recent addition to the genus is the inclusion of *Xa. dyei* with its three pathovars, namely, *eucalypti*, *laureliae* and *dysoxyli* classified on the basis of multi-locus sequence analysis (Young *et al.*, 2010). The strains of this species have a unique fatty acid methyl ester profile.

Type species–*Xanthomonas campestris*

Systematic position: Domain-*Bacteria*; Phylum-*Proteobacteria*; Class-"*Gammaproteobacteria*"; Order-"*Xanthomonadales*"; Family-"*Xanthomonadaceae*"; Genus-*Xanthomonas*; Species-*Xa. albilineans, Xa. alfalfae*, *Xa. alfalfae* subsp. *alfalfae*, *Xa. alfalfae* subsp. *citrumelonis*, *Xa. arboricola*, *Xa. arboricola*- pv. *celebensis*, pv. *corylina*, pv. *fragariae,* pv. *juglandis*, pv. *populi*, pv. *pruni*; *Xa. axonopodis*, *Xa. axonopodis*- pv. *allii*, pv. *axonopodis*, pv. *bauhiniae*, pv. *begoniae*, pv. *betlicola*, pv. *biophyti*, pv. *cajani*, pv. *cassiae*, pv. *clitoriae*, pv. *coracanae*, pv. *cyamopsidis*, pv. *desmodii*, pv. *desmodiigangetici*, pv. *desmodiilaxiflori*, pv. *desmodiirotundifolii*, pv. *dieffenbachiae*, pv. *erythrinae*, pv. *fascicularis*, pv. *glycines*, pv. *khayae*, pv. *lespedezae*, pv. *maculifoliigardeniae*, pv. *manihotis*, pv. *martyniicola*, pv. *melhusii*, pv. *nakataecorchori*, pv. *patelii*, pv. *pedalii*, pv. *phyllanthi*, pv. *physalidicola*, pv. *poinsettiicola*, pv. *punicae*, pv. *rhynchosiae*, pv. *ricini*, pv. *sesbaniae*, pv. *tamarindi*, pv. *vasculorum*, pv. *vignaeradiatae*, pv. *vignicola*, pv. *vitians*; *Xa. bromi*, *Xa. campestris*, *Xa. campestris*- pv. *aberrans*, pv. *alangii**, pv. *amaranthicola**, pv. *amorphophalli**, pv. *aracearum**, pv. *arecae**, pv. *argemones**, pv. *armoraciae*, pv. *arracaciae**, pv. *asclepiadis**, pv. *azadirachtae**, pv. *badrii**, pv. *barbareae*, pv. *betae**, pv. *bilvae**, pv. *blepharidis**, pv. *boerhaaviae**, pv. *brunneivaginae**, pv. *campestris,* pv. *cannabis**, pv. *cannae**, pv. *carissae**, pv. *centellae**, pv. *clerodendri**, pv. *convolvuli**, pv. *coriandri**, pv. *daturae**, pv. *durantae**, pv. *esculenti**, pv. *euphorbiae**, pv. *fici**, pv. *guizotiae**, pv. *gummisudans**, pv. *heliotropii**, pv. *incanae,* pv. *ionidii**, pv. *lantanae**, pv. *laureliae**, pv. *lawsoniae**, pv. *leeana**, pv. *leersiae**, pv. *malloti**, pv. *merremiae**, pv. *mirabilis**, pv. *mori**, pv. *musacearum**, pv. *nigromaculans**, pv. *obscurae**, pv. *olitorii**, pv. *papavericola**, pv. *parthenii**, pv. *passiflorae,* pv. *paulliniae**, pv. *pennamericanum**, pv. *phormiicola**, pv. *physalidis**, pv. *plantaginis,* pv. *raphani,* pv. *sesami**, pv. *spermacoces**, pv. *syngonii**, pv. *tardicrescens**, pv. *thespesiae**, pv. *thirumalacharii**, pv. *tribuli**, pv. *trichodesmae**, pv. *uppalii**, pv. *vernoniae**, pv. *viegasii**, pv. *viticola**, pv. *vitiscarnosae**, pv. *vitistrifoliae**, pv. *vitiswoodrowii**, pv. *zantedeschiae**, pv. *zingibericola**, pv. *zinniae**; *Xa. cassavae, Xa. citri, Xa. citri* pv. *anacardii, Xa. citri* subsp. *citri, Xa. citri* subsp. *malvacearum, Xa. citri* pv. *mangiferaeindicae, Xa. codiaei, Xa. cucurbitae, Xa. cynarae, Xa. dyei-* pv. *dysoxyli*, pv. *eucalypti*, pv. *laureliae*; *Xa. euvesicatoria, Xa. fragariae, Xa. fuscans, Xa. fuscans* subsp. *aurantifolii, Xa. fuscans* subsp. *fuscans, Xa. gardneri, Xa. hortorum, Xa. hortorum-* pv. *carotae*, pv. *hederae*, pv. *pelargonii*, pv. *taraxaci*; *Xa. hyacinthi, Xa. maltophilia, Xa. melonis, Xa. oryzae, Xa. oryzae-* pv. *oryzae*, pv. *oryzicola*; *Xa. perforans, Xa. phaseoli, Xa. pisi, Xa. populi, Xa. sacchari, Xa. theicola, Xa. translucens, Xa. translucens-* pv. *arrhenatheri*, pv. *cerealis*, pv. *graminis*, pv. *phlei*, pv. *phleipratensis*, pv. *poae*, pv. *secalis*, pv. *translucens*, pv. *undulosa*; *Xa. vasicola, Xa. vasicola* pv. *holcicola, Xa. vesicatoria.*

5.3.1.3.39. Xylella Wells *et al.* 1987: The genus was established by Wells *et al.* (1987) to accommodate xylem-limited procaryotes responsible for causing Pierce's disease of grapevines, phony peach disease, leaf scorches of oak, mulberry, sycamore, elm, almond and plum, and periwinkle wilt. The below given description of the genus is taken from Wells *et al.* (1987).

Cells are Gram-negative, predominantly single straight rods measuring 0.25 - 0.35 × 0.9 - 3.5 μm. Under certain cultural conditions, long filamentous strands are formed. Nonmotile and nonsporing. Colonies are creamy to white and are generally of two types: convex to pulvinate, smooth, opalescent with entire margins, and umbonate, rough with finely undulated margins. The bacteria are strictly aerobic and growth is inhibited by 2.5% carbon dioxide. Metabolism is respiratory in which the oxygen is used as terminal electron acceptor. Non-fermentative, non-halophilic and non-pigmented; oxidase test negative and catalase test positive. Optimal temperature for growth is 26 - 28°C, and optimum pH 6.5 - 6.9. The bacteria are nutritionally fastidious, requiring a specialized medium such as buffered cysteine-yeast extract agar containing charcoal or glutamine-peptone medium containing serum albumin. The G + C content of the DNA is 51.0 - 52.4 mol% (T_m) or 52.0 - 53.1 mol% (Bd). Members are found mainly in the xylem of plant tissue.

Type species- *Xylella fastidiosa*

Although Wells *et al.* (1987) recognized only one species to include all the strains, there are definitely different pathotypes, most of which have

---

* These pathovars were not examined by Vauterin *et al.* (1995) nor reclassified by others since 1995 and therefore are *Xa. campestris sensu lato* and may not belong to the species as emended by Vauterin *et al.* (1995) (Bull *et al.*, 2010).

not been characterized and compared with each other (Hopkins and Purcell, 2002). Schaad *et al.* (2004) proposed three new subspecies, namely, *Xyl. fastidiosa* subsp. *piercei, Xyl. fastidiosa* subsp. *multiplex* and *Xyl. fastidiosa* subsp. *pauca. Xyl. fastidiosa* subsp. *pauca* has not been validated due to the lack of deposition of a culture of the type strain. As the type strain designated for subsp. *piercei* is the type strain of the species, the name *Xyl. fastidiosa* subsp. *fastidiosa* takes priority over *Xyl. fastidiosa* subsp. *piercei*.

Systematic position: Domain-*Bacteria*; Phylum-*Proteobacteria*; Class-*"Gammaproteobacteria"*; Order-*"Xanthomonadales"*; Family-*"Xanthomonadaceae"*; Genus-*Xylella*; Species-*Xyl. fastidiosa, Xyl. fastidiosa* subsp. *fastidiosa, Xyl. fastidiosa* subsp. *multiplex.*

5.3.1.3.40. Xylophilus Willems *et al.* 1987: Willems *et al.* (1987) found that the causal agent of bacterial necrosis and canker of grapevines, earlier classified as *Xanthomonas ampelina*, was not related to the genus *Xanthomonas*. They created a new genus *Xylophilus* and transferred *Xa. ampelina* to the new genus as *Xylophilus ampelinus*. This being the only species in the genus was designed as type species.

The below given description of the genus is taken from Willems *et al.* (1987). Cells are Gram-negative, straight to slightly curved rods, measure 0.4 - 0.8 × 0.6 - 3.3 μm and are motile by a single polar flagellum. Long filamentous cells (length ≥ 30 μm) may appear in older cultures. Cells occur singly, in pairs, or in short chains. Strictly aerobic, having a respiratory type of metabolism using oxygen as the terminal electron acceptor. The bacteria are chemoorganotrophs with oxidative carbohydrate metabolism; oxidase test negative and catalase test positive. The growth is generally very slow and poor even at the optimal temperature of 24°C. The G + C content of the DNA ranges from 68.0 - 69.0 mol%. The genus belongs to the acidovorans rRNA complex in rRNA superfamily III and is equidistantly related to the other taxa in this complex. *Xylophilus* strains can grow on L-glutamine but not on calcium lactate, while the reverse is true for *Xanthomonas* strains.

Type species- *Xylophilus ampelinus*

Systematic position: Domain-*Bacteria*; Phylum-*Proteobacteria*; Class-*"Betaproteobacteria"*; Order-*"Burkholderiales"*; Family-*Comamonadaceae*; Genus-*Xylophilus*; Species-*Xyp. ampelinus.*

REFERENCES

Ah-You, N., Gagnevin, L., Grimont, P.A.D. *et al.* (2009). Polyphasic characterization of xanthomonads pathogenic to members of the Anacardiaceae and their relatedness to species of *Xanthomonas. Int. J. Syst. Evol. Microbiol.* 59: 306 - 18.

Al-Saady, N.A., Khan, A.M., Calari, A. *et al.* (2008). '*Candidatus* Phytoplasma omanense', associated with witches'-broom of *Cassia italica* (Mill.) Spreng. in Oman. *Int. J. Syst. Evol. Microbiol.* 58: 461 - 66.

Arocha, Y., Antesana, O., Montellano, E. *et al.* (2007). '*Candidatus* Phytoplasma lycopersici', a phytoplasma associated with 'hoja de perejil' disease in Bolivia. *Int. J. Syst. Evol. Microbiol.* 57: 1704 – 10.

Arocha, Y., Lopez, M., Pinol, B. *et al.* (2005). '*Candidatus* Phytoplasma graminis' and '*Candidatus* Phytoplasma caricae', two novel phytoplasmas associated with diseases of sugarcane, weeds and papaya in Cuba. *Int. J. Syst. Evol. Microbiol.* 55: 2451 - 63.

Baldani, J.I., Baldani, V.L.D., Seldin, L. *et al.* (1986). Characterization of *Herbaspirillum seropedicae* gen. nov., sp. nov., a root-associated nitrogen-fixing bacterium. *Int. J. Syst. Bacteriol.* 36: 86 – 93.

Baldani J.I., Pot, B., Kirchhof, G. *et al.* (1996). Emended description of *Herbaspirillum*; inclusion of [*Pseudomonas*] *rubrisubalbicans*, a mild plant pathogen, as *Herbaspirillum rubrisubalbicans* comb. nov.; and classification of a group of clinical isolates (EF Group 1) as *Herbaspirillum* species 3. *Int. J. Syst. Bacteriol.* 46: 802 - 10.

Beijerinck, M.W. and Folpmers, T. (1916). Verslagen van de gewone vergadering der wis-en naturkundige afdeeling. Konenklijke Akademie van Westenchappen te Amsterdam, 18: 1198 - 1200.

Bouchek-Mechiche, K., Gardan, L., Andrivon, D. *et al.* (2006). *Streptomyces turgidiscabies* and *Streptomyces reticulsiscabiei*: One genomic species, two pathogenic groups. *Int. J. Syst. Evol. Microbiol.* 56: 2771 - 76.

Bouchek-Mechiche, K., Gardan, L., Normand, P. *et al.* (2000). DNA relatedness among strains of *Streptomyces* pathogenic to potato in France: description of three new species, *S. europaeiscabiei* sp. nov. and *S. steliiscabiei* sp. nov. associated with common scab, and *S. reticulsiscabiei* sp. nov. associated with netted scab. *Int. J. Syst. Evol. Microbiol.* 50: 91 -99.

Bouzar, H. and Jones, J.B. (2001). *Agrobacterium larrymoorei* sp. nov., a pathogen isolated from aerial tumours of *Ficus benjamina. Int. J. Syst. Evol. Microbiol.* 51: 1023 - 26.

Brady, C.L., Venter, S.N., Cleenwerck, I. *et al.* (2009). *Pantoea vagans* sp. nov., *Pantoea eucalypti* sp. nov., *Pantoea deleyi* sp. nov. and *Pantoea anthophila* sp. nov. *Int. J. Syst. Evol. Microbiol.* 59: 2339 - 45.

Breed, R.S., Murray E.G.D. and Smith, N.R. (eds) (1957). *Bergey's Manual of Determinative Bacteriology*. 7th Ed. Williams and Wilkins Co. Baltimore, 1094 pp.

Brenner, D.J., Staley, J.T. and Krieg, N.R. (2001). Classification of prokaryotic organisms and the concept of bacterial speciation. In: D.R. Boone and R.W. Castenholz (volume editors), G.M. Garrity (editor-in-chief). *Bergey's Manual of Systematic Bacteriology*. 2nd Ed. Vol. 1, *The Archaea and the Deeply Branching and Phototrophic Bacteria*. Springer-Verlag, New York, pp. 27 - 31.

Buchanan, R.E. and Gibbons, N.E. (eds) (1974). *Bergey's Manual of Determinative Bacteriology*. 8th Ed. Williams and Wilkins Co., Baltimore, 1268 pp.

Bull, C.T., De Boer, S.H., Denny, T.P. *et al.* (2008). Demystifying the nomenclature of bacterial plant pathogens. *J. Plant Pathol.* 90: 403 – 17.

Bull, C.T., De Boer, S.H., Denny, T.P. *et al.* (2010). Comprehensive list of names of plant pathogenic bacteria, 1980-2007. *J. Plant Pathol.* 92: 551 - 92.

Buonaurio, R., Stravato, V.M., Kosako, Y. *et al.* (2002). *Sphingomonas melonis* sp. nov., a novel pathogen that causes brown spots on yellow Spanish melon fruits. *Int. J. Syst. Evol. Microbiol.* 52: 2081 - 87.

Dastager, S.G., Lee, J.C., Ju, Y.J. *et al.* (2009). *Leifsonia kribbensis* sp. nov., isolated from soil. *Int. J. Syst. Evol. Microbiol.* 59: 18 - 21.

Davis, M.J., Gillaspie, A.G, Jr., Vidaver, A.K. *et al.* (1984). *Clavibacter*: a new genus containing some phytopathogenic coryneform bacteria, including *Clavibacter xyli* subsp. *xyli* sp. nov., subsp. nov. and *Clavibacter xyli* subsp. *cynodontis* subsp. nov., pathogens that cause ratoon stunting disease of sugarcane and Bermudagrass stunting disease. *Int. J. Syst. Bacteriol.* 34: 107 - 17.

De Ley, J., Segers, P. and Gillis, M. (1978). Intra- and intergeneric similarities of *Chromobacterium* and *Janthinobacterium* ribosomal ribonucleic acid cistrons. *Int. J. Syst. Bacteriol.* 28: 154 - 68.

Dye, D.W., Bradbury, J.F., Goto, M. *et al.* (1980). International standards for naming pathovars of phytopathogenic bacteria and a list of pathovar names and pathotype strains. *Rev. Plant. Pathol.* 59: 153 - 68.

Evtushenko, L.I., Dorofeeva, L.V., Subbotin, S.A. *et al.* (2000). *Leifsonia poae* gen. nov., sp. nov., isolated from nematode galls on *Poa annua*, and reclassification of '*Corynebacterium aquaticum*' Leifson 1962 as *Leifsonia aquatica* (ex Leifson 1962) gen. nov., nom. rev., comb. nov. and *Clavibacter xyli* Davis *et al.* 1984 with two subspecies as *Leifsonia xyli* (Davis *et al.* 1984) gen. nov., comb. nov. *Int. J. Syst. Evol. Microbiol.* 50: 371 - 80.

Gardan, L., Bella, P., Meyer, J.M. *et al.* (2002). *Pseudomonas salomonii* sp. nov., pathogenic on garlic, and *Pseudomonas palleroniana* sp. nov., isolated from rice. *Int. J. Syst. Evol. Microbiol.* 52: 2065 - 74.

Gardan, L., Christen, R., Achouak, W. *et al.* (2004). *Erwinia papayae* sp. nov., a pathogen of papaya (*Carica papaya*). *Int. J. Syst. Evol. Microbiol.* 54: 107-13.

Gardan, L., Dauga, C., Prior, P. *et al.* (2000). *Acidovorax anthurii* sp. nov., a new phytopathogenic bacterium which causes bacterial leaf-spot of anthurium. *Int. J. Syst. Evol. Microbiol.* 50: 235 – 46.

Gardan, L., Gouy, C., Christen, R. *et al.* (2003). Elevation of three subspecies of *Pectobacterium carotovorum* to species level: *Pectobacterium atrosepticum* sp. nov., *Pectobacterium betavasculorum* sp. nov. and *Pectobacterium wasabiae* sp. nov. *Int. J. Syst. Evol. Microbiol.* 53: 381 - 91.

Gardan. L., Stead, D.E., Dauga, C. *et al.* (2003a). *Acidovorax valerianellae* sp. nov., a novel pathogen of Lamb's lettuce [*Valerianella locusta* (L.) Laterr.]. *Int. J. Syst. Evol. Microbiol.* 53: 795 – 800.

Garrity, G.M., Bell, J.A. and Lilburn, T. (2005). The revised road map to the manual. In: D.J. Brenner, N. R. Krieg and J.T. Staley (volume editors), G.M.

Garrity (editor-in-chief). *Bergey's Manual of Systematic Bacteriology*. 2nd Ed. Vol. 2, *The Proteobacteria*, Part A, *Introductory Essays*. Springer-Verlag, New York, pp. 159 - 200.

Gauthier, G., Gauthier, M. and Christen, R. (1995). Phylogenetic analysis of the genera *Alteromonas*, *Shewanella*, & *Moritella* using genes coding for small-subunit rRNA sequences and division of the genus *Alteromonas* into two genera, *Alteromonas* (emended) and *Pseudoalteromonas* gen. nov., and proposal of twelve new species combinations. *Int. J. Syst. Bacteriol.* 45: 755 - 61.

Gavini, F., Mergaert, J., Beji, A. *et al.* (1989). Transfer of *Enterobacter agglomerans* (Beijerinck 1888) Ewing and Fife 1972 to *Pantoea* gen. nov. as *Pantoea agglomerans* comb. nov. and description of *Pantoea dispersa* sp. nov. *Int. J. Syst. Bacteriol.* 39: 337 - 45.

Gillis, M., Van Van, T., Bardin, R. *et al.* (1995). Polyphasic taxonomy in the genus *Burkholderia* leading to an emended description of the genus and proposition of *Burkholderia vietnamiensis* sp. nov. for N_2-fixing isolates from rice in Vietnam. *Int. J. Syst. Bacteriol.* 45: 274 - 89.

Goncalves, E.R. and Rosato, Y.B. (2000). Genotypic characterization of xanthomonad strains isolated from passion fruit plants (*Passiflora* spp.) and their relatedness to different *Xanthomonas* species. *Int. J. Syst. Evol. Microbiol.* 50: 811 - 21.

Goodfellow, M. (1984). Reclassification of *Corynebacteirum fascians* (Tilford) Dowson in the genus *Rhodococcus*, as *Rhodococcus fascians* comb. nov. *Syst. Appl. Microbiol.* 5: 225 - 29.

Goto, M. and Kuwata, H. (1988). *Rhizobacter daucus* gen. nov., sp. nov., the causal agent of carrot bacterial gall. *Int. J. Syst. Bacteirol.* 38: 233 - 39.

Goyer, C., Faucher, E. and Beaulieu, C. (1996). *Streptomyces caviscabies* sp. nov., from deep-pitted lesions in potatoes in Quebec, Canada. *Int. J. Syst. Bacteriol.* 46: 635 - 39.

Guo, Y., Zheng, W., Rong, X. *et al.* (2008). A multilocus phylogeny of the *Streptomyces griseus* 16S rRNA gene clade: use of multilocus sequence analysis for streptomycete systematics. *Int. J. Syst. Evol. Microbiol.* 58: 149 - 59.

Hauben, L., Moore, E.R.B., Vauterin, L. *et al.* (1999). Validation of publication of new names and new combinations previously effectively published outside the IJSB. *Int. J. Syst. Bacteriol.* 49: 1–3. Effective publication : Hauben, L., Moore, E.R.B., Vauterin, L. *et al.* (1998). Phylogenetic position of phytopathogens within the *Enterobacteriaceae*. *Syst. Appl. Microbiol.* 21: 384 - 97.

Hiruki, C. and Wang, K. (2004). Clover proliferation phytoplasma: '*Candidatus* Phytoplasma trifolii'. *Int. J. Syst. Evol. Microbiol.* 54: 1349 - 53.

Hodgetts, J., Boonham, N., Mumford, R. *et al.* (2008). Phytoplasma phylogenetics based on analysis of *secA* and 23S rRNA gene sequences for improved resolution of candidate species of '*Candidatus* Phytoplasma'. *Int. J. Syst. Evol. Microbiol.* 58: 1826 – 37.

Holt, J.G., Krieg, N.R., Sneath P.H.A., Staley, J.T. and Williams, S.T. (eds) (1994). *Bergey's Manual of Determinative Bacteriology*. 9th Ed. Williams and Wilkins, Baltimore, 787 pp.

Hopkins, D.L. and Purcell, A.H. (2002). *Xylella fastidiosa*: Cause of Pierce's disease of grapevine and other emergent diseases. *Plant Dis.* 86: 1056 - 66.

Hormaeche, E. and Edwards, P.R. (1960). A proposed genus *Enterobacter*. *Int. Bull. Bacteriol. Nomen. Taxon.* 10: 71 – 74.

Hulse, M., Johnson, S. and Ferrieri, P. (1993). Agrobacterium infections in humans: experience at one hospital and review. *Clin. Infect. Dis.* 16: 112 – 17.

IRPCM Phytoplasma/Spiroplasma Working Team - Phytoplasma taxonomy group (2004). '*Candidatus* Phytoplasma', a taxon for the wall-less, non-helical prokaryotes that colonize plant phloem and insects. *Int. J. Syst. Evol. Microbiol.* 54: 1243 - 55.

Jagoueix-Eveillard, S., Bove, J.M. and Garnier, M. (1994). The phloem-limited bacterium of greening disease of citrus is a member of the ∝ subdivision of the *Proteobacteria*. *Int. J. Syst. Bacteriol.* 44: 379 - 86.

Jones, J.B., Lacy, G.H., Bouzar, H. *et al.* (2004). Reclassification of the *xanthomonads* associated with bacterial spot disease of tomato and pepper. *Syst. Appl. Microbiol.* 27: 755 - 62.

Judicial Commission of the International Committee on Systematics of Prokaryotes. (2008). *Corynebacterium ilicis* is typified by ICMP 2608 =ICPB CI144, *Arthrobacter ilicis* is typified by DSM 20138 =ATCC 14264 =NCPPB 1228 and the two are not homotypic synonyms, and clarification of the authorship of these two species. Opinion 87. *Int. J. Syst. Evol. Microbiol.* 58: 1976 - 78.

Jumas-Bilak, E., Carlier, J.P., Jean-Pierre, H. *et al.* (2007). *Jonquetella anthropi* gen. nov., sp. nov., the first member of the candidate phylum '*Synergistetes*' isolated from man. *Int. J. Syst. Evol. Microbiol.* 57: 2743 – 48.

Jung, H.Y., Sawayanagi, T., Kakizawa, S. *et al.* (2002). '*Candidatus* Phytoplalsma castaneae', a novel phytoplasma taxon associated with chestnut witches' broom disease. *Int. J. Syst. Evol. Microbiol.* 52: 1543 - 49.

Jung, H.Y., Sawayanagi, T., Kakizawa, S. *et al.* (2003). '*Candidatus* Phytoplasma ziziphi', a novel phytoplasma taxon associated with jujube witches'-broom disease. *Int. J. Syst. Evol. Microbiol.* 53: 1037 - 41.

Jung, H.Y., Sawayanagi, T., Wongkaew, P. *et al.* (2003a). '*Candidatus* Phytoplasma oryzae', a novel phytoplasma taxon associated with rice yellow dwarf disease. *Int. J. Syst. Evol. Microbiol.* 53: 1925 - 29.

Kageyama, B., Nakae, M., Yagi, S., *et al.* (1992). *Pantoea punctata* sp. nov., *Pantoea citrea* sp. nov., and *Pantoea terrea* sp. nov. isolated from fruit and soil samples. *Int. J. Syst. Bacteriol.* 42: 203 - 10.

Kim, W.S., Gardan, L., Rhim, S.L. *et al.* (1999). *Erwinia pyrifoliae* sp. nov., a novel pathogen that affects Asian pear trees (*Pyrus pyrifolia* Nakai). *Int. J. Syst. Bacteriol.* 49: 899 - 906.

Lapage, S.P., Sneath, P.H.A., Lessel, E.F., Skerman, V.B.D., Seeliger, H.P.R. and Clark, W.A. (1975). *International Code of Nomenclature of Bacteria*: 1976 Revision. American Society for Microbiology, Washington, DC, USA.

Lee, I.M., Bottner, K.D., Secor, G. *et al.* (2006). '*Candidatus* Phytoplasma americanum', a phytoplasma associated with a potato purple top wilt disease complex. *Int. J. Syst. Evol. Microbiol.* 56: 1593 – 97.

Lee, I.M., Davis, R.E. and Gundersen-Rindal, D.E. (2000). Phytoplasma: Phytopathogenic mollicutes. *Annu. Rev. Microbiol.* 54: 221 - 55.

Lee, I.M., Gundersen-Rindal, D.E., Davis, R.E. *et al.* (2004). '*Candidatus* Phytoplasma asteris', a novel phytoplalsma taxon associated with aster yellows and related diseases. *Int. J. Syst. Evol. Microbiol.* 54: 1037 - 48.

Lee, I.M., Martini, M., Marcone, C. *et al.* (2004a). Classification of phytoplasma strains in the elm yellows group (16SrV) and proposal of '*Candidatus* Phytoplalsma ulmi' for the phytoplasma associated with elm yellows. *Int. J. Syst. Evol. Microbiol.* 54: 337 - 47.

Liefting, L.W., Weir, B.S., Pennycook, S.R. *et al.* (2009). '*Candidatus* Liberibacter solanacearum', associated with plants in the family Solanaceae. *Int. J. Syst. Evol. Microbiol.* 59: 2274 – 76.

Lincoln, S.P., Fermor, T.R. and Tindall, B.J. (1999). *Janthinobacterium agaricidamnosum* sp. nov., a soft rot pathogen of *Agaricus bisporus. Int. J. Syst. Bacteriol.* 49: 1577 - 89.

Marcone, C., Gibb, K.S., Streten, C. *et al.* (2004). '*Candidatus* Phytoplasma spartii', '*Candidatus* Phytoplasma rhamni' and '*Candidatus* Phytoplasma allocasuarinae', respectively associated with spartium witches'-broom, buckthorn witches'-broom and allocasuarina yellows diseases. *Int. J. Syst. Evol. Microbiol.* 54: 1025 - 29.

Marcone, C., Schneider, B. and Seemuller, E. (2004a). '*Candidatus* Phytoplasma cynodontis', the phytoplasma associated with Bermuda grass white leaf disease. *Int. J. Syst. Evol. Microbiol.* 54: 1077 - 82.

Martin, K., Schumann, P., Rainey, F.A. *et al.* (1997). *Janibacter limosus* gen. nov., sp. nov., a new actinomycete with *meso*-diaminopimelic acid in the cell wall. *Int. J. Syst. Bacteriol.* 47: 529 - 34.

Martini, M., Lee, I.M. Bottner, K.D. *et al.* (2007). Ribosomal protein gene-based phylogeny for finer differentiation and classification of phytoplasmas. *Int. J. Syst. Evol. Microbiol.* 57: 2037 – 51.

McLeod, M.P., Warren, R.L., Hsiao, W.W.L. *et al.* (2006). The complete genome of *Rhodococcus* sp. RHA1 provides insights into a catabolic powerhouse. *Proc. Natl. Acad. Sci.* USA, 103: 15582 - 87.

Mergaert, J., Verdonck, L. and Kersters, K. (1993). Transfer of *Erwinia ananas* (synonym, *Erwinia uredovora*) and *Erwinia stewartii* to the genus *Pantoea* emend. as *Pantoea ananas* (Serrano 1928) comb. nov. and *Pantoea stewartii* (Smith 1898) comb. nov., respectively, and description of *Pantoea stewartii* subsp. *indologenes* subsp. nov. *Int. J. Syst. Bacteriol.* 43: 162 - 73.

Mori, K., Yamaguchi, K., Sakiyama, Y. *et al.* (2009). *Caldisericum exile* gen. nov., sp. nov., an anaerobic, thermophilic, filamentous bacterium of a novel bacterial phylum, *Caldiserica* phyl. nov., originally called the candidate phylum OP5, and description of *Caldisericaceae* fam. nov., *Caldisericales* ord. nov. and *Caldisericia* classis nov. *Int. J. Syst. Evol. Microbiol.* 59: 2894 – 98.

Murray, R.G.E. and Stackebrandt, E. (1995). Taxonomic Note: Implementation of the provisional status *Candidatus* for incompletely described procaryotes. *Int. J. Syst. Bacteriol.* 45: 186 - 87.

Park, D.H., Kim J.S., Kwon S.W. *et al.* (2003). *Streptomyces luridiscabiei* sp. nov., *Streptomyces puniciscabiei* sp. nov. and *Streptomyces niveiscabiei* sp. nov., which cause potato common scab disease in Korea. *Int. J. Syst. Evol. Microbiol.* 53: 2049 - 54.

Parkinson, N., Aritua, V., Heeney, J. *et al.* (2007). Phylogenetic analysis of *Xanthomonas* species by comparison of partial gyrase B gene sequences. *Int. J. Syst. Evol. Microbiol.* 57: 2881 – 87.

Parkinson, N., Stead, D., Bew, J. *et al.* (2009). *Dickeya* species relatedness and clade structure determined by comparison of *recA* sequences. *Int. J. Syst. Evol. Microbiol.* 59: 2388 – 93.

Roumagnac, P., Gagnevin, L., Gardan, L. *et al.* (2004). Polyphasic characterization of xanthomonads isolated from onion, garlic and Welsh onion (*Allium spp.*) and their relatedness to different *Xanthomonas* species. *Int. J. Syst. Evol. Microbiol.* 54: 15 - 24.

Saglio, P., Lhospital, M., Lafleche, D. *et al.* (1973). *Spiroplasma citri* gen. and sp. n.: A mycoplalsma-like organism associated with "Stubborn" disease of citrus. *Int. J. Syst. Bacteriol.* 23: 191 - 204.

Samson, R., Legendre, J.B., Christen, R., *et al.* (2005). Transfer of *Pectobacterium chrysanthemi* (Burkholder *et al.* 1953) Brenner *et al.* 1973 and *Brenneria paradisiaca* to the genus *Dickeya* gen. nov. as *Dickeya chrysanthemi* comb. nov. and *Dickeya paradisiaca* comb. nov. and delineation of four novel species, *Dickeya dadantii* sp. nov., *Dickeya dianthicola* sp. nov., *Dickeya dieffenbachiae* sp. nov. and *Dickeya zeae* sp. nov. *Int. J. Syst. Evol. Microbiol.* 55: 1415 - 27.

Sasaki, J., Chijimatsu, M. and Suzuki, K.-I. (1998). Taxonomic significance of 2, 4-diaminobutyric acid isomers in the cell wall peptidoglycan of actinomycetes and reclassification of *Clavibacter toxicus* as *Rathayibacter toxicus* comb. nov. *Int. J. Syst. Bacteriol.* 48: 403 - 10.

Sawada, H., Ieki, H., Oyaizu, H. *et al.* (1993). Proposal for rejection of *Agrobacterium tumefaciens* and revised descriptions for the genus *Agrobacterium* and for *Agrobacterium radiobacter* and *Agrobacterium rhizogenes. Int. J. Syst. Bacteriol.* 43: 694 - 702.

Schaad, N.W., Postnikova, E., Lacy, G. *et al.* (2004). *Xylella fastidiosa* subspecies: *X. fastidiosa* subsp. *piercei,* subsp. nov., *X. fastidiosa* subsp. *multiplex*, subsp. nov., and *X. fastidiosa* subsp. *pauca* subsp. nov. *Syst. Appl. Microbiol.* 27: 290 - 300.

Schaad, N.W., Postnikova, E., Sechler, A. *et al.* (2009). List of new names and new combinations previously effectively, but not validly, published. *Int. J. Syst. Evol. Microbiol.* 59: 923 - 25. Effective publication: Schaad, N.W., Postnikova, E., Sechler, A. *et al.* (2008). Reclassification of subspecies of *Acidovorax avenae* as *A. avenae* (Manns 1905) emend., *A. cattleyae* (Pavarino, 1911) comb. nov., *A. citrulli* (Schaad *et al.*, 1978) comb. nov., and proposal of *A. oryzae* sp. nov. *Syst. Appl. Microbiol.* 31: 434 – 46.

Schneider, B., Torres, E., Martin, M.P. *et al.* (2005). '*Candidatus* phytoplasma pini', a novel taxon from *Pinus silvestris* and *Pinus halepensis. Int. J. Syst. Evol. Microbiol.* 55: 303 - 307.

Sechler, A., Schuenzel, E.L., Cooke, P. *et al.* (2009). Cultivation of '*Candidatus* Liberibacter asiaticus', '*Ca.* L. africanus' and '*Ca.* L. americanus' associated with huanglongbing. *Phytopathology* 99: 480 – 86.

Seemuller, E. and Schneider, B. (2004). '*Candidatus* Phytoplasma mali', '*Candidatus* Phytoplasma pyri' and '*Candidatus* Phytoplasma prunorum', the causal agents of apple proliferation, pear decline and European stone fruit yellows, respectively. *Int. J. Syst. Evol. Microbiol.* 54: 1217 - 26.

Skerman, V.B.D., McGowan, V. and Sneath, P.H.A. (eds) (1980). Approved lists of bacterial names. *Int. J. Syst. Bacteriol.* 30: 225 - 420.

Sutra, L., Christen, R., Bollet, C. *et al.* (2001). *Samsonia erythrinae* gen. nov., sp. nov., isolated from bark necrotic lesions of *Erythrina* sp., and discrimination of plant-pathogenic *Enterobacteriaceae* by phenotypic features. *Int. J. Syst. Evol. Microbiol.* 51: 1291 - 1304.

Takahashi, K., Nishiyama, K. and Sato, M. (1996). *Pseudomonas syringae* pv. *broussonetiae* pv. nov., the causal agent of bacterial blight of paper mulberry (*Broussonetia kazinoki* x *B. papyrifera*). *Ann. Phytopathol. Soc. Japan* 62: 17 - 22.

Takeuchi, M., Hamana, K. and Hiraishi, A. (2001). Proposal of the genus Sphingomonas sensu stricto and three new genera, Sphingobium, *Novosphingobium* and *Sphingopyxis*, on the basis of phylogenetic and chemotaxonomic analyses. *Int. J. Sys. Evol. Microbiol.* 51: 1405 – 17.

Takeuchi, M., Sakane, T., Yanagi, M. *et al.* (1995). Taxonomic study of bacteria isolated from plants: Proposal of *Sphingomonas rosa* sp. nov., *Sphingomonas pruni* sp. nov., *Sphingomonas asaccharolytica* sp. nov. and *Sphingomonas mali* sp. nov. *Int. J. Syst. Bacteriol.* 45: 334 - 41.

Teixeira, D.C., Saillard, C., Eveillard, S. *et al.* (2005). '*Candidatus* Liberibacter americanus', associated with citrus huanglongbing (greening disease) in Sao Paulo State, Brazil. *Int. J. Syst. Evol. Microbiol.* 55: 1857 - 62.

Trebaol, G., Gardan, L., Manceau, C. *et al.* (2000). Genomic and phenotypic characterization of *Xanthomonas cynarae* sp. nov., a new species that causes bacterial bract spot of artichoke (*Cynara scolymus* L.) *Int. J. Syst. Evo. Microbiol.* 50: 1471 - 78.

Urakami, T., Ito-Yoshida, C., Araki, H. *et al.* (1994). Transfer of *Pseudomonas plantarii* and *Pseudomonas glumae* to *Burkholderia* as *Burkholderia* spp. and description of *Burkholderia vandii* sp. nov. *Int. J. Syst. Bacteriol.* 44: 235 - 45.

Valiunas, D., Staniulis, J. and Davis, R.E. (2006). '*Candidatus* Phytoplasma fragariae', a novel phytoplasma taxon discovered in yellows diseased strawberry, *Fragaria* x *ananassa*. *Int. J. Syst. Evol. Microbiol.* 56: 277 - 81.

Vaneechoutte, M., Kampfer, P., Baere, T.D. *et al.* (2004). *Wautersia* gen. nov., a novel genus accommodating the phylogenetic lineage including *Ralstonia eutropha* and related species, and proposal of *Ralstonia* [*Pseudomonas*] *syzygii* (Roberts *et al.* 1990) comb. nov. *Int. J. Syst. Evol. Microbiol.* 54: 317 - 27.

Vauterin, L., Hoste, B., Kersters, K. *et al.* (1995). Reclassification of *Xanthomonas*. *Int. J. Syst. Bacteriol.* 45: 472 - 89.

Verdin, E., Salar, P., Danet, J.L., *et al.* (2003). '*Candidatus* Phytoplasma phoenicium' sp. nov., a novel phytoplasma associated with an emerging lethal disease of almond trees in Lebanon and Iran. *Int. J. Syst. Evol. Microbiol.* 53: 833 - 38.

Waldee, E.L. (1945). Comparative studies of some peritrichous phytopathogenic bacteria. *Iowa State Coll. J. Sci.* 19: 435 - 84.

Wei, W., Davis, R.E., Lee, I.-M. *et al.* (2007). Computer-simulated RFLP analysis of 16S rRNA genes: identification of ten new phytoplasma groups. *Int. J. Syst. Evol. Microbiol.* 57: 1855 – 67.

Wei, W., Lee, I.-M., Davis, R.E. *et al.* (2008). Automated RFLP pattern comparison and similarity coefficient calculation for rapid delineation of new and distinct phytoplasma 16Sr subgroup lineages. *Int. J. Syst. Evol. Microbiol.* 58: 2368 –77.

Wells, J.M., Raju, B.C., Hung, H.Y. *et al.* (1987). *Xylella fastidiosa* gen. nov., sp. nov.: Gram-negative, xylem-limited, fastidious plant bacteria related to *Xanthomonas* spp. *Int. J. Syst. Bacteriol.* 37: 136 - 43.

Willems, A., Falsen, E., Pot, B. *et al.* (1990). *Acidovorax*, a new genus for *Pseudomonas facilis, Pseudomonas delafieldii,* E. Falsen (EF) group 13, EF group 16, and several clinical isolates, with the species *Acidovorax facilis* comb. nov., *Acidovorax delafieldii* comb. nov., and *Acidovorax temperans* sp. nov. *Int. J. Syst. Bacteriol.* 40: 384 - 98.

Willems, A., Gillis, M., Kersters, K. *et al.* (1987). Transfer of *Xanthomonas ampelina* Panagopoulos 1969 to a new genus, *Xylophilus* gen. nov., as *Xylophilus ampelinus* (Panagopoulos 1969) comb. nov. *Int. J. Syst. Bacteriol.* 37: 422 - 30.

Willems, A., Goor, M., Thielemans, S. *et al.* (1992). Transfer of several phytopathogenic *Pseudomonas* species to *Acidovorax* as *Acidovorax avenae* subsp. *avenae* subsp. nov., comb. nov., *Acidovorax avenae* subsp. *citrulli, Acidovorax avenae* subsp. *cattleyae*, and *Acidovorax konjaci*. *Int. J. Syst. Bacteriol.* 42: 107 - 19.

Winslow, C.E.A., Broadhurst, J., Buchanan, R.E. *et al.* (1920). The families and genera of the bacteria. Final report of the Committee of the Society of American Bacteriologists on characterization and classification of bacterial types. *J. Bacteriol.* 5: 191 - 229.

Yabuuchi, E., Kosako, Y., Fujiwara, N. *et al.* (2002). Emendation of the genus *Sphingomonas* Yabuuchi *et al.* 1990 and junior objective synonymy of the species of three genera, *Sphingobium, Novosphingobium* and *Sphingopyxis*, in conjuction with *Blastomonas ursincola*. *Int. J. Syst. Evol. Microbiol.* 52: 1485 - 96.

Yabuuchi, E., Kosako, Y., Naka, T. *et al.* (1999). Validation of publication of new names and new combinations previously effectively published outside the IJSB. *Int. J. Syst. Bacteriol.* 49: 935 - 36. Effective publication: Yabuuchi, E., Kosako, Y., Naka, T. *et al.* (1999). Proposal of *Sphingomonas suberifaciens* (van Bruggen, Jochimsen and Brown 1990) comb. nov., *Sphingomonas natatoria* (Sly 1985) comb. nov., *Sphingomonas ursincola* (Yurkov *et al.*

1997) comb. nov., and emendation of the genus *Sphingomonas*. *Microbiol. Immunol.* 43: 339 - 49.

Yabuuchi, E., Kosako, Y., Oyaizu, H. *et al.* (1993). Validation of the publication of new names and new combinations previously effectively published outside the IJSB. *Int. J. Syst. Bacteriol.* 43: 398 - 99. Effective publication: Yabuuchi, E., Kosako, Y., Oyaizu, H. *et al.* (1992). Proposal of *Burkholderia* gen. nov. and transfer of seven species of the genus *Pseudomonas* homology group II to the new genus, with the type species *Burkholderia cepacia* (Palleroni and Holmes 1981) comb. nov. *Microbiol. Immunol.* 36: 1251 - 75.

Yabuuchi, E., Kosako, Y., Yano, I. *et al.* (1996). Validation of the publication of new names and new combinations previously effectively published outside the IJSB. *Int. J. Syst. Bacteriol.* 46: 625 - 26. Effective publication: Yabuuchi, E., Kosako, Y., Yano, I. *et al.* (1995). Transfer of two *Burkholderia* and an *Alcaligenes* species to *Ralstonia* gen. nov.: Proposal of *Ralstonia pickettii* (Ralston, Palleroni and Doudoroff 1973) comb. nov., *Ralstonia solanacearum* (Smith 1896) comb. nov. and *Ralstonia eutropha* (Davis 1969) comb. nov. *Microbiol. Immunol.* 39: 897 - 904.

Yabuuchi, E., Yano, I., Oyaizu, H. *et al.* (1990). Validation of the publication of new names and new combinations previously effectively published outside the IJSB. *Int. J. Syst. Bacteriol.* 40: 320 - 21. Effective publication: Yabuuchi, E., Yano, I., Oyaizu, H. *et al.* (1990). Proposals of *Sphingomonas paucimobilis* gen. nov. and comb. nov., *Sphingomonas parapaucimobilis* sp. nov., *Sphingomonas yanoikuyae* sp. nov., *Sphingomonas adhaesiva* sp. nov., *Sphingomonas capsulata* comb. nov., and two genospecies of the genus *Sphingomonas*. *Microbiol. Immunol.* 34: 99 - 119.

Yamada, K. and Komagata, K. (1972). Taxonomic studies on coryneform bacteria V. Classification of coryneform bacteria. *J. Gen. Appl. Microbiol.* 18: 417 - 31.

Yoon, J.H., Lee, H.B., Yeo, S.H. *et al.* (2004). *Janibacter melonis* sp. nov., isolated from abnormally spoiled oriental melon in Korea. *Int. J. Syst. Evol. Microbiol.* 54: 1975 - 80.

Young, J.M. (2008). An overview of bacterial nomenclature with special reference to plant pathogenic bacteria. *Syst. Appl. Microbiol.* 31: 405 - 24.

Young, J.M., Bull, C.T., De Boer S.H. *et al.* (2004). Names of plant pathogenic bacteria, 1864 - 2004. Online [http://www.isppweb.org/names_bacterial_revised.asp].

Young, J.M., Dye, D.W., Bradbury, J.F. *et al.* (1978). A proposed nomenclature and classification for plant pathogenic bacteria. *N. Z. J. Agric. Res.* 21: 153 - 77.

Young, J.M., Wilkie, J.P., Park, D.C. *et al.* (2010). List of new names and new combinations previously effectively, but not validly, published. *Int. J. Syst. Evol. Microbiol.* 60: 1009 – 1010. Effective Publication: Young, J.M., Wilkie, J.P., Park, D.C. *et al.* (2010). New Zealand strains of plant pathogenic bacteria classified by multi-locus sequence analysis; proposal of *Xanthomonas dyei* sp. nov. *Plant Pathol.* 59: 270 – 81.

Zgurskaya, H.I., Evtushenko, L.I., Akimov, V.N. *et al.* (1993). *Rathayibacter* gen. nov., including the species *Rathayibacter rathayi* comb. nov., *Rathayibacter tritici* comb. nov., *Rathayibacter iranicus* comb. nov., and six strains from annual grasses. *Int. J. Syst. Bacteriol.* 43: 143 - 149.

Zhao, Y., Sun, Q., Wei, W., *et al.* (2009). '*Candidatus* Phytoplasma tamaricis', a novel taxon discovered in witches'-broom-diseased salt cedar (*Tamarix chinensis* Lour.). *Int. J. Syst. Evol. Microbiol.* 59: 2496 - 2504.

Zhao, Y., Wei, W., Lee, I.M. *et al.* (2009a). Construction of an interactive online phytoplasma classification tool, *i*PhyClassifier, and its application in analysis of the peach X-disease phytoplasma group (16SrIII). *Int. J. Syst. Evol. Microbiol.* 59: 2582 - 93.

Zreik, L., Bove, J.M. and Garnier, M. (1998). Phylogenetic characterization of the bacterium-like organism associated with marginal chlorosis of strawberry and proposition of a *Candidatus* taxon for the organism, '*Candidatus* Phlomobacter fragariae'. *Int. J. Syst. Bacteriol.* 48: 257 - 261.

Chapter 6

GENOME EVOLUTION OF PHYTOPATHOGENIC PROCARYOTES

A great amount of biochemical and genetic diversity exists within the procaryotic organisms. Due to their ability to utilize diverse energy sources, electron donors and acceptors, they are most widely distributed and found in every conceivable environment. As the higher organisms on earth evolved in the presence of procaryotes, the latter are intimately associated with the former, including plants. The procaryotes, on account of their intimate association with plants, interact with plants and affect directly or indirectly their overall health. Some procaryotes improve the plant health while others cause harmful effects and in some cases may even lead to plant death. The latter group includes plant pathogenic procaryotes/bacteria that cause a large number of plant diseases and are of direct interest to plant pathologists.

6.1. BACTERIAL GENOME

The genomes of bacteria contain deoxyribose nucleic acid (DNA) that is composed of four nucleic acid bases, namely, adenine (A), guanine (G), thymine (T) and cytosine (C). It is the specific number and order of these bases that uniquely identifies an organism. Bacterial genomes are made up of large DNA molecules arranged generally in a single chromosome. The chromosomes are generally circular in shape but may be linear in some cases. Besides chromosomes, bacterial genomes often contain extra chromosomal DNA molecules, called plasmids. The plasmids may vary in length from a few kilo bases to over 400 kilo bases (kb).

The genome size of plant pathogenic bacteria varies with different genera and species. Even different strains of the same species show a high degree of variability (Vinatzer and Yan, 2008). The smaller genomes are estimated to be around 2.5 million base pairs (mbp), the larger can be greater than 5 mbp (Pierson and Ishimaru, 2000). Among the phytopathogenic bacteria, phytoplasmas have genomes of smaller sizes, ranging from 530-1350 kilo base pairs (kbp), than the other bacteria. '*Candidatus* Phytoplasma cynodontis', the causal agent of Bermuda grass white leaf disease, has the smallest genome size (530 kbp) and may represent the smallest chromosome known for any living organism. An exact proportion and distribution of each of the four bases in the genome is obtained by sequencing the genome completely. The relative amount of each base, reported as the G + C content, is a characteristic of a genus. Some genera like *Erwinia* has a low G + C content ranging from 49.8 - 54.1 mol% while others such as, *Clavibacter* (70 ± 5 mol%), *Curtobacterium* (66.0 - 71.0 mol%) and *Streptomyces* (69.0 - 78.0 mol%) have a relatively higher G + C content.

6.2. PATHOGENICITY ISLANDS

Plant pathogenic bacteria have large segments of DNA associated with pathogenicity, called 'pathogenicity islands'. These pathogenicity islands (PAIs) can be located using genetic tools like transposon mutagenesis and complementation. The G + C contents of some of these PAIs are distinct from the overall G + C content of the organism. From this and other relevant information, it is postulated that the presence of PAIs can indicate how some bacteria may have evolved to become pathogens. PAIs are clusters of virulence genes that have been acquired through transfer and recombinant events, and contribute to a pathogenic phenotype. As these genes often come from phylogenetically diverse genomes, G + C content varies among genes in a PAI. PAIs have a mosaic structure and are subject to a continuous process of gene acquisition, gene loss and gene decay (Loria *et al.,* 2006).

Emergence of new plant pathogens or increase in virulence of the existing pathogens has always been a matter of great concern to plant pathologists. An understanding of the evolutionary mechanisms, leading to emergence of new pathogens or of aggressive virulent strains of the existing ones, is essential to reduce or minimize the losses caused by plant pathogens. The development of analytical tools and availability of genomes' sequenced data have greatly increased our understanding of the mechanisms affecting the evolution of plant pathogens. In contrast to eucaryotes, which evolved mostly through modifications of existing genetic information, and in which horizontal gene transfer played a minor role, bacteria have acquired a significant proportion of their genetic diversity through the acquisition of DNA sequences from distantly related organisms, including eucaryotes.

6.3. HORIZONTAL GENE TRANSFER

Horizontal gene transfer (HGT), also called **lateral gene transfer**, is defined to be the movement of genetic material between bacteria other than by descent in which the information travels through the generations as the cell divides. By contrast, vertical transfer occurs when an organism receives genetic material from its ancestor, i.e., its parent or species from which it is evolved. HGT is most often thought of as a sexual process that requires a mechanism for the mobilization of a chromosomal DNA among bacterial cells. As the bacteria do not reproduce sexually, they have acquired several mechanisms, namely, transformation, conjugation and transduction for the exchange of genetic material. Natural genetic transformation is believed to be the essential mechanism for the attainment of genetic plasticity in many species of bacteria. During bacterial evolution, the ability of bacteria and the archaea to adapt to new environments most often results from the acquision of new genes through horizontal transfer rather than by the alteration of gene functions through numerous point mutations.

HGT is influenced more by physical than phylogenetic proximity, and that genes carried by HGT are often associated with pathogenicity and adaptation to a changing environment or to new environmental niches. Thus, coding sequences that are retained following HGT are more likely to be derived from bacteria sharing a particular niche and thus coding for functions related to survival in that niche.

HGT is very important in the sense that it may cause almost instant emergence of new pathogens, in comparisons to the hundreds or even thousands of years needed for emergence of new pathogens by other mechanisms. No doubt, HGT is an important driving force for the evolution of bacterial genome and for rapid adaptation to changing environments; pathoadaptive mutations, which increase virulence by means of minor genetic alterations such as point mutation, and gene rearrangement or deletion, also contribute to the evolution of virulence.

Earlier, all bacteria were assumed to be clonal, i.e., the variation between strains of their species is caused more by mutation than by recombination. Smith *et al.* (1993) revolutionized the field of microbial evolution by developing molecular evolutionary tests that showed that the recombination occurs frequently in some species. However, with more publications coming in favour of recombination, it is clearly established that recombination is an important evolutionary mechanism in numerous bacterial species. A notable example is *Escherichia coli,* which was previously considered to be clonal. Extensive sequence analyses of many isolates of *Es. coli* has now confirmed the earlier evidence of recombination in some *Es. coli* clades (Wirth *et al.,* 2006).

Recombinations occur frequently only when DNA sequences of the donor and the recipient are very similar. However, these can be detected only when DNA sequences of the donor and recipient are sufficiently different from each other for recombination events to be inferred from sequence data with statistical significance. Multilocus sequence typing

(MLST) allows for the determination of contribution of homologous recombination resulting from conjugation, transduction, or transformation, to variation between strains of a species. Recombination in many bacterial species was overlooked before the advent of MLST. Sarkar and Guttman (2004) found little recombination in *Pseudomonas syringae* on the basis of analysis of distantly and closely related isolates and mutation contributed approximately four times more than the recombination to variation. The basis for the relatively small contribution of recombination to variation between the analyzed isolates was probably due to low percentage (91%) of DNA sequence identity between distantly related isolates. Also, the adaptation of *Psm. syringae* pathovars to different plant species probably makes physical contact between distantly related isolates more difficult. Yan *et al.* (2008) reported that recombination contributed 5.8 times more than mutation to variation between isolates when closely related isolates with a DNA sequence identity of approximately 99% were analyzed. Many isolates studied had a partially overlapping host range, thus giving an ample opportunity for recombination. Even in cases where a common host between isolates was not identified, isolates may have evolved only recently from ancestors with overlapping host ranges. They were also able to confirm high rates of recombination between closely related *Psm. syringac* isolates by identifying recombination breakpoints within several genes and by applying various recombination tests. However, for some gene fragments, the DNA identity between isolates was more than 99.3% and it was too high to allow detection of recombination.

The analysis of a mix of closely and distantly related isolates of *Ralstonia solanacearum* (Castillo and Greenberg, 2007) and *Xylella fastidiosa* (Scally *et al.*, 2005) revealed that both the pathogens evolved mainly by mutation rather than recombination. However, the analysis of a large number of closely related isolates of *Ral. solanacearum* (Bertolla *et al.*, 1999) and *Xanthomonas axonopodis* pv. *vesicatoria* (Basim *et al.*, 1999) showed that *Ral. solanacearum* was found to be naturally competent during infection, potentially allowing a very high rate of recombination; and recombination rates were also high enough in *Xa. axonopodis* pv. *vesicatoria* to be detected in controlled coinfections. Thieme *et al.* (2005) sequenced complete genome of *Xa. campestris* pv. *vesicatoria* and found novel insights into the genome plasticity and pathogenicity of the bacterium. They further revealed that several regions of atypical G + C content were associated with both insertion sequence elements and tRNA genes, and contained putative pathogenicity determinants which might have been acquired by horizontal gene transfer. Goss *et al.* (2005) also found high recombination rates in *Psm. viridiflava* after analyzing a larger number of closely related isolates. The results of majority of the above mentioned studies clearly show that recombination probably plays an important role in the evolution of many bacterial plant pathogenic species but it is easily overlooked when an insufficient number of closely related isolates is sampled.

Toth *et al.* (2006), in their review article, compared the genome of *Er. carotovora* subsp. *atroseptica* (recent name, *Pectobacterium atrosepticum)* with those of human and animal pathogenic eterobacteria, namely, *Es. coli, Salmonella enterica, Yersinia pestis, Y. pseudotuberculosis and Shigella flexneri,* and suggested HGT in *Er. carotovora* subsp. *atroseptica* may directly have led to the acquisition of number of determinants that mediate its interactions, pathogenic or otherwise with plants, offering a glimpse into its evolutionary divergence from animal pathogenic enterobacteria. *Er. carotovora* subsp. *atroseptica* genome contains 17 major putative horizontally acquired islands and several minor horizontally acquired islands. In both the cases, the genes involved in pathogensis and/or ecological adaptation are found. Two of the horizontally acquired islands represent complete prophages, whereas the others show evidence of phage remnants and transposase, and integrase genes that are often associated with HGT.

Immediate public access to the data of complete genome sequences opens up a new biological age. It is considered the birth of a new science: **Genome-based biology.** In addition to the implications for medicine, the knowledge of microbial complete genome sequence also provides a wealth of information for tracing evolutionary networks. Hundreds of bacterial genomes including the genomes of dozens of plant pathogenic bacteria have been sequenced. The genome-sequenced data have provided valuable information pertaining to sources of genomic variation and evolution of new strains. Some of the phytopathogenic bacteria, whose genomes have been sequenced, are given below.

1. *Xyl. fastidiosa* clone 9a5C
2. *Ral. solanacearum* strains GMI 1000, MOIK 2 and 1609
3. *Psm. syringae* pv. *tomato* strains DC 3000 and T 1
4. *Xa. axonopodis* pv. *citri* (recent name, *Xa. citri* subsp. *citri)*
5. *Xa. campestris* pv. *campestris* strains 85-10, ATCC 33913 (CFBP 5241) and 8004 (CFBP 6650)
6. *Xa. campestris* pv. *vesicatoria* (recent name, *Xa. vesicatoria)*
7. *Xa. oryzae* pv. *oryzae* strains MAFF 311018, KACC 10331 and PXO 99A
8. *'Candidatus* Liberibacter asiaticus'
9. *Er. carotovora* subsp. *atroseptica* (recent name, *Pectobacterium atrosepticum)* strain 1043

6.4. GALL-FORMING *PANTOEA AGGLOMERANS*- A CASE STUDY

The most classical example of the evolution of a new bacterial plant pathogen *is Pa. agglomerans*. This bacterium, which is widespread in nature as a commensal and epiphytic bacterium associated with many plants, was transformed into a host-specific gall-forming pathogen on gypsophila and beet. Barash and Manulis-Sasson (2009), in their most recent review article, have given an excellent account of transformation of *Pa. agglomerans* into an hrp-dependent and host-specific pathogen. The transformation was achieved by the evolution of a pathogenicity plasmid that harbours most, if not all of the genes, required for gall formation and host specificity. The pathogenicity plasmid designated as pPATH is unique in structure and contains a pathogenicity island (PAI). The PAI was evolved on an interon plasmid of the IncN family, which is distributed among genetically diverse population of *Pa. agglomerans*. The horizontal gene transfer was the major contributing force in the creation of PAI, although a pathoadaptive mechanism might also be involved. Plant pathogenic strains of *Pa. agglomerans* isolated from gypsophila contain a pPATH. The presence of six insertion sequence (IS) elements belonging to five families (i.e., IS*1*, IS*3*, IS*5*, IS*6* and IS*630*) along the PAI is a strong evidence that an intensive HGT may have taken place during the evolution of the pPATH. The IS*3* family is one of the largest families of IS elements, with members found in diverse bacterial groups. Several IS elements found in phytopathogenic bacterial plasmids are members of the IS*3* family. These elements were indentified in *Xa. campestris, Psm. syringae, Er. amylovora* (Sundin, 2007) and *Agrobacterium tumefaciens*.

The pathogenicity plasmid of *Pa. agglomerans* pv. *gypsophilae* (*Pag*), designated as $pPATH_{Pag}$, is approximately 135 kb and it contains a pathogenicity island of ~75 kb. The pPATH harbours an *hrp*/*hrc* gene cluster, type III effectors, a gene cluster encoding for biosynthesis of indole acetic acid (IAA) and cytokinins, and many diverse insertion sequence elements and partial sequences from genes of other bacteria. Mutations in genes encoding components of the type III secretion system or the *hrp*/*hrc* regulatory gene cascade completely abolished gall formation while mutations in genes encoding type III effectors or biosynthesis of IAA and cytokinins reduced gall development substantially or almost totally. These findings suggest that pathogenicity of *Pa. agglomerans* pv. *gypsophilae* and *Pa. agglomerans* pv. *betae* is hrp-dependent, and that acquisition of the plasmid-borne PAI was a crucial step in the transformation of *Pa. agglomerans* into a gall-forming and host-specific plant pathogen.

The pathogenicity plasmid of *Pa. agglomerans* pv. *gypsophilae* is nonconjugative which is supported by the lack of putative genes that show significant homology to *tra* genes. Plant synthesized IAA and cytokinins are absolutely required for gall initiation, whereas the pathogen-secreted phytohormones are not needed. However, pathogen-secreted IAA and cytokinins do have a significant effect on gall size and might be important for expression of the *hrp* regulatory genes as signalling molecules.

Plasmids are common residents of phytopathogenic bacteria and contribute significantly to host evolution in a multifaceted manner. Plasmids tend to encode determinants of virulence and ecological fitness that can enhance adaptation to a specific niche or can influence niche expansion. Many of these determinants appear to have been acquired from other bacteria via horizontal transfer, illustrating an important function of these plasmids in the acquisition of sequences that enable rapid evolution. These genes can ultimately be delivered to the host chromosome through plasmid integration events, thus stabilizing important acquired determinants within the genome. Sundin (2007) has given an excellent account of contribution of phytopathogenic bacterial plasmids to the evolutionary history of their hosts.

6.5. EVOLUTION IN *STREPTOMYCES* SPECIES

Streptomyces are multicellular, Gram-positive, high G + C content bacteria, with a complex development cycle, large linear chromosomes and a high capacity for catabolism and secondary metabolism. In addition to linear chromosomes, *Streptomyces* species contain both circular and linear plasmids that add to adaptability and enhance horizontal gene transfer. Minorities of *Streptomyces* species are plant pathogens and the emergence of new pathogenic species has occurred many times in agricultural systems. Plant pathogenic species contain a large PAI that contains the thaxtomin biosynthetic genes as well as other virulence determinants.

The emergence of new plant pathogenic species of *Streptomyces* suggests that the *Streptomyces* PAI is moblizable, at least in some strains. The appearance of *St. turgidiscabies* in Japan is the latest example of a newly emerged species. Due to its recent emergence, *St. turgidiscabies* was used to test the hypothesis that the PAI could be transferred to other *Streptomyces* species via conjugation. *St. coelicolor* and *St. diastatochromogenes* were used as recipients. Analysis of transconjugates indicated that all known pathogenicity loci were mobilized during the conjugation events. Additional analysis indicated that a DNA element of approximately 660 kb integrated into the *St. coelicolor* chromosome in most of the transconjugates. Mobilization of the 660 kb version of the PAI to *St. coelicolor* did not confer a pathogenicity phenotype, while its mobilization to *St. diastatochromogenes* enabled the transconjugates to produce thaxtomin (toxin involved in pathogenesis) and colonize excised tuber tissue (Kers *et al.,* 2005). It is evident from these results that PAI is not the sole pathogenicity determinant.

The unhindered transfer of *St. turgidiscabies* PAI to saprophytic *Streptomyces* species *in vitro* suggests that mobilization of pathogenicity determinants among *Streptomyces* species occurs frequently in nature. However, the occurrence of a limited number of pathogenic *Streptomyces* species in contrast to a large number of saprophytic *Streptomyces* species in nature suggests that there may be a genetic barrier in most *Streptomyces* species that results in instability of PAI or prevents the expression of the pathogenicity genes in the recipient strains. It is further confirmed by the fact that mobilization of PAI into *St. coelicolor* did not confer pathogenic ability to this well known saprophytic species.

6.6. *XA. ORYZAE* PV. *ORYZAE*- A CASE OF RAPID EVOLUTION

Complete genome sequences have been published for three strains of *Xa. oryzae* pv. *oryzae* (*Xoo*), namely, MAFF 311018 (MAFF), a Japanese race 1 strain, also referred to as T 7174 (Ochiai *et al.,* 2005), KACC 10331 (KACC), a Korean race 1 isolate, also known as KXO 85 (Lee *et al.,* 2005) and PXO 99^{A}, a strain of Philippines race 6 (Salzberg *et al.,* 2008). The genomes of MAFF and KACC overall are highly similar to each other in gene content and organization. PXO 99^{A} is considerably more distant from either of MAFF or KACC. In contrast to MAFF and KACC, PXO 99^{A} is virulent on large number of rice varieties representing diverse genetic sources of resistance, including the broad-spectrum recessive resistance gene *xa5*. Relatively few resistance genes including *xa13* are effective against PXO 99^{A} while *xa13* is ineffective against MAFF and KACC. As MAFF and KACC are highly similar in genome content and organization, Salzberg *et al.* (2008) have compared PXO 99^{A} and MAFF and have drawn the following conclusions.

Xoo genome has a remarkable plasticity and this plasticity is most strikingly evident in the large number of major rearrangements and indels between these strains. The differences suggest that the *Xoo* genome evolves rapidly and this conclusion is perhaps best supported by the 212 kb sequence duplication in PXO 99^{A}. The duplication represents a remarkable recent event, with only a single nucleotide difference differentiating between two copies in PXO 99^{A}. Gene duplication contributes to gene diversification, permitting for unconstrained evolution of otherwise indispensable sequences. The abundant duplications in PXO 99^{A} can act as an important source of genomic variation in *Xoo*. It is clear from the analysis of 212 repeat kb that IS elements play an important role in generating duplications. Comparison of MAFF and PXO 99^{A} provided clear evidence of the involvement of homologous recombination among transcription activator like (TAL) effector genes in generating differences in their structure and copy number at genomic locations that were otherwise conserved, indicating that the sequences themselves play a major role in generating that diversity.

The analysis of sequenced genomes of *Xanthomonas* species reveals that the genus as a whole shows a high degree of genomic variation. Notably, *Xoo* strains contain the greatest number and diversity of IS elements of all the sequenced xanthomonads. The genome sequence of PXO 99A and its comparison with those of MAFF and KACC have clearly shown that the *Xoo* genome is highly plastic and rapidly evolving. These findings help to explain the extraordinary diversity of *Xoo* genotypes and development of a large number of races isolated from different parts of the world.

The development of microbial genomic poses a new challenge for plant pathologists to make the best use of this available information. They should apply this information most meticulously for reducing threats and losses from plant pathogenic bacteria and also improve the viable use of beneficial bacteria. The limited number of effective chemicals and sources of disease resistance compel to think and evolve other strategies to control bacterial diseases of plants. Plant virologists have already utilized their knowledge of complete plant viral genomes to develop transgenic virus-resistant plants. The efforts should also be made to do the same with bacterial genomics. The possibilities to develop methods of disease control in a manner analogous to pharmaceutical approaches for veterinary and clinical medicine should also be explored. Knowing the complete genome sequences of many plant pathogenic bacteria may allow us to identify the Achilles' heel of a plant pathogenic bacterium.

REFERENCES

Barash, I. and Manulis-Sasson, S. (2009). Recent evolution of bacterial pathogens: The gall-forming *Pantoea agglomerans* case. *Annu. Rev. Phytopathol.* 47: 133 - 52.

Basim, H., Stall, R.E., Minsavage, G.V. *et al.* (1999). Chromosomal gene transfer by conjugation in the plant pathogen *Xanthomonas axonopodis* pv. *vesicatoria*. *Phytopathology* 89: 1044 - 49.

Bertolla, F., Frostegard, A., Brito, B. *et al.* (1999). During infection of its host, the plant pathogen *Ralstonia solanacearum* naturally develops a state of competence and exchanges genetic material. *Mol. Plant-Microbe Interact.* 12: 467 - 72

Castillo, J.A. and Greenberg, J.T. (2007). Evolutionary dynamics of *Ralstonia solanacearum*. *Appl. Environ. Microbiol.* 73: 1225 - 38.

Goss, E.M., Kreitman M. and Bergelson, J. (2005). Genetic diversity, recombination and cryptic clades in *Pseudomonas viridiflava* infecting natural populations of *Arabidopsis thaliana*. *Genetics* 169: 21 - 35.

Kers, J.A., Cameron, K.D., Joshi, M.V. *et al.* (2005). A large, mobile pathogenicity island confers plant pathogenicity on *Streptomyces* species. *Mol. Microbiol.* 55: 1025 - 33.

Lee, B.-M., Park, Y.-J., Park, D.-S. *et al.* (2005). The genome sequence of *Xanthomonas oryzae* pathovar *oryzae* KACC 10331, the bacterial blight pathogen of rice. *Nucleic Acids Res.* 33: 577 - 86.

Loria, R., Kers, J. and Joshi, M.V. (2006). Evolution of plant pathogenicity in *Streptomyces. Annu. Rev. Phytopathol.* 44: 469 - 87.

Ochiai, H., Inoue, Y., Takeya, M. *et al.* (2005). Genome sequence of *Xanthomonas oryzae* pv. *oryzae* suggests contribution of large numbers of effectors genes and insertion sequences to its race diversity. *Jpn. Agric. Res. Q.* 39 (4): 275 - 87.

Pierson III, L.S. and Ishimaru, C.A. (2000). Genomics of plant-associated bacteria : A glimpse of the future that has become reality. APSnet Feature Story, [http://www.apsnet.org/online/feature/Genomics/].

Salzberg, S.L., Sommer, D.D., Schatz, M.C. *et al.* (2008) Genome sequence and rapid evolution of the rice pathogen *Xanthomonas oryzae* pv. *oryzae* PXO 99A. *BMC Genomics* 9: 204.

Sarkar, S.F. and Guttman, D.S. (2004). Evolution of the core genome of *Pseudomonas syringae,* a highly clonal, endemic plant pathogen. *Appl. Environ. Microbiol.* 70: 1999 - 2012.

Scally, M., Schuenzel, E.L., Stouthamer, R. *et al.* (2005). Multilocus sequence type system for the plant pathogen *Xylella fastidiosa* and relative contributions of recombination and point mutation to clonal diversity. *Appl. Environ. Microbiol.* **71:** 8491 - 99.

Smith, J.M., Smith, N.H., O'Rourke, M. *et al.* (1993). How clonal are bacteria? *Proc. Natl. Acad. Sci.* USA, 90: 4384 - 88.

Sundin, G.W. (2007). Genomic insights into the contribution of phytopathogenic bacterial plasmids to the evolutionary history of their hosts. *Annu. Rev. Phytopathol.* 45: 129 - 51.

Thieme, F., Koebnik, R. and Bekel, T. (2005). Insights into genome plasticity and pathogenicity of the plant pathogenic bacterium *Xanthomonas campestris* pv. *vesicatoria* revealed by the complete genome sequence. *J. Bacteriol.* 187: 7254 - 66.

Toth, I.K., Pritchard, L. and Birch, P.R.J. (2006). Comparative genomics reveals what makes an enterobacterial plant pathogen. *Annu. Rev. Phytopathol.* 44: 305 - 36.

Vinatzer, B.A. and Yan, S. (2008). Mining the genomes of plant pathogenic bacteria. How not to down into gigabases of sequence. *Mol. Plant Pathol.* 9: 105 - 18.

Wirth, T., Falush, D., Lan, R. *et al.* (2006). Sex and virulence in *Escherichia coli*: An evolutionary perspective. *Mol. Microbiol.* 60: 1136 - 51.

Yan, S., Liu, H., Mohr, T.J. *et al.* (2008). Role of recombination in the evolution of model plant pathogen *Pseudomonas syringae* pv. *tomato* DC3000, a very atypical tomato strain. *Appl. Environ. Microbiol.* 74: 3171 - 81.

Chapter 7

FASTIDIOUS PHYTOPATHOGENIC PROCARYOTES: THEIR CULTIVATION AND CHARACTERIZATION, AND PLANT DISEASES CAUSED BY THEM

7.1. FASTIDIOUS PHYTOPATHOGENIC PROCARYOTES

Most of the phytopathogenic procaryotes are capable of growing on the routine culture media. However, a few of them are fastidious in their cultural requirements. These include spiroplasmas, phytoplasmas, *Ralstonia syzygii, Leifsonia xyli* subspp. *xyli* and *cynodontis, Xylella fastidiosa* subspp. *fastidiosa* and *multiplex*, *Serratia marcescens*, some pathovars of *Curtobacterium flaccumfaciens* and a few unidentified bacteria. These bacteria colonize vascular tissue of their host plants. The vascular-colonizing fastidious phytopathogenic procaryotes/bacteria can be divided into following three categories.

1. Wall-less mollicutes (phytoplasmas and spiroplasmas)
2. Walled phloem-inhabiting bacteria
3. Walled xylem-limited bacteria

Until 1967, yellows type plant diseases were thought to be caused by plant viruses. An inability to isolate any microbe from the infected plants, the ability of the causal agents to pass through filters which retained bacteria and the transmission of these agents by grafting and/or by insect vectors led support to the fact that the plant yellows are caused by viruses. Doi *et al.* (1967) reported the presence of **mycoplasma-like organisms** (MLOs) in the phloem elements of plants infected with mulberry dwarf, potato witches' broom and aster yellows. From 1967 to 1994, these microorganisms were called mycoplasma-like organisms due to their superficial similarity to that group.

7.2. PHYTOPLASMA

In 1994, the trivial name 'phytoplasma' was adopted by the Phytoplasma Working Team at the 10th Congress of International Organization of Mycoplasmology, replacing the term "mycoplasma-like organisms" (ICSB, 1993, 1997). After establishing the etiology of the mulberry dwarf, many other diseases considered to be of viral etiology but in which no viral agent was found, were shown to be caused by phytoplasmas. The phytoplasmas has not been cultured so far and it has limited their characterization. However, availability of recent molecular techniques has resulted in a rapid understanding of phytoplasmas leading to the establishment of a "candidate" genus, ***'Candidatus* Phytoplasma'.**

7.3. SPIROPLASMAS

Corn stunt is an important disease in topics and subtropics. Investigations by researchers on this disease around 1970, led to the discovery of another related group of plant pathogens. Davis and Worley found tiny, helical cells lacking cell wall in the filtrate of stunted corn plants. They later named these microbes as spiroplasmas. After the discovery of *Spiroplasma kunkelii,* the causal agent of corn stunt, the spiroplasmas were implicated in other important plant diseases. The spiroplasma associated with citrus stubborn disease was designated as *Sp. citri*. The same species was later isolated from horseradish plants exhibiting symptoms of "brittleroot" disease. The characteristic symptoms of the diseases are stunting, phloem necrosis and fragility of roots from which the name of the disease is derived. The third phytopathogenic spiroplasma, *Sp. phoeniceum*, was isolated from periwinkle plants, planted as traps for *Sp. citri* among the trees in a Syrian citrus grove. Besides phytopathogenic spiroplasmas, spiroplasmas occur as epiphytes and pathogens of animals including humans.

Mycoplasmas, spiroplasmas, phytoplasmas and their relatives are procaryotes lacking cell walls and they belong to class Mollicutes (molli-soft and cute-skin). They are the smallest, simplest and self-replicating cellular organisms and measure 0.3 - 1.2 μm in diameter. They lack a rigid cell wall and are bounded only by a plasmalemma; hence they are pleomorphic in shape. Mycoplasmas and spiroplasmas can be cultured on artificial media but their inability to grow on most routine culture media and the requirements for sterols and other unusual ingredients in growth media for many of them, has led to call them fastidious. The mollicutes have the smallest known genomes (10^{8-9} daltons) and this characteristic has prompted the scientists to opt favourably for sequencing their genomes. Bermuda grass phytoplasma has the smallest genome of 530 kb in contrast to about 4600 kb genome of *Escherichia coli.* The mollicutes also have a low G + C content ranging from 23 - 41 mol%. Spiroplasmas, as the name indicates, are usually helical in shape and measure 0.08 - 0.2 μm in width and 2 - 4 μm in length. When grown in liquid media and viewed using dark field microscopy, spiroplasmas exhibit a distinct helical morphology and motility that allows them to move through liquid media. No apparent flagella or other appendages have been observed that could explain how they move through liquid media. When present in their insect vectors, phytopathogenic spiroplasmas may lose their helical shape.

Phytoplasmas and spiroplasmas cause large number of plant diseases collectively called **"yellows"**. According to Calavan and Oldfield (1979) more than 600 plant diseases, infecting several hundred plant species, are caused by them. On the basis of symptoms, the yellows are divided into two broad categories. In the first category, caused by spiroplasmas and some phytoplasmas, the predominant symptoms are foliar chlorosis, stunting, wilting, die-back, phloem necrosis and collapse of the sieve tubes and companion cells leading to plant decline and death. These symptoms result mainly from phloem dysfunction and possibly also due to the competition between the pathogen and the host plant for the photosynthates present in the phloem. In the second category, caused by another group of phytoplasma, imbalance of plant growth regulators causes deformities and overgrowths of the host tissues. The overgrowths include early development and proliferation of axillary shoots leading to witches' broom, bunchy top growth and swollen veins. Flower deformities consist of partial or complete sterility, phyllody, virescence, asymmetrical shape and colour-break. Economic losses result from reduction in product quality or partial to complete loss in yield (Fletcher and Wayadande, 2002).

For an effective management of a disease, proper identification of the disease and an accurate etiology are very important. Symptomatology can be effective in identifying certain diseases showing characteristic external symptoms such as stunting and organ deformities associated with internal symptoms like phloem discolouration and/or necrosis. More definite method to identify a disease is the isolation/detection of the causal agent and its identification, which can be achieved with spiroplasmas but not with phytoplasmas. In the 1970s and 1980s, convincing evidences for association of many phytoplasmas with plant diseases were obtained using electron microscopy. The phytoplasmas were present in high numbers in sieve elements of the symptomatic plants but absent in healthy plants. Light and fluorescence microscopies were also used for this purpose. In the light microscopy, the staining of sections of infected plant tissues with Dienes' stain colours the mollicute-colonized sieve tubes. In the fluorescence microscopy, nucleic acid stain, DAPI is used to stain the plant tissue. Later on, with the availability of antisera to a number of phytopathogenic mollicutes, ELISA and imunoblotting techniques were successfully used. Currently, molecular tagging methods, i.e., DNA probes and primers of varying sensitivity and specificity are often used. In case of phytoplasmas, these methods are often supplemented with RFLP pattern analysis to know their phylogenetic placement and to determine the relationship of the causal agent with the other known mollicutes.

Spiroplasmas are transmitted by leafhoppers, planthoppers and psyllids. These mollicutes multiply in their insect vectors (Fletcher *et al.*, 1998). After being acquired by the feeding insect during phloem ingestion, they move into the body cavity via the midgut regions of the intestine. The mollicutes multiply in the gut epithelial cells and they move across the plasmalemma and basal lamina into the haemocoel. Using haemolymph as a multiplication and transport medium, the pathogens migrate to the salivary glands, and from there they are ejected with saliva into phloem during subsequent phloem probing. Depending upon the titre of the infected host and duration of the acquisition period, the entire transmission process may take 2 - 4 weeks. The transmission is not transovarial (Fletcher and Wayadande, 2002).

Corn leafhopper, *Dalbulus maidis* is the main vector of *Sp. kunkelii*. Transmission efficiency of this vector is quite high, which can go as high as 90 – 100%. *D. elimatus* is another vector of this pathogen. *Sp. citri*, the cause of citrus stubborn is found in all major citrus growing regions of the world. In the United States, beet leafhopper, *Circulifer tenellus* is the main vector of this pathogen while in Europe, *C. congeners* and *Scaphitopius* spp. are important field vectors. *C. tenellus* has a wide host range, which probably may be the reason why *Sp. citri* infects large number of host plants belonging to different families. The additional information on citrus stubborn is given under the description of this disease.

Leafhoppers are the main vectors of phytoplasmas but other insects like cixiid planthoppers and psyllids transmit causal agents of coconut lethal yellowing and pear decline, respectively. *D. maidis*, vector of maize bushy stunt phytoplasma has 60 - 80% transmission efficiency. Other leafhoppers, *Macrosteles quadrilineatus* and *Scaphoides titanus* are efficient vectors of the aster yellows and flavescence doree phytoplasmas, respectively. The higher survival rate of insects carrying the maize bushy stunt or aster yellow phytoplasmas than those of mollicute-free insects, especially at lower temperatures, shows that phytoplasma infection may be beneficial to the vector insects. It is possible that phytoplasmas may have evolved first in insects and then moved to the plant hosts.

The following strategies can be adopted for the management of plant diseases caused by wall-less mollicutes.

1. Use of resistant varieties is the best option but these are available for a very few host plants. Maylayan dwarf palm is

recommended against coconut lethal yellowing and ash genotypes tolerant to ash yellow against ash yellow.

2. Tetracyclines are effective against these pathogens but these are not recommended for food plants because of high cost and pesticide residues in food products. However, tetracycline relatives have been used to preserve specimen trees and shrubs in landscapes.
3. Meristem tip culture and heat or CO_2 therapies are effective in eliminating these pathogens from certain plant materials.
4. Control of insect vectors with pesticides and these should be applied before the arrival of the insect in the field.

7.4. FASTIDIOUS PHLOEM-COLONIZING BACTERIA

Some of the plant diseases caused by walled-bacteria also posed a problem in their identification. Inability to isolate the causal agents of these diseases often led to believe, first that they are caused by viruses and after the discovery of the phytoplasmas, to be caused by them. Phloem-inhabiting walled-bacteria are generally small, rod-shaped, Gram-negative bacteria. The outer membrane of some of these bacteria is wavy or rippled in appearance, due to which they were earlier referred to **"Rickettsia-like organisms"** (RLOs). They were also called rickettsia-like because they resemble in their morphology, ultra structure, insect transmission and fastidious nature to those of human pathogens in the Rickettsiaceae, but this resemblance is only superficial in most cases. However, at least one of these fastidious procaryotes, i.e., papaya bunchy top bacterium, is truly rickettsia-like. It is related to members of the α-1 subgroup of the *Proteobacteria* in the genus *Rickettsia* on the basis of the DNA sequence homology of several different genes, including 16S ribosomal RNA gene (Davis, 2001a). Due to the advent of physical and molecular tools like electron microscopy, PCR and 16S rDNA sequences analysis, the number of known diseases caused by phloem-inhabiting walled-bacteria have increased considerably. Davis (2001), in his review listed 24 diseases infecting vegetables, fruits, grain crops, ornamental plants and trees. These have been reported from different regions including the Unites States, Canada, South America, the Caribbean, Asia, Africa, Europe and the Middle East countries.

Among these pathogens, the causal agents of citrus huanglongbing (greening disease) are of great economic importance. The other pathogens included in this group are the causal agents of papaya bunchy top, yellow vine of cucurbits and clover club leaf. '*Candidatus* Liberibacter asiaticus', '*Ca.* Li. africanus' and '*Ca.* Li. americanus', the causal agents of Asian, African and American citrus huanglongbing, respectively have been classified and belong to alpha-2 subgroup of *Proteobacteria*. In 200(

another Liberibacter has been found infecting an ornamental rutaceous tree, cape-chestnut (*Calodendrum capense*) in South Africa. This has been classified as '*Ca.* Li. africanus subsp. capensis'. The symptoms, host range, dissemination and management approaches of citrus huanglongbing pathogens are given under the description of the disease in chapter 13 (13.44). The bacterium causing papaya bunchy top falls into K-1 subgroup of the same family. It is a rod-shaped bacterium, measures 0.25 - 0.35 × 0.8 - 1.6 μm and is surrounded by a Gram-negative type cell wall that apparently lacks a peptidoglycan layer in the periplasmic space. The pathogen of clover club leaf has not yet been classified.

Another economically important pathogen of this group is *Se. marcescens,* which causes yellow vine of cucurbits. It can be readily cultured on common bacteriological media. The strains isolated from cucurbits differ significantly from other strains of this bacterium. *Se. marcescens* is generally found in food products, stored wastes, water bodies and soil as an innocuous saprophyte, but may cause illness in immunocompromised vertebrates, including humans and insects species. Strains of *Se. marcescens* also act as plant growth-promoting rhizobacteria while inhabiting the rhizosphere of plants. Due to their antifungal properties, certain *Se. marcescens* strains are also being evaluated as potential biocontrol agents against plant pathogens. All the above-mentioned phloem-inhabiting pathogens, except *Se. marcescens* and '*Ca.* Liberibacter spp.', have not yet been cultured. Sechler *et al.* (2009) have reported for the first time the culturing and pathogenicity of '*Ca.* Li. asiaticus' and '*Ca.* Li. americanus' associated with citrus huanglongbing. Two strains of '*Ca.* Li. asiaticus' and one of '*Ca.* Li. americanus', grown on a medium (designed by them) called Liber A, proved pathogenic to citrus and these strains were isolated from noninoculated tissues of inoculated trees and seedlings 9 and 2 months later, respectively. The identity of the bacteria was confirmed by RT-PCR and 16S rDNA sequencing.

Until the 1980s, clover club leaf was the best-studied disease among the diseases caused by phloem-inhabiting bacteria. The name of the disease is based on the delayed opening of young leaflets, as it happens in crimson clover (*Trifolium incarnatum*), which remain curled in a cup-like shape. Artificially, the pathogen infects a large number of host plants belonging to at least 7 families and the additional symptoms often include a slight floral virescence. The chain of cells ("thalli") having undulatory motility, distinctive of this group, are found in the extracts of infected plants. Yellow vine of cucurbits was discovered in late 1980s. It causes considerable losses to watermelon, pumpkin, squash and cantaloupe but has not been found infecting cucumber in the field. Infected plants turn yellow, wilt and often show a rapid total plant collapse. Phloem necrosis of affected plants is evident as a honey-brown discolouration. The disease was originally found in southern Oklahoma and northern Texas but now it occurs in several other midwestern states and in East in Tennessee.

Leafhopper, *Empoasca stevensii* transmits the papaya bunchy top bacterium but the pathogen-vector relationship is not well worked out. Clover club leaf bacterium is transmitted by a leafhopper, *Agalliopsis novella* and the bacterium also multiplies inside the leafhopper. Squash bug, *Anasa tristis*, a true bug, acts as a vector of *Se. marcescens*. This relationship seems to be unusual because true bugs generally do not act as vectors of phloem-associated pathogens. Squash bugs acquire the bacterium after 24 hours of feeding and transmit it in 1 - 21 days after acquisition. The bacterium is present in the haemolymph of some but not all the transmitters, indicating a non-circulative relation of the pathogen and its vector.

The reduction in pathogen dissemination by the use of vector-resistant varieties is the best approach for long term control of these pathogens. However, in case of heavy vector infestation, the application of insecticide is often the best alternative. The use of parasitoids and predators, if effective, should be preferred over the use of insecticides. Use of resistant varieties of watermelon, exclusion of the squash bug from susceptible plants and use of trap crops to prevent or reduce squash bug infestation are some other options for the control of yellow vine of cucurbits.

7.5. FASTIDIOUS XYLEM-LIMITED BACTERIA (XLB)

Walled bacterial pathogens also colonize the water transporting cells of the xylem of their host plants. Earlier these diseases were thought to be caused by viruses and later on by the RLOs due to the superficial similarities of the rippled cell walls of these pathogens with those of RLOs. Although these bacteria are termed fastidious, many of them have been grown in culture. Important pathogens included in this group are *L. xyli* subspp. *xyli* and *cynodontis*, different strains and subspecies of *Xyl. fastidiosa, Ral. syzygii* and several pathovars of *Cur. flaccumfaciens*. These pathovars of *Cu. flaccumfaciens* cause vascular diseases of beets, beans and soybean, tulips and poinsettias (Davis and Vidaver, 2001).

Among the XLB, the best-studied pathogen is *Xyl. fastidiosa*. It has a wide host range and causes considerable economic losses to its most host plants. Besides causing different diseases, *Xyl. fastidiosa* colonizes numerous plant species without causing symptoms or any noticeable stress. Studies during 1950s demonstrated that more than 100 plant species harboured *Xyl. fastidiosa*. The presence of the pathogen in the symptomless plants was demonstrated through its vector transmission to grapevine or alfalfa indicator plants. *In vitro* culturing of the pathogen and use of serological assays also expanded the list of symptomless plants.

Xyl. fastidiosa strains are difficult to isolate and grow in axenic culture. An important factor for the successful isolation is the selection of the host tissue for isolation that contains the largest population of the bacterium. These bacteria do not grow on most common bacteriological media but require specialized media. Growth is slow even on these specialized media with doubling times of nine hours to two days (Hopkins, 2001). Different strains and subspecies of *Xyl. fastidiosa* cause different diseases, namely, Pierce's disease of grapevines, Phony peach disease, leaf scorches of oleander, almond, pear and several forest trees, leaf scald of plum and coffee, citrus variegated chlorosis etc. In 1955, Hopkins and associates suggested the causal role of *Xyl. fastidiosa* in citrus blight. Among the different hosts attacked by the bacterium, maximum economic loss has been caused to citrus and grapes.

Different immunodiagnostic techniques such as fluorescent antibody, immunogold labeling and ELISA have been used to detect *Xyl. fastidiosa* in insect vectors and plant hosts. Molecular techniques for detection include DNA-DNA hybridization, rep-PCR, RAPD-PCR and RFLPs. PCR can detect less than 100 cells per sample and this number is similar to thresh hold values obtained by *in vitro* culture. PCR primers specific to *Xyl. fastidiosa* strains that cause citrus variegated chlorosis have been developed using cloned and sequence-specific RAPPD-PCR products. The main drawback of the PCR is that it does not differentiate between living and dead cells (Purcell and Hopkins, 1996)

Xyl. fastidiosa is the first plant pathogenic bacterium for which the complete genome has been sequenced. *Xyl. fastidiosa* is a heterogeneous species and consists of different strains. Schaad *et al.* (2004) proposed three new subspecies, namely, *Xyl. fastidiosa* subsp. *piercei*, *Xyl. fastidiosa* subsp. *multiplex* and *Xyl. fastidiosa* subsp. *pauca*. As the type strain designated for subsp. *piercei* is the type strain of the species, the name *Xyl. fastidiosa* subsp. *fastidiosa* takes priority over *Xyl. fastidiosa* subsp. *piercei*. The authors (Schaad *et al.,* 2004a) have also provided a correction for this. In their collection of isolates (Schaad *et al.,* 2004), *Xyl. fastidiosa* subsp. *fastidiosa* represented strains from grapevine, alfalfa, almond and maple, *Xyl. fastidiosa* subsp. *multiplex* from peach, elm, plum, pigeon grape, scyamore and almond, and *Xyl. fastidiosa* subsp. *pauca* from citrus only. *Xyl. fastidiosa* subsp. *pauca* has not been validated due to the non-deposition of a culture of its type strain (Young *et al.*, 2004).

Xyl. fastidiosa is transmitted by xylem-feeding homopterans. Important vectors include members of the leafhopper sub-family Cicadellinae (sharpshooters) and of the spittle bug family Cercopidae. Much progress has been made in understanding the process by which it spreads within the xylem vessels of susceptible plants as well as the traits that contribute to its acquisition and transmission by sharpshooter vectors (Chatterjee *et al.,* 2008). *Xyl. fastidiosa* cells ingested by leafhopper sharpshooters colonize the food canal within the straw-like stylets, which are used to pierce the plant tissues. The bacteria attached to the insect cuticle grow into biofilms that line the inner surface of the food canal. After acquiring the bacterium, both adults and nymphs can transmit the bacterium immediately but nymphs are unable to transmit it after molting. However, an adult after acquisition can transmit the bacterium from several months to its total life span. The symptoms of Pierce's disease, its management and insect transmission are given under the description of the disease in chapter 13 (13.48).

The symptoms of citrus variegated chlorosis, caused by *Xyl. fastidiosa,* are found on twigs, leaves and fruits. The symptoms are more pronounced on 3- to 6-year-old trees and mainly on sweet orange cultivars. The characteristic symptoms are small, interveinal chlorotic yellow spots, resembling zinc deficiency on upper side of mature leaves, which correspond to slightly raised, light to dark brown, gummy lesions on the underside of the leaves. Fruits of affected trees are much smaller than normal, very firm and ripen earlier. The growth rate of infected plants is greatly reduced and twigs and branches may wilt but usually the plants do not die.

In coffee leaf scorch, the symptoms appear as large marginal and apical scorched areas on recently matured leaves. The symptoms generally appear on young flushes of plants. Symptoms are more pronounced during winter, particularly during water stress. Fruit size and yield are considerably reduced. Plant death occurs after several years of infection.

L. xyli subsp. *xyli* (syn. *Clavibacter xyli* subsp. *xyli*) and *L. xyli* subsp. *cynodontis* (syn. *Cla. xyli* subsp. *cynodontis*) are two other important plant pathogens of this group. *L. xyli* subsp. *xyli*, the causal agent of ratoon stunting disease of sugarcane is economically more important than *L. xyli* subsp. *cynodontis*, the causal agent of Bermuda grass stunting disease. Immunogold labeling coupled with electron microscopy has been applied to detect *L. xyli* subsp. *xyli* in the infected host plants. Tissue blotting by immunostaining of bacterial cells on filters to assess sugarcane for colonization by this bacterium was faster and more sensitive than ELISA. The details of this pathogen and other relevant information is available under the description of ratoon stunting disease of sugarcane in chapter 13 (13.9).

L. xyli subsp. *cynodontis* infects maize plants systemically and develops large populations of cells without causing any visible symptoms. It also affects many other plants if mechanically inoculated. The lack of development of symptoms in maize plants systemically colonized by this bacterium led to the development of its genetically transformed strains containing *Cry*IA(*c*) gene from *Bacillus thuringiensis* to control insects which feed on xylem tissue. (Lampel *et al.,* 1994). Colonization of maize seed or seedlings by its recombinant strains reduced tunneling by the European corn borer by 85% but did not significantly increase the yield. *Ral. syzygii* causes Sumatra disease of cloves. Its presence in xylem cells have been confirmed on the basis of microscopic examination of diseased plants. Epidemic clues suggest that it may have originated from forest plants.

REFERENCES

Calavan, E.C. and Oldfield, G.N. (1979). Symptomatology of spiroplasmal plant diseases. In: R.F. Whitcomb and J.G. Tully (eds). *Mycoplasmas.* Vol. III, Academic Press, Inc., New York, pp. 37 - 64.

Chatterjee, S., Almeida, R.P.P. and Lindow, S. (2008). Living in two worlds: The plant and insect life styles of *Xylella fastidiosa. Annu. Rev. Phytopathol.* 46: 243 - 71.

Davis, M.J. (2001). Fastidious phloem-limited bacteria. In: N.W. Schaad, J.B. Jones and W. Chun (eds). *Laboratory Manual for the Identification of Plant Pathogenic Bacteria.* 3rd Ed. APS Press, Inc., St. Paul, Minnesota, pp. 275 - 82.

Davis, M.J. (2001a). Fastidious phloem-limited bacteria. In: O.C. Maloy and T.D. Murray (eds). *Encyclopedia of Plant Pathology.* Vol. 1, John Wiley and Sons, Inc., New York, pp. 449 - 52.

Davis, M.J. and Vidaver, A.K. (2001). Coryneform plant pathogens. In: N.W. Schaad, J.B. Jones and W. Chun (eds). *Laboratory Manual for the Identification of Plant Pathogenic Bacteria.* 3rd Ed. APS Press, Inc., St. Paul, Minnesota, pp. 218 - 35.

Doi, Y.M., Teranaka, M., Yora, K. *et al.* (1967). Mycoplasma- or PLT group-like microorganisms found in the phloem elements of plants infected with mulberry dwarf, potato witches' broom, aster yellows, or paulownia witches' broom. *Ann. Phytopathol. Soc. Jpn.* 33: 259 - 66.

Fletcher, J. and Wayadande, A. (2002). Fastidious vascular-colonizing bacteria. APSnet Education Center, [http://www.apsnet.org/education/introplantpath/pathogengroups/fastidious/].

Fletcher, J., Wayadande, A., Melcher, U. *et al.* (1998). The phytopathogenic mollicute-vector interface: A closer look. *Phytopathology* 88: 1351 - 58.

Hopkins, D.L. (1989). *Xylella fastidiosa*: a xylem-limited bacterial pathogen of plants. *Annu. Rev. Phytopathol.* 27: 271 - 90.

Hopkins, D.L. (2001). Xylem-limited bacteria. In: O.C. Maloy and T.D. Murray (eds). *Encyclopedia of Plant Pathology.* Vol. 2, John Wiley and Sons, Inc., New York, pp. 1212 - 13.

Int. Comm. Syst. Bacteriol., Subcomm. Taxon. Mollicutes. (1993). Minutes of the interim meetings, 1 and 2 August 1992, Ames, Iowa. *Int. J. Syst. Bacteriol.* 43: 394 - 97.

Int. Comm. Syst. Bacteriol., Subcomm. Taxon. Mollicutes. (1997). Minutes of the interim meetings, 12 and 18 July 1996, Orlando, Florida, USA. *Int. J. Syst. Bacteriol.* 47: 911 - 14.

Lampel. J.S., Canter, G.L., Dimock, M.B. *et al.* (1994). Integrative cloning, expression and stability of the *Cry*IA(*c*) gene from *Bacillus thuringiensis* subsp. *kurstaki* in a recombinant strain of *Clavibacter xyli* subsp. *cynodontis*. *Appl. Environ. Microbiol.* 60: 501 - 08.

Lee, I.-M., Davis, R.E. and Gundersen-Rindal, D.E. (2000). Phytoplasma: Phytopathogenic mollicutes. *Annu. Rev. Microbiol.* 54: 221 - 55.

Purcell, A.H. (2006). *Xylella fastidiosa*: A scientific and community internet resource on plant disease caused by the bacterium *Xylella fastidiosa*. Online [http://nature.berkeley.edu/xylella/].

Purcell, A.H. and Hopkins, D.L. (1996). Fastidious xylem-limited bacterial plant pathogens. *Annu. Rev. Phytopathol.* 34: 131 - 51.

Schaad, N.W., Postnikova, E., Lacy, G. *et al.* (2004). *Xylella fastidiosa* subspecies: *X. fastidiosa* subsp. *piercei* subsp. nov., *X. fastidiosa* subsp. *multiplex* subsp. nov., and *X. fastidiosa* subsp. *pauca* subsp. nov. *Syst. Appl. Microbiol.* 27: 290 - 300.

Schaad, N.W., Postnikova, E., Lacy, G. *et al.* (2004a). Erratum. *Xylella fastidiosa* subspecies: *X. fastidiosa* subsp. *piercei* subsp. nov., *X. fastidiosa* subsp. *multiplex* subsp. nov., and *X. fastidiosa* subsp. *pauca* subsp. nov. *Syst. Appl. Microbiol.* 27: 763.

Sechler, A., Schuenzel, E.L., Cooke, P. *et al.* (2009). Cultivation of '*Candidatus* Liberibacter asiaticus', '*Ca.* L. africanus' and '*Ca.* L. americanus' associated with huanglongbing. *Phytopathology* 99: 480 – 86.

Young, J.M., Bull, C.T., De Boer S.H. *et al.* (2004). Names of plant pathogenic bacteria, 1864 - 2004. Online [http://www.isppweb.org/names_bacterial_revised.asp].

Chapter 8

INTERACTION AMONG PLANT PATHOGENIC BACTERIA AND OTHER MICROORGANISMS

8.1. Soil inhabitants/invaders
8.2. Plant growth-promoting rhizobacteria
8.3. Phylloplane
8.4. Endophytes
8.5. Interaction between fungal and bacterial plant pathogens
8.6. Interaction between bacterial plant pathogens
8.7. Interaction between plant pathogenic bacteria and nematodes
8.8. Interaction between bacterial and viral plant pathogens

Plant pathogenic bacteria inhabit diverse habitats. They are present in soil where they live as free-living bacteria or associated with plant disease debris. Plant pathogenic bacteria also inhabit plants where they can occur as pathogens causing various diseases, as epiphytes or endophytes. Many plant pathogenic bacteria are associated with seeds; the association may be external infestation or internal infection or may be both. In majority of these habitats, they live in association with other microorganisms, namely, bacteria, filamentous fungi, yeasts, viruses, nematodes, protozoa, algae etc. The presence of these microorganisms influences the survival, multiplication and disease causing ability of the phytopathogenic bacteria. The interaction among these microorganisms may result in variation in symptoms produced, length of incubation period, rate of disease development and disease severity or even suppression of disease.

Soil is a complex matrix in which numerous types of microorganisms, including plant pathogenic bacteria, survive and multiply. These microorganisms are subjected to external stresses such as temperature and moisture extremes. They are also in constant competition for nutrients, subject to predation and parasitism and exposed to toxic chemicals. Different soil types, climatic regions, seasons and plant species grown also influence the type of microorganisms and their population density. Exudates and lysates released by the plant roots greatly influence the microbial population in the rhizosphere. In contrast to phyllosphere, rhizosphere is comparatively less affected by UV light, dew, rain and variation in temperature.

8.1. SOIL INHABITANTS/INVADERS

Some of the phytopathogenic procaryotes are soil inhabitants while most of them are soil invaders. Perhaps the most truly soil-inhabiting phytopathogenic procaryotes are *Streptomyces* spp., the causal agents of scabs of potato and other hosts. These species are found worldwide in arable lands. *Pseudomonas marginalis* and *Clostridium* spp., opportunistic bacteria associated with soft-rotting diseases are also common in many soils. *Agrobacterium tumefaciens* and *Ralstonia solanacearum* are other plant pathogens endemic in soil. *Pectobacterium carotovorum* subsp. *carotovorum*, earlier considered to be a soil bacterium, has a poor surviving ability in soil. Most other plant pathogenic bacteria are generally considered as soil invaders because they have poor competitive saprophytic ability. They are able to survive in soil mainly in association with diseased plant debris. The survival in association with plant debris is generally longer under cold and dry conditions. Under wet conditions, microorganisms cause rapid decomposition of the plant debris, thereby reducing the survival period of phytopathogenic bacteria. In addition to entering the soil in association with diseased host tissue, soil invaders may also be introduced into the soil as free cells released by the diseased plants or after the decomposition of the infected tissue. The survival of these free cells in soil may vary from a few days to several months, governed mainly by soil conditions. *Psm. savastanoi* pv. *glycinea* and *Xanthomonas phaseoli* survived for less than 3 months in field soil at 25°C while they survived for 6 - 9 months at 5°C.

Most plant pathogenic bacteria survive for longer periods of time in sterile than non-sterile soil. This clearly indicates that other microorganisms present in soil influence directly or indirectly the survival of plant pathogenic bacteria in soil. Various types of interactions occur among plant pathogenic bacteria and soil microorganisms. The production of bacteriocins by *Ag. radiobacter* strains K 84 and K 1026 and the use of these strains for controlling crown gall caused by *Ag. tumefaciens* is well documented. The commercial formulations of these strains for biological control of crown gall are available.

Many soil microorganisms, particularly actinomycetes, produce a variety of antibiotics. Plant pathogenic bacteria vary greatly in their relative sensitively to antibiotics produced by soil microorganisms. Plant pathogens like xanthomonads that survive poorly in soil tend to be more sensitive to these antibiotics than those that survive for longer periods of time in soil. *Psm. putida* M 17 produces an antibiotic inhibitory to *Pectobacterium* spp., capable of controlling soft rot of potato tubers but an antibiotic-negative mutant of M 74 is less effective. Many strains of *Psm. aeruginosa* are cyanogenic and inhibit the growth of many fungal and bacterial plant pathogens with the production of hydrogen cyanide. Strong cyanogenic strains are capable of completely inhibiting the growth of *Ag. tumefaciens*, *Ral. solanacearum*, *Pe. chrysanthemi* and *Xa. vesicatoria*.

Some free-living nematodes feed on bacteria and their number may go up to 2×10^7 individuals/m^2 in soil, capable of consuming 800 kg of bacterial cells/ha/year. In a laboratory experiment, the nematode *Plectus parietinus* ingested 5000 cells/min when allowed to feed on an *Acinetobacter* sp., while another nematode, *Pristionchus lheritieri* also thrived on pure cultures of *Erwinia amylovora, Pe. carotovorum* subsp. *carotovorum* and *Psm. syringae* pv. *phaseolicola*. Protozoa also feed on bacteria in soil and according to an estimate ingest about 9×10^{14} cells/m^2/year. The decline in populations of *Xa. campestris* pv. *campestris* and rhizobia in field soil due to protozoa has been reported. The information of amoeba feeding on a *Pseudomonas* sp. on pea roots is also available in the literature (De Boer, 1982).

8.2. PLANT GROWTH-PROMOTING RHIZOBACTERIA

Bacillus spp. and *Pseudomonas* spp., particularly *Psm. fluorescens* and *Psm. putida* are the common inhabitants of soil. These bacteria are called plant growth-promoting rhizobacteria (PGPR) because they promote plant growth by suppressing the growth of major and minor soil borne plant pathogens and also of the nonparasitizing deleterious rhizosphere microorganisms. In some cases, PGPR also directly promote the plant growth.

Some of the rhizosphere bacteria or rhizobacteria are capable of inducing **systemic acquired resistance** against a wide variety of plant pathogens. Rhizosphere-mediated **induced systemic resistance** (ISR) has been demonstrated against fungi, bacteria and viruses in *Arabidopsis,* bean, carnation, cucumber, radish, tobacco, and tomato. Bacterial strains differ in their ability to induce resistance in different plant species, and plants also show variation in the expression of ISR upon induction by specific bacterial strains (Van Loon *et al.,* 1998). Induction of systemic resistance has been demonstrated by *Psm. putida* WCS 358 against *Psm. syringae* pv. *tomato* in *Arabidopsis* (bacterial speck), by *Psm. fluorescens* S 97 against *Psm. syringae* pv. *phaseolicola* in bean (halo blight), by *Psm. putida* 89B-27 and *Serratia marcescens* 90-166 against *Er. tracheiphila* in cucumber (bacterial wilt) and by *Se. marcescens* 90-166 against *Psm. syringae* pv. *tabaci* in tobacco (wildfire).

8.3. PHYLLOPLANE

The aerial surface (above-ground parts) of plants is called the **phyllosphere** or more correctly **phylloplane**. The phylloplane is colonized mainly by a variety of bacteria, yeasts and fungi, and less frequently by algae, protozoa and nematodes. However, bacteria form a characteristic component of the phylloplane microflora of the majority of plants. The number and kind of microorganisms on phylloplane vary with plant species, season, age of plants, and perhaps, most importantly, with changes in the **microclimate** at the plant surface. While there have been some investigations of the colonists of buds and flowers, most work on phylloplane microbiology has been focused on leaves, a more dominant aerial plant part. Bacteria are by far the most numerous colonists of leaves, often being found in numbers averaging 10^{6-7} cells/cm^2 (up to 10^8 cells/g) of leaf. The microorganisms inhabiting the phylloplane are called **epiphytes** and most of them live as saprophytes.

The inoculum of saprophytic bacteria, which inhabit the leaves, is believed to come from the seeds, the roots and the air. From the numerous microorganisms, which arrive on the leaf surface, only those get established which get adapted to this habitat and are able to multiply. These organisms are termed as **residents** while those which fail to get established but remain viable on leaves for only short period are termed as **casuals** (Blakeman, 1982).

The majority of bacteria found on aerial plant surfaces are Gram-negative and belong to *Erwinia, Pseudomonas, Xanthomonas, Flavobacterium* and *Pantoea*. Pigmented bacteria, which are rarely found in the rhizosphere, dominate the leaf surface. It is mainly due to the reason that they can withstand solar irradiation. *Pa. agglomerans,* a yellow-pigmented bacterium, has been isolated frequently from leaves and flowers of different plants. Strains of *Psm. fluorescens* also occur widely on leaf surfaces of various plants. Gram-positive bacteria are much less frequently found on leaf surfaces.

Some of the plant pathogenic bacteria grow as epiphytes in a nonpathogenic **"resident phase"** on their hosts, and in some cases on nonhosts also, before entering into the host. Because of the greater economic importance of this phase, there is much more information on the behaviour of these pathogens as epiphytes than that of the true saprophytes. *Psm. syringae* pv. *syringae* occurs widely as an epiphyte on its hosts, namely, bean, tomato, peach, pear and sour cherry. It also occurs as an epiphyte on various nonhosts including hairy vetch, weed grasses and broad-leafed herbs. From the hairy vetch, the bacterium spreads to the bean host by water splashes during rainstorms. *Psm. syringae* pv. *morsprunorum* also occurs as an epiphyte in association with *Psm. syringae* pv. *syringae* on the leaves of sour cherry but not on the nonhost plants.

Psm. savastanoi pv. *glycinea* grows epiphytically on soybean and epiphytic growth on primary leaves may provide the inoculum for the subsequent infection of trifoliate leaves. Epiphytic population of *Xa. phaseoli* varies with different varieties, it being more on susceptible varieties compared to resistant varieties.

Prevention of blossom blight is a key step in the management of fire blight of pome fruits, because the cankers and bacterial ooze originating from blossom infections provide much of the inoculum for other phases of the disease. Blossom blight phase of fire blight is the result of epiphytic growth of *Er. amylovora* on floral structures including stigmas, anthers and hypanthium. Stigmas have been demonstrated to be the primary site of epiphytic colonization by *Er. amylovora*. Epiphytic populations of *Er. amylovora* attain sizes of 10^{6-7} cfu/stigma. Considering the weight of a pear stigma around 0.15 mg, the carrying capacity of a pear stigma on a fresh weight basis comes to 10^{9-10} cfu/g (Johnson and Stockwell, 1998). Other plant pathogenic bacteria, which grow as epiphytes on their hosts include *Psm. syringae* pv. *lachrymans* on cumber, *Burkholderia cepacia* on onion, *Psm. syringae* pv. *morsprunorum* and *Xa. arboricola* pv. *pruni* on peach, *Xa. vesicatoria* on tomato and *Pe. carotovorum* subsp. *carotovorum* on potato.

Various types of interactions occur among microorganisms present on the leaf surface. There may be competition for space and nutrients. Reports of production of antibacterial antibiotics and bacteriocins are also available in the literature. Much of the information on interaction among bacteria on leaves has come from the investigations with the goal of exploiting such interactions to obtain biological control of plant diseases or to control frost injury caused by epiphytic bacteria.

Pa. agglomerans, a pigmented bacterium occurs as a common epiphyte on fruit trees and crop plants. Its epiphytic growth on host results in creating acidic conditions to the levels, which are unfavorable to some pathogens. *Er. amylovora* was unable to grow in pear nectar in the presence of this bacterium (Riggle and Klos, 1972). Using culture made up to stimulate nutrient conditions in the nectar, *Pa. agglomerans* consumed all the organic nitrogen and reduced the pH to levels inhibitory for *Er. amylovora*. *Pa. agglomerans* also inhibits the growth of *Xa. oryzae* pv. *oryzae* on rice leaves by creating highly acidic conditions.

Psm. fluorescens, an epiphyte produces an antibiotic capable of inhibiting the growth of *Psm. syringae* pv. *phaseolicola* both in culture and on bean plants. The antibiotic readily translocates into the seedlings. *Psm. syringae* pv. *syringae* is a widespread epiphyte on aerial plant surfaces. Most strains of this bacterium produce bacteriocins, that too having wide spectrum. Relatively few strains of *Psm. syringae* pv. *phaseolicola* produce bacteriocins having a narrow spectrum of activity (Vidaver *et al.,* 1972). Bacteriocin-like substances inhibitory to *Er. amylovora* are also produced by *Pa. agglomerans*.

The composition of epiphytic bacterial communities can be strongly manipulated by establishment of certain antagonistic microorganisms at appropriate time during plant growth. Perhaps the best-studied interactions are the achievements of biological control of plant pathogens and bacteria causing frost injury. Certain epiphytic bacteria act as ice-nucleating agents and cause frost injury on sensitive plants, such as deciduous fruit trees (on flower), sensitive herbaceous plants (on leaves) and tropical plant species. The strains of *Psm. syringae* pv. *syringae* and *Pa. agglomerans* are among the most active naturally occurring ice-nuclei bacteria, catalyzing ice formation at temperatures as warm as -1°C. Since the ice-nucleation temperature of the plants increases with increasing population sizes of ice$^+$ bacteria, a strategy of frost control is to reduce the number of ice$^+$ bacteria on these plants at the onset of cold temperatures. Fortunately, the population size of ice$^+$ bacteria on young tissues such as flowers or young leaves of most plants is very small and increases with time due to arrival of bacterial cells followed by their multiplication on plants (Lindow, 1982). Pre-emptive competitive exclusion of ice$^+$ bacteria with naturally, occurring non-ice nucleation-active bacteria has been found to be an effective and practical mean of frost control. When applied to plants before their colonization by other bacteria, such antagonists rapidly multiply and compete for limited resources, presumably carbon compounds. However, the antagonism via the production of antimicrobial compounds seems to be an uncommon phenomenon on the surface of leaves. Recombinant Ice$^-$ bacteria, the first recombinant microorganism released into the open environment in field experiments, were able to competitively exclude the Ice$^+$ bacteria (Skirvin *et al.*, 2000). Recombinant Ice$^-$ bacteria can be obtained by inactivating the gene responsible for the cell wall protein involved in crystal formation. The bacteria differ in their ability to exclude other bacteria from a leaf surface but similarity in resource requirements maximizes the competitive exclusion of immigrant bacteria. Management of frost injury by reducing population of Ice$^+$ bacteria has become an established method of forest control. For this purpose, lyophilized preparation of *Psm. fluorescens* strain A 506 are now commercially available (Blightban A506; Nufarm Americas, Inc., Sugar Land, Tex.) for spray application to a variety of crop plants (Lindow and Brandl, 2003).

Nonchemical management of fire blight disease is another great achievement of exploitation of microbial interactions, based on detailed studies of the ecology of both the pathogen and antagonists on plants (Johnson and Stockwell, 1998). Biocontrol of fire blight disease is very effective and at par with chemical control; additionally it reduces fruit russetting and frost injury to the plants. Prior colonization of the stigmatic surface of flowers with nonpathogenic bacteria such as *Psm. fluorescens* strain A 506 and *Pa. agglomerans* strains can greatly inhibit colonization by *Er. amylovora* leading to considerable reduction in disease. These antagonists also spread readily from inoculated to noninoculated flowers, thereby giving protection to these flowers. *Psm. fluorescens* strain A 506 is commercial available (Blight-Ban A 506, Plant Health Technologies, Boise, ID) for biological control of fire blight.

8.4. ENDOPHYTES

Certain microorganisms grow as endophytes inside the plants. Endophytes are microorganisms that do not visibly harm the host plant but can be isolated from surface-disinfected plant tissue or the inner parts of the plants. These microorganisms are different from those, which cause latent infection of the host but do not elicit disease symptoms under any given environments. They colonize an ecological niche similar to that of phytopathogens. This characteristic may favour them as candidates for biocontrol and can be exploited for this purpose. Available information suggests that some of the endophytic microorganisms have the capacity to control plant pathogens. In some cases, they can also accelerate seedling emergence and promote plant establishment under adverse conditions and enhance plant growth and development. Some of the endophytic bacteria commonly found associated with different plant species include *Bacillus* spp., *Enterobacter cloacae, En. aerogenes, Methylobacterium* spp. *Nocardia* spp., *Pa. agglomerans, Bu. cepacia, Psm. fluorescens, Psm. aeruginosa, Curtobacterium flaccumfaciens* etc.

Araujo *et al.* (2002) isolated *Cur. flaccumfaciens* significantly more frequently from asymptomatic plants of citrus growing in an orchard infected with citrus variegated chlorosis (CVC). As the symptomatic and asymptomatic plants were of the same genotype and both were growing under the same condition, they postulated that the association of *Cur. flaccumfaciens* with asymptomatic plants may be imparting resistance to these plants against CVC. Denaturing gradients gel electrophoresis is a very useful technique for studying population of culturable and nonculturable plant endophytes and soil microorganisms. This method provides a fingerprint of the microbial community structure of the sampled habitat in which each band of the fingerprint represents a group of bacteria having 16S rDNA sequences with similar denaturation points.

Endosymbionts are the endophytic bacteria, which contribute to the health of a plant and in turn acquire certain benefits in the plant niche. True endosymbionts differ from endophytes by virtue of the mutualistic benefits of the association. They differ from ectosymbionts by their internal association with plant cells. The classical example of an endosymbiotic relationship is that which occurs between legumes and species of *Rhizobium* and *Bradyrhizobium,* where bacteria fix atmospheric nitrogen in the nodules. Although rhizobial endosymbionts are best known for their association with the root hairs of legumes and for their ability to fix nitrogen, there are also *Rhizobium* species that infect leaves through invasion of the epidermal cells of hosts such as the aquatic legume, *Neptunia oleracea*. Endosymbionts are also found associated with non-leguminous plants. There are more than 370 species belonging to five genera included in families Myriaceae and Rubiaceae that possess nonrhizobial endosymbionts. These represent members of the enterobacteria and of the genus *Phyllobacterium,* many of which produce nodules on leaves. Recent work in several laboratories has also shown that enteric bacteria are common inhabitants of the interior of plants. The entry of enteric human pathogens such as *Salmonella* spp. *Escherichia coli* and *Es. coli* O157: H7 into plants has been demonstrated. The extent of endophytic colonization by strains is regulated by plant defences, and several genetic determinants necessary for this interior colonization in endophytic bacteria have been identified. Common virulence factors in plant and animal pathogens have also been described in bacteria that can infect both plant and animal models (Tyler and Triplett, 2008).

In nature, many times more than one pathogen infect the same plant. Therefore, various types of interactions occur among these pathogens on the surface of the plant or inside the plant tissue. These pathogens may belong to the same group or to the different groups like fungi, bacteria, viruses, nematodes etc.

8.5. INTERACTION BETWEEN FUNGAL AND BACTERIAL PLANT PATHOGENS

Simultaneous infection of fungal and bacterial pathogens occurs on plants. The severity of bacterial wilt of potato and tomato caused by *Ral. solanacearum* increases if these plants are already infected by wilt caused by *Fusarium* and *Verticillium* spp. Moreover, the root lesions caused by these fungal pathogens also serve as avenues for the entry of *Ral. solanacearum.*

Inoculation of *F. oxysporum*, *F. solani* and *V. dahliae* on the newly emerged potato plants that were inoculated before planting with *Pe. atrosepticum* significantly increases the incidence of blackleg compared to inoculation of *Pe. atrosepticum* alone. However, the inoculation of *Pe. carotovorum* subsp. *carotovorum* under similar conditions does not cause any increase in the incidence of seed tuber decay. The presence of *Bu. cepacia* in wounded citrus fruits promotes the infection by *Penicillium digitatum* because the bacterium suppresses the phenylalanine ammonia lyase activity and also the wound healing.

Most recently, Kobayashi and Crouch (2009) have reviewed the information on diverse type of bacterial-fungal interactions including biological control systems, endobiotic relationships and diseases of cultivated mushrooms, and have also emphasised the benefits of studying these systems at the molecular level. There are few descriptions of naturally occurring interactions between pathogenic bacteria and fungal hosts. However, the diseases of cultivated mushrooms caused by bacteria have received much more attention and a number of diseases and their causal agents have been described.

Mushroom blotch diseases are caused by different bacteria. Blotch symptoms typically occur on basidiocarps (or caps) and occasionally on mushroom stipes. They start as discoloured lesions, which eventually become sunken or pitted at later stages depending on the type of blotch disease. Two main blotch diseases, namely, **brown blotch** and **ginger blotch** have been described and among them, brown blotch caused by *Psm. tolaasii* has been studied in detail. These two diseases can be distinguished from each other on the basis of symptoms and physical nature of the damage caused on the basidiocarps. The main difference is in the intensity of colour produced in the symptom expression. Typical brown blotch symptoms are characterized by dark brown discolourations that ultimately develop into deep sunken lesions. In ginger blotch, the lesions have a lighter or typical ginger-coloured appearance and these lesions may show no or only mild sunken appearance. The colour of the lesions can vary with the strain and species involved. *Psm. tolaasii* produces tolaasin, a cylic lipopeptide toxin that is essential for the production of symptoms. The application of purified toxin alone produces symptoms and a mutant strain blocked in tolaasin production does not cause the disease. Besides tolaasin, other factors are also involved in blotch symptom production because nontolaasin-producing bacteria can also cause blotch disease. *Psm. reactans*, a species related to *Psm. tolaasii*, produces a related lipodepsipeptide compound, WLIP. However, neither *Psm. reactans* nor its toxin produces the severity level of the blotch disease, similar to that of produced by *Psm. tolaasii* or tolaasin. The cylic lipopeptide toxins are important contributors to blotch disease symptom production and these toxins are produced by a diverse group of bacteria. Even the strains of *Psm. tolaasii* differ in the types of cylic lipopeptide molecules they produce.

The same bacterial species may also cause different diseases of mushrooms as found in case of *Psm. agaricus*. *Psm. agaricus* is one of the bacterial species reported to be associated with causing milder blotch-like symptoms on *Agaricus bisporus* (Lo Cantore and Iacobellis, 2004). It also causes **yellow disease** of oyster mushrooms, which is characterized as yellow droplets on the surface of infected sporocarps. *Psm. agaricus* also causes another disease called **dipy gill**. The characteristic feature of the disease is bacterial ooze originating from gills of sporocarps and from longitudinal splits within the stripe where bacteria colonize host tissue both intercellularly and intracellularly. The virulence determinants of *Psm. agaricus* have not been determined.

Soft rots are another group of mushroom diseases caused by bacteria. The symptoms comprise of pitting and sticky yellowish or brownish blotches on caps leading to complete dissolution of mushrooms within a few days. *Bu. gladioli* pv. *agaricola* causes a **rapid soft rot**, previously called cavity disease, which occurs on the sporocarps and stipes of *Agaricus* spp. *Janthinobacterium agaricidamnosum* has been reported to cause symptoms similar to those of rapid soft rot on *A. bisporus*. Recently, Kim *et al.* (2007) reported a *Pantoea* sp. to cause a different kind of rot on caps and stipes of the king oyster mushroom, *Pleurotus eryngii*. In this disease, water-soaked lesions accompanied by soft rot appear on caps and stipes of the infected sporocarps. Among the soft rots, only the disease caused by *Bu. gladioli* pv. *agaricola* has been investigated to study the host-pathogen interaction. Type II secretion system (TIISS) and secreted enzymes play a role in this disease. Mutations within TIISS produce mutant strains that do not degrade mushrooms and produce significantly less or no chitinase or proteinase activity compared to wild-type parent strains. TIISS mutants are able to colonize mushroom tissue in a biofilm-like manner but are unable to degrade fungal hyphae in comparison to wild-type parent strain. *In vitro* antifungal activity of mutants reveals that type II secretions are involved in mushroom degradation but not in hyphal growth inhibition.

Internal stipe necrosis is caused by an enteric bacterium, *Ewinella americana*. The characteristic symptom of the disease is browning or severe necrosis of the central core tissue of the stipe. In more severe cases, the symptoms spread outwardly. The mechanism(s) of the pathogenesis has not been worked but the chitinolytic activity of the bacterium, attributed to a single gene, has been determined. Although the exact role of chitinase in pathogenesis is not known, the high susceptibility of stipe tissue to chitinase degradation indicates an important role of this enzyme as a virulence factor in the disease development. In **mummy disease** of mushrooms, the mushrooms infected in early stage fail to reach maturity and instead of decaying, they become mummified. A *Pseudomonas* sp. was implicated as the cause of this disease and in another independent study; intracellular colonization of mushroom spore cells by the bacteria was observed and confirmed. *Psm. aeruginosa* and *Psm. fluorescens* have also been independently implicated as causal agents of the disease but none of these investigations have been verified so far. It appears that etiology of the mummy disease has not been finally settled.

8.6. INTERACTION BETWEEN BACTERIAL PLANT PATHOGENS

In many cases, the same plant species acts as a host of more than one pathogenic bacterium and in some of these cases; the same host may be attacked by more than one bacterial pathogen at the same time. In majority of these cases, the combined infection leads to aggravate the disease situation resulting in more losses.

In crucifers, particularly the cabbage and cauliflower, combined infection of black rot (*Xa. campestris* pv. *campestris*) and stump rot (*Pectobacterium* spp.) commonly occurs resulting in considerable losses.

Bashir *et al.* (2007) reported increased disease severity due to the combined infection of *Xa. campestris* pv. *sesami* and *Psm. syringae* pv. *sesami*, the causal agents of bacterial blight and bacterial leaf spot of sesame, respectively. They found that leaf infection due to combined infection was 80.6% corresponding to the disease score of 4.03 while corresponding figure for bacterial blight was 70.6% (disease score 3.78) and for bacterial leaf spot was 50.0% (disease score 2.50). The combined infection also caused higher defoliation of 50.0% in comparison to 36.66% and 38.0% defoliation caused by bacterial blight and leaf spot, respectively. The figures for stem infection for combined infection, bacterial blight and leaf spot were 47.16, 43.16 and 26.66%, respectively.

The common blight and fuscous blight of bean caused by *Xa. phaseoli* and *Xa. fuscans* subsp. *fuscans*, respectively often occur simultaneously in many cases. As most symptoms of both the diseases are also similar, the identification of these diseases on the basis of symptoms is difficult.

8.7. INTERACTION BETWEEN PLANT PATHOGENIC BACTERIA AND NEMATODES

The association of plant parasitic nematodes and plant pathogenic bacteria is found in many plant diseases. In some of these diseases, the nematodes act as carrier or vectors of bacteria while in some cases synergistic effect on disease development is found. In yellow ear rot of wheat, *Anguina tritici*, the causal agent of ear cockle of wheat, acts as a carrier of *Rathayibacter tritici*, but it also produces nematode galls in ears if environmental conditions are favourable for its multiplication, in addition to initial symptoms of ear cockle of wheat. *Rat. toxicus* capable of potentially colonizing and producing toxin in a wide range of cereals and fodder grasses including annual ryegrass (*Lolium rigidum*) is vectored by *A. funesta* and another undescribed species of *Anguina*. The infection of tobacco, tomato and brinjal by root knot nematodes, *Meloidogyne incognita* and other species of this genus, increases the severity of bacterial wilt caused by *Ral. solanacearum*. Moreover, the cultivars which are resistant to *Ral. solanacearum* infection alone, show the susceptible reaction after the infection caused by root knot nematodes. This may be due to the overcoming of the penetration barrier meant for the bacterium, by the nematode. The attack of ring nematode, *Criconemoides xeroplax* on French prune trees results in considerable increase in the incidence and severity of bacterial canker caused by *Psm. syringae* pv. *syringae*.

8.8. INTERACTION BETWEEN BACTERIAL AND VIRAL PLANT PATHOGENS

The infection of *Chrysanthemum morifolium* by chrysanthemum stunt viroid (CSV) reduces the pith maceration of chrysanthemum cuttings caused by *Pe. chrysanthemi* pv. *chrysanthemi*. In CSV infected cuttings, the bacterium remains confined to xylem vessels, while in the absence of viral infection, it spreads from xylem vessels to the adjacent vascular cells and into pith. Inoculation of cowpea seedlings by cowpea mosaic virus even 1 day after the inoculation of *Ag. tumefaciens* inhibited the development of tumours. The failure of tumour formation may be due to the production of an inhibitory substance by viral infection and its subsequent translocation to the site of bacterial infection.

The combined infection of potato plants with potato viruses X and S, and *Clavibacter michiganensis* subsp. *sepedonicus* results in a significant decrease in yield in comparison to invidual infection of each pathogen. Moreover, the typical symptoms of both the viral diseases fail to develop (Goto, 1992).

REFERENCES

Araujo, W.L., Marcon, J., Maccheroni, W. Jr. *et al.* (2002). Diversity of endophytic bacterial populations and their interaction with *Xylella fastidiosa* in citrus plants. *Appl. Environ. Microbiol.* 68: 4906 - 14.

Bashir, S., Irfan ul-Haque, M., Mukhtar, T. *et al.* (2007). Pathogenic variation in *Pseudomonas syringae* and *Xanthomonas campestris* pv. *sesami* associated with blight of sesame. *Pak. J. Bot.* 39: 939 - 43.

Blakeman, J.P. (1982). Phylloplane interactions. In: M.S. Mount and G.H. Lacy (eds). *Phytopathogenic Prokaryotes*. Vol. 1, Academic Press, New York, pp. 307 - 33.

De Boer, S.H. (1982). Survival of phytopathogenic bacteria in soil. In: M.S. Mount and G.H. Lacy (eds). *Phytopathogenic Prokaryotes*. Vol. 1, Academic Press, New York, pp. 285 - 306.

Goto, M. (1992). *Fundamentals of Bacterial Plant Pathology.* Academic Press, Inc., San Diego, 342 pp.

Johnson, K.B. and Stockwell, V.O. (1998). Management of fire blight: A case study in microbial ecology. *Annu. Rev. Phytopathol.* 36: 227 - 48.

Kim, M.K., Ryu, J.S. and Lee, Y.H. (2007). First report of *Pantoea* sp. induced soft rot disease of *Pleurotus eryngii* in Korea. *Plant Dis.* 91: 109.

Kobayashi, D.Y. and Crouch, J.A. (2009). Bacterial/fungal interactions: From pathogens to mutualistic endosympbionts. *Annu. Rev. Phytopathol.* 47: 63 - 82.

Lindow, S.E. (1982). Epiphytic ice nucleation-active bacteria. In: M.S. Mount and G.H. Lacy (eds). *Phytopathogenic Prokaryotes*. Vol. 1, Academic Press, New York, pp. 335 - 62.

Lindow, S.E. and Brandl M.T. (2003). Microbiology of the phyllosphere. *Appl. Environ. Microbiol.* 69: 1875 - 83.

Lo Cantore, P. and Iacobellis, N.S. (2004). First report of brown discoloration of *Agaricus bisporus* caused by *Pseudomonas agarici* in southern Italy. *Phytopathol. Medit.* 43: 35 - 38.

Riggle, J.H. and Klos, E.J. (1972). Relationship of *Erwinia herbicola* to *Erwinia amylovora*. *Can. J. Bot.* 50: 1077 - 83.

Skirvin, R.M., Kohler, E., Steiner, H. *et al.* (2000). The use of genetically engineered bacteria to control frost on strawberries and potatoes. Whatever happened to all of that research? *Sci. Hortic.* (Amst.) 84: 179 - 89.

Tyler, H.L. and Triplett, E.W. (2008). Plants as a habitat for beneficial and/or human pathogenic bacteria. *Annu. Rev. Phytopathol.* 46: 53 - 73.

Van Loon, L.C., Bakker, P.A.H.M. and Pieterse, C.M.J. (1998). Systemic resistance induced by rhizosphere bacteria. *Annu. Rev. Phytopathol.* 36: 453 - 83.

Vidaver, A.K., Mathys, M.L., Thomas, M.E. *et al.* (1972). Bacteriocins of the phytopathogens *Pseudomonas syringae, P. glycinea* and *P. phaseolicola*. *Can. J. Microbiol.* 18: 705 - 13.

Chapter 9

BACTERIOPHAGES

9.1. DEFINITION

Bacteriophages, shortly called phages, are the viruses, which kill bacteria. These were independently discovered by Federick Twort in 1915 and by Felix d' Herelle in 1917. D' Herelle gave the name bacteriophages, which means bacteria eaters and is derived from two words, bacteria and greek word 'phagein', meaning to eat. Even before the discovery of phages, there have been documented reports of river water having the ability to cure infectious diseases like leprosy. In 1986, E.H. Hankin reported that waters of Ganga and Jamuna rivers of India contain some agent, which had a marked antibacterial activity against cholera bacterium and that agent could pass through a very fine porcelain filter.

Since their discovery, the phages have been used for various purposes. They have been used as therapeutic agents, for the identification of bacteria and forecasting of bacterial plant diseases. Starting from 1930, pioneering virologists like Luria, Delbruck and others studied bacteriophages as model systems to investigate viral structure, replication of viruses, molecular genetics etc. The study of phages has greatly enhanced our understanding of gene structure replication, transcription and regulation. Phage Lambda has been extensively used as vector in recombinant DNA studies.

Phages are found wherever their hosts, i.e., bacteria are found. These are most widely distributed and are commonly found in soil, intestines of animals, seawater etc. One of the densest natural sources for phages and other viruses is seawater, where virion population up to 9×10^8 particles per millilitre have been found in microbial mats at the surface (Wommack and Colwell, 2000) and up to 70% of marine bacteria may be infected by phages (Prescott, 1993).

Bacteriophages are obligate parasites of bacteria. They can exist in nature but cannot replicate outside the host. Phages or phage particles consist of an outer protein coat or capsid enclosing the nucleic acid. The nucleic acid can be DNA or less often RNA. The most common type of nucleic acid found in phages is double-stranded linear DNA, but double-stranded circular DNA, single-stranded linear and circular DNA and single- and double-stranded linear RNA are also found. The length of nucleic acid varies from 5 - 500 kilo base pairs. As the phages having larger nucleic acid molecules contain more genes, they have more complex life cycles and are less dependent on bacterial enzymes for their replication. Phages having less than 10 genes depend almost entirely on cellular function, while those possessing 30 - 100 genes depend largely on proteins encoded by their genetic material. A few of phages having largest particles contain so many genes that they need no host genes for replication of DNA. Most phages contain genes that encode a variety of proteins. All known phages utilize protein-synthesizing system, amino acids and energy-generating systems of their host cells, therefore, the phages can multiply only in the metabolizing bacteria (Maloy *et al.,* 1994).

The nucleic acids of phages often contain unusual or modified bases. In phage T4, cytosine is replaced by glucosylated 5-hydroxy-methylcytosine and in phage SPO1, hydroxymethyl uracil replaces thymine. These modified bases protect phage nucleic acid from nucleases that break down host nucleic acids during phage infection. The number of different kinds of proteins and the amount of each kind of protein in the phage particle will vary depending upon the phage. The simplest phages have many copies of only one or two different types of proteins while more complex phages have many different kinds of proteins. The capsid or protein coat plays below mentioned three important roles in the life cycle of a phage.

1. Protects the phage genome from DNA-degrading enzymes during the extracellular search.
2. Adsorption of phage particle to the bacterial cell.
3. Delivery of the phage genome into the cytoplasm of the host cell.

The phages are of different sizes and shapes. Most phages range in size from 24 - 200 nm long. T4 phage is among the largest phages. It is approximately 200 nm long and 80 - 100 nm wide.

9.2. TYPES OF PHAGES

On the basis of morphology, phages are classified into three categories.

(a) Tailed icosahedral

(b) Tailless icosahedral

(c) Filamentous

In tailed icosahedral phages, the head generally measures 50 - 100 nm in diameter and the tail about 15 - 170 × 10 - 20 nm. phi X174, a typical icosahedral tailless phage has a maximum diameter of 225 A°. The filamentous phages are generally 800 - 1400 nm long and 4 - 7 nm broad. The tail is a hollow tube through which nucleic acid passes during injection. In some phages like T4, the tail is surrounded by a contractile sheath, which contracts at the time of injection of the nucleic acid. Some phages have a base plate at the end of the tail and one to six tail fibres attached to it (Fig. 9.1). These structures are involved in the attachment of the phage to the host. The length of the tail varies with different phages. Phage P1 of *Escherichia coli*, described in 1955, has an icosahedral head, an intricate tail complete with tube, contractile sheath, base plate and tail fibres. Phage P22 of *Salmonella typhimurium* is much simpler in comparison to P1; its tail is very much reduced and the head appears to be resting on the base plate. The phages that do not have a base plate and tail fibres, involve other structures for binding the phage to the host.

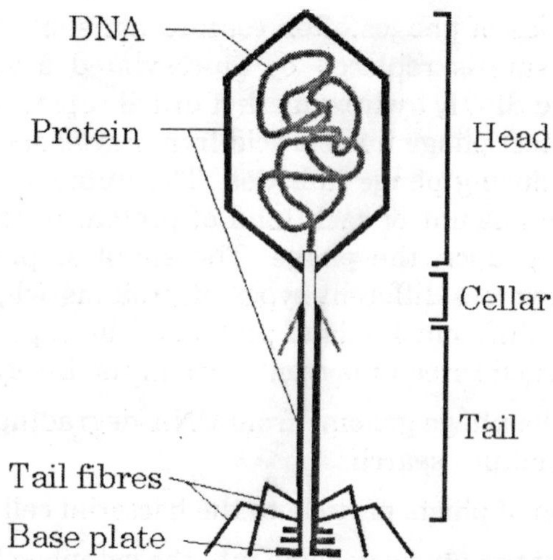

Figure 9.1 Tailed bacteriophage

Fiers *et al.* (1976) described for the first time complete nucleotide sequence of the genome of phage MS2 and a year later, Sanger *et al.* (1977) sequenced the genome of phi X174. The complete nucleotide sequence of the phage RSA1 genomic DNA (Fujiwara *et al.*, 2008) and of phages RSM1 and RSS1 (Kawasaki *et al.*, 2007) have also been determined.

Some important discoveries made while studying phages include genetic transduction (Zinder and Lederberg, 1952), acquisition and loss of genes from genome (Campbell, 1962), molecular basis of DNA recombination (Meselson and Weigle, 1961), triplet nature of DNA code (Crick *et al.*, 1961) and repression and activation of genes (Ptashne, 2004).

9.3. CLASSIFICATION OF PHAGES

The double-stranded DNA tailed phages belonging to order Caudovirales, account for 95% of all the phages reported in the literature and possibly make up the majority of the phages present on the planet. The classification of the phages by the International Committee on Taxonomy of Viruses, based on morphology and nucleic acid of phages, is given in Table 9.1.

Table 9.1. Classification of phages based on morphology and nucleic acid by International Committee on Taxonomy of Viruses

Family	Morphology	Nucleic acid
Myoviridae	Non-enveloped, contractile tail	Linear ds DNA
Siphoviridae	Non-enveloped, long, non-contractile tail	Linear ds DNA
Podoviridae	Non-enveloped, short, non-contractile tail	Linear ds DNA
Tectiviridae	Non-enveloped, isometric	Linear ds DNA
Corticoviridae	Non-enveloped, isometric	Circular ds DNA
Lipothrixviridae	Enveloped, rod-shaped	Linear ds DNA
Plasmaviridae	Enveloped, pleomorphic	Circular ds DNA
Rudiviridae	Non-enveloped, rod-shaped	Linear ds DNA
Fuselloviridae	Non-enveloped, lemon-shaped	Circular ds DNA
Inoviridae	Non-enveloped, filamentous	Circular ss DNA
Microviridae	Non-enveloped, isometric	Circular ss DNA
Leviviridae	Non-enveloped, isometric	Linear ss RNA
Cystoviridae	Enveloped, spherical	Segmented ds RNA

ds means double-stranded and ss means single-stranded.

The phages are generally of two types, i.e., **virulent** and **temperate**. Virulent phages, also called **lytic phages,** start multiplying inside the host cell immediately after entry and cause lysis of the cell in a short period. T4 and other T-phages of *Es. coli* fall in this category. Temperate or **avirulent phages** (phage P22 of *Sal. typhimurium* and phages P1 and Lambda of *Es. coli*) do not cause lysis of the host cell immediately after the infection. After entry into the host cell, viral genome gets integrated into the host chromosome and replicates along with the multiplication of host cell. The inserted phage genome is called **prophage** and the bacterial cell carrying the prophage is called **lysogen**. The replication of prophage along with the host cell contributes indirectly to the phage growth because every cell carrying prophage is a potential source for production of phage particles. The process of establishment of a prophage is called lysogenization. The phenomenon of **lysogenization** was first discovered in 1920s. However, the prophage is not always a part of chromosome in all the lysogenic cycles. The genome of phage P1 does not integrate with the bacterial chromosome but remains in free form called P1 plasmid. Lambda phage of *Es. coli* may cause lysis of the host cells but more frequently causes **lysogeny** (Fig. 9.2).

As lysis of the host cell occurs in both lytic and temperate (although at a later stage) phages, it is more appropriate to classify the phages as **temperate** and **obligately lytic** (a.k.a., virulent) **phages** (Gill and Abedon, 2003). A temperate phage does not always enter into a lysogenic relationship with a host bacterium. In fact many temperate phage infections result in the immediate lysis of the host cells, i.e., lytic cycle rather than lysogeny depending upon the characteristics of the infecting phage and metabolic stage of the host. The prophage remains dormant during active metabolic state of the host cell. When the host conditions deteriorate, possibly due to the depletion of nutrients, than the prophage becomes active, initiates the reproductive cycle and causes lysis of the host cell. In the lysogenic cycle, the prophage may also be occasionally released from the host chromosome in some host cells and initiates lytic cycle as in virulent phages. These cells are finally lysed and liberate free phage particles. This phenomenon is called **spontaneous induction**. Prophages can also be induced to undergo a lytic cycle by ultraviolet irradiation or mitomycin C treatment. Lysogenic bacteria carrying prophages become resistant to same or similar/related phages. Phage mutants that are able to bypass this resistance are described as *vir* (for virulence) mutants. The advantage associated with such virulence is the ability to actively infect homologous lysogens, i.e., bacteria lysogenized by the same phage. In nature, mutations occur in phages and phage mutants differing in plaque morphology, host range, UV sensitivity etc. appear. Mutants differing in plaque morphology are readily discovered on agar plate cultures. The major disadvantage due to phage mutation is the development of non-virulent strains of phages used in phage therapy.

9.4. PHAGE OR LYSOGENIC CONVERSIONS

It is the phenomenon by which sometimes, prophages provide benefits to the host bacterium by adding new functions to the bacterial genome. A well known example is a conversion of an avirulent strain of

Vibrio cholerae into a virulent strain capable of causing cholera. Due to this reason, temperate phages are not used for therapy.

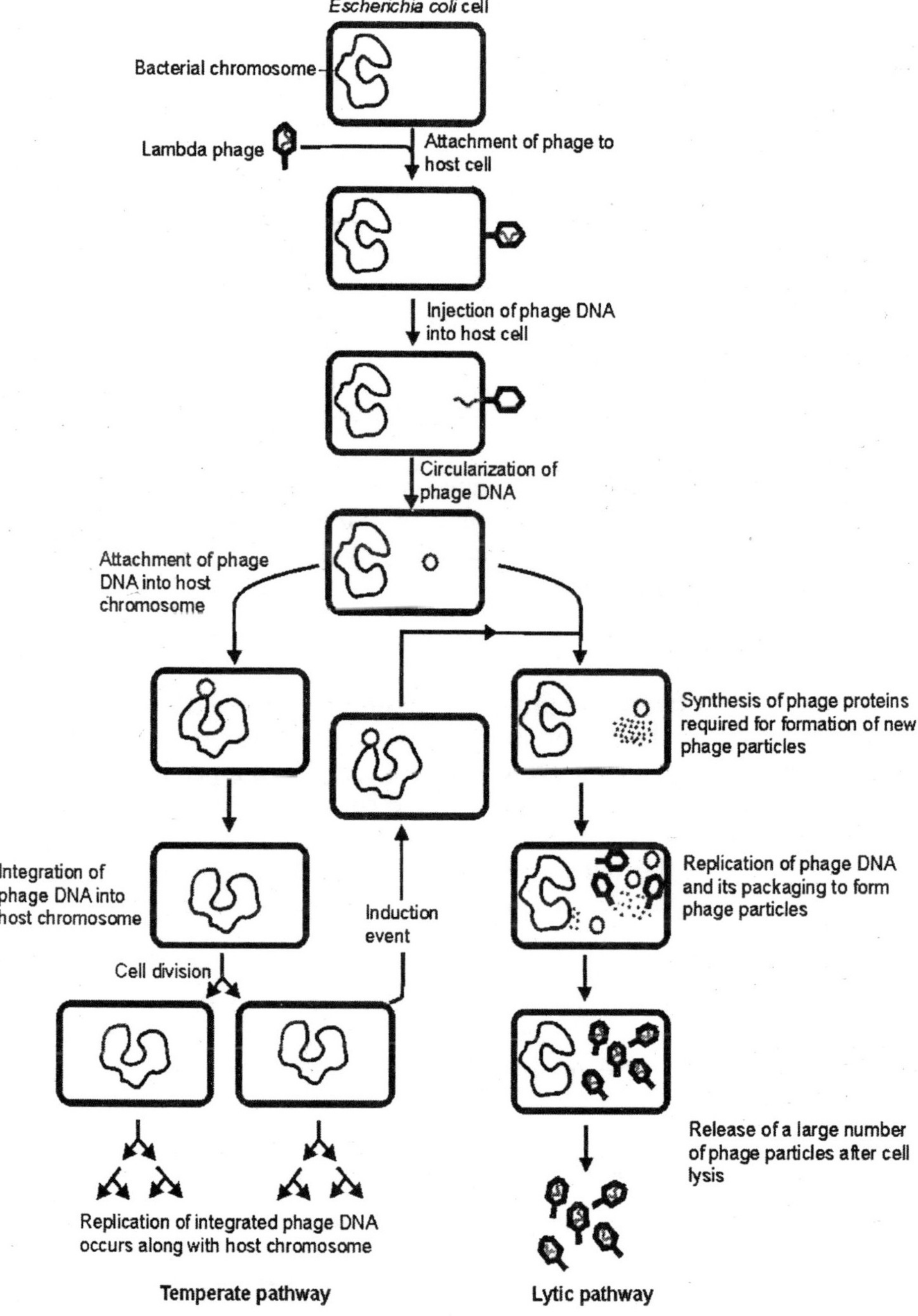

Figure 9.2 Life cycle of bacteriophage Lambda

9.5. REPLICATION OF PHAGES

Most phages undergo a lytic life cycle. The following steps are involved in the replication of lytic phages.

9.5.1. Adsorption or attachment of phage to host cell

The attachment of phage particles to the bacteria is influenced by the population density of both bacteria and phages. More number of bacterial cells and phage particles increase the frequency of attachment. There are specific receptors or adsorption sites on the surface of bacterial cells including lipopolysaccharides, teichoic acids, proteins or even flagella to which phages attach. This specificity means that a phage can infect only the bacterium bearing those particular receptor(s) to which it can bind. Therefore, the receptor specificity in turn determines the host range of a phage. The tail fibres locate receptors/adsorption sites and spikes fix the phage particles to the receptors. The filamentous phages adsorb to the tips of pili, the attachment being end-to-end. After the adsorption, pili are depolymerized and the phage particles come in contact with the cell surface. The adsorption of mycoplasma phages occurs to the specific receptor proteins on the unit membrane (Goto, 1992). The phage receptors also serve the other purposes.

9.5.2. Transfer of phage nucleic acid into bacterial cell

After contacting the appropriate receptor, the tail fibers bring the base plate close to the surface of host cell. Lysozyme present in the tail causes lysis of bacterial cell wall. In more complex phages like T4, tail is surrounded by a contractile sheath, which contracts and the core with an internal hole penetrates the cell membrane and injects the nucleic acid. The contraction of the tail occurs, possibly with the help of ATP present in the tail (Prescott, 1993). The coat protein of the phage particles, called **'ghost'**, is left outside the bacterial cell. Not much is known about the mechanism how the other phages transfer the nucleic acid. Some phages with short tails probably transfer the nucleic acid to the periplasm. In case of tailless phages, it seems that the phage coat breaks open and releases the nucleic acid on the cell wall from where it is transferred across the cell wall.

9.5.3. Replication of phage nucleic acid and proteins

After the transfer of the phage nucleic acid into the bacterium, the shutdown of the host DNA or RNA synthesis is achieved by different ways by different phages and one way is degradation of the host DNA. During this process, the bacteria often lose the ability either to replicate or transcribe their own DNA. Phages containing single-stranded DNA or RNA cause comparatively less shutdown. The phage takes the control of host cell and redirects its biosynthetic pathways to synthesize phage nucleic acid and proteins. This reprogramming is achieved by transcribing and then translating phage genes. The host RNA polymerase generally initiates the transcription of phage genes. After the first set of transcription, the bacterial polymerase is usually modified to recognize phage-specific promoters, or a phage-specific RNA polymerase is synthesized. Phage proteins are synthesized as per requirement by regulating the transcription. Messenger RNA made, in early stage, usually encodes the enzymes required for takeover of the bacterium for phage DNA replication, and for the synthesis of late mRNA. Late mRNA encodes the components of the phage particles, i.e., proteins needed for making protein coat and enzyme required to break the cell wall (Malacinski, 2003).

9.5.4. Assembly of phage particles

The process of assembling phage particles is also called **morphogenesis.** Two types of proteins, namely, structural and catalytic proteins are required for this process. Structural proteins constitute the phage particles while catalytic proteins participate in the assembly process but do not become a part of the phage particles. Maturation proteins, a subgroup of catalytic proteins, shape the phage DNA into an appropriate form for packing into the phage particles. The assembly of icosahedral phages comprises the following stages.

1. Structural proteins aggregate to form a phage head and a tail, if the latter is present.
2. The nucleic acid condenses and enters into the preformed head.
3. The tail attaches to the filled head.

In filamentous phages, the formation of a phage particle from the nucleic acid and the protein is a single step process (Maloy *et al.*, 1994).

9.5.5. Release of phage particles

At the later stage of infection cycle, most phages synthesize enzymes, which cause lysis of the host cells. Two most commonly synthesized enzymes are: (a) holin, which causes disruption of the cytoplasmic membrane and (b) lysozyme which hydrolyses hydrogen bonds of peptidoglycan layer of the cell wall. The cell lysis occurs due to the action of these enzymes releasing the phage particles in the surrounding medium. The suspension containing the newly released phage particles is called **lysate**. The number of phage particles released per cell varies with the type of phage and the physiological condition of the host cell. Generally, 50 - 100 phage particles are produced per cell. The number of phage particles released from a single cell is called the **average burst size**. The release of phage particles leads to reduction in turbidity of liquid cultures and formation of plaques in agar plates. The plaque morphology is specific in a particular combination of a bacterium and a phage and is used as a dist inguishing character. Filamentous phages are generally released in large numbers, i.e., several hundred per cell and are released without lysis of the host cell. The phage progeny is released continuously by out folding of the cell wall. This process is called **extrusion** and it does not result in a major damage to the cell. The cells infected with such filamentous phages can continue to produce phage particles for long periods of time. As the cells are not lysed, the turbidity of the liquid cultures does not change. The life cycle of phage T4 of *Es. coli* at 37°C is completed in 25 minutes. However, most phages complete their life cycles in 20 - 60 minutes.

Most phages of plant pathogenic bacteria contain DNA but some contain RNA. Although the DNA of most phages contains the normal bases, namely, adenine, guanine, cytosine and thymine, the DNA of some phages contains different bases. In phage Xp-12 of *Xanthomonas oryzae* pv. *oryzae,* cytosine is completely replaced by 5-methylcytosine and this replacement imparts unusual physical properties to the DNA. The filamentous phage Xf of the same bacterium has a protein coat unusually high in hydrophobic amino acids, which may account for its liability to organic solvents and to low concentration of sodium lauryl sulphate. Filamentous phage Cf of *Xa. citri* subsp. *citri* (syn. *Xa. axonopodis* pv. *citri*) contains a single-stranded DNA and is 1008 ± 99 nm long. It has a narrow host range and forms small clear plaques. It gets integrated into the host chromosome as a prophage. The double-stranded RNA lipid-containing phage ϕ 6 of *Pseudomonas syringae* pv. *phaseolicola* is approximately 60 nm in diam and contains a three-segmented unique RNA molecule (Goto, 1992). It is the most extensively characterized phage of the phytopathogenic bacteria (Vidaver, 1976).

9.6. USES OF PHAGES

The phages have been employed for various purposes for the benefit of mankind. Some of the important uses are given below:

9.6.1. Phage therapy

Since the discovery of phages, these had been extensively used for treating various ailments of humans and in some cases success rates were extremely high. The first report of use of phages for controlling *Bacillus anthracis* and staphylococci infections was made by Brunoghe and Maisin in 1921 from France. A year later, the phages were evaluated for the control of typhoid fever and bacillary dysentery. The phages were also used extensively to control cholera outbreaks in India (Morrison, 1932). There are several hundred reports of phage therapy in the literature but most of them pertain to the work carried on in former Soviet Union and eastern European countries. After the First World War, the work on phage therapy was largely abandoned in the West but it continued in former Soviet Union, Georgia and eastern Europe. As much of the literature was either in Russian or Georgian, it remained unavailable to the West for many years. According to Kutter (1997), more than 800 papers were published concerning disease control or prevention with phages. A report from the Institute of Immunology and Experimental Medicine of the Polish Academy of Sciences clearly illustrates the success of phage therapy. A total of 550 patients were treated at chemical and hospital departments in three different cities against major pathogens belonging to *Staphylococcus, Pseudomonas, Escherichia, Klebsiella* and *Salmonella* genera. The success rate for phage therapy ranged from 80 - 95%. The overall success rate was even higher (94%) for 518 patients where antibiotic therapy was found ineffective.

Phages are considered to be very safe for therapeutic use because a very few cases of side-effects have been reported. A case of liver pain, lasting for several hours, was reported in one study conducted in Poland. This might have occurred due to the excessive release of endotoxins resulting from the rapid death of bacterial cells. However, the side-effects are also associated with the use of many antibiotics. In the past, some failures were associated with phage therapy, which might have occurred due to the following reasons.

(a) Failure to select phages specific to target bacteria.

(b) Improper selection of phages due to the wrong identification of bacteria.

(c) Use of phages for infections involving mixtures of different bacteria.

(d) Use of temperate phages instead of lytic phages.

(e) Failure to neutralize gastric acidity prior to oral phage administration.

Antibiotics are also widely used for controlling bacterial infections. A comparison of phages and antibiotics as therapeutic agents, listing the advantages and disadvantages of both, is given in Table 9.2.

Table 9.2. Comparison between phages and antibiotics as antibacterial agents.

	Bacteriophages	Antibiotics
1	Are specific or very specific in killing the bacteria. Hence, correct identification of the bacterium is required.	Generally broad-spectrum activity. Hence, the exact identification of the bacterium is not necessary.
2	In case of mixed/multiple infections, different phages are required for different bacteria.	A single antibiotic usually works against multiple infections.

3	Very few or no side-effects.	Multiple side-effects including intestinal disorders, weakness etc.
4	The receptors targeted by the phages are generally present on bacterial membrane or capsule, which are important virulence determinants. Development of phage-resistance usually means change in these structures and may, therefore, lead to the development of attenuated bacterial strains.	Bacterial strains resistant to antibiotics usually remain pathogenic.
5	Production is simple and relatively cheap.	Production is expensive.
6	Selection of a phage against a phage-resistant bacterium is a rapid process and frequently can be accomplished in a short period. It is relatively much less expensive	The development of an antibiotic against a resistant bacterium is a lengthy process and may take years. It is also highly expensive.

The efficacy of phages in controlling bacterial infections and contaminations prompted some countries to grant licenses for the sale of phage products. During 1930s, phage therapy products were licensed for sale in the United States. Health Ministry of former Soviet Union routinely licensed active phage preparations for use in humans. Probably, the therapeutic phages sold and used in eastern Europe were also licensed.

9.6.2. Use in food industry

Phages are also used in food industry to check food spoilage. *Listeria monocytogenes* causes listeriosis, a food-borne disease recognized as an important health hazard. Outbreaks of the disease lead to the death of about 25% infected ones making it as the highest fatality rate of all food-borne diseases. *Li. monocytogenes* is found in raw and ready-to-eat meat, poultry and seafood, and dairy products. The bacterium is a remarkably tough organism having high resistance to salt, nitrite, dry conditions and acidity. But its ability to multiply under refrigerated conditions is a major concern for food processors.

The phage technology centre of EBI Food Safety in the Netherlands is producing bacteriophages, including the one for the control of *Li. monocytogenes* on the industrial scale. These phages are supplied to the customers in the European Union and the United States.

On 18th August 2006, the Food and Drug Administration (FDA) of the United States approved the use of bacteriophages on ready-to-eat meat and poultry products to kill *Li. monocytogenes* by amending its food additive regulations. These phages were also given a GRAS (Generally

Recognized as Safe) status. In July 2007, the same phages were approved for use on all food products. Phage LMP-102, manufactured by Intralytix, a company based in Baltimore, USA, is used as food additive. This viral additive consists of six individually purified phages in equal proportion and is used @ of about one millilitre solution per 500 sq. centimetres of food product surface area just prior to package. Each phage is effective against various strains of *Li. monocytogenes.* The basis for using the mixture of phages is to minimize the chances of the bacterium to develop resistance to the phages.

9.6.3. Identifications of bacteria

Phages can be used for the identification of bacteria including phytopathogenic ones. The phages used for this purpose should have a high specificity for the species or the strains to be identified. The host range of the phages varies greatly. The host range may be narrow, confined to a particular species or a strain or a group of strains within a species. The non-specific or polyvirulent phages have a wide host range, which can infect bacteria belonging to different species, genera or even families. Lambda phage has a narrow host range and infects only some strains of *Es. coli* while phage P1 has a wide host range and can infect all enterobacteria and even myxobacteria.

The host range of phages may vary with the source of their occurrence. In general, the phages isolated from diseased plant material exhibit a narrow host range and produce large plaques. On the contrary, phages obtained from soil or sewage have a wide host range and form small plaques. The phages isolated from sewage have been found to attack enteric bacteria of humans as well as plant-associated bacteria. The phages attacking different groups of bacteria also differ in their host range. Phages infecting *Pectobacterium carotovorum* subsp. *carotovorum* generally have a narrow host range. Similarly, ϕ RSM1 and ϕ RSS1 of *Ralstonia solanacearum* have a narrow host range while two other phages, namely, ϕ RSA1 and ϕ RSL1 of the same bacterium have a relatively wide host range and the phage ϕ RSA1 was able to infect all the 15 strains of the bacterium belonging to different races and biovars tested in the study (Yamada *et al.,* 2007). The ability of two phages namely, Cp1 and Cp2 to infect roughly 98% strains of *Xa. citri* subsp. *citri* (syn. *Xa. axonopodis* pv. *citri*) occurring in Japan indicates the wide host range of the phages of this pathogen.

In some phages, phage virulence is correlated to pathogenic variability or certain biochemical properties of the host bacterium. These phages can be used for quick identification of pathogenic races. Certain phages of *Ral. solanacearum* attack strains of the bacterium characterized by lactose oxidation and pathogenicity to tobacco. Phages specific to a lactose-utilizing strain of *Xa. citri* subsp. *malvacearum* (syn. *Xa.*

axonopodis pv. *malvacearum*) and a mannitol-utilizing strain of *Xa. citri* subsp. *citri* occur in nature and are used for quick identification of these bacteria.

9.6.4. Detection of bacteria from substrate

Phytopathogenic bacteria occur in various ecological situations/niches and in majority of these cases they are accompanied by complex microbial populations. Hence, the isolation of these bacteria is difficult due to the presence of other microorganisms. However, the phages have been used to detect phytopathogenic bacteria from soil, diseased plant debris, seeds, phylloplane etc. and success achieved in many cases. The increase in population of phage added to the substrate indicates the presence of the host bacterium in the substrate. The propagated phage is separated from the microbial complex either by centrifugation or by chloroform treatment. In case of low population of the target bacterium in the substrate, the selective enrichment of the bacterium is required.

Phages have been successfully used to detect phytopathogenic bacteria from seeds of cowpea and green gram (Soni and Thind, 1991), rice, soybean, pea and kidney bean.

The presence of bacteria in seed can also be indirectly detected from the seedlings raised from the infected seeds. The detection of *Clavibacter michiganensis* subsp. *michiganensis* from infected tomato seeds is generally more effective from the seedlings raised from these seeds. Two phages, ϕ RSM1 and ϕ RSS1 of *Ral. solanacearum* are highly strain specific and can be effectively used for the detection of specific strains (Yamada *et al.,* 2007). Kawasaki *et al.* (2007) reported that bacterial cells of *Ral. solanacearum,* transformed with a green fluorescent protein (GFP)-expressing plasmid pRSS12, constructed from a filamentous phage ϕ RSS1, emitted a strong green fluorescence and the presence of these cells was easily monitored in infected potato tissues and soil samples. These results suggest that this plasmid can serve as an easy-to-use GFP-tagging tool for any given strain of *Ral. solanacearum* in cytological and field studies. A novel technique called Sensing of Phage-Triggered Ion Cascades has been developed by Dobozi-King *et al.* (2005) for the rapid detection and identification of live pathogenic bacteria on the scale of minutes, with an unparalleled specificity. The method has a potential ultimate sensitivity of 1 bacterium per microlitre (1 bacterium/mm^3) and is applicable to only physiological viable cells infected by mycophages and siphophages. The method combines the specificity and avidity of bacteriophages with fluctuation analysis of electrical noise. The method is based on the massive transitory ion leakage that occurs at the moment of phage DNA injection into the host cell. The ion fluxes require only that the cells should be physiologically viable (i.e., having energized membranes) and can occur within seconds after mixing the cells with sufficient concentration of phage particles.

9.6.5. Disease forecasting

Phages have been used for forecasting bacterial blight of rice. The forecasting is more accurate under normal conditions. The time when phage population in irrigation water exceeds 200 particles per ml of water corresponds to the time 10 days before the initial disease development. For an accurate forecasting, the water sample should be taken in the morning at the outflow.

9.6.6. Control of bacterial plant diseases

Phages have been successfully used for the control of bacterial plant pathogens. This topic is covered in chapter 12.

SELECTED REFERENCES

Campbell, A. (1962). Episomes. *Adv. Genet.* 11: 101 - 45.

Crick, F.H.C., Leslie Barnett, F.R.S., Brenner, S. *et al.* (1961). General nature of the genetic code for proteins. *Nature* 192: 1227 - 32.

Dobozi-King, M., Seo, S., Kim, J.U. *et al.* (2005). Rapid detection and identification of bacteria: SEnsing of Phage-Triggered Ion Cascade (SEPTIC). *J. Biol. Phys. & Chem.* 5: 3 - 7.

Fiers, W., Contreras, R., Duerinck, F. *et al.* (1976). Complete nucleotide sequence of bacteriophage MS2 RNA: primary and secondary structure of the replicase gene. *Nature* 260: 500 - 07.

Gill, J. and Abedon, S.T. (2003). Bacteriophage ecology and plants. APSnet Feature Story, November 2003.

Goto, M. (1992). *Fundamentals of Bacterial Plant Pathology.* Academic press, Inc, San Diego, 342 pp.

Kawasaki, T., Satsuma, H., Fujie, M. *et al.* (2007). Monitoring of phytopathogenic *Ralstonia solanacearum* cells using green fluorescent protein-expressing plasmid derived from bacteriophage ϕ RSS1. *J. Biosci. & Bioengin.* 104: 451 - 56.

Kutter, E. (1997). Phage therapy: Bacteriophages as antibiotics. [http://www.evergreen.edu/phage/phagetherapy/phagetherapy.htm.]

Malacinski, G.M. (2003). *Freifelder's Essentials of Molecular Biology.* 4th Ed. Narosa Publishing House, New Delhi, 491 pp.

Maloy, S.R., Cronan, J.E. Jr. and Freifelder, D. (1994). *Microbial Genetics.* 2nd Ed. Narosa Publishing House, New Delhi, 524 pp.

Meselson, M. and Weigle, J.J. (1961). Chromosome breakage accompanying genetic recombination in bacteriophages. *Proc. Natl. Acad. Sci.* USA, 47: 857 - 68.

Morrison, J. (1932). *Bacteriophage in the Treatment and Prevention of Cholera* Lewis, London.

Prescott, L. (1993). *Microbiology.* W.C. Brown Publishers, ISBN 0 - 697 - 01372 3.

Ptashne, M. (2004). *A Genetic Switch: Phage Lambda Revisited.* 3rd Ed. Col Spring Harbor Laboratory Press.

Sanger, F., Air, B.M., Barrell, B.G. *et al.* (1977). Nucleotide sequence o bacteriophage ϕX174 DNA. *Nature* 265: 687 - 95.

Soni, P.S. and Thind, B.S. (1991). Detection of *Xanthomonas campestris* pv *vignaeradiatae* (Sabet *et al.*) Dye from green gram seeds and *X. campestri* pv. *vignicola* (Burkh.) Dye from cowpea seeds with the help of bacteriophages. *Pl. Dis. Res.* 6: 6 - 11.

Vidaver, A.K. (1976). Prospects for control of phytopathogenic bacteria by bacteriophages and bacteriocins. *Annu. Rev. Phytopathol.* 14: 451 - 65.

Wommack, K.E. and Colwell, R.R. (2000). Virioplankton: Viruses in aquati ecosystems. (http://mmbr.asm.org/cgi/pmidlookup?view=long&pmid=10704 475). *Microbiol. Mol. Biol. Rev.* 64: 69 - 114.

Yamada, T., Kawasaki, T., Nagata, S. *et al.* (2007). New bacteriophages tha infect the phytopathogen *Ralstonia solanacearum. Microbiology* 153: 2630 39.

Zinder, N.D. and Lederberg, J. (1952). Genetic exchange in *Salmonella. J Bacteriol.* 64: 679 - 99.

Chapter 10

HOST-PATHOGEN RELATIONSHIP

Plants in nature are generally resistant to most pathogens. The ability of a pathogen to cause disease in a host plant is usually the exception, not the rule. This is because the plants have an inbuilt ability to recognize potential invading pathogens and to mount successful defences. Moreover, it is also evident from the fact that a given plant species is attacked only by a limited number of pathogens out of a large number of plant pathogens running into several thousands. However, in contrast, successful pathogens do cause diseases because they are able to evade recognition by the host or suppress the host defence mechanisms or both (Staskawicz, 2001).

The resistance is broadly classified into two categories, i.e., **nonhost resistance** and **actual/real resistance**. Nonhost resistance is due to the fact that a given plant species is not a host of a particular pathogen and this is the most common form of resistance in nature. Most plant pathogenic bacteria except *Agrobacterium tumefaciens*, *Ralstonia solanacearum* and *Rhodococcus fascians* have a limited or specific host range. Actual or real resistance is due to the inherent properties of the host. Resistance or susceptibility of the host is the outcome of the interaction, direct or indirect, between the genomes of host plant and pathogen. Even within a plant species that is susceptible to a particular pathogen, different cultivars may show different disease reactions ranging from susceptible to resistant to a particular strain or race of the pathogen. In the same way, different races or strains of the pathogen also vary in virulence toward a specific plant variety (Fig. 10.1). Resistance to a pathogen is often correlated with a **hypersensitive response** (HR). HR is the result of localised induced cell defences in the host plant at the site of infection caused by a pathogen. It results from quick mobilization of defence signals by the affected and surrounding cells and the subsequent release of toxic compounds that often kill both the invaded and surrounding cells, and the invading pathogen. HR thus, limits the growth and spread of the pathogen and imparts resistance to the host against the pathogen. An HR may not always be visible because in some cases only a single cell or very few cells are affected and visible symptoms do not appear. Resistance gene *Xa21* of rice does not induce visible HR against *Xanthomonas oryzae* pv. *oryzae*. Injection of many plant pathogenic bacteria into leaf tissues of nonhost plants, particularly tobacco, results in the development of HR. The artificially induced HR consists of large leaf sectors becoming water-soaked at first and then necrotic and finally collapsing within 8 to 12 hours after inoculation. The bacteria injected into the leaf tissues are generally killed rapidly. The HR may also occur when virulent strains of plant pathogenic bacteria are injected into resistant cultivars of susceptible host or when avirulent strains are injected into susceptible cultivars. It shows that in case of plant pathogenic bacteria, HR is the result of an incompatible host-pathogen relationship. HR in plant diseases caused by bacteria was first recognised by Klement (1963) and then it was subsequently elaborated by him and his associates in a series of papers. Host response evoked in plants by a few incompatible bacterial cells is normally not readily discernible. The application of **injection-infiltration method** has made the recognition of HR quite easy. In the injection-infiltration method, bacterial cell suspension containing 10^7 cfu/

ml is injected into the tobacco leaf tissue with the help of a hypodermic syringe. Different cultures can be inoculated into different interveinal sections of the same tobacco leaf.

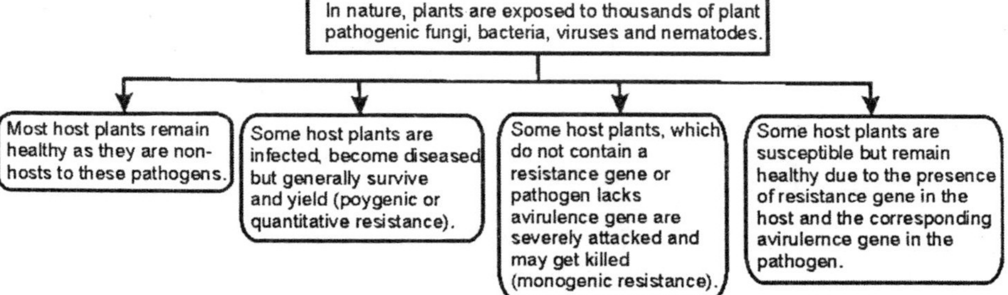

Figure 10.1 Host responses to pathogen attack

HR is considered a central component of plant resistance responses. Several effector proteins of *Pseudomonas syringae* pathovars are known to inhibit HR, though in most of these cases the molecular basis of this inhibition is yet not known. It shows that certain bacterial effectors have evolved the ability to suppress programmed cell-death responses. Some effectors seem to interfere with recognition events responsible for an HR rather than actively suppressing the HR. For instance, AvrRpt2 is not known to suppress cell death in general but it interferes with the HR triggered by AvrRpm1.

10.1. GENE-FOR-GENE HYPOTHESIS

During 1940s, H.H. Flor published his insightful and classical work on the genetics of interaction between flax and its obligate rust pathogen, *Malampsora lini*, which led to the development of gene-for-gene hypothesis. In its most simplest form, the gene-for-gene hypothesis states that for each gene that confers virulence to the pathogen, there is a corresponding gene in the host that confers resistance to the host and vice versa. Each gene in the host can be identified only by its counterpart gene in the pathogen, and vice versa. However, if the plant host does not contain resistance gene, the pathogen is still capable of causing disease on that plant even though it still contains the avirulence gene. After the publication of Flor's gene-for-gene concept, an outstanding progress toward understanding the molecular basis for resistance has been made notably on two aspects, i.e., the identification and cloning of the *avr* (avirulence) and *R* (resistance) genes and the characterization of cellular defence responses that are induced after recognition. Induced plant defence responses include enhanced production of active oxygen species, phytoalexin biosynthesis, cell wall reinforcement, the release of hydrolytic enzymes or other inhibitory proteins and the HR (Leach and White, 1996).

The gene-for-gene concept has been found to operate in many bacterial plant diseases. Some of the examples include soybean and *Psm.*

savastanoi pv. *glycinea*, tomato and *Psm. syringae* pv. *tomato*, common bean and *Psm. syringae* pv. *phaseolicola*, rice and *Xanthomonas oryzae* pv. *oryzae*, pea and *Psm. syringae* pv. *pisi* and cotton and *Xa. citri* subsp. *malvacearum*. However, unlike other phytopathogenic bacteria, race system of *Ral. solanacearum* is not based on gene-for-gene interaction, i.e., different cultivars carrying different *R* gene(s). Instead, the races are based on their pathogenicity on different host species.

10.2. AVIRULENCE GENES

Avirulence (*avr*) genes are found in plant pathogens, including plant pathogenic bacteria. Avirulence genes can be defined as genes in the pathogen that encode a protein product that is conditionally recognised directly or indirectly only by those plants that contain the complementary *R* gene. Specific recognition results in the induction of defence gene expression and the inhibition of pathogen growth. Avirulence genes generally elicit a hypersensitive reaction on a resistant host. A pathogen strain may contain many avirulence genes, and the combination of avirulence genes within a particular strain specifies the physiological race of the strain. The first avirulence gene characterized, *avrA* was cloned from race 6 of *Psm. savastanoi* pv. *glycinea,* and when transferred to other races of the pathogen, conferred the ability to elicit resistant response only on cultivars of soybean having resistance gene *Rpg2* (Staskawicz *et al.*. 1984). Leach and White (1996) have given a comprehensive account of avirulence genes, their structural features, functions, race specificity and role in pathogenicity and interaction of *avr* and *hrp* genes.

More than 40 bacterial avirulence genes have been cloned and characterized (Leach and White, 1996). However, the function of these gene products is known in few cases including that of *avrD* of *Psm. savastnoi* pv. *glycinea*. The product of *avrD* likely functions indirectly to elicit resistance in soybean as the evidence suggests the gene product is an enzyme involved in elicitor production. In most cases, bacterial avirulence gene function is dependent on interactions with the hypersensitive response and pathogenicity *(hrp*, pronounced as harp) genes. Strains with mutations in *hrp* genes cannot elicit their normal avirulence gene-media-ted, race-specific interactions with host plants. Although the intracellular enzymatic functioning of AvrD protein is clearly established, analyses of structure-function relationship of other *avr* genes, particularly the *avrBs3* family, are suggestive of an extrabacterial role for the Avr protein.

Increasing evidence suggests that *avr* genes play roles other than the race-specificity also. *Xa. vesicatoria* strains containing an inactivated *avrBs2* gene show less aggressiveness to a susceptible pepper cultivar than wild-type strains. Mutants in *avrBs2* homologs in *Xa. alfalfae* subsp. *alfalfae* also show reduced multiplication in alfalfa plants, suggesting that *avrBs2* appears to be essential for the growth of these pathogens in these hosts. *Xa. citri* subsp. *malvacearum* containing *avrb6* induces more water-soaking on cotton than the strains without *avrb6*. Other avirulence genes that function in aggressiveness or fitness to host plants are the *avrE* and *avrA* genes from *Psm. syringae* pv. *tomato, avrRpm1* from *Psm. syringae* pv. *maculicola* and *avrXa7* from *Xa. oryzae* pv. *oryzae*.

10.3. RECOGNITION OF AVR PROTEINS OF PATHOGENS BY THE HOST PLANT

Interaction between *R* genes and their matching *avr* genes has been investigated in many host pathogen combinations. The direct interaction between *R* and *avr* gene products of these investigated host-pathogen relationships have been found in few cases, while in majority of the cases there is no physical interaction between *R* and *avr* genc products. It appears that in majority of the cases, the recognition of Avr proteins by R proteins is indirect, i.e., through at least one-third component to which the Avr protein binds and is recognised. The binding of the Avr protein to its virulence target acts as a signal for the *R* gene, which then initiates the defence responses. The direct or indirect recognition of Avr proteins by the R proteins leads to activation of signalling networks by the latter, resulting in resistant responses. The *avrPto* of *Psm. syringae* pv. *tomato* encodes AvrPto protein that is delivered by the bacterium directly into the host cell cytoplasm. It was earlier thought that *PtoR* gene of tomato, which confers resistance against *Psm. syringae* pv. *tomato*, encodes for a cytoplasmic protein kinase that appears to interact directly with the bacterial AvrPto protein. Later on, it was shown that the expression of *Pto* requires the presence and expression of another gene, *Prf*, which is located within the *Pto* gene cluster. However, the recent work shows that perhaps, *Pto* is not the true resistance gene, but encodes the virulence target of AvrPto. Now it is considered an indirect recognition of AvrPto by *Prf* gene rather than a direct recognition of AvrPto by Pto. Hence, it is not a direct interaction of an *avr* with its *R* gene but an interaction between an Avr protein and its virulence target (Agrios, 2004).

AvrBs2 protein produced by *Xa. vesicatoria*, casual agent of bacterial spot of tomato and pepper, and several other Avr proteins produced by various pathovars of *Psm. syringae* are among the best studied proteins so far. In *Xa. vesicatoria*/pepper combination, *Bs2R* gene confers resistance against the pathogen by producing an NBS-LRR protein (has a hydrophobic N terminus), which recognizes avrBs2 protein encoded by *avrBs2* gene. The *avrBs2* gene, highly conserved among different strains of *Xa. vesicatoria* and also among pathovars of *Xa. campestris*, is needed for the full virulence of the bacterium on susceptible hosts. A mutant strain of *Xa. vesicatoria* failed to trigger a resistant response in plants carrying *Bs2R* gene but did not show any reduction in its virulence in susceptible plants. As this observation does not separate virulent and avirulent functions of *avrBs2*, it is not likely that recognition of AvrBs2 occurs after binding to its virulence target (Agrios, 2004)

Most of the *avr* genes investigated so far do not play any role in the pathogenicity or virulence of the bacteria. Even the inactivation of these genes by mutations does not affect the virulence or pathogenicity of the pathogen. However, some *avr* genes including *avrBs2* of *Xa. vesicatoria* play a role in the pathogenicity. The gene *avrBs2* encodes a protein, which is necessary for pathogenicity and this is evident from the fact that this gene is present in all strains of the pathogen, and the mutants lacking this gene do not infect the previously susceptible hosts. The other examples of *avr* genes enhancing virulence or pathogenicity include *pthA* of *Xa. citri* subsp. *citri* and *avrb6* of *Xa. citri* subsp. *malvacearum*. Both these genes enhance the virulence of a weakly pathogenic strain of *Xa. alfalfae* subsp. *citrumelonis*, enabling it to cause canker lesions on its host. *pthA* is required for the pathogenicity of *Xa. citri* subsp. *citri* to cause Asiatic citrus canker on citrus while it can also act as *avr* gene and cause HR on cotton and bean when transferred to *Xa. citri* subsp. *malvacearum* and *Xa. phaseoli*, the pathogens of cotton and bean, respectively. The examples of *avr* genes, namely, *avrE* and *avrA* of *Psm. syringae* pv. *tomato*, *avrRpm1* of *Psm. syringae* pv. *maculicola* and *avrXa7* of *Xa. oryzae* pv. *oryzae* that enhance aggressiveness or fitness to the host plants, have already been cited above.

10.4. HYPERSENSITIVE RESPONSE AND PATHOGENICITY GENES

The hypersensitive response and pathogenicity genes (*hrp* genes) are those genes, which control the ability of pathogenic bacteria to cause disease on the susceptible hosts and to elicit hypersensitive response on resistant hosts. They may be one of the most important groups of genes found in phytopathogenic bacteria in relation to pathogenicity and host range. The characteristic phenotype associated with *hrp* mutants is the inability to cause disease on susceptible host plants, as well as the inability to elicit HR on resistant cultivars of susceptible hosts and nonhost plants. *Hrp* genes were first described for *Psm. syringae* pv. *phaseolicola* and subsequently have been described in many Gram–negative bacteria, including *Ral. solanacearum*, *Xa. oryzae* pv. *oryzae*, *Erwinia amylovora*, *Pectobacterium carotovorum* subsp. *carotovorum*, *Dickeya chrysanthemi*, *Pantoea stewartii* and pathovars of *Xa. campestris* and *Psm. syringae*. In most bacterial species, *hrp* genes are located in two distinct clusters. The large *hrp* gene cluster consists of six to nine transcriptional units, with each transcription unit coding for several proteins. The *hrp* gene cluster of *Psm. syringae* pv. *phaseolicola* is at least 22 kb in size and is composed of at least eight transcriptional units. In most bacteria, *hrp* clusters are presumably located in the chromosome. However the *hrp* cluster of *Ral. solanacearum* is found on an approximately 200 kb mega-plasmid.

Other important groups of genes, in addition to *hrp* genes, are needed by phytopathogenic bacteria to cause disease. The *eps* genes of *Ral. solanacearum* and the *pel* genes of *Pectobacterium* soft-rot bacteria are a few of the many genes encoding pathogenicity or virulence factors that contribute to the disease and symptom development. Significantly, the expression of these genes occurs independently of hrp systems. It appears that the bacteria have evolved multiple mechanisms for the secretion of pathogenicity and virulence factors.

Studies of *hrp* genes and their gene products and their related secretion systems have demonstrated that certain fundamental determinants of pathogenicity are conserved among bacterial pathogens of plants and animals. *hrp* genes encode a secretory and regulatory system dedicated to the transport of pathogenicity and virulence proteins, as well as Avr proteins. The other important roles, which the hrp genes play, are the transcriptional regulation of *avr* genes and the expression of *avr-R* gene-mediated resistance (Lindgren, 1997).

10.5. HARPINS

The *hrp* gene clusters of phytopathogenic bacteria contain genes that encode elicitors of the HR. These *hrp*-encoded elicitors are called harpins. In general, harpins are glycine-rich, hydrophilic molecules that lack cysteine. They also lack amino-terminal signal peptides, which is consistent with their secretion by a Type III pathway. *Psm. syringae* pv. *syringae* produces an hrp-dependent harpin protein, which differs from most Avr proteins, in that it causes cell death, a hypersensitive-like response, when injected in the leaf apoplast. It shows that this harpin functions on the outside of the plant cell. Harpin may associate with liposomes and with bilayer membranes on which it apparently forms pores. Through these pores, water and nutrients from the cell move to the apoplast, where they serve as food for the bacteria. These pores also serve as openings for the movement of other types of effectors into the cells. This harpin-mediated HR also differs from HRs induced by *avr* genes in that it does not depend on a matching *R* gene.

10.6. BACTERIAL SECRETION SYSTEMS

Most determinants of pathogenicity need to be secreted to interact with host cells. Six secretion systems are found in Gram-negative bacteria, which deliver various chemicals, including proteins and nucleic acids, to perform a multitude of biochemical and physiological functions. The secretion systems of plant pathogenic bacteria play a major role in their pathogenicity as they transport bacterial proteins and other molecules, required for pathogenicity, into the host cells. *Pe. atrosepticum* strain 1043 contains genes encoding all the six secretion systems, suggesting a complex secretome and an impressive ability to interact with its environ-ment(s). All the six secretion systems are found in plant pathogenic bacteria.

10.6.1. Type I secretion system (TISS)

It is present nearly in all plant pathogenic bacteria. It secretes proteins of various sizes, often with calcium-binding glycine rich repeats, from the cytoplasm to the extracellular matrix in a single step. Various toxins like hemolysins, rhizobiocin and cyclolysin are also secreted. They consist of ATP-binding cassette proteins and carry out in and out movement of a variety of compounds for which the energy is provided by the hydrolysis of ATP. In the tribe erwiniae, the TISS is required for secretion of proteases, which apparently play only a minor role in pathogenesis. The genome sequence of *Pe. atrosepticum* strain 1043 has revealed additional TISSs. On the whole, TISS plays a minor role in pathogenicity.

10.6.2. Type II secretion system (TIISS)

It is common among Gram-negative bacteria and shares many features with the type IV pilus system. TIISSs are found in several phytopathogens, including xanthomonads, pseudomonads, *Xylella fastidiosa, Ral. solanacearum* and the erwiniae. In *Xa. campestris* pv. *campestris* and *Xa. oryzae* pv. *oryzae,* it is involved in the secretion of plant cell wall-degrading enzymes (PCWDEs), polysaccharide and proteases, and hence plays an important role in virulence. TIISS has been studied in detail in soft rot erwiniae (SRE). It is responsible for the secretion of PCWDEs and is, therefore, essential for soft rot disease development. TIISS is involved in the translocation of various virulence factors including toxins, enzymes and proteins. The translocation of proteins involves a two-step process. In the first step, unfolded proteins move to the periplasm via the Sec pathway across the inner membrane. In the second step, processed and folded proteins move through the periplasm and across the outer membrane via an apparatus consisting of 12 - 14 proteins encoded by a cluster of genes. *Xanthomonas and Ralstonia* species possess two TIISS per cell and use them for secretion of virulence factors such as pectinolytic and cellulolytic enzymes. *Xylella* and *Agrobacterium* species possess only one TIISS per cell. *Agrobacterium* species have the genes only for the first step of protein transport across the inner membrane and these use the TIVSS for the remaining process. TIISS is regulated in part by a quorum-sensing mechanism (Agrios, 2004).

10.6.3. Type III secretion system (TIIISS)

It plays a most important role in the pathogenicity of bacteria belonging to the genera *Xanthomonas, Pseudomonas* and *Ralstonia.* Its primary function is the transport of effector proteins across the bacterial membrane and further into the plant cell. In addition to the above-mentioned three genera, TIIISS also occurs in most Gram-negative bacteria *(Erwinia* and *Pantoea* species) including those causing diseases in humans and animals. The translocation of the bacterial effector proteins into the host cells interferes with host cell signal transduction and other cellular processes, leading to the enhancement of bacterial virulence in the susceptible hosts. Without the TIIISS, phytopathogens cannot defeat basal defences of the host, grow in plants, produce disease symptoms on host or elicit hypersensitive response in nonhosts. TIIISS mutants fail to suppress the basal plant defences. Some effectors also act as virulence proteins that are recognized by corresponding host resistance (R) proteins, triggering the HR that restricts pathogen spread. Alfano and Collmer (2004) have presented a conceptual overview of TIIISS effectors functioning as both elicitors and suppressors of plant defences. As the TIIISS in phytopathogens was first recognized through its association with the HR, the genes encoding it have been named as hypersensitive response and pathogenicity (*hrp*) genes. There are a number of genes in the *hrp* cluster, which show a high level of similarity to those in the animal pathogens, and these have been termed hrp conserved (*hrc*) genes. The *hrp* cluster in the SRE has been demonstrated to play an important role in pathogenesis. *Er. amylovora* (*Eam*), a close relative of SRE, has been best studied of the erwiniae in terms of TIIISS and its effectors. Three TIIISS secreted proteins, namely, DspE/A_{Eam}, $HrpN_{Eam}$ (harpin) and $HrpW_{Eam}$ have been characterized in *Er. amylovora.* DspE/A_{Eam} appears to be widely conserved in many plant pathogens and is currently the only demonstrated effector protein in the erwiniae. $HrpW_{Eam}$ and $HrpN_{Eam}$, on the other hand, have been termed helper proteins that may assist translocation of effectors (such as DspE/A), but are believed to remain outside the host cell (Alfano and Collmer, 2004).

10.6.4. Type IV secretion system (TIVSS)

This secretion system transports macromolecules between bacteria and from bacteria into their eucaryotic hosts. These transferred proteins are very similar to those, which mobilize plasmids among bacteria. The *virB* operon of *Ag. tumefaciens* encodes 11 proteins that form an organized structure. These proteins are involved in the transfer of T-DNA strand from the bacterium to the cytoplasm of the plant cell. The transporting structure originating from bacterial inner membrane stretches through the outer membrane and terminates into a pilus-like structure that protrudes from the bacterial cell (Agrios, 2004). Three other phytopathogenic bacteria, namely, *Pe. atrosepticum* strain 1043, *Xa. citri* subsp. *citri* and *Xyl. fastidiosa* possess gene clusters homologous to *VirB*. The presence of *virB* cluster of *Pe. atrosepticum* strain 1043 within horizontally acquired island 7 suggests that it was acquired through horizontal gene transfer.

10.6.5. Type V (autotransporters) and VI (two-partner secretion) secretion systems

Both of these systems are found in a wide range of Gram-negative bacteria and are involved in transport of large proteins across the outer membrane. Unlike the TIISS and TIIISS, each transporter system secretes only a single protein type although different systems can transport different proteins. *Xanthomonas* and *Xylella* species contain genes for TVSS (autotransporters), which encode surface associated adhesins.

TVISS or two-partner secretion system has been found associated with secretion of hemolysins and cytolysins in *Serratia marcescens*. A homologous system is present in *Xyl. fastidiosa* and *Ral. solancearum*. Two-partner system also occurs in *Pe. atrosepticum* strain 1043, *Er. amylovora* and *D. chrysanthemi* (Toth *et al.*, 2006).

10.7. PATHOGEN EFFECTORS

After penetrating the cell wall, pathogens come in contact with host plasma membrane, where they encounter extracellular surface receptors that recognise pathogen-associated molecular patterns. The term **'pathogen-associated molecular patterns'** (PAMP) was developed by researchers of mammalian innate immune system to describe this type of defence-activating compounds. According to Bent and Mackey (2007), the term **'microbe-associated molecular patterns'** (MAMP) is gaining favour because nonpathogenic microorganisms also possess PAMPs. Well-known examples of MAMPs that are detected by plants include bacterial flagellins, lipopolysaccharides or elongation factor-Tu, fungal chitin, or oomycete Pep-13 or heptaglucosides. Perception of a microorganism at the plasma membrane initiates **MAMP-triggered immunity** (MTI), which usually halts infection before the microbe gains a hold in the plant. However, pathogenic microorganisms have evolved the ways to suppress MIT by interfering with recognition at the plasma membrane or by secreting effector proteins into the plant cell cytosol that presumably alter resistance response. The ability to deliver pathogen proteins directly into plant host cells to alter plant defence is considered a unifying theme among phytopathogens. In a consequence to pathogens' ability to suppress primary defences, plants developed a more specialised mechanism to detect microbes, referred here as **effector-triggered immunity** (ETI). ETI involves the direct or indirect recognition of the microbial proteins used to subvert MTI by plant resistance (R) proteins.

Activation of protein-mediated resistance also suppresses microbial growth, but usually not before limited proliferation/growth of the invading pathogen. To overcome this kind of resistance, pathogens have evolved to adapt effectors to interfere with ETI (Chisholm *et al.,* 2006).

There is growing evidence that during infection, pathogens actively suppress the plant's MAMP-triggered defences. Almost all the pathogenic microbes encode effectors that suppress MAMP-triggered defences but the effectors produced by phytopathogenic bacteria are best characterized. Gram-negative bacterial pathogens have acquired a TIIISS through either horizontal gene transfer or adaptation of the flagellar apparatus. TIIISS enabled bacteria directly deliver effector proteins into plant cells to suppress MAMP defence responses. *Psm. syringae* can secrete about 20 - 30 effectors during infection (Chang *et al.*, 2005). Effectors promote pathogenicity and TIIISS is essential for the development of disease symptoms and bacterial multiplication. It is hypothesized that collective action of effectors alters plant physiology of susceptible hosts to support pathogen growth. Bacterial effector proteins that are delivered to plants can possess enzymatic activity (Table 10.1).

Psm. syringae effectors, AvrPto, AvrRpt2 and AvrRpm1 inhibit defence responses elicited by MAMP recognition. Also wild-type *Xa. campestris* effectors suppress the plant defence responses to the bacterial MAMP lipopolysaccharide. *Xa. campestris* TIIISS deficient bacteria have no effect on MAMP perception. These results show that bacterial effectors delivered by TIIISS may play a crucial role in suppressing MAMP-triggered defence responses elicited by MAMPs.

Bacterial effector proteins have also been shown to activate plant transcription. AvrBs3, AvrXa10 and AvrXa7, members of the *Xanthomonas* AvrBs3 effector family, contain a C-terminal nuclear localization signal (NLS) and an acidic transcriptional activation domain (AAD). These features suggest that this family of effectors functions in the plant nucleus to alter transcription during infection. In fact the NLS of AvrBs3 family is functional, and the AAD of AvrXa10 is capable of transcriptional activation of reporter genes in *Arabidopsis* and yeast. These findings suggest that the AvrBs3 effector family alters plant nuclear gene transcription during pathogenic infection, likely as a means to down regulate host defences.

Three plant signalling molecules, namely, salicylic acid (SA), jasmonic acid (JA) and ethylene regulate plant defence against microbial attack. SA-mediated signalling is critical in establishing local and systemic bacterial resistance, while JA-mediated signalling is induced in response to mechanical wounding and herbivore attack. JA and SA defence pathways are mutually antagonistic and bacterial pathogens exploit this fact to overcome SA-dependent defence responses. The toxin, coronatine, produced by *Pseudomonas* species during infection, acts as a

Table 10.1. Enzymatic activity of biochemically characterized effectors and selected elicitors

Effector	Organism	Biochemical function	Plant target(s)	*R* gene	Phenotype	Reference
AvrRpt2	*Pseudomonas syringae*	Protease[a]	RIN4	RPS2	Cleaves RIN4, interferes with *R* gene-mediated defence, inhibits basal defence and manipulates JA pathway	Reviewed by Mudgett (2005)
AvrB	*Psm. syringae*	_	RIN4	RPM1	RIN4 phosphorylation, manipulates JA pathway	Reviewed by Mudgett (2005)
AvrRpm1	*Psm. syringae*	_	RIN4	RPM1	RIN4 phosphorylation, inhibits basal defence	Reviewed by Mudgett (2005)
HopPtoD2	*Psm. syringae*	Protein phosphatase[a]	_	_	Suppresses programmed cell death and PR expression	Reviewed by Mudgett (2005)
AvrPphB	*Psm. syringae*	Protease[a]	PBS1	RPS5	Cleaves PBS1, manipulates JA pathway	Reviewed by Mudgett (2005)
AvrPtoB	*Psm. syringae*	E3 ligase[a], ubiquitin-conjugating enzyme	_	Pto	_	Janjusevic *et al.* (2005
XopD	*Xa. campestris*	Cysteine[a] protease	SUMO		_	Reviewed by Mudget (2005)
AvrXv4	*Xa. campestris*	Cysteine protease	SUMO	XV4	_	Reviewed by Mudget (2005)
AvrBsT	*Xa. campestris*	Cysteine protease	SUMO	_	_	Reviewed by Mudget (2005)

[a]Biochemical function has been demonstrated *in vitro*.
RIN4 = Resistance regulator
- = Information not available
PR= Pathogenesis-related

JA mimic and contributes to the virulence by suppressing SA-mediated host responses. Coronatine is not the only bacterial factor that interferes with SA-mediated defence responses. Multiple effector proteins such as AvrB, AvrRpt2, AvrPphB, HopPtok and AvrPphE$_{pto}$ have been shown to manipulate the JA pathway in a combined action. Bacterial effectors also prevent cell wall fortification achieved by callous deposition in cell wall appositions (papillae). Three effectors produced by *Psm. syringae,* namely, AvrPto suppresses papillae formation and AvrE and HopPtoM suppress callose deposition during infection.

Of the large number of bacterial effector proteins known todate, only a few have been biochemically characterized. The characterized effectors show enzyme activity and modify host proteins to promote bacterial virulence. Information on their enzymatic functions has come from analyses of their protein structure prediction programmes to uncover conserved catalytic residues and protein folds. The elucidation of biochemical functions of yet uncharacterized effectors and their targets will greatly help in understanding the role of effectors in disease development.

Psm. syringae effector AvrPtoB has been shown to inhibit ETI associated programmed cell death in susceptible tomato plants, although this phenotype did not enable prediction of enzymatic activity. Crystallization of the C-terminal domain of AvrPtoB, which is responsible for inhibiting programmed cell death, reveals homology to components of eucaryotic E3 ubiquitin ligases. Moreover, AvrPtoB was demonstrated to possess ubiquitin ligase activity *in vitro*. Mutation of key residues eliminated the ubiquitin ligase activity of AvrPtoB *in vitro* and its ability to inhibit cell-death *in vivo*. Hence, it is evident that AvrPtoB acts as a mimic of host ubiquitin ligases, transferring ubiquitin to plant proteins involved in the regulation of programmed cell-death.

Effectors XopD, AvrXv4 and AvrBsT produced by *Xanthomonas* species, being cysteine proteases, interfere with the plant SUMO protein conjugation pathway (Table 10.1). SUMO is post-translationally linked to proteins in a way analogous to the ubiquitin conjugation system. In contrast to ubiquitination, sumoylation does not promote protein degradation but instead alters a number of different functional parameters, depending on the protein substrate. Sumoylation alters protein properties like subcellular localization, partnering, DNA binding and activation of transcriptional factors. Host proteins are desumoylated in the presence of these effectors.

Effector AvrRpt2 of *Psm. syringae* is also a cysteine protease and it cleaves the peptide sequence VP × FG × W. The protease activity of this effector is essential for its self-processing activity and also for virulence function within the plant cell. During the infection of *Arabidopsis*, AvrRpt2 cleaves resistance regulator RIN4. Genetic evidence shows that AvrRpt2 also targets additional proteins to promote pathogenesis and several *Arabidopsis* proteins that contain variations of the peptide sequence VP × FG × W, were cleaved by it in a transient expression system. Thus, it is likely that AvrRpt2 eliminates multiple proteins during infection and these are its virulence targets.

Alfano and Collmer (2004) reported that many effectors could act as molecular double agents that betray the pathogens to plant defences in some interactions and suppress host defences in others. AvrPtoB of *Psm. syringae* pv. *tomato* strain DC 3000 is a novel example of an effector acting as a double agent in promoting pathogenesis. AvrPtoB is able to suppress programmed cell-death in tomato plants lacking the *Pto* resistance gene. However, AvrPtoB possesses an N-terminal domain that can elicit the HR in these otherwise susceptible tomato plants and a C-terminal domain that is required to suppress the HR triggered by the N-terminal domain. Therefore, a mutant of DC 3000, which produces only the N-terminal domain of AvrPtoB is avirulent on tomato unless complemented with full-length AvrPtoB. It is worth mentioning that several of the effectors demonstrated to have HR suppressor activity are members of effector families that were previously shown to have avirulence activity. Therefore, many effectors may be double agents, but whether they act as elicitors or suppressors will depend on the complement of R proteins in the host, on the complement of effectors in the bacterium, but most likely also on quantitative factors associated with the timing and level of delivery relative to other effectors.

10.8. QUORUM SENSING

Quorum sensing (QS) is a mechanism by which bacteria communicate from cell to cell via the secretion and detection of small, diffusible signal molecules. There are certain traits that are only expressed when bacteria are crowded together. Traits regulated in phytopathogenic bacteria by QS include the production of extracellular polysaccharides, degradative enzymes, antibiotics, siderophores and pigments, as well as hypersensitive reaction and pathogenicity protein secretion, Ti-plasmid transfer, motility, biofilm formation and epiphytic fitness. Since QS regulatory systems are often required for pathogenesis, interference with QS signalling may offer a means of controlling bacterial diseases of plants. *N*-acyl-homoserine lactones (AHLs) are the most commonly reported type of QS signals. Over 50 species of Proteobacteria, including plant-associated species, produce AHLs. *Ral. solanacearum* and *Xa. campestris* use unique fatty acid and butyrolactone derivatives as QS signals to regulate production of pathogenicity factors. The fact that the expression of pathogenicity factors in a variety of plant pathogenic bacteria is critically dependent on QS is of particular interest to plant pathologists. A brief account of the QS mechanisms operating in four bacterial plant pathogenic species is given below.

10.8.1. *Ag. tumefaciens*

In crown gall, tumours are formed after the transfer of T-DNA from the tumour inducing (Ti) plasmid of *Ag. tumefaciens* into the chromosome of the plant cell. The opines produced in the tumours serve as nutrients for the pathogen. Both AHLs and opines act as chemical signals to induce the transfer of the resident Ti-plasmid to plasmidless agrobacteria present in the galls. QS control is mediated by TraR, a LuxR-type AHL receptor that responds most strongly to *N*-3-oxo-octanoyl-HL. This AHL is synthesized by the LuxI homolog TraI. Transfer of the nopaline-type Ti-plasmid, pTiC58 is mediated by the phosphodiester opines, agrocinopines A and B, while the transfer of octopine-type Ti-plasmid needs octopine. The agrocinopines interact with the AccR repressor to derepress both the coregulated agrocinopine catabolic locus (*acc)* and the linked, divergently transcribed *arc* (*a*grocinopine *r*egulation of *c*onjugation) operon that contains $traR_{nop}$. In the octopine strains, octopine interacts with the OccR transcription factor to activate an operon comprised of the octopine catabolic genes (*occ*), an ABC-type transport system and $traR_{oct}$. TraR regulon of both types of plasmids includes two linked *tra* operons, a separate *trb* operon and the *rep* operon for the replication of Ti-plasmid. The structural components for Ti-plasmid conjugal transfer are encoded by the combined action of *tra* and *trb* operons. In each system, *traR* is part of a separate transcription unit near the *tra* gene cluster, while *traI* is the first gene in the *trb* operon.

Considerable information is available on the molecular and structural basis of QS control of Ti plasmid conjugal transfer. In the nopaline strain C 58, ApoTraR appears to be plasma membrane-associated but disassociates from the membrane when complexed with AHL. AHL-TraR dimerizes and positions the C-terminal helix-turn-helix domain to attach to the major groove of the 18-bp *tra*-box recognition site; *tra*-boxes are located immediately upstream of the *traAFB*, *traCDG*, *traI-trb* and *rep* operons. The activation of *traI-trb* operon by TraR results in creation of a positive feedback loop for *traI*. TraR activity also stimulates *traM* gene expression, which encodes a small anti-activator with affinity for the C-terminal domain of TraR. It is assumed that TraM titrates TraR to prevent Ti plasmid transfer until TraR levels exceed than those of TraM.

In young tumours, sufficient nutrients are present, which favour rapid growth of colonizing agrobacteria over Ti-plasmid maintenance. At this stage, the galls contain a significant population of avirulent agrobacteria due to plasmid loss and these bacteria appear to have a definite growth advantage. At later stages of gall formation, nutrients may be limiting and the bacterial cell density may be sufficiently high to trigger QS to result in Ti-plasmid transfer to more cells in the galls. These cells then would be capable of causing new infections. The higher plasmid copy number would also increase the expression of the opine catabolic genes resulting in more efficient use of the available opines.

10.8.2. *Pe. carotovorum* subsp. *carotovorum*

QS in *Pe. carotovorum* subsp. *carotovorum* controls the population density-dependent expression of pathogenicity factors such as extracellular enzymes and the Hrp secretion system, as well as carbapenem antibiotic production. The primary AHL, 3-oxoC6HL, is produced by the LuxI-like signal synthase CarI. Mutants defective in *carI* do not produce carbapenem, pectolytic enzymes, endoglucanases and proteases, and fail to secrete harpin. Therefore, they are completely nonpathogenic. The Car QS system directly regulates the genes for the production of carbapenem, which constitute the *carA-H* biosynthetic operon. The QS mechanism of soft rot *Pectobacterium* species has been called the **"mob attack"** mechanism of pathogenesis. The bacteria avoid production of pectic enzymes in early stages of infection possibly to avoid triggering a defence response. Early production of cell wall degrading enzymes can release pectic wall fragments that can elicit host defense responses prematurely. The pathogen thus requires to build a sufficient population before risking detection. In an experiment, *carI* was introduced into transgenic tobacco plants. These plants produced 3-oxoC6HL and presumably activated the *Pe. carotovorum* subsp. *carotovorum* to produce pectic enzymes early in infection leading to triggering of a defence response. This increased resistance of transgenic plants to the pathogen suggests that such an early enzyme production might be harmful to the pathogen. The production of carbapenem late in infection avoids the tissue colonization by other bacteria and thus protects the food supply of the pathogen.

10.8.3. *Ral. solanacearum*

Ral. solanacearum causes vascular wilt of many solanaceous and a few non-solanaceous plants. After entering the vascular system, it colonizes the entire plant and its population exceeds 10^{10} cells/cm length of stem tissue. Most of its traits related to infection and virulence are regulated by the Phc (*ph*enotype *c*onversion) regulatory system in a population density-dependent manner. PhcA, a LysR-type transcriptional regulator is the most important component of a complex regulatory system and its activity is regulated by a unique, volatile signal molecule, 3-OH palmitic acid methyl ester (3-OH PAME). *phcA* mutants do not produce extracellular polysaccharide (EPS), pectin methyl esterase and endoglucanase but they have an increased motility and higher production of polygalacturonase and siderophore. 3-OH PAME acts as a QS signal and is used to sense population density and to regulate the production of EPS and degradative enzymes such as pectin methyl esterase, endoglucanase, polygalacturonase etc. Low levels of 3-OH PAME lead to reduced

EPS and exoenzyme synthesis but increased levels of 3-OH PAME promote PhcA activity, resulting in enhanced production of EPS and exoenzymes but decreased motility and siderophore synthesis. Thus, the Phc regulatory system serves as a master control switch to turn off behaviours suited to free-living survival of the bacterium and to turn on these when needed for initial host contact, microcolony/biofilm formation and pathogenesis.

10.8.4. *Pa. stewartii* subsp. *stewartii*

The bacterium causes Stewart's wilt of sweet corn and maize. Inside the plant, the bacterium grows in the apoplast (intercellular spaces) of young leaves causing "water-soaked" lesions and colonizes the xylem vessels leading to subsequent wilting. Vascular occlusion is due to the accumulation of large amounts of EPS on pit membranes. EPS synthesis in *Pa. stewartii* subsp. *stewartii* strain DC 283 is regulated in part by the EsaI/EsaR QS system. The AHL synthase, EsaI, catalyzes the production of 3-oxoC6HL and small amounts of 3-oxoC8HL. The *esaI* gene is constitutively expressed and not subject to EsaR-mediated autoregulation. Mutations in the *esaI* gene cause pleiotropic effects, eliminating ALH production, EPS synthesis and virulence, while the mutations in *esaR* gene result in constitutive, growth-independent hypermucoidy. In contrast, the wild-type strain produces EPS in a population density-dependent manner, with measurable levels detected primarily at population densities > 10^8 cells/ml. The inoculation of *esaI* mutant strain on sweet corn seedlings fails to induce wilting and causes only slight water-soaking as this strain has a functional copy of EsaR that represses EPS synthesis and Hrp-mediated water-soaking. In contrast, the inoculation of wild-type strains results in the death of seedlings. The *esaR* mutant and *esaI* and *esaR* double-deletion mutants are significantly less virulent than the wild-type strains in spite of the fact that they produce large amounts of EPS and express *hrp* genes. These findings clearly show that QS control of pathogenicity genes is important. The reduced virulence of the mutants indicates that early production of some virulence factors such as EPS may be counterproductive unless they are produced at the proper time and location of infection. *In vitro* studies with the wild-type and QS mutants showed that EPS synthesis interferes with the ability of the pathogen to attach to plastic surfaces. The same attachment deficiency could be artificially induced in the *esaI* mutant strain through exogenous addition of AHL, the degree of attachment being inversely proportional to the amount of AHL supplied. In plant essays, the similar behaviour for bacterial attachment to plants was observed; the *esaI* mutants remained localized at the site of infection, whereas the wild-type strains moved at an impressive rate through the vascular system. Another species of *Pantoea*, *Pa. agglomerans* pv. *gypsophilae*, the causal agent of gypsophyla galls, produces C4HL.

Do the plants have a mechanism to counteract QS signals? Plant pathogenic bacteria depend greatly on QS regulation to coordinate their colonization and infection of plant hosts. At present, there are no reports to indicate that the plants have the ability to produce compounds that specifically inhibit the production/expression of bacterial QS signals. However, a significant percentage of soil bacteria do produce enzymes for degrading or inactivating AHLs. The best studied of these enzymes is AiiA, a lactonase from *Bacillus cereus* that opens the lactone ring of all AHL tested, reducing their potency as signals by about 1000-fold. Interestingly when AiiA was expressed in transgenic tobacco and potato plants, the plants showed a high degree of resistance to infection by *Pe. carotovorum* subsp. *carotovorum*, which depends heavily on 3-oxo-C6-HSL-mediated QS for pathogenicity. Therefore, destruction or inactivation of AHL signals in situ could be useful strategy for engineering disease-resistant plants. The information available so far suggests that AHLs are not degraded substantially by plant enzymes, at least in the rhizosphere of soil-grown plants.

Available information shows that some higher plants namely, pea, rice, tomato, soybean and *Medicago truncatula* secrete various compounds that act like (mimic) bacterial QS signals. Such "signal-mimic" compounds may provide plants with important tools to disrupt or manipulate QS regulation in associated bacteria. First of all, QS signal-mimic compounds were found in marine red alga, *Delisera pulchra*. The active compounds were a set of halogenated furanones that are structurally similar to AHLs and interact specifically with AHL receptors. These compounds also inhibit AHL-mediated QS in many bacteria, interfere with biofilm formation and substantially alter the structure of natural bacterial communities on the algal surface in marine environments. Higher plants also produce signal-mimic compounds that can inhibit bacterial responses to an added AHL signal. However, in contrast to the strictly inhibitory activities of the furanone mimics of the *D. pulchra*, many of the signal mimics from higher plants stimulate AHL-induced QS behaviours (Von Bodman *et al.*, 2003).

10.9. ROLE OF TOXINS, ENZYMES AND AUXINS IN PLANT DISEASE DEVELOPMENT

Various chemical compounds produced by plant pathogenic procaryotes play a role in disease development. They may produce all or a part of the disease syndrome. These compounds can be broadly categorized as toxins, enzymes, auxins etc.

10.9.1. Toxins

Several phytopathogenic bacteria produce toxins that increase the severity of the disease caused by the pathogen, i.e., they affect the virulence of the pathogen but are not essential for the pathogen to cause disease. None of the toxins produced by plant pathogenic bacteria is host-specific, i.e., they also affect other host plants, which are not normally attacked by the pathogen in nature. A brief account of some of the important toxins is given below.

10.9.1.1. Tabtoxin: Tabtoxin is produced by *Psm. syringae* pv. *tabaci,* the causal agent of wildfire of tobacco; by strains of this bacterium occurring on other hosts such as bean and soybean and also by other pathovars of *Psm. syringae* attacking maize, oats and coffee. Necrotic spots on leaves produced by toxigenic strains are typically surrounded by yellow haloes. Purified toxin and culture filtrate of the bacterium produce symptoms identical to those of wildfire of tobacco on many plant species belonging to different families, including nonhost species, in addition to tobacco, indicating nonhost specificity of the toxin. The mutants of *Psm. syringae* pv. *tabaci,* which have lost the ability to produce the toxin (Tox$^-$) show reduced virulence and produce necrotic leaf spots without yellow haloes. Tox$^-$ strains are indistinguishable from *Psm. angulata,* the causal agent of angular leaf sport of tobacco, but now the latter is considered the non-toxigenic form of *Psm. syringae* pv. *tabaci.*

Tabtoxin is a dipeptide and consists of two amino acids, namely, L-threonine and tabtoxinine (Fig. 10.2). Tabtoxin as such is not toxic to the plants, but in the plant cells peptidases of bacterial and plant origin hydrolyse it in the presence of zinc and release tabtoxinine, which is the active moiety. Tabtoxinine inactivates the enzyme glutamine synthetase, which results in reduction of glutamine levels and as a consequence of it ammonia accumulates to the toxic concentrations. Accumulation of ammonia results in uncoupling of photorespiration and destruction of thylakoid membrane of the chloroplasts leading to chlorosis and ultimately necrosis of host tissue.

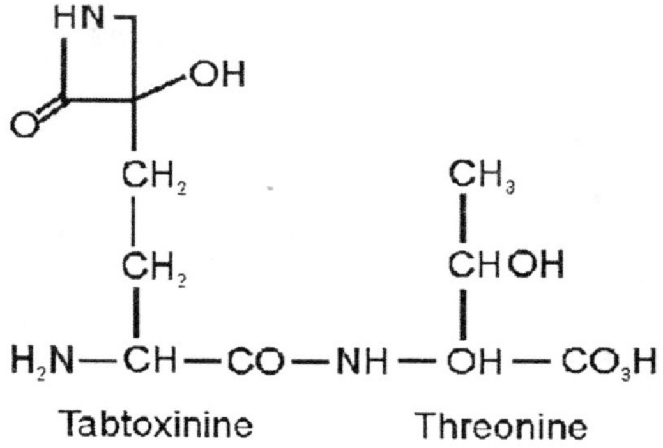

HN
OH
O
CH_2 CH_3
CH_2 CHOH
$H_2N—CH—CO—NH—OH—CO_3H$
Tabtoxinine Threonine

Figure 10.2 Structure of tabtoxin

10.9.1.2. Phaseolotoxin: It is produced by *Psm. syringae* pv. *phaseolicola,* the causal agent of halo blight of bean and some other legumes. The toxin causes localized and systemic chlorosis and in the latter case, the toxin moves from the lesions to the lamina of trifoliate leaves through phloem. The chlorotic symptoms produced on plants treated with toxin alone are identical to the symptoms produced on naturally infected plants indicating that they are the result of toxin produced by the bacterium. In addition to chlorosis, infected plants and plants treated with purified toxin also show reduced growth of newly formed leaves, disruption of apical dominance and accumulation of ornithine.

Phaseolotoxin is a modified tripeptide consisting of ornithine, alanine, arginine and a phosphosulfinyl group. In the plant tissue, plant enzymes cleave the peptide bonds of the toxin and release alanine, arginine and phosphosulfinylornithine. Phosphosulfinylornithine is the active toxin. The toxin causes a reversible inhibition of enzyme ornithine carbamoyltransferase. As this enzyme normally converts ornithine to citrulline, a precursor of arginine, the inactivation of the enzyme results in the accumulation of ornithine and reduced levels of citrulline and arginine. It seems that toxin also interferes with lipid synthesis, reduces the activity of ribosomes, inhibits pyrimidine nucleotide biosynthesis and changes the permeability of membranes. The toxin is responsible for a major loss to the host because it lowers the resistance of the host towards other fungal, bacterial and viral pathogens in addition to the halo blight pathogen (Agrios, 2004). Toxin-producing strains of the bacterium are also sensitive to it but they acquire tolerance to it by producing the toxin insensitive ornithine carbamoyltransferase, which has a markedly different kinetic.

10.9.1.3. Tagetitoxin: Tagetitoxin is produced by *Psm. syringae* pv. *tagetis,* which infects zinnia, marigold, sunflower and artichoke. The application of as little as 20 ng of toxin to marigold and zinnia plants causes apical chlorosis after 2 - 3 days of treatment. It appears that the toxin directly inhibits RNA synthesis by affecting RNA polymerase.

Other nonhost-specific toxins produced by plant pathogenic bacteria include coronatine, syringomycins, syringotoxins, rhizobitoxine, albcidins, thaxtomins etc. Coronatine is produced by *Psm. syringae* pv. *atropurpurea,* the pathogen of chocolate spot disease of Italian Ryegrass and also by *Psm. syringae* pvs. *maculicola*, *morsprunorum* and *tomato*, and *Psm. savastanoi* pv. *glycinea*. The toxin functions by suppressing the induction of defence-related genes. However, it appears that the toxin is not essential for pathogenicity of all the strains of the bacteria. Coronatine causes chlorosis and is also a major virulence factor in *Psm. syringae* pathovars. It is formed by conjugation of the polyketide coronafacic acid to coronamic acid. Coronafacic acid may also be independently coupled to other amino acids, forming a variety of coronafacoyl conjugates. *Cfa* gene cluster and *cma* operon synthesize coronafacic acid and coronamic acid, respectively. Coronafacic ligase enables the conjugation of coronafacic acid to coronamic acid (Bender *et al.,* 1999). Syringomycin is a major virulence factor in *Psm. syringae* and it promotes necrosis by pore formation in the host plasma membrane.

Bell *et al.* (2004) have sequenced the genome of *Er. carotovora* subsp. *atroseptica* (recent name, *Pectobacterium atrosepticum*), the first enterobacterial plant pathogen genome, and have characterized the virulence factors. The genome of *Pe. atrosepticum* contains *syr* genes similar to those required for syringomycin production, which suggests that it may also attack the host plasma membrane to form ion channels and cause autolysis. *Pe. atrosepticum* also possesses *cfa* genes that are highly similar in structure and organization to those in *Psm. syringae* but lacks genes similar to those in the *cma* operon. It also contains the *cfl* gene necessary for ligation of coronafacic acid to amino acids and may thus synthesize coronafacoyl conjugates. Mutations in the genes *cfa6* and *cfa7* significantly reduce pathogenicity in potato, suggesting that coronafacic acid is an important virulence factor of *Pe. atrosepticum.* In *Pe. atrosepticum, cfa* and syringomycin genes are found on horizontally acquired islands 2 and 6, respectively indicating that these have been acquired by horizontal gene transfer during the evolution of the bacterium.

Rhizobitoxine is produced by *Burkholderia andropogonis,* the causal agent of bacterial stripe of sorghum. Albicidins are produced by *Xa. albilineans,* the causal agent of leaf scald of sugarcane. Albicidins block the replication of eucaryotic DNA and development of plastids resulting in chlorosis of emerging leaves. Syringomycins, syringotoxins, syringostatins and syringopeptins are a group of lipodepsipeptides produced by different strains of *Psm. syringae* pv. *syringae.* These toxins are responsible for the antimicrobial and phytotoxic activity of the bacterium. Syringomycins are 30 times more active than syringopeptins in antimicrobial activity but at least 40 times less active in electrolyte leakage assay with carrot tissue. The mode of action of these toxins is related to destruction of plasma membrane associated functions such as K^+ efflux, Ca^{2+} influx, H^+-ATPase activity, membrane potential and protein phosphorilation. Among these functions, Ca^{2+} transport associated with elevation of membrane potential seems to be most important concerning ion channel formation in the lipid bilayer of host cell membrane (Goto, 1992). Thaxtomins are a group of toxins produced by *Streptomyces* species, which cause scab of potato, sugarbeet and radish. These toxins cause cell hypertrophy of affected plant tissue and stunting of seedlings.

10.9.2. Growth regulators

The growth regulators like auxins and cytokinins also play a role in the pathogenesis of certain plant pathogenic bacteria. The role of auxins in plant disease development has been studied more extensively in some bacterial diseases of plants. *Ral. solanacearum,* causal agent of wilt of solanaceous plants produces 100 times more indole-3-acetic acid (IAA) in diseased plants in comparison to healthy plants. The high levels of IAA increase the plasticity of cell walls and this increased plasticity of cell walls renders the protein, pectin and cellulose components of cell walls more accessible to their respective enzymes secreted by the pathogen and also facilitates their degradation by these enzymes. Increased levels of IAA may also be responsible for increased respiration and transpiration of infected plants.

Psm. savastanoi pv. *savastanoi,* the causal agent of knot disease of olives and oleander also produces IAA, which induces infected plants to form galls or knots. The production of IAA is positively correlated with the severity of the disease and the strains that do not produce IAA fail to induce the gall formation. The bacterial genes that control IAA production are present in a plasmid of the bacterium. However, a gene present in the chromosome of the bacterium also synthesizes some IAA. In crown gall disease, IAA and cytokinins are produced in tumour cells. The increased levels of IAA and cytokinins are sufficient to cause autonomous enlargement and division of transformed tumour cells. The cytokinins are also involved in leafy gall and shoot proliferations of herbaceous plants caused by *Rho. fascians* (Putnam and Miller, 2007). The multiple cytokinins can be isolated from supernatants of *Rho. fascians* cell cultures but the amounts of classical cytokinins produced by *Rho. fascians* in cultures are too low to account for leafy gall symptoms. Moreover, the level of cytokinins produced in axenic cultures of virulent isolates are only moderately greater than those produced by avirulent isolates, thus suggesting that the increased level is not sufficient for pathogenicity. It is presumed that a cytokinin-like compound is produced by the bacteria, or that in planta, the close proximity of cytokinins produced by the bacteria have a greater impact on plant cells than those isolated from culture supernatants and applied in bioassays (Eason *et al.,* 1996). The cytokinins also inhibit the degradation of nucleic acids and proteins resulting in delayed onset of senescence.

Virulent isolates of *Rho. fascians* possess a large conjugative linear plasmid of about 200 kb that bears at least three loci required for pathogenicity. The determinants of fasciation induction on the plasmid pFiD188 (of strain D 188) reside within the *fas, att* and *hyp* operons. The *fas* locus contains six genes that encode an isophenyl transferase, plus enzymes that modify the product, and overall produce an unknown molecule that stimulates symptom development. A fourth locus involved in pathogenicity, *vic,* has been found on the bacterial chromosome and is associated with the ability of the bacterium to utilize glyoxylate shunt of the Krebs cycle as an alternate pathway for catabolism of specific nutrients (Vereecke *et al.,* 2002).

10.9.3. Enzymes

Various enzymes, namely, pectinases, cellulases, proteases, esterases and cutinases are involved in the pathogenesis of plant pathogenic bacteria. Among them, role of pectinases has been studied more extensively. Pectinases or pectolytic enzymes cause degradation of pectic substances. Pectic substances are the main components of the middle lamella and act as cementing material between the cell walls. Pectic substances also constitute a major part of the primary cell walls, in which they form an amorphous gel filling the spaces between the cellulose microfibrils. Pectic substances are polysaccharides consisting mostly of chains of galacturonan molecules interspersed with a much smaller number of rhamnose molecules and small side chains of galacturonan, xylan and some other five carbon sugars.

10.9.3.1. Pectinases: Pectinases are involved mainly in plant diseases where soft rotting of affected plants or plant parts occurs. Soft rot of fleshy fruits and vegetables is caused by many bacteria belonging to *Pectobacterium, Xanthomonas, Pseudomonas, Bacillus, Clostridium* and *Dickeya*. Among them, most important are members of *Pe. carotovorum* subsp. *carotovorum, Pe. atrosepticum* and *D. chrysanthemi*. These bacteria generally attack a wide range of dicotyledonous but a limited number of monocotyledonous and woody plants. *Pectobacterium* soft rot bacteria produce a variety of pectinases, namely, pectate lyase or pectic acid transeliminase (endo-PL), pectin lyase or pectin transeliminase (endo-PNL), pectin methylesterase (PME) and polygalacturonase (endo- and exo-PG). The activity of pectolytic enzymes determined using sliced tissues of storage organs such as potato tubers, roots of carrot, radish and turnips, and succulent leaves often does not correlate with the activity determined on intact parts of the actively growing plants due to active defence reactions. The pectolytic activity of these bacteria also varies with different strains.

Among the pectolytic enzymes mentioned above, endo-PL is the major virulence factor but endo-PG of *Pe. carotovorum* subsp. *carotovorum* is also capable of causing maceration of potato tissues. Nonivasion of the host (*Saintpaulia* sp.) leaves by PME-deficient mutant of *D. chrysanthemi* also indicates that PME is involved in the pathogenesis of this pathogen. A common feature of extracellular pectic enzymes produced by *Pectobacterium* spp. is the presence of a single hydrolytic enzyme and multiple isozymes of pectate lyase. Many strains of *D. chrysanthemi* produce five endo-PL isozymes designated as PLa to e. PLa is acidic, PLb and PLc are neutral, and PLd and PLe are alkaline in reaction. These isozymes are different in their isoelectric point but are similar in certain properties and functions. PLa, PLb, PLc, PLd, and PLe are encoded as independent transcriptional units on the loci *pelA, pelB, pelC, pelD,* and *pelE,* respectively. The endo-PL of *Pe. carotovorum* subsp. *carotovorum* is also composed of five isozymes, having different isoelectic points and are designated as PL1 to 5. The endo-PL of *Pe. atrosepticum* comprises of 6 isozymes.

The pectobacteria produce more endo-PL and the enzyme is rapidly induced in these bacteria by pectic compounds. Its production is repressed by glucose. Endo-PNL is induced in many strains of *D. chrysanthemi* and *Pe. carotovorum* subsp. *carotovorum* by UV light, mitomycin C and other DNA-damaging agents. Pectic compounds do not induce this enzyme. Its coordinate induction with bacteriocin or a temperate phage further suggests that it is controlled by the SOS regulatory system (Collmer and Keen, 1986).

A constitutive protein export system controls the secretion of pectinases. In *D. chrysanthemi,* the production of endo-PL is controlled by two constitutive enzymes encoded on two loci *out1* and *out2.* The locus *out1* controls the export of endo-PL from the cytoplasm to the periplasm and it synchronizes with the translation of mRNA. The complete enzyme protein after excision from the leader peptide is released outside the cytoplasmic membrane. Endo-Pl accumulated in the periplasm across the outer membrane is transferred by the locus *out2.* In *D. chrysanthemi,* the *out* genes are scattered on the chromosome, while in *Pe. carotovorum* subsp. *carotovorum,* they are present in a cluster. The export deficient mutants *(out⁻)* are defective in secretion of cellulase and protease, and impaired in motility and virulence (Goto, 1992).

Isozyme PLe plays a major role in the pathogenesis of *D. chrysanthemi* because the amount of this isozyme produced accounts for 50 - 60% of the total isozyme proteins produced in the plant tissue. The maceration activity of PLa is roughly 1/1000 that of PLe and it alone cannot macerate potato tissues. The low maceration activity of PLa is due to its low isoelectric point, which prevents binding of PLa proteins to the negatively charged plant cell walls. However, *D. chrysanthemi* requires the participation of PLa along with the PLe to express its pathogenicity fully for the infection of *Saintpaulia* sp. leaves. There seems to be synergistic effect between these isozymes. Similarly, the combined activity of both endo-PL and endo-PNL also increases the maceration activity (Goto, 1992).

The production of extracellular pectolytic enzymes by pathogens is both constitutive and induced. The availability of pectin polymers and the released galacturonan units regulates the production of these enzymes. It seems that the pathogens at all times produce small amounts of pectolytic enzymes constitutively, which release from the pectin a small number of galacturonan monomers, dimers, or oligomers. These molecules act as inducers for the enhanced synthesis and release of pectolytic enzymes (substrate induction). But at higher concentration, these molecules also repress the synthesis of these enzymes (catabolite repression). In some resistant host-pathogen combinations, pectic fragments are released from the cell wall due to the pectolytic activity, which function as endogenous elicitors of the defence mechanism of the host.

In addition to soft rot bacteria, certain other phytopathogenic bacteria, including *Ral. solanacearum,* produce pectolytic enzymes. The production of pectolytic and cellulolytic enzymes by *Ral. solanacearum* and their role in the pathogenesis of the bacterium has been reviewed by Schell (2000). *Ral. solanacearum* produces one pectin methylesterase (PME) and three polygalacturonases (PG). Three polygalacturonases include one endo-PG, PglA also called PehA(52 kDa) and two exo-PG, PehB (PglB) and PehC (PglC). PglA produces oligogalaturonides as well as mono-, di- and tri-galacturonic acid molecules. The exo-poly-α-D-galacturonosidase, PehB (74 kDa) releases only digalacturonic acid, while PehC produces only monogalacturonic acid. In pectin rich regions of plants, these four enzymes break pectin oligomers and also supply monogalacturonic acid molecules for the growth of the bacterium. PehB-deficient mutants show only slight reduction in their virulence, while PehA (PglA)-deficient mutants take at least double the time than the wild-type to cause wilting and killing of plants. This reduction in virulence is even more striking in soil infestation in comparison to wound inoculation of petioles or stems. Markedly slow invasion and colonization by PehA$^-$ and PehB$^-$ mutants than the wild-types suggests that, in addition to a nutritional role, polygalacturonases enhance movement of the bacteria through pectin rich areas such as middle lamella, pit membranes between xylem vessels and sites of secondary root emergence. As the PME$^-$ mutants show full virulence on certain hosts, the role of PME in wilt is not clear.

10.9.3.2. Cellulases: Some phytopathogenic bacteria, namely, *Ral. solanacearum, Xa. campestris* pv. *campestris* and *D. chrysanthemi* produce cellulases or cellulolytic enzymes. *Ral. solanacearum* produces two extracellular glucanases that can attack β-1,4-glycosidic linkages of cellulose. These enzymes are Egl (42 kDa) that releases cellobiose, cellotriose and cellotetrose from soluble cellulose and CbhA (66 kDa), a β-1 - 4 - exocellobiohydrolase that releases cellobiose from the reducing ends of crystalline cellulose. Egl-deficient mutants of *Ral. solanaearum* are slower to cause disease symptoms than wild-type. The inoculation of Egl-deficient mutants through soil infestation also results in reduced speed and efficiency for root invasion and colonization of stems in comparison to wound inoculation of petioles and stems. It indicates the possibility that Egl facilitates the invasion of roots and/or penetration of xylem vessels by degrading cellulosic glucans in the cell walls. A nutritional role in planta for Egl in biovars 2 and 3 is possible because they oxidize cellobiose but doubtful in biovars 1 and 4 as they do not oxidize cellobiose.

10.9.4. Extracellular polysaccharides

Extracellular polysaccharides or exopolysaccharides (EPS) are produced by many plant pathogenic bacteria and they form a capsule or slime layer around the bacterial cells. EPS protects the bacteria under adverse environmental conditions and also contributes to the virulence of pathogenic bacteria. *Ral. solanacearum, Er. amylovora, Pa. stewartii, Xyl. fastidiosa* and other plant pathogenic bacteria are known to produce EPS.

Polysaccharides play an important role in the virulence of members of the tribe erwiniae. Exopolysaccharides, namely, amylovoran and stewartan produced by *Er. amylovora* and *Pa. stewartii*, respectively block xylem vessels of the host plants to cause their wilting. EPS in these pathogens is produced by genes closely related to *wza* gene cluster in *Escherichia coli*. EPS in *D. chrysanthemi* appears to be coregulated with other virulence factors and mutations in lipopolysaccharide/enterobacterial common antigen structure in both *D. chrysanthemi* and *Pe. atrosepticum* are associated with reduced virulence. *Pe. atrosepticum* strain 1043 contains many genes, which encode enzymes that produce a range of polysaccharides including lipopolysaccharides (e.g. *rfa* and *waa* genes), EPS/O-antigen (e.g. *wza* and *rfb* genes), enterobacterial common antigen (e.g. *wec* and *rff* genes) and membrane-derived oligosaccharides (periplasmic glucans, e.g. *opg* genes), all of which are common among the enterobacteria (Toth *et al.,* 2006). The production of exopolysaccharides and their role in virulence of *Ral. solanacearum* has been reviewed by Schell (2000). *Ral. solanacearum* produces exopolysaccharide I (EPS I), which is a main virulence factor of the bacterium. EPS I is a long polymer ($>10^6$ Da) with a trimeric repeat unit of N-acetyl galactosamine, 2-N-acetyl-2-deoxy-L-galacturonic acid and 2-N-acetyl-4-N-(3-hydroxybutanoyl)-2-4-6-trideoxy-D-glucose. On various culture media and in plants, *Ral. solanacearum* produces large amounts of EPS I, comprising more than 90% of the total exopolysaccharide. Many diverse strains of the bacterium produce EPS I, of which 85% is released by the cells, while only 15% is retained in the form of cell surface bound capsule.

Mutants of *Ral. solanacearum* lacking synthesis of only EPS I rarely wilt or kill plants, even after injection of large number of cells directly into the stems. The stem colonization by EPS I-deficient mutants at a much slower rate and to a much lesser extent than the wild-type suggests that EPS I promotes stem colonization. The accumulation of EPS I in xylem vessels leads to wilting by vascular occlusion or rupture of vessel elements due to excessive hydrostatic pressure. It has also been found that EPS I-deficient mutants do not readily move into the xylem vessels, but rather agglutinate and degenerate in the intercellular spaces of root cortex. A 16 kb *eps* operon encodes many proteins comprising the biosynthetic pathway of EPS I, since mutations throughout *eps* abrogate EPS I production. The *eps* operon contains more than 12 genes; its transcription is directed by a promoter that requires the products of 10 regulatory genes and more than 3 signals for full activation.

The wilting and/or decline symptoms induced by *Xyl. fastidiosa*, another xylem-limited bacteria, also result from mild but chronic water deficiency of the host plant. This is caused by blocking of xylem vessels with bacterial cells or pectic substances. The symptoms of leaf scorch develop when plants suffer from acute water deficiency on account of increased transpiration due to sunlight.

It is evident from the above given information that considerable progress has been made on certain aspects of the host-pathogen relationship. Many *avr* and *R* genes have been identified and cloned and considerable progress has been made in understanding the molecular basis for resistance. The role of QS signals in regulating the genes for epiphytic fitness and for major pathogenicity factors has been highlighted. However, the information on many aspects and mechanisms is scanty and they need to be investigated more thoroughly. The direct interaction between *R* and *avr* gene products of the investigated host pathogen relationships have been found in very few cases. What kind of relationship occurs in other cases, and if of different type, what is the basis for it? As indirectly recognised effectors in many cases can escape detection only by ceasing virulence activity (Bent and Mackcy, 2007), hence their thorough and detailed investigation may give some clues to exploit this act of loss of virulence for the plant disease management. There is a need to translate the knowledge of elicitors, effectors and *R* genes to confer durable disease resistance to obtain an effective and practical disease control. The future work on QS should aim to make changes that are specific to particular pathogens, tissues and developmental windows without impacting beneficial microbes.

REFERENCES

Agrios, G.N. (2004). *Plant Pathology.* 5th ed. Elsevier Academic Press, San Diego, 922 pp.

Alfano, J.R. and Collmer, A. (1997). The type III (Hrp) secretion pathway of plant pathogenic bacteria: trafficking harpins, Avr proteins and death. *J. Bacteriol.* 179: 5655 – 62.

Alfano, J.R. and Collmer, A. (2004). Type III secretion system effector proteins: Double agents in bacterial disease and plant defense. *Annu. Rev. Phytopathol.* 42 : 385 – 414.

Bell, K.S., Sebaihia, M., Pritchard, L. *et al.* (2004). Genome sequence of the enterobacterial phytopathogen *Erwinia carotovora* subsp. *atroseptica* and characterization of virulence factors. *Proc. Natl. Acad. Sci.* USA, 101: 11105 – 10.

Bender, C.L., Alarcon-Chaidez, F. and Gross, D.C. (1999). *Pseudomonas syringae* phytotoxins: mode of action, regulation and biosynthesis by peptide and polyketide synthetases. *Microbiol. Mol. Biol. Rev.* 63: 266 - 92.

Bent, A.F. and Mackey, D. (2007). Elicitors, effectors and *R* genes: The new paradigm and a lifetime supply of questions. *Annu. Rev. Phytopathol.* 45: 399 - 436

Chang, J.H., Urbach, J.M., Law, T.F. *et al.* (2005). A high-throughput, near-saturating screen for type III effector genes from *Pseudomonas syringae. Proc. Natl. Acad. Sci.* USA, 102: 2549 - 54.

Chisholm, S.T., Coaker, G., Day, B. *et al.* (2006). Host-microbe interactions: Shaping the evolution of the plant immune response. *Cell* 124: 803 - 14.

Collmer, A. and Keen, N.T. (1986). The role of pectic enzymes in plant pathogenesis. *Annu. Rev. Phytopathol.* 24: 383 - 409.

Eason, J.R., Morris, R.O. and Jameson, P. (1996). The relationship between virulence and cytokinin production by *Rhodococcus fascians* (Tilford 1936) Goodfellow 1984. *Plant Pathol.* 45: 323 - 31.

Goto, M. (1992). *Fundamentals of Bacterial Plant Pathology*. Academic Press, Inc. San Diego, 342 pp.

Janjusevic, R., Abramovitch, R.B., Martin, G.B. *et al.* (2005). A bacterial inhibitor of host programmed cell death defenses is an E3 ubiquitin ligase. *Science* 311: 222 - 26.

Keen, N.T. (2000). A century of plant pathology: A retrospective view on understanding host-parasite interactions. *Annu. Rev. Phytopathol* .38: 31 - 48.

Klement, Z. (1963). Rapid detection of the pathogenicity of phytopathogenic Pseudomonads. *Nature* (London) 199: 299 - 300.

Leach, J.E. and White, F.F. (1996). Bacterial avirulence genes. *Annu. Rev. Phytopathol.* 34: 153 - 79.

Lindgren, P.B. (1997). The role of *hrp* genes during plant-bacterial interactions. *Annu. Rev. Phytopathol.* 35: 129 - 52.

Mudgett, M.B. (2005). New insights into the function of phytopathogenic bacterial type III effectors in plants. *Annu. Rev. Plant Biol.* 56: 509 - 31

Nester, E. (2008). *Agrobacterium*: The natural genetic engineer: 100 years later. APSnet Feature Story, (http://www.apsnet.org/online/feature/Agrobacterium/)

Putnam, M.L. and Miller, M.L. (2007). *Rhodococcus fascians* in herbaceous perennials. *Plant Dis* . 91: 1064 - 76.

Schell, M.A. (2000). Control of virulence and pathogenicity genes of *Ralstonia solanacearum* by an elaborate sensory network. *Annu. Rev. Phytopathol.* 38: 263 - 92.

Staskawicz, B.J. (2001). Genetics of plant-pathogen interactions specifying plant disease resistance. *Plant Physiol.* 125: 73 - 76.

Staskawicz, B., Dahlbeck, D. and Keen, N.T. (1984). Cloned avirulence gene of *Pseudomonas syringae* pv. *glycinea* determines race-specific incompatibility on *Glycines max* (L.) Merr. *Proc. Natl. Acad. Sci.* USA, 81: 6024 – 28.

Toth, I.K., Pritchard, L. and Birch, P.R.J. (2006). Comparative genomics reveals what makes an enterobacterial plant pathogen. *Annu. Rev. Phytopathol.* 44: 305 – 36.

Vereecke, D., Cornelius, K., Temmerman, W. *et al.* (2002). Chromosomal locus that affects pathogenicity of *Rhodococcus fascians. J. Bacteriol.* 184: 1112–20.

Von Bodman. S.B., Bauer, W.D. and Coplin, D.L. (2003). Quorum sensing in plant-pathogenic bacteria. *Annu. Rev. Phytopathol.* 41: 455 – 82.

Chapter 11

DIAGNOSIS OF BACTERIAL DISEASES OF PLANTS

11.1. DIAGNOSIS

For an effective disease management, the correct diagnosis of a disease based on the identification of its causal agent is essential. The diagnosis of some diseases on the basis of symptoms can be made to a satisfactory level but the identification of the causal agent is necessary for confirmation. For the diagnosis of a plant disease, a comprehensive host list that covers a known disease, its typical symptoms and known potential pathogens for a specific host are required. The compendia of different crop diseases published by American Phytopathological Society are very valuable publications for this purpose. A list of plant diseases found on the American Phytopathological Society website (http://www.apsnet.org/online/common/top.asp) is also a source of valuable information. *Westcott's Plant Disease Handbook* is also useful because specific symptoms associated with each disease are given (Horst, 2001). The photographs of symptoms, especially the coloured ones, also aid in the identification of plant diseases. A list of known pathogens for a given crop greatly reduces the choices to one or two suspected genera for the given symptoms.

The symptoms of plant diseases caused by different groups of pathogens, namely, fungi, bacteria and viruses have many similarities. However, two signs, i.e., water-soaking and bacterial exudation are associated only with most but not all the bacterial diseases. The amount of bacterial ooze varies with the nature of disease and the environmental conditions. The ooze is generally more forceful in vascular diseases and in freshly caused lesions. The intensity of ooze is also more under humid conditions. In a dry atmosphere, the bacterial ooze is converted into scales, flakes or beads, often becoming detectable only with the help of a magnifying glass, especially when it is sparse. Riley *et al.* (2006) have also stressed the importance of signs by stating that signs are much more specific to disease-causing agents than are symptoms and are extremely useful in the diagnosis of a disease and identification of the agent causing the disease.

In bacterial wilt of solanaceous plants, the surface of the transverse section is moister in appearance in comparison to a fungal wilt. White droplets of bacterial exudates generally appear on the cut surface and the ooze can be intensified by squeezing the infected plant tissue. A diagnostic feature of bacterial wilt of cucurbits is coming out of the bacterial ooze from the cut ends of the infected stems. This sticky bacterial mass can be drawn into fine threads by touching the cut end of the stem with a finger (or knife blade) and gently pulling the finger away from the cut end. The fine threads of bacterial ooze can also be seen by pulling away the cut ends of the infected stem (Fig. 13.22.b).

Microscopic examination of diseased plants helps greatly in establishing bacterial nature of the diseases. The bacterial streaming is found in almost all the bacterial diseases except the crown gall. A portion of the lesion cut from the border adjacent to the healthy tissue is mounted in a drop of water and placed under the microscope, without covering it with a cover slip. Clouds of bacteria streaming out of the cut end of the mounted tissue can be seen. The ooze is more forceful in vascular diseases like bacterial blight of rice. Phase contrast microscopy is more suitable for this purpose because starch or latex particles can be distinguished from the bacterial cells. Bacterial ooze test can also be performed in the field by cutting an infected leaf (bacterial blight of rice) or stem (wilt of solanaceous plants) and placing it in a beaker of clean water. After a few minutes, threads of bacterial ooze start coming out of the cut end. In diseases caused by fastidious procaryotes such as spiroplasmas, phytoplasmas and *Xylella fastidiosa*, the vascular bundles of the sectioned tissues are examined with an electron microscope to detect the presence of these pathogens.

The confirmation of the diagnosis should be made by proving the Koch's postulates. To fulfil the Koch's postulates, the causal bacterium should be isolated and purified. After inoculation, it should produce the typical symptoms of the disease on the host, identical to those found in the nature. The bacterium should be reisolated from the inoculated plants and it should be identical to the inoculated pathogen in all respects. Generally, the isolation of the bacterium should be made on a general, non-selective medium instead of a semi-selective medium. A general purpose, non-selective but differential medium containing tetrazolium chloride is useful for distinguishing a potential pathogen from saprophytes during the initial isolation as long as the concentration of the tetrazolium chloride is reduced to 0.001% to avoid inhibitory effects on members of *Xanthomonas* genus. However, the semi-selective media are useful for confirming the presence of suspected pathogens. Semi-selective media are more helpful for isolations from soil but generally not from plant tissues because the latter are generally surface sterilized before isolation. A few of the bacterial plant pathogens such as phytoplasmas and papaya bunchy top pathogen have not been cultured on artificial media. The presence of such pathogens in the host plants should be confirmed by other methods such as electron microscopy.

The ultimate confirmation of a plant pathogen should be done by proving its pathogenicity. However, this requires time, availability of host plants and optimum environmental conditions for disease development. Moreover, in some bacterial pathogens, the time required to produce the disease symptoms is too long. Some strains of *Clavibacter michiganensis* subsp. *michiganensis* require 21 - 28 days to produce the symptoms. On the other hand, the confirmation by pathogenicity tests is often quick for leaf spotting pathogens as they take only 3 - 4 days to produce the symptoms under optimum conditions. For some bacterial pathogens, hypersensitivity test on tobacco, as an alternative to pathogenicity test, is quite useful while it does not work with others. Lelliot and Stead (1987) have given a detailed account for the diagnosis of bacterial diseases of plants by giving their detailed symptoms, diagnostic procedures, host inoculation tests, composition of culture media, and procedures for staining techniques and biochemical tests.

The diagnosis of an undescribed or a new disease is certainly more complicated and requires more skill and expertise than confirming the presence of an already known disease. Therefore, the entire set of diagnostic principles should be used to narrow down the choices, as no simple tests are suitable. More emphasis should be put on microscopic examination and the initial association of the pathogen with the disease syndrome. Before spending lot of time on numerous phenotypic and genotypic tests for pathogen identification, some basic information about the pathogen should be obtained to decide which tests are to be used for further identification. The tests for basic information include Gram-staining reaction, oxidative versus fermentative metabolism, presence or absence of spores and flagella. To further characterize the bacterium, the relatively inexpensive tests include metabolic tests (API strip tests, bio Merieux, Inc.), metabolic substrate analysis (Microlog TM, Biolog, Inc., Hayward, CA), analysis of fatty methyl esters (MIDI, Newark, DE), or 16S rDNA sequence analysis. An extreme caution must be exercised in interpreting results of any one test, especially for identification of an unknown disease caused by an unidentified bacterial pathogen. The past experience emphasizes the importance of performing basic bacteriological tests before accepting a name that comes from a database, regardless of the refined analytically methods used to generate a similarity index.

A polyphasic analysis still appears to be the most reliable approach for identification of new pathogens. After the pathogenicity has been proved and the genus determined with relatively simple basic bacteriological tests, the final identification of the bacterium is interpreted from the results of various genotypic methods. Hu *et al.* (2001) reported that comparison of 16S rDNA sequence data with phenotypic data for type strains assisted in selection of determinative tests that may discriminate distinct taxa for simplified laboratory analyses.

11.2. MONOCLONAL ANTIBODIES

With the development of hybridoma technology, considerable improvement has been made in immunodiagnostic techniques. Monoclonal antibodies (MAbs) produced by this method have a defined specificity to a single epitope. Hybridomas produce consistent antibodies, which give consistent result in different laboratories. Several MAbs produced by plant pathogenic bacteria and used in their epidemiological studies are listed in Table 11.1. Many of these antibodies have been characterized by testing specificities with large number of strains of the target pathogens. Some bacterial taxa are relatively homozygous in the sense that they possess common antigenic determinants and one antibody generally reacts with all or nearly all strains of the taxon. *Cla. michiganensis* subsp. *michiganensis* and *Xa. axonopodis* pv. *pelargonii* fall in this category. On the other hand, many plant pathogenic bacteria like *Pectobacterium atrosepticum* and *Pe. carotovorum* subsp. *carotovorum* are serologically heterogeneous, therefore, not all members of the population react with a polyclonal antiserum or a taxon-specific MAb. Pathogen specific MAbs that react with heat stable lipopolysaccharides enable development of robust detection kits for use in rapid laboratory and field diagnosis. MAbs directed towards capsule and/or extracellular polysaccharides in pathogens, such as *Cla. michiganensis* subsp. *michiganensis* and *Ral. solanacearum*, are useful in a number of immunodiagnostic formats.

Table 11.1. Taxon-specific monoclonal antibodies produced for plant pathogenic bacteria[a]

Genus, species/ subspecies or pathovar	MAb designation	No. of target strains tested	Reference*
Acidovorax avenae subsp. *citrulli*	3D1F3	26	M. Bandla (personal communication)
Agrobacterium tumefaciens biovar 3	A6F21-1D3G7C8		Bishop *et al.* (1989). *Phytopathology* 79: 995 - 98.
Clavibacter	Cmm1	88	Alvarez *et al.* (1993).

michiganensis subsp. *michiganensis*			*Phytopathology* 83: 1405.
Cla. michiganensis subsp. *sepedonicus*	McAb1 to McAb-5	19	De Boer and Wieczørek (1984). *Phytopathology* 74: 1431 - 34.
Erwinia carotovora subsp. *atroseptica*	14/18.6, 14/2, 14/8.6	3	Vernon-Shirley and Burns (1992). *J. Appl. Bacteriol.* 72: 97 - 102.
Er. chrysanthemi	6A6	36	Singh *et al.* (1999). *Plant Dis.* 84: 443 - 48.
Er. stewartii	C/G7/B2	43	Lamka *et al.* (1991). *Phytopathology* 81: 839 - 46.
Pseudomonas avenae	Pal to Pa-5	20	Alvarez *et al.* (1993). *Proc. Symp. Seed Health Test. 1st,* Ottawa, Ont., Canada.
Psm. syringae pv. *phaseolicola*	Six MAbs[b]	223	Ovod *et al.* (1995). *Phytopathology* 85: 226 - 32.
	AG-1, AG-2	9	Wong, W.C. (1990). *Lett. Appl. Microbiol.* 10: 241 - 44.
Ralstonia solanacearum species-specific	Rs1, Rs1a	75	Alvarez *et al.* (1992). Presented at Bact. Wilt Int. Conf. Kaohsiung, Taiwan.
strain-specific	MAb		Griep *et al.* (1998). *Phytopathology* 88: 795 - 803.
Xanthomonas albilineans	12 MAbs	1	Tsai *et al.* (1990). *Plant Prot. Bull.* 32: 125 - 35.
	7 MAbs	38	Alvarez *et al.* (1996). *Plant Pathol.*45: 358 - 66.
Xa. axonopodis pv. *begoniae*	Xcb-1	26	Benedict *et al.* (1990). *Appl. Environ. Microbiol.* 56: 572 - 74
Xa. axonopodis pv. *citri*	X-4600	30	Permar and Gottwald (1989). *Phytopathology* 79: 780 - 83.
Xa. axonopodis pv. *dieffenbachiae*	Xcd 1, Xcd 3, Xcd7	329	Lipp *et al.* (1992). *Phytopathology* 82: 677 - 82.
	Xcd 108	302	Norman and Alvarez (1994). *Plant Dis.* 78: 954 - 58.
Xa. campestris pv. *campestris*	X9, X13, X17	200	Alvarez *et al.* (1985). *Phytopathology* 75: 722 - 28.
	10C5, 20H6, 16B5, 17C12, 10H12, 11B6	37	Franken A.A.J.M. (1990). *Proc. Symp. Perspect. Monoclon. Antib. Agric.* PUDOC Wageningen, Neth.

Xa. hortorum pv. *pelargonii*	Xcp-1	76	Benedict *et al.* (1990). *Appl. Environ. Microbiol.* 56: 572 - 74.
Xa. hortorum pv. *phaseoli*	XP2	18	Wong, W.C. (1990). *Lett. Appl. Microbiol.* 10: 241 - 44.
Xa. oryzae pv. *oryzae*	Xco1, Xco2, Xco5	178	Benedict *et al.* (1989). *Phytopathology* 79: 322- 28.
Xa. oryzae pv. *oryzicola*	Xco1a	8	Benedict *et al.* (1989). *Phytopathology* 79: 322 - 28.
Xa. translucens pv. *undulosa*	AB3-B6	44	Bragard *et al.* (1993). *Fitopathol. Bras.* 18: 42 - 50.
Xa. campestris pv. *mangiferaeindicae*	XCM-1-XCM-6	4	Sanders *et al.* (1994). *J. Appl. Bacteriol.* 77: 509 – 18.
Xylophilus ampelinus		63	Gorris *et al.* (1989) *Proc. 7th Int. Conf. Plant Path. Bact.* Budapest, pp. 913 - 21.

[a]Data summarized from Alvarez (2004)
[b] Two MAbs (Ps core-1 and Ps core-2), specific to core lipopolysaccharide and four O-chain-specific MAbs (Ps-O:2 - 1, Ps-O:2 - 2, Ps-O:2 - 3 and Ps-O:3 - 1) were used to classify 223 strains belonging to 19 pathovars of *Psm. syringae.*
*References included in the table are not given under the references.

Most of the MAbs produced for different bacterial plant pathogens have been tested using ELISA. The ability of ELISA assays to detect 10^{5-6} cfu ml^{-1} is adequate for the identification of bacterial pathogens from symptomatic plants and colonies on the selective media. The sensitivity of ELISA can be enhanced ten-fold by using an extraction buffer containing ethylenediaminetetra acetic acid and lysozyme, which releases lipopolysaccharides into solution, thereby enhancing the antibody-antigen reaction without increasing background readings. ELISA techniques using both polyclonal and monoclonal antibodies are available for several taxa of phytopathogenic bacteria, and rapid detection kits are also commercially available. Multitarget ELISA, which can detect more than one species in the same ELISA plate well using different enzyme labels, have been developed by Agdia, Inc. A multiple ELISA for detection of *Cla. michiganensis* subsp. *michiganensis* and *Xa. axonopodis* pv. *vesicatoria* is currently available (Alvarez, 2004).

11.3. FLOW CYTOMETRY

The development of flow cytometry has enhanced the immunodiagnostic detection of bacteria. In this technique, as the bacterial cells or other particles pass individually through a sensor in a liquid stream, they are rapidly identified and quantified. Cells are identified by conjugation of fluorescent dyes to specific antibodies and multiple cellular parameters are determined simultaneously based on cell's fluorescence and its ability to scatter light. The cells may be sorted electronically, permitting purification and/or culture of subpopulations of selected cells for further confirmatory tests (Alvarez, 2001; Alvarez and Adams, 1999). Flow cytometry has been used for the detection of *Cla. michiganensis* subsp. *michiganensis* in tomato seed extracts, of *Xa. axonopodis* pv. *dieffenbachiae* in anthurium (Alvarez and Adams, 1999), of *Xa. campestris* pv. *campestris* in seed extracts of *Brassica* sp. (Chitarra *et al.*, 2002) and to determine the viability of *Ral. solanacearum* in seed potatoes (Van der Wolf *et al.*, 2004).

Other methods such as immunofluorescence, immunomagnetic separation and lateral flow devices are also used for the detection of plant pathogenic bacteria. Immunofluorescence is widely used in Europe to detect bacterial pathogens in seed and propagating materials. In the Netherlands, it is used to screen 60,000 potato seed pieces annually to determine the presence of *Ral. solanacearum* (Van der Wolf and Schoen, 2004). It is also used to detect *Cla. michiganensis* subsp. *sepedonicus* in potato seed pieces and in France, it is used to screen tomato seed lots for the presence of *Cla. michiganensis* subsp. *michiganensis*. Lateral flow devices use the principles, primarily those of ELISA, but various types of filters used act as solid support for the initial binding reaction (Danks and Barker, 2000). A lateral flow device test kit, developed by Central Science Laboratory, UK, detects *Ral. solanacearum* in a 3-minute single step process. Rapid ImmunoStrip® kits for detection of *Ral. solanacearum, Cla. michiganensis* subsp. *michiganensis* and *Xa. hortorum* pv. *pelargonii* are sold by Agdia, Inc. In immunomagnetic separations, target cells are isolated from a mixed solution using paramagnetic polystyrene beads coated with specific antibodies. After washing, bound cells can be used for polymerase chain reaction or they can be grown on semi-selective media as they are viable.

11.4. GENOMIC TECHNIQUES

During the last two decades, many genomic techniques have been developed for the detection of known pathogens in plant samples. Several workers have reported the development and/or application of pathogen-specific primers for detection of bacterial plant pathogens in heterogeneous mixtures. Most of these involve amplification by PCR coupled with one or more other techniques. Multiplex PCR is used to identify several pathogens simultaneously (Bertolini *et al.*, 2003). A new cooperational PCR (Co-PCR) has also been developed to detect *Ral. solanacearum* in water (Caruso *et al.*, 2003). Another PCR technique, real-time PCR is rapid and more accurate for the detection of bacterial pathogens and has several other advantages. Other genomic methods include DNA-hybridization, dot-blots and nucleic acid sequence-based amplifications.

The main emphasis has been on the development of specific probes for the detection of target pathogens. In 1994, Maulis and associates developed pathovar specific probes and PCR tests for *Xa. campestris* pv. *pelargonii*, which could detect 10 - 50 cfu of the bacterium per sample. Therefore, these tests should be able to detect the pathogens in asymptomatic plant tissue also. The primers and a probe were also developed for the detection of *Psm. syringae* pv. *actinidiae* in asymptomatic kiwi fruits (Koh and Nou, 2002). Cubero and Graham (2002) developed primer sets for *Xa. axonopodis* pv. *citri* that can distinguish pathotype A from *Xa. aurantifolii* pathotypes B and C. Primers were based on sequence differences in the ITS region and *pthA* gene. Primer sets based on ribosomal sequences had high level of specificity for *Xa. axonopodis* pv. *citri* A (of citrus canker type A), whereas those based on *pthA* were universal for pathogens of all types of citrus bacterial cankers.

Different methods developed for the detection and identification of plant pathogenic bacteria have merits and demerits. Genomic techniques are quick and have more specificity and sensitivity but they have not been adopted in all the laboratories due to certain limitations. For some laboratories, the cost of chemicals and equipment, and the requirement of trained personnel for DNA extraction are the limitations for using PCR in the routine. Therefore, keeping in view the merits and demerits of all the available methods, Alvarez (2004) has rightly advocated the adoption of integrated approaches for the detection of plant pathogenic bacteria and diagnosis of bacterial diseases.

11.5. ENRICHMENT TECHNIQUES

The enrichment technique has frequently been used to increase the sensitivity of IF, ELISA, and/or PCR. An enrichment-ELISA assay using pathogen-specific MAbs was used to detect *Xa. axonopodis* pv. *dieffenbachiae* in leaf samples, *Pe. atrosepticum* or *Ral. solanacearum* in potato seed pieces. Alvarez *et al.* (1997) reported that even a limited culturing step, which is still insufficient for colony identification, increases the sensitivity of ELISA by about 10,000-fold. Enrichment assays have the added advantage that only the viable cells multiply to give a positive signal. Immunofluorescence colony-staining (IFC) technique combines an enrichment step in pour plates for colony detection with fluorochrome labeled antibodies and also increases the sensitivity of IF approximately 10^4 times (Veena and Van Vuurde, 2002). The microcolonies formed due to the multiplication of viable cells can be seen with an epifluorescence stereomicroscope at 40 - 60X whereas individual (dead) cells are not visible at this low magnification. IFC is used mainly for epidemiological studies and for confirmation of results achieved with other detection methods, which do not distinguish between living and dead cells. Optimization of buffers and conjugates also improves test results.

Antibodies of subclass IgG are highly suitable for IFC formats because their molecules are smaller than those of IgAs (dimers) and IgMs (pentamers) and they diffuse readily through agarose gel.

The enrichment of target bacteria on semi-selective media also improves the sensitivity of PCR reaction. Such tests are often termed "BIO-PCR" because living cells of colonies harvested from culture plates are used in the PCR reaction. The major bands result from amplifications of DNA from living cells, although small amplicons may also be formed from dead bacteria. Sakthivel *et al.* (2001) were able to detect 55 fg of DNA per reaction tube which was equivalent to 70 cfu of *Xa. oryzae* pv. *oryzae* ml^{-1} of original sample and about 7 cells per reaction tube. To detect *Burkholderia cepacia* by a "BIO-PCR" assay, contaminated samples were incubated only for 24 hr in broth; DNA was extracted and added to PCR beads with specific primers. The entire process was completed in 27 hr in comparison to 5 - 6 days required for isolation and identification of the target bacterium.

Certain compounds that interfere with DNA amplification have limited the use of direct PCR for the detection of target bacteria from natural samples. In a number of such cases, immunocapture or immunomagnetic separation has been employed to solve the problem. The optimum time for immunocapture of *Psm. syringae* pv. *syringae* with Advanced Magnetics™ beads coated with a polyclonal antiserum was only one hour incubation. The detection limits were significantly improved when immunocapture was followed by a specific PCR test. Walcott and Gitaitis (2000) reported that immunomagnetic separation plus PCR resulted in a 100-fold increase in detection of *Aci. avenae* subsp. *citrulli* over direct PCR. Van Overbeek *et al.* (2002) also supported the polyphasic approach for studying the interaction between *Ral. solanacearum* and its biocontrol agent, *Psm. corrugata*.

Schaad *et al.* (2003) have reviewed the advances made in molecular-based diagnostic techniques used for the detection of plant pathogens including bacteria. They have compared the efficiency and sensitivity of conventional PCR, real-time PCR, BIO-PCR and immunocapture PCR. Real-time BIO-PCR is much more sensitive than conventional real-time PCR and it can detect 2 cfu/ml of *Psm. syringae* pv. *phaseolicola* in seed extracts and of *Cla. michiganensis* pv. *sepedonicus* in potato tuber extracts. Multiplex PCR can be used to detect more than one species of bacteria in the same reaction tube using probes labeled with different fluorescent dyes. To avoid cross absorption, the wavelength of each dye must be well separated.

Most recently, Miller *et al.* (2009) have stressed the importance of early and accurate diagnoses and pathogen surveillance on local, regional and global scales to predict outbreaks and allow time for development and application of mitigation strategies. Plant disease diagnostic networks have developed worldwide to address the problems of efficient and effective disease diagnosis and pathogen detection, engendering cooperation of institutes and experts within countries and across national borders. The internet has opened global avenues for access to databases, communication and cooperation not thought possible decades ago. Without this innovation, the organization of effective networks to meet the challenge of invasive pathogens was unlikely. Case studies of regional, national and international diagnostic networks are presented. The authors have also emphasized the importance of morphological identification and systematics for the detection and identification of the pathogens in spite of frequent use of molecular methods in developed countries.

REFERENCES

Alvarez, A.M. (2001). Differentiation of bacterial populations in seed extracts by flow cytometry. In: S.H. De Boer (ed.). *Plant Pathogenic Bacteria*. Kluwer, Dordrecht, pp. 393 - 96.

Alvarez, A.M. (2004). Integrated approaches for detection of plant pathogenic bacteria and diagnosis of bacterial diseases. *Annu. Rev. Phytopathol.* 42: 339 - 66.

Alvarez, A.M. and Adams, P.D. (1999) Flow cytometry: A promising tool for seed health testing. *Proc. Third Int. Seed Health Symp.* Ames, Iowa, Int. Seed Test. Assoc. pp. 110 - 14.

Alvarez, A.M., Rehman, F.U. and Leach, J.E. (1997). *Comparison of serological and molecular methods for detection of Xanthomonas oryzae pv. oryzae in rice seed*. Presented at Seed Health Test: Progr. Towards the 21st Century, Cambridge, UK.

Bertolini, E., Olmas, A., Lopez, M.M. *et al.* (2003). Multiplex nested RT-PCR in a single closed tube for sensitive and simultaneous detection of four RNA viruses and *Pseudomonas savastanoi* pv. *savastanoi* in olive trees. *Phytopathology* 93: 286 - 92.

Caruso, P., Bertolini, E., Cambra, M. *et al.* (2003). A new and sensitive Co-operational polymerase chain reaction (Co-PCR) for rapid detection of *Ralstonia solanacearum* in water. *J. Microbiol. Methods* 55: 257 - 72.

Chitarra, L.G., Langerak, C.J., Bergervoet, J.H. *et al.* (2002). Detection of the plant pathogenic bacterium, *Xanthomonas campestris* pv. *campestris* in seed extracts of *Brassica* sp. applying fluorescent antibodies and flow cytometry. *Cytometry* 47: 118 - 26.

Cubero, J. and Graham, J.H. (2002). Genetic relationship among worldwide strains of *Xanthomonas* causing canker in citrus species and design of new primers for their identification by PCR. *Appl. Environ. Microbiol.* 68: 157 - 64.

Danks, C. and Barker, I. (2000). On-site detection of plant pathogens using lateral flow devices. *Bull. OEPP/EPPO* 30: 421 - 26.

Horst, R.K. (2001). *Westcott's Plant Disease Handbook.* 6th Ed. Kluwer Academic Publishers, Boston, M. A.

Hu F.P., Young, J.M., Triggs, C.M. *et al.* (2001). Relationships within the Proteobacteria of plant pathogenic *Acidovorax* species and subspecies, *Burkholderia* species and *Herbaspirillum rubrisubalbicans* by sequence analysis of 16S rDNA, numerical analysis and determinative tests. *Antonie Van Leeuwenhoek J. Microbiol. Serol.* 80: 201 - 14.

Koh, Y.J. and Nou, I.S. (2002). DNA markers for identification of *Pseudomonas syringae* pv. *actinidiae. Mol. Cells* 13: 309 - 14.

Lelliot, R.A. and Stead, D.E. (1987). *Methods for the Diagnosis of Bacterial Diseases of Plants.* Blackwell Scientific Publications, Oxford, 215 pp.

Miller, S.A., Beed, F.D. and Harmon, C.L. (2009). Plant disease diagnostic capabilities and networks. *Annu. Rev. Phytopathol.* 47: 15 - 38.

Riley, M.B., Williamson, M.R. and Maloy, O. (2006). Plant disease diagnosis. APSnet Feature Story, [http://www.apsnet.org/online/feature/plant disease/].

Sakthivel, N., Mortensen, C.N. and Mathur, S.B. (2001). Detection of *Xanthomonas oryzae* pv. *oryzae* in artificially inoculated and naturally infected rice seeds and plants by molecular techniques. *Appl. Microbiol. Biotechnol.* 56: 435 - 41.

Schaad, N.W., Frederick, R.D., Shaw, J. *et al.* (2003). Advances in molecular-based diagnostics in meeting crop biosecurity and phytosanitary issues. *Annu. Rev. Phytopathol.* 41: 305 - 24.

Van der Wolf, J.M. and Schoen, C.D. (2004). *Bacterial Pathogens: Detection and Identification Methods.* Marcel Dekker, New York, pp.1 - 5.

Van der Wolf, J.M., Sledz, V., Van Elsas, J.D. *et al.* (2004). Flow cytometry to detect *Ralstonia solanacearum* and to assess viability. In: C. Allen, P. Prior and A.C. Hayward (eds). *Bacterial Wilt: The Disease and the Ralstonia solanacearum Species Complex.* APS Press, St. Paul, Minnesota, p 6.

Van Overbeek, L.S., Cassidy, M., Kozdroz, J. *et al.* (2002). A polyphasic approach for studying the interaction between *Ralstonia solanacearum* and potential control agents in the tomato phytosphere. *J. Microbiol. Methods* 48: 69 - 86.

Veena, M.S. and van Vuurde, J.W. (2002). Indirect immunofluorescence colony staining method for detecting bacterial pathogens of tomato. *J. Microbiol. Methods* 49: 11 - 17.

Walcott, R.R. and Gitaitis, R.D. (2000). Detection of *Acidovorax avenae* subsp. *citrulli* in watermelon seed using immunomagnetic separation and the polymerase chain reaction. *Plant Dis.* 84: 470 - 74.

Chapter 12

DISEASE MANAGEMENT WITH BIOCONTROL AGENTS, BACTERIOPHAGES, BACTERIOCINS AND SIDEROPHORES

Plant pathogenic bacteria cause devastating losses to crop plants and ultimately huge sufferings to the mankind. Bacterial plant pathogens are very difficult to control due to their rapid spread in the crops. The problem is further complicated due to the limited number of chemicals available for their control. At present, only antibiotics like streptomycin, Streptocycline etc. and copper fungicides are mainly used for their management. The use of antibiotics is restricted in many countries due to the risk of development of bacterial strains resistant to these antibiotics and transmission of this character to the human bacterial pathogens. The reports of plant pathogenic bacteria developing resistance to copper fungicides have also appeared in the literature. Hence, there is an urgent need to develop new management strategies free from these drawbacks. Nowadays, more emphasis is laid on biological control and the use of bacteriophages, bacteriocins and siderophores for the management of bacterial plant pathogens. Bacteriophages, bacteriocins and siderophores are also components of the biological control. A brief account of management of bacterial plant pathogens achieved through the use of biological control, bacteriophages, bacteriocins and siderophores is given in this chapter.

12.1. BIOLOGICAL CONTROL

Bacterial diseases of plants are managed using various approaches including use of chemicals. There are many drawbacks associated with chemical use. The major problem is the development of pathogen strains resistant to the chemicals. The occurrence of resistant strains of *Erwinia amylovora*, cause of fire blight of pome fruits and of *Xanthomonas vesicatoria*, cause of bacterial spot of tomato and pepper, to streptomycin is well known. The strains of plant pathogenic bacteria, resistant to copper fungicides are also found where these fungicides have been repeatedly used for a long time. The use of chemicals also causes environmental pollution and residue problems. Moreover, nowadays, a demand for pesticide free products is also increasing rapidly. In certain cases, the use of chemicals is not permitted (on peach fruits before harvest) due to residue problem or if the product is to be certified as organic.

Keeping in view the above-mentioned limitations associated with chemical use, there is an urgent need to develop an ecofriendly alternative to chemical use. Biological control offers an attractive alternative to chemical use. Biocontrol is fundamentally an applied ecology. More specifically, the goal is to manage a microbial community to favour the biocontrol agent (s) and disfavour the pathogens. Exclusion of latecomers by a pioneering species is a general principal of community ecology and a fundamental tenet of biological control (Andrews, 1992).

Biological control of plant diseases is gaining importance. It has been tried successfully in many diseases and some of the diseases are being controlled on commercial scale. Its success can also be gauzed from the number of publications and review papers published on the topic. Important review papers published during the last two decades include Weller (1988), Wilson and Wisniewski (1989), Andrews (1992), Sutton and Peng (1993), van Loon *et al.* (1998), Paulitz and Belanger (2001), Fravel (2005) and Compant *et al.* (2005).

For a successful biological control, biocontrol agent must colonize the host surface. The establishment of a biocontrol agent is necessary before the arrival of the pathogen when the antagonist is not an aggressive colonizer. Colonization by an antagonist is a general requirement for biological control, and thus microorganisms that grow well on the host surface are usually better candidates than those that do not. It is generally believed that resident microorganisms generally work better as biocontrol agents. However, in certain cases, the introduced microorganisms do equally well. Soil-borne bacteria can colonize pear flowers and give control of fire blight almost as good as commercial bactericides. The deciding factor for residency vs. transiency is whether the invaders can survive the unfavourable weather conditions and not whether they can grow under favourable conditions.

The search for biocontrol agents should be made in the suppressive soils, from the rhizosphere of healthy plants in an infected crop or from a healthy crop growing in a pathogen infected area. For the control of post-harvest diseases of fruits, microorganisms, which quickly and thoroughly colonize fruit surface, should be selected. Biocontrol agents are generally screened *in vitro* before testing them in the greenhouse or in the field. *In vitro* tests are generally poor predictors of field performance. Andrews (1992) has given a scheme for screening of biocontrol agents.

The understanding of the mechanism(s) of a biocontrol agent can facilitate optimization of control, as well as help to screen for more efficient strains of the agent. Biocontrol agents employ different mechanisms to check the growth of the pathogens. These mechanisms include compet-ition for nutrients and host space, production of antibiotics, bacteriocins and siderophores, induction of systemic resistance etc. Most of the biocontrol agents employ more than one mechanism to antagonise the target pathogen.

12.2. PLANT GROWTH-PROMOTING RHIZOBACTERIA

Many plant-associated bacteria improve plant health by inhibiting the growth of major and minor (does not attack plants directly but produce metabolites harmful to the plants) pathogens. These bacteria are called **plant growth-promoting bacteria** (PGPB). The most widely studied group of PGPB are the plant growth-promoting rhizobacteria (PGPR) that colonize the root surfaces and the rhizosphere. Kloepper *et al.* (1999) and Gray and Smith (2005) have reported that some of these PGPR can also enter inside the roots and establish as endophytes. Many of them are able to cross the endodermis barrier, moving from the root cortex to the vascular system and ultimately live as endophytes in stem, leaves, tubers and other organs. The major mechanisms of biocontrol employed by PGPB include competition for an ecological niche or a substrate, production of inhibitory metabolites and induction of systemic resistance in host plants to a broad spectrum of pathogens. Compant *et al.* (2005) has published an excellent review on biocontrol of plant diseases through the use of PGPB, highlighting the principles, mechanisms of action and future prospects.

PGPB control the plant pathogens by the synthesis of allelochemicals, including iron-chelating siderophores, antibiotics, biocidal volatiles, lytic enzymes and detoxification compounds. A variety of antibiotics including amphisin, 2,4-diacetylphloroglucinol (DAPG), oomycin A, phenazine, HCN, tropolone, pyoluteorin, pyrrolnitrin, tensin and cyclic lipopetides are produced by pseudomonads, and oligomycin A, Zwittermicin A, xanthobaccin and kanosamine are produced by *Bacillus, Streptomyces* and *Stenotrophomonas* spp. (Compant *et al.*, 2005). An esterase produced by *Pantoea dispersa* causes an irreversible detoxification of albicidin produced by *Xa. albilineans*, cause of leaf scald of sugarcane. The detoxification coupled with biocontrol efficacy leads to an excellent control of the disease (Amusa and Odunbaku, 2007). Velusamy *et al.* (2006) reported the production of DAPG inhibitory to *Xa. oryzae* pv. *oryzae,* by plant-associated *Pseudomonas fluorescens* strains and the production was detected on the basis of PCR screening and confirmed by HPLC, HNMR and IR analyses. They further reported that DAPG suppressed bacterial blight of rice up to 59 - 64% in nethouse and field experiments and Tn5 mutants defective in DAPG production (Phl$^-$) were much less effective in suppressing the disease. Biocontrol activity through the synthesis of alellochemicals, commonly found in free-living rhizobacteria, have also been found in endophytic bacteria. Sessitsch *et al.* (2004) reported that certain endophytic bacteria isolated from field-grown potato plants reduced *in vitro* the growth of *St. scabiei* and *Xa. campestris* through production of siderophores and antibiotics. An endophytic *Psm. fluorescens* strain FPT 9601 has been reported to synthesize DAPG and deposit DAPG crystals in and around roots of tomato.

12.3. INDUCED SYSTEMIC RESISTANCE

Certain PGPB trigger a phenomenon known as induced systemic resistance (ISR) to protect the plants from different types of pathogens, i.e., fungal, bacterial and viral. Manifestation of ISR is dependant on the combination of host plant and bacterial strain. Most reports of PGPB-mediated ISR pertain to free-living rhizobacteria, but endophytic bacteria have also been reported to trigger ISR. **Systemic acquired resistance** (SAR), a phenomenon similar to ISR, is triggered in plants by pathogens or weakly virulent strains of pathogens. In this case, the defence mechanisms of the host react to the invasion of the pathogen leading to hypersensitive reaction due to which pathogen remains confined to local necrotic lesions of brown necrotic desiccated tissue. A major distinction often drawn between ISR and SAR, is that the latter is dependant on accumulation of salicylic acid (SA). Majority of PGPB that activate ISR appear to do it via a SA-independent pathway involving jasmonate and ethylene signals. However, some PGPB do trigger an SA-dependent signalling pathway by producing very small amounts (nanograms) of SA in the rhizosphere.

Ba. subtilis GBO 3 and *Ba. amyloquefaciens* IN 937a activate an ISR pathway by secreting volatiles in *Arabidopsis* seedlings challenged with the soft rot pathogen, *Pectobacterium carotovorum* subsp. *carotovorum*. PGPB activated ISR alters the host physiology and metabolic processes resulting in strengthening of cell walls and increased synthesis of plant defence chemicals upon challenge by pathogens. Inoculation of tomato plants with an endophytic *Psm. fluorescens* WCS 417r and subsequent colonization of epidermal or hypodermal cells result in thickening of the outer tangential and outer most part of the radial side of the first layer of the cortical cell walls. Compant *et al.* (2005) observed that endophytic colonization by *Burkholderia phytofirmans* Ps JN in the grapevine resulted in a host defence reaction consisting of phenolic compound accumulation and strengthening of cell walls in the endodermis and in several cortical cell layers.

Fravel (2005), in his review article, has emphasized the need to intensify the research in production, formulation and delivery of biocontrol agents to boost their commercialization. He has also suggested combining biocontrol agents with each other and with other control methods, integrating biocontrol into an overall system. The first biocontrol agent used commercially, *Agrobacterium radiobacter* str. K 84, was registered with the United States Environmental Protection Agency (EPA) in 1979 for the control of crown gall. Ten years later, the first fungus, *Trichoderma harzianum* ATCC 20476 was registered with EPA for the control of a plant disease. Till 2001, a total of 14 bacteria and 12 fungi were registered with EPA for the control of plant diseases and one bacterium has been removed from the EPA list. Commercial products used for biocontrol of plant diseases account about 1% of agricultural chemical sales (Lidert, 2001) while the fungicides represent approximately 15% of pesticide sales (http://www.epa.gov.). However, in spite of their extremely low percentage in pesticide sales, the biopesticides have contributed immensely to the control of plant diseases.

The formulation of a biocontrol agent can affect its performance. The populations of a *Streptomyces* sp. remain stable in talcum powder and starch granules for 10 - 14 weeks and the stability is more at 4°C than at 24°C. The methods and strategies for introducing and maintaining of biocontrol agents also affect the outcome of biocontrol. The ideal biocontrol introduces or promotes the antagonists only when and where they are needed or are most effective, and minimizes wasteful application of biocontrol agents to nontargets.

Biocontrol agents can be applied with bactericides, SAR inducers or other methods of plant disease management. A mixture of six bacteriophages has been tried singly and in combination with acibenzolar-S-methyl (ASM) or copper hydroxide plus mancozeb. Johnson *et al.* (2004) optimized timings of sprays of *Psm. fluorescens* (Blight Ban A506) and *Psm. agglomerans* to control fire blight on apple and pear blossoms by adopting fire blight forecasting concepts. Moss *et al.* (2007) found 75-3S hrpG mutant of *Xa. campestris* pv. *vesicatoria* as effective as ASM in controlling bacterial spot of tomato under field conditions. They further reported that it caused a mean reduction in foliar disease severity of ~76% compared to the mean of ~29% for *Psm. syringae* Cit7, a most effective biocontrol agent found previously. In chapter 13, the descriptions of crown gall of plants, fire blight of pome fruit trees and bacterial spot of tomato and pepper contain additional information on the biocontrol of these diseases.

12.4. BACTERIOPHAGES

Bacteriophages (shortly called phages) are viruses, which kill bacteria. Their ability to kill bacteria has prompted the scientists to use them for the control of bacteria pathogenic to humans and plants, since their discovery in the second decade of 20th century. The use of phages to kill human pathogens has been dealt in chapter 9.

Among the earliest investigations on the use of phages for plant disease control are those of Coons and Kotila (1925), Kotila and Coons (1925) and Moore (1926). Coons and Kotila (1925) controlled rotting of carrot discs by combined inoculation of *Er. carotovora* subsp. *carotovora* and its phage. In the same year, Kotila and Coons prevented the rotting of potato tubers by co-inoculation of *Er. carotovora* subsp. *atroseptica* and its phage. In 1934, came a report by Massey, indicating a low incidence of bacterial blight of cotton in fields flooded by Nile river water. He suggested that the low incidence of the disease was due to the presence of a phage in river water. There is also a report of control of Stewart's wilt of corn with phages. Phage treatment of infected seed reduced the disease incidence from 18% to 14%.

In spite of these and other successful reports indicating the success of phages for plant disease control, their use on field scale did not become a reality for many years. Several workers questioned their reliability and effectiveness for controlling bacterial disease on field scale. Okabe and Goto (1963) stated that in general, the phages seem to be ineffective for controlling the plant diseases. Later on, Vidaver (1976) also expressed similar views and went a step further by not recommending the phages for plant disease control, citing the following reasons: (1) As in most cases, purified phages were not used for plant disease control, the control may be due to the combined effect of phages, media components and metabolic products. (2) The significant control of disease was achieved only with application of phage prior to pathogen inoculation but not with simultaneous or post applications. (3) The phage may mutate from a virulent to temperate type. (4) A virulent phage may enter into lysogeny, hence making the infected cell immune to the attack of another phage. (5) A phage may transduce characters like virulence or drug resistance, from one bacterium to another bacterium. Okabe and Goto's report of 1955 substantiates the transduction of virulence in *Psm. solanacearum*. (6) Phages are affected by environmental conditions like temperature, UV light irradiation etc. Goto (1992) also concluded that the practical use of phages for control of bacterial plant diseases in the field was not successful.

Recently, Jones *et al.* (2007) have given a comprehensive account of work done on the use of phages for disease control and have clearly showed the practicability of phage use for controlling bacterial plant diseases citing some important success stories. Enhanced knowledge and development of advanced research techniques have adequately solved the problems faced earlier. The major problem of controlling phage-resistant bacterial strains has been solved by development and selection of host range (h) mutant phages. The techniques to develop protective phage formulations and integration of biocontrol agents and systemically acquired resistance inducers with phage use have further enhanced the efficacy of phages for plant disease management. Moreover, the remarkable success achieved in controlling many important bacterial human pathogens has also given impetus to successful plant disease management. Besides achieving highly effective disease control, there are several below given potential advantages in using phages in disease control.

1. Phages are self-replicating; hence there is no need of repeated applications. In many cases, one application will do the job.
2. Phages are natural components of biosphere; they can be readily isolated from wherever their hosts, i.e., bacteria, are present.
3. Phage receptors on bacteria are generally pathogenicity determinants; the development of phage resistant mutants usually leads to attenuation in bacterial virulence.
4. As phages are non-toxic to eucaryotic cells, they can be used in cases where use of chemical is prohibited such as treatment of peach fruits before harvest.
5. Phages are specific and they eliminate only target bacteria in contrast to broad-spectrum antibiotics. Therefore, they do not harm other members, including beneficial of the indigenous flora.
6. Phages are not affected by most agrochemicals; hence they can be used in combination with these chemicals and also with biocontrol agents.
7. It is much easier and less expensive to develop a mutant phage against a phage resistant bacterium than to develop an antibiotic against an antibiotic resistant bacterium.
8. Phage preparations are fairly easy and inexpensive to produce and can be stored at 4°C for months without significant reduction in titer.

The development of bacterial strains resistant to phages was a major limiting factor in using phages for plant disease control. This aspect has been adequately highlighted by earlier workers (Katznelson, 1937; Okabe and Goto, 1963; Vidaver, 1976). Jackson (1989) developed a process to deal with occurrence of phage-resistant mutants by preparing a mixture of host range mutant (h-mutants) phages for disease control. H-mutants (h=host range or extended host range) possess the ability to lyse bacterial strains that are resistant to the parent phage, while still being capable of lysing the wild-type bacterium. Therefore, they have an extended host range compared to the parent phage. The application of a mixture of four phages, including wild-type and h-mutant phages, twice a week in the early morning prior to sunrise, gave significantly better control of bacterial spot of tomato. The phage mixture caused 17% reduction in disease severity while copper-mancozeb application caused 11% reduction. Moreover, the phage application also gave significantly higher yield of extra large fruits than the copper-mancozeb treatment (Flaherty *et al.* 2000). Flaherty *et al.* (2001) used a mixture of h-mutants, developed from five phages, exhibiting the broadest host range, which reduced the incidence of bacterial blight of geranium caused by *Xa. campestris* pv. *pelargonii* by 50% or more compared to control, and the disease incidence was also significantly less than those of recommended bactericides.

The phyllosphere is a relatively harsh environment compared to rhizosphere for the persistence of phages. Air temperature, sunlight, UV irradiation, rains etc. affect the persistence of phages in the phyllosphere. Various formulations have been tested to increase the persistence of phages on the plants. Phages applied in three formulations, namely, PCF (0.5% pregelatinized corn flour + 0.5% sucrose), skim milk (0.75% skim milk powder + 0.5% sucrose) and Casecrete (0.5% Casecrete NH-400, a water-soluble casein polymer + 0.25% pregelatinized corn flour + 0.5% sucrose), increased 4,700, 100,000 and 38,500 times in number, respectively than non-formulated phages, two days after application. These formulations improved disease control due to phage application and the effect was more pronounced in the field compared to greenhouse conditions.

Another strategy to enhance plant disease control with phages is to apply them through bacteria that are able to persist in the plant environment and that are also host to these phages. A strain of *Pa. agglomerans* was used to deliver and sustain a mixture of four phages to control fire blight of pear caused by *Er. amylovora*. *Pa. agglomerans* is also a biological control agent of *Er. amylovora.* The strains of both the bacteria were susceptible to these phages (Svircev *et al.* 2006). Tanaka *et al.* (1990) also used the similar strategy to control tobacco bacterial wilt by applying a strain of *Ralstonia solanacearum*. The phages that are better adapted to targeted environment and multiply rapidly are more suitable for disease control. Balogh (2006) found that two of the three phages that were able to multiply rapidly in *Xa. axonopodis* pv. *citri* in *in vitro* were not able to multiply in the same bacterium on grapefruit leaf surface and declined rapidly within hours, hence no control of citrus canker. A third phage, originally isolated from aerial plant tissue, multiplied rapidly *in vivo,* increased in numbers more than three-fold within 9 hours and gave effective control of the disease.

Timings of phage application in relation to arrival of the pathogen on the host also influence the disease control in several instances. The application of phage simultaneously with the pathogen inoculation or shortly before usually gives better results. Phage application one hour or one day before inoculation of peach bacterial spot pathogen gave best control while the fire blight of apple was best controlled when the phage mixture and *Er. amylovora* were applied at the same time. Black rot of cabbage and bacterial spot of pepper are best controlled when the phages are applied on the same day of pathogen inoculation.

The field application of phage during the daytime should be made when the sunlight irradiation is minimum to achieve better results. The control of bacterial spot of tomato was better with phage applications in the evening than morning (Balogh *et al.,* 2003). While in another study, control was better when phage was applied at dawn in comparison to late morning application. For an effective control of tomato bacterial leaf spot, the phage mixture should contain $10^{6\text{-}8}$ plaque-forming units per ml.

Various approaches have been tried to incorporate phages to develop integrated management strategies for bacterial plant diseases. The combined application of phages with avirulant strain of *Ral. solanacearum* for the control of tobacco bacterial wilt and with *Pa. agglomerans* for the control of fire blight of pear have already been mentioned above. Tomato has been used as a model system for developing an integrated management strategy to control foliar bacterial diseases. Various combinations including bacterial antagonists, phages, plant growth-promoting rhizobacteria and systemically acquired resistance (SAR) inducers have been compared in greenhouse experiments. Acibenzolar-S-methyl (ASM), a SAR inducer, significantly reduced bacterial spot of tomato in field but the combination of ASM and phage provided an additional reduction in disease pressure and resulted in more effective foliar disease control than ASM, phage or copper-mancozeb alone (Obradovic *et al.,* 2004). Phage treatment integrated with other practices is currently widely used in greenhouse and production fields in Florida as a part of a standard integrated management programme for control of bacterial spot of tomato.

The successful use of phages for controlling bacterial plant diseases is evident from the fact that the phages are commercially produced and used on the large scale in the greenhouse and production fields. Agri Phi, Inc., established by L. E. Jackson was the first company to commercially produce phages, specifically for control of bacterial plant diseases (Jackson, 1989). Based on greenhouse and field trials, OmniLytics, Inc., in Salt Lake City, UT (formerly Agri Phi, Inc.) was the first to receive Environmental Protection Agency's registration (registration No. 67986 - 1) to use phages in agriculture. The registration is for using host-specific phages on tomatoes in greenhouse and production fields as a part of standard integrated management programme to control tomato bacterial spot (Jones *et al.,* 2006). Recently, Lang *et al.* (2007) have recommended the integration of phage mixtures with ASM for the management of onion leaf blight caused by *Xa. axonopodis* pv. *allii* to reduce the growers' reliance on conventional copper bactericides applied with mancozeb.

Another attractive and encouraging area is the use of recombinant phage lysins. Endolysins are bacteriophage-encoded peptidoglycan hydrolases that accumulate in the cytosol of phage-infected bacterial cells. These endolysins cause the cell lysis at the end of the lytic cycle. Due to their ability to recognize unique receptors present in the bacterial cell wall, they show a highly specific targeting of their host cells while leaving the normal flora unaffected. Another positive feature of endolysins is their modular structure, which allows it to be manipulated by exchanging domains to alter the host spectrum or to improve the catalytic activity. Several *in vitro* and *in vivo* investigations show that these enzymes possess many desirable properties which make them highly suitable to act as antimicrobials. The trials conducted in the past have demonstrated the potential of the endolysins to control *Ba. anthracis,* bacteria present in the food and feed and plant pathogenic bacteria.

12.5. BACTERIOCINS

As defined by Nomura in 1967, bacteriocins are non-replicating, bactericidal protein-containing substances which are produced by certain strains of the same or closely related species. This definition includes heterogeneous substances ranging from low-molecular-weight particles resembling bacteriophage protein components. A. Gratia was the first person to discover bacteriocins in 1925. As the first compound discovered by him killed *Escherichia coli,* he called it colicine. The **term bacteriocine** was coined in 1953; the terminal "e" is omitted in the current terminology of bacteriocins.

The bacteriocins differ in their molecular weight, microbial targets, mode of action and immunity mechanism. They differ from traditional antibiotics that they have a narrow killing range and are toxic to bacteria closely related to the producing strain. Bacteriocins have been found in almost every bacterial species examined todate, and within a species tens or even hundreds of different kinds of bacteriocins are produced (James *et al.,* 1991). Halobacteria universally produce their own version of bacteriocins, called **halocins.** The *Archaea* produce their own distinct family of bacteriocin-like antimicrobials, called as **archaeocins** (Riley and Wertz, 2002). According to Klaenhammer (1988), 99% of all bacteria may produce at least one bacteriocin and the only reason we have not isolated more is that few researchers have looked for them. Bacteriocinogenic strains produce bacteriocins in nature but their production can also be induced by treatment with ultraviolet irradiation or mitomycin C.

A uniform system of naming bacteriocins does not exist. Agrocin 84 is derived from the name of producing genus *Agrobacterium* while syringacin is derived from the species name of *Psm. syringae* pv. *syringae.*

The bacteriocins are classified or categorized in several ways based on method of killing, genetics, method of production and molecular weight and chemistry. Another method of classification places the bacteriocins into class I, class IIa/b/c and class III.

The production of bacteriocins is controlled by plasmids as in case of agrocin 84, and the bacteriocin produced by *Pa. agglomerans*. In some bacteria, chromosomal genes code for the production of bacteriocins. Several bacteriocins require specific sites on the bacterial cell wall, called receptors, for attachment and killing. Agrocin 84 binds to a specific protein of the periplasmic space of the crown gall bacterium and the specific sensitivity is determined by the Ti-plasmid of the host bacterium. This specificity is lost in case cell wall is lost due to the action of enzymes or chemicals. In that case, the number of bacteriocins can bind to the cytoplasmic membrane and kill the bacteria in different ways. It shows that many bacteriocins have two stages of adsorption; one based on specific receptor or attachment site in the outer cell membrane and the other independent of a specific site. This also indicates that the bacteriocins can also be effective against wall-less bacteria such as phytoplasmas (Vidaver, 1983). Bacteriocins kill bacteria in several ways, i.e., by affecting protein synthesis, DNA stability, energy flux or membrane integrity. Agrocin 84 specifically inhibits DNA synthesis because of its being a structural analog of adenine. Carotovorocin produced by *Pe. carotovorum* subsp. *carotovorum* has phospholipase activity and thus acts on the cytoplasmic membrane.

Although the application of bacteriocins for controlling bacterial plant diseases seems to be an attractive proposition, very little is known about the occurrence and roles of these compounds in plant pathogenic bacteria. Recently, Holtsmark *et al.* (2008) have given an excellent and exhaustive account of work done on production of bacteriocins by plant pathogenic bacteria, indicating the possibility of controlling bacterial plant pathogens with the bacteriocins. Almost all known bacteriocins produced by gram-negative plant pathogenic bacteria are proteins. The bacteriocins produced by *Psm. syringae* pv. *syringae* are S-type pyocins which are high molecular-weight compounds (65 - 80 kDa) and are related to colicins. S-type pyocins are complexes of two components: a large component with killing activity and a smaller immunity protein and they act mainly on other species of *Pseudomonas*. Recently, Chuang *et al.* (2007) reported the production of a smaller colicin- and pyocin-like antimicrobial protein of about 55 kDa, called carocin S1 by *Pe. carotovorum* (syn. *Er. carotovora* subsp. *carotovora*). It was shown to inhibit other strains of the same bacterium, probably by exerting DNase activity. Colicin-like proteins are also found in the genomes of both the sequenced strains of *Xylella fastidiosa* and of *Xa. oryzae* pv. *oryzae*. Hert *et al.* (2005) showed that production of protein bacteriocins, not yet characterized, by

Xa. perforans, enabled the producer strain to exert inhibitory effect on *Xa. euvesicatoria*, both in the laboratory and field trials.

Gram-negative plant pathogenic bacteria also produce bacteriocins that exert their antimicrobial activity by self-assembling into cytotoxic phage tail-like fibres. Among them, the best studied bacteriocins are carotovorocins produced by *Pe. carotovorum* subsp. *carotovorum*. The carotovorocins produced by one strain kill other strains of *Pe. carotovorum* subsp. *carotovorum* and the killing spectra of these bacteriocins are determined by the structure of the tail fibres. Phage tail-like bacteriocins are also produced by *Ral. solanacearum* and avirulent bacteriocin producing strains of *Ral. solanacearum* reduce the development of bacterial wilt of tobacco. A phage tail-like bacteriocin produced by *Serratia plymithicum* inhibits the growth of *Er. amylovora*.

Several plant pathogenic bacteria produce protein bacteriocins, which are different from the antimicrobial proteins mentioned above and/or that have not been sufficiently characterized. *Xa. axonopodis* pv. *glycines* (syn. *Xa. campestris* pv. *glycines*) produces a heterodimeric bacteriocin called glycinecin, which is encoded by two separate genes, *glyA* and *glyB* that give rise to 39 and 14 kDa subunits, respectively. The inhibitory range of the bacteriocin includes mainly other pathovars of *Xa. campestris* and *Xa. oryzae* pv. *oryzae* (Heu *et al.*, 2001). Glycinecin shares no significant similarity with any protein sequences known todate and acts by permeablizing the membrane of target cells (Pham *et al.*, 2004). The 30 kDa lectin-like putidacin produced by a rhizosphere isolate BW11M1 of *Pseudomonas* sp. is another bacteriocin of this category. This bacteriocin contains regions that resemble mannose-binding domains of lectins in monocotyledonous plants and it inhibits the growth of strains of a number of *Pseudomonas* species, including pathovars of *Psm. syringae*. Parret *et al.* (2005) reported the production of two lectin-like bacteriocins, having similar inhibitory spectra, by *Psm. fluorescens* str. Pf-5, a well-known biocontrol agent. A purified bacteriocin from *Psm. syringae* pv. *ciccaronei* consists of three proteins (45 - 76 kDa). It was found inhibitory to *Psm. savastanoi* pv. *savastanoi* in the field trials.

Very little is known about protein bacteriocins produced by Gram-positive plant pathogenic bacteria. The best-studied one is impomicin, a heat-sensitive 10 kDa protein produced by *St. ipomoea*. It inhibits closely related strains, primarily other strains of *St. ipomoea*.

Peptide bacteriocins are abundantly produced in nature but very little is known about their production by plant pathogenic bacteria. *Ag. tumefaciens* and *Rhizobium leguminosarum* produce peptide bacteriocins called trifolitoxins, which consist of 11 amino acids carrying several (partly unknown) posttranslational modifications. Producer species compete with closely related species.

Holtsmark *et al.* (2006) were the first to report a plant pathogen producing a known type of peptide bacteriocin, namely, a lantibiotic belonging to class I, type B. This bacteriocin called michiganin A; produced by *Clavibacter michiganensis* subsp. *michiganensis* inhibits the growth of another closely related pathogen of potato, *Cla. michiganensis* subsp. *sepedonicus* (Holtsmark *et al.*, 2006, 2007). It inhibited *Cla. michiganensis* subsp. *sepedonicus* at a concentration range of 20 - 200 ng ml^{-1}, which is a normal inhibitory range for peptide bacteriocins. Michiganin A resembles other type B lantibiotics such as actagardine and mersacidin. Recently Bonini *et al.* (2007) have reported the bacteriocin production by *Xa. axonopodis* pv. *citri* and *Xa. campestris* pv. *passiflorae* strains, characterized them and named them citricin and passifloricin, respectively.

In addition to control of bacterial plant diseases, bacteriocins can be employed for the following purposes.

1. Forecasting of bacterial plant diseases.
2. Use in epidemiological studies to type or differentiate strains that are otherwise indistinguishable. It has acquired a wide spread application and various typing schemes based either on the production of or on sensitivity to a range of different bacteriocins are used.
3. Identification of bacterial strains, because production of particular groups of bacteriocins is confined to certain taxonomic groups.

The control of phytopathogenic bacteria with bacteriocins has received the attention of many research workers and has been tried for many diseases. Among them, the control of crown gall of stone fruits has been investigated in detail and it is being practised on commercial scale in many countries (see 13.40- crown gall of plants). Although, the control of crown gall is not solely due to the production of bacteriocin as stated by Vidaver (1976), the application of living strain K 1026 of *Ag. radiobacter* gives highly effective control of the disease. Purified syringacin 4-A have been tried to control *Psm. syringae* pv. *phaseolicola* on bean. Its pre-inoculation spray @ 3 ng per leaf reduced the lesion count from 250 to zero. Its application to soybean seed protected the seed against *Psm. savastanoi* pv. *glycinea* infection and also increased the seed germination by 20%. The cost of treatment was 0.49 US dollar per hectare (Vidaver, 1976). Guven (2000) found that all the 86 isolates of *Psm. syringae* pv. *phaseolicola,* collected from different bean growing areas of Eskisehir, Turkey, produced bacteriocins, which formed 24 different groups. Lavermicocca *et al.* (2002) reported that application of crude bacteriocin produced by *Psm. syringae* pv. *ciccaronei* @ 6000 arbitrary units ml^{-1}, reduced the severity of olive knots caused by *Psm. savastanoi* pv. *savastanoi* by reducing the knot weight by 81 and 51% on leaves and leaf scar sites, respectively. Bacteriocin application also reduced the epiphytic population of the pathogen, which was at least 350 and 20 times lower than the control populations on twigs and on leaves, respectively.

In spite of some limitations associated with bacteriocins for their use in plant disease control, they have a great potential to be used as prophylactic treatment for seed-or tuber-borne pathogens, prevention of secondary spread of bacteria from infected plants and protection of high value crops like apple and pear against fire blight and stone fruits against crown gall. An important recommendation in the use of bacteriocins is that at least two and preferably three serologically unrelated bacteriocins must be used simultaneously. It will reduce the chances of development of bacterial strains resistant to the bacteriocins.

The future research on bacteriocins may reveal new regulatory processes and pathways resulting in their economical and viable mass production. In such cases, bacteriocins can be applied directly in some kind of semi-pure or pure form. Otherwise, one may deliver bacteriocins using nonvirulent producer strains.

Montesinos (2007) has given a comprehensive account of **anti-microbial peptides** (AMPs) dealing with their production and role in plant disease control, and suggesting a strong possibility of developing transgenic plants based on their expression, for plant disease control. AMPs are short sequence peptides with generally fewer than 50 amino acid residues reported in living systems, which are a first line of defence in plants and animals. Around 900 AMPs have been reported in the literature.

Due to their strong antimicrobial activity, AMPs have a potential for the control of plant pathogens. Unfortunately, AMPs from microbial origin like lipidic cyclopeptides and peptaibols have a high toxicity that limits their direct use for plant disease control. Therefore, several AMPs have been used to develop shorter and less toxic analogues by synthetic producers. The fact that plant and animal AMPs are produced by ribosomal synthesis, has provided tools for developing transgenic plants expressing genes coding for the synthesis of these compounds conferring partial or total resistance to plant pathogens.

Synthetic AMPs have been developed using solid-phase methods and those that facilitate the performance of combinatorial libraries. Combinatorial chemistry is very effective for developing new molecules by improving their antimicrobial activity against selected target pathogens and minimizing their toxicity to plants and animals and susceptibility to proteases. MB39, an analogue of cecropin B, inhibits *Pe. betavasculorum, Cla. michiganensis*, three pathovars of *Psm. syringae* and two pathovars of *Xa. campestris*. D4E1, another analogue of cecropin B inhibits *Psm. syringae* pv. *tabaci* and *Xa. citri* subsp. *malvacearum*.

Gene constructions including sequences coding for AMPs have been expressed in many crop plants and these plants exhibit varying degree of protection against different plant pathogens. Montesinos (2007) has listed more than 20 AMPs of animal, fungal and plant origins and 10 synthetic AMPs along with their sources, and plant species transformed expressing these compounds. These transformed plants exhibit resistance to various fungal and bacterial pathogens. Transgenic plants, obtained by transferring various AMPs encoding genes and showing resistance to different bacterial plant pathogens are given in Table 12.1. Vidal *et al.* (2006)

Table 12.1. Antimicrobial peptides (AMPs) expressed in transgenic plants that confer partial resistance to pathogens.

Origin	AMP	Source	Plant transformed	Pathogen	Reference
Animal	Cecropin A, B	Moth haemolymph	Rice	*Xanthomonas oryzae* pv. *oryzae*	*Sharma *et al.* (2000). *FEBS Lett.* 484: 7 - 11.
	Tachyplesin	Crab haemolymph	Potato	*Erwinia carotovora*	*Allefs *et al.* (1996). *Mol. Breed.* 2: 97 - 105.
	Sarcotoxin IA	Fruit fly Haemolymph	Tobacco	*Pseudomonas syringae* pv. *tabaci*, *Er. carotovora* subsp. *carotovora*	*Ohshima *et al.* (1999). *J. Biochem.* 125: 431 - 35.
Plant	Hordothionin	Barley	Tobacco	*Clavibacter michiganensis, Psm. syringae* pv. *tabaci*	*Carmona *et al.* (1993). *Plant J.* 3: 3457.
Synthetic	SB - 37	Cecropin analogue	Potato, apple	*Er. carotovora.* subsp. *carotovora*	*Arce *et al.* (1999). *Am. J. Potato Res.* 76: 169 - 77.
	Shiva - 1	Cecropin analogue	*Anthurium*	*Xa. axonopodis* pv. *dieffenbachiae*	*Kuehnle *et al.* (2004). *Hort Sci.* 39: 1327 - 31.
	SB - 37, Shiva - 1	Cecropin analogues	Tobacco	*Ralstonia solanacearum,Psm. syringae* pv. *tabaci*	*Jaynes *et al.* (1993). *Plant Sci.* 89: 43 - 53. *Huang *et al.* (1997). *Phytopathology* 87: 494 - 99.
	MB - 39	Cecropin analogue	Apple	*Er. amylovora*	*Liu *et al.* (2001). *Plant Cell Rept.* 20: 306 - 12.
	MSI - 99	Magainin analogue	Grapevine	*Agrobacterium tumefaciens*	Vidal *et al.* (2006)
	Rev 4	Indolicidin analogue	Tobacco/ *Arabidopsis*	*Psm. syringae* pv. *tabaci, Er. carotovora*	Xing *et al.* (2006)

*References not included under the references

reported that MSI - 99 expressed in grapevine and protected the plants against *Ag. tumefaciens.* The analogue of indolicidin Rev 4 has been expressed in tobacco and is effective against *Psm. syringae* pv. *tabaci* and *Er. carotovora* infections (Xing *et al.*, 2006). Transgenic plants given in Table 12.1 and also those listed by Montesinos (2007) have been developed only up to the experimental stage and no commercial cultivars have been marketed due to regulatory limitations and social concerns. Moreover, exploitation of large number of AMPs as active ingredients of pesticides has also not been achieved. Development of these compounds as pesticides for agricultural use has several constraints like intrinsic toxicity, low stability of some compounds, high cost of production and the need to develop suitable formulations. Therefore, the future tasks consist of developing less toxic and more stable compounds having lower production costs by improving preparative synthesis and biotechnological techniques using microbial systems or transgenic crops.

12.6. SIDEROPHORES

It is a Greek word meaning iron carrier. Siderophores are low molecular weight, high affinity iron (III) chelators secreted by bacteria, fungi and many plants. Iron Fe^{3+} ions have a very low solubility at neutral pH and therefore, cannot be utilized by organisms. Siderophores dissolve these ions by chelation as soluble Fe^{3+} complexes that can be taken up by active transport mechanisms. One major group of siderophores, i.e., derivatives of hydroxamic acid, chelates ferric iron very strongly. Some examples of siderophores produced by different bacteria and fungi include bacillibactin by *Ba. subtilis,* enterobactin by *Es. coli*, pseudobactin by *Pseudomonas* B10, fusarinine C by *Fusarium roseum* and ferrichrome by *Ustilago sphaerogena*.

When grown under low-iron conditions, fluorescent pseudomonads produce the high affinity peptidic siderophore, pyoverdine but in many cases, a second siderophore having less affinity for iron is also produced. A considerable progress has been made in the elucidation of siderophore-mediated high-affinity iron uptake by *Pseudomonas* spp., particularly *Psm. aeruginosa.* Some of the genes for the biosynthesis and uptake of these siderophores have been identified and functions of the encoded proteins are known. Iron uptake through siderophores is regulated at several levels, i.e., via the general iron-sensitive repressor FUR (Ferric Uptake Regulator), via extracytoplasmic sigma factors/anti-sigma factors or via other regulators. As pseudomonads are ubiquitous bacteria, their genomes contain large number of genes encoding receptors for the uptake of heterologous ferrisiderophores or heme indicating their better adaptability to diverse iron resources. An exciting recent development is the evidence provided by Cornelis (2008) for a cross talk between the iron regulon and other regulatory networks, including the diffusible signal molecule-mediated quorum sensing in *Psm. aeruginosa*.

Kloepper *et al.* (1980) were the first to demonstrate the importance of siderophores production as a mechanism of biological control. As siderophores sequester the limited supply of iron (III) in the rhizosphere, they limit its availability to the pathogens and ultimately suppress their growth. Fluorescent pseudomonads have been shown to act as biocontrol agents against certain soil-borne plant pathogens. They produce yellow-green pigments (pyoverdines), which fluoresce under UV light and function as siderophores. Pyoverdines chelate iron in the rhizosphere and deprive pathogens of iron, which is required for their growth and pathogenesis.

The availability of iron (III) in the soil declines logarithmically with increasing soil pH. Therefore, the siderophore-mediated suppression of plant pathogens is greater in neutral or alkaline soils than in acid soils. The following factors account for the suppression of plant pathogens by siderophores.

(a) The pathogen may or may not produce siderophores; if they are produced, the amount of siderophores produced is small or the siderophores have a lower affinity for iron than those of produced by the antagonist(s).

(b) Pathogens are unable to use siderophores produced by antagonists or by other microorganisms in their immediate environment.

(c) Pathogens produce a siderophore that can be used by the antagonist(s). The most convincing evidence supporting the role of siderophores in biocontrol is the fact that siderophore-deficient mutants are less suppressive to the pathogens in the rhizosphere than parental strains.

Jagadeesh *et al.* (2001) found that siderophore-deficient mutants of a fluorescent *Pseudomonas* sp. did not control bacterial wilt of tomato caused by *Ral. solanacearum* as much as the wild-type strain, thus confirming the role of siderophores in the biocontrol mechanism. Siderophore-deficient mutants gave 18.75% disease control in comparison to 75.00% given by wild-type strain.

REFERENCES

Amusa, N.A. and Odunbaku, O.A. (2007). Biological control of bacterial diseases of plants in Nigeria: Problems and prospects. *Res. J. Agric. & Biolog. Sci.* 3: 979 - 82.

Andrews, J.H. (1992). Biological control in the phyllosphere. *Annu. Rev. Phytopathol.* 30: 603 - 35.

Balogh, B. (2006). Characterization and Use of Bacteriophages Associated with Citrus Bacterial Pathogens for Disease Control. PhD. Thesis. Univ. FL., Gainesville.

Balogh, B., Jones, J.B., Momol, M.T. *et al.* (2003). Improved efficacy of newly formulated bacteriophages for management of bacterial spot on tomato. *Plant Dis.* 87: 949 - 54.

Bonini, M., Maringoni, A.C. and Rodrigues Neto, R. (2007). Characterization of *Xanthomonas* spp. strains by bacteriocins. *Summa Phytopathol.* 33: 24 - 29.

Chuang, D.Y., Chien, Y.C. and Wu, H.P. (2007). Cloning and expression of the *Erwinia carotovora* subsp. *carotovora* gene encoding the low-molecular-weight bacteriocin, carocin S1. *J. Bacteriol.* 189: 620 - 26.

Compant, S., Duffy, B., Nowak, J. *et al.* (2005). Use of plant growth-promoting bacteria for biocontrol of plant diseases: Principles, mechanisms of action and future prospects. *Appl. Environ. Microbiol.* 71: 4951 - 59.

Coons, G.H. and Kotila, J.E. (1925). The transmissible lytic principle (bacteriophage) in relation to plant pathogens. *Phytopathology* 15: 357 - 70.

Cornelis, P. (ed.) (2008). *Pseudomonas: Genomics and Molecular Biology.* 1st Ed., Caister Academic Press. (http://www.horizonpress.com/pseudo).

Flaherty, J.E., Harbaugh, B.K., Jones, J.B. *et al.* (2001). H-mutant bacteriophages as a potential biocontrol of bacterial blight of geranium. *Hort. Sci.* 36: 98 - 100.

Flaherty, J.E., Jones, J.B., Harbaugh, B.K. *et al.* (2000). Control of bacterial spot on tomato in the greenhouse and field with H-mutant bacteriophages. *Hort. Sci.* 35: 882 - 84.

Fravel, D.R. (2005). Commercialization and implementation of biocontrol. *Annu. Rev. Phytopathol.* 43: 337 - 59.

Goto, M. (1992). *Fundamentals of Bacterial Plant Pathology.* Academic Press, Inc. San Diego, 342 pp.

Gray, E.J. and Smith, D.L. (2005). Intracellular and extracellular PGPR: Commonalities and distinctions in the plant-bacterium signalling processes. *Soil Biol. Biochem.* 37: 395 - 412.

Guven, K. (2000). Bacteroicin typing of some Turkish isolates of *Pseudomonas syringae* pv. *phaseolicola. Turk. J. Biol.* 24: 795 - 801.

Hert, A.P., Roberts, P.D., Momal, M.T. *et al.* (2005). Relative importance of bacteriocin-like genes in antagonism of *Xanthomonas perforans* tomato race 3 to *Xanthomonas euvesicatoria* tomato race 1 strain 259. *Appl. Environ. Microbiol.* 71: 3581 - 88.

Heu, S., Oh, J., Kang, Y. *et al.* (2001). *gly* gene cloning and expression and purification of glycinecin A, a bacteriocin produced by *Xanthomonas campestris* pv. *glycines* 8ra. *Appl. Environ. Microbiol.* 67: 4105 - 10.

Holtsmark, I., Eijsink, V.G.H. and Bruberg, M.B. (2008). Bacteriocins from plant pathogenic bacteria. *FEMS Microbiol. Lett.* 280: 1 - 7.

Ioltsmark, I., Mantzilas, D., Eijsink, V.G.H. *et al.* (2006). Purification, characterization and gene sequence of Michiganin A, an actagardine-like lantibiotic produced by the tomato pathogen *Clavibacter michiganensis* subsp. *michiganensis*. *Appl. Environ. Microbiol.* 72: 5814 - 21.

Ioltsmark, I., Mantzilas, D., Eijsink, V.G.H. *et al.* (2007). The tomato pathogen *Clavibacter michiganensis* subsp. *michiganensis*: Producer of several antimicrobial substances. *J. Appl. Microbiol.* 102: 416 - 23.

ackson, L. E. (1989). *U. S. Patent No. 4828999.*

agadeesh, K.S., Kulkarni, J.H. and Krishnaraj, P.U. (2001). Evaluation of the role of fluorescent siderophore in the biological control of bacterial wilt in tomato using Tn^5 mutants of fluorescent *Pseudomonas* sp. *Curr. Sci.* 81: 882 - 83.

ames, R., Lazdunski, C. and Pattus, F. (eds) (1991). *Bacteriocins, Microcins and Lantibiotics.* Vol. 65, Springer-Verlag, New York, 519 pp.

ohnson, K.B., Stockwell, V.O. and Sawyer, T.L. (2004). Adaptation of fire blight forecasting to optimize the use of biological controls. *Plant Dis.* 88: 41 - 48.

ones, J.B., Iriarte, F.B., Obradovic, A. *et al.* (2006). Management of bacterial spot on tomatoes with bacteriophages. *Proc. Int. Symp. Biol. Control Bac. Plant Dis., 1st* Darmstadt, Germany, 408: 154, Land-Forstwirtsch: Mitt. Biol. Bundesanst.

ones, J.B., Jackson, L.E. Balogh, B. *et al.* (2007). Bacteriophages for plant disease control. *Annu. Rev. Phytopathol.* 45: 245 - 62.

Katznelson, H. (1937). Bacteriophages in relation to plant diseases. *Bot. Rev.* 3: 499 - 521.

Klaenhammer, T R. (1988). Bacteriocins of lactic acid bacteria. *Biochimie* 70: 337 - 49.

Kloepper, J.W., Leng, J., Teintze, M. *et al.* (1980). Enhanced plant growth by siderophores produced by plant growth-promoting rhizobacteria. *Nature* 286: 885 - 86.

Kloepper, J.W., Rodriguez-Ubana, R., Zehnder, G.W. *et al.* (1999). Plant root-bacterial interactions in biological control of soil borne diseases and potential extension to systemic and foliar diseases. *Austral. Plant. Pathol.* 28: 21 - 26.

Kotila, J.E. and Coons, G.H. (1925). Investigations on the black leg disease of potato. *Mich. Agric. Exp. Sta. Tech. Bull.* 3 - 29.

ang, J.M., Gent, D.H. and Schwartz, H.F. (2007). Management of *Xanthomonas* leaf blight of onion with bacteriophages and a plant activator. *Plant Dis.* 91: 871 - 78.

avermicocca, P., Lonigro, S.L., Valerio, F. *et al.* (2002). Reduction of olive knot disease by a bacteriocin from *Pseudomonas syringae* pv. *ciccaronei*. *Appl. Environ. Microbiol.* 68: 1403 - 07.

idert, Z. (2001). Biopesticides: Is there a path to commercial success? In: M. Vurro, J. Gressel, T. Butt *et al.* (eds). *Enhancing Biocontrol Agents and Handling Risks.* IOS Press, Amsterdam, P. 283.

Montesinos, E. (2007). Antimicrobial peptides and plant disease control. *FEMS Microbiol. Lett.* 270: 1 - 11.

Moore, E.S. (1926). d' Herelle's bacteriophage in relation to plant parasites. *South Afr. J. Sci.* 23: 306.

Moss, W.P., Byrne, J.M., Campbell, H.L. *et al.* (2007). Biological control of bacterial spot of tomato using *hrp* mutants of *Xanthomonas campestris* pv. *vesicatoria. Biol. Control* 41: 199 - 206.

Obradovic, A., Jones, J.B., Momol, M.T. *et al.* (2004). Management of tomato bacterial spot in the field by foliar applications of bacteriophages and SAR inducers. *Plant Dis.* 88: 736 - 40.

Okabe, N. and Goto, M. (1963). Bacteriophages of plant pathogens. *Annu. Rev. Phytopathol.* 1: 397 - 418.

Parret, A. H., Temmerman, K. and De Mot, R. (2005). Novel lectin-like bacteriocins of biocontrol strain *Pseudomonas fluorescens* pf-5. *Appl. Environ. Microbiol.* 71: 5197 - 5207.

Paulitz, T.C. and Belanger, R.R. (2001). Biological control in greenhouse systems. *Annu. Rev. Phytopathol.* 39: 103 - 33.

Pham, H.T., Riu, K.Z., Jang, K.M. *et al.* (2004). Bactericidal activity of glycinecin A, a bacteriocin derived from *Xanthomonas campestris* pv. *glycines*, on phytopathogenic *Xanthomonas campestris* pv. *vesicatoria* cells. *Appl. Environ. Microbiol.* 70: 4486 - 90.

Riley, M.A. and Wertz, J.E. (2002). Bacteriocins: Evolution, ecology and application. *Annu. Rev. Microbiol.* 56: 117 - 37.

Sessitsch, A., Reiter, B. and Berg, G. *et al.* (2004). Endophytic bacterial communities of field-grown potato plants and their plant growth-promoting and antagonistic abilities. *Can. J. Microbiol.* 50: 239 - 49.

Sutton, J.C. and Peng, G. (1993). Manipulation and vectoring of biocontrol organisms to manage foliage and fruit diseases in cropping systems. *Annu. Rev. Phytopathol.* 31: 473 - 93.

Svircev, A.M., Lehman, S.M., Kim, W. *et al.* (2006). Control of the fire blight pathogen with bacteriophages. *Proc. Int. Symp. Biol. Control Bact. Plant Dis., 1st*, Seeheim/Darmstadt, Germany, 259, Berlin: Dtsch. Bibl., CIP-Einh. aufn.

Tanaka, H., Negishi, H. and Maeda, H. (1990). Control of tobacco bacterial wilt by an avirulent strain of *Pseudomonas solanacearum* M4S and its bacteriophage. *Ann. Phytopathol. Soc. Jpn.* 56: 243 - 46.

Van Loon, L.C., Bakker, P.A.H.M. and Pieterse, C.M.J. (1998). Systemic resistance induced by rhizosphere bacteria. *Annu. Rev. Phytopathol.* 36: 453 - 83.

Velusamy, P., Immanuel, J.E., Gnanamanickam, S.S. *et al.* (2006). Biocontrol of rice bacterial blight by plant-associated bacteria producing 2,4-diacetylphloroglucinol. *Can. J. Microbiol.* 52: 56 - 65.

Vidal, J.R., Kikkert, J.R., Malnoy, M.A. *et al.* (2006). Evaluation of transgenic 'chardonnay' (*Vitis vinifera*) containing magainin genes for resistance to crown gall and powdery mildew. *Transgenic Res.* 15: 69 - 82.

Vidaver, A.K. (1976). Prospects for control of phytopathogenic bacteria by bacteriophages and bacteriocins. *Annu. Rev. Phytopathol.* 14: 451 - 65.

Vidaver, A.K. (1983). Bacteriocins: The lure and the reality. *Plant Dis.* 67: 471-75.

Weller, D.M. (1988). Biological control of soil borne plant pathogens in the rhizosphere with bacteria. *Annu. Rev. Phytopathol.* 26: 379 - 407.

Wilson, C.L. and Wisniewski, M.E. (1989). Biological control of post harvest diseases of fruits and vegetables: An emerging technology. *Annu. Rev. Phytopathol.* 27: 425 - 41.

Xing, H., Lawrence, C.B., Chambers, O. *et al.* (2006). Increased pathogen resistance and yield in transgenic plants expressing combinations of the modified antimicrobial peptides based on indolicidin and magainin. *Planta* 223: 1024 - 32.

Chapter 13

DESCRIPTION OF PLANT DISEASES

13.1. BACTERIAL BLIGHT OF RICE

Bacterial blight (hereafter referred to as BB) is one of the mos devastating diseases of rice (*Oryza sativa* L.) due to its high epiphytoti potential and its destructiveness to nitrogen-responsive high yieldin cultivars in both temperate and tropical climates. Its occurrence in 197 in Africa has added a new dimension to concerns about its transmissio and dissemination. It has also been a serious problem on hybrid rice. I the 1980s when hybrid rice was first released in China, epiphytotic occurred widely because of lack of resistance in the hybrid rice, resultin in severe yield losses.

It is probably one of the most extensively studied diseases and large number of research publications, dealing with different aspects o the pathogen and the disease, have appeared in the literature. Severa review papers containing valuable information on the disease have als been published by workers from different countries. Two such reviev papers have also appeared from India (Srivastava, 1972; Thind, 2002). A recent review paper by Nino-Liu *et al.* (2006) provides very usefu information on different aspects of the casual agent of the disease. Fo additional information, the readers may also see annual reports o International Rice Research Institute, Philippines and Central Ric Research Institute, Cuttack, India.

The disease was first seen by farmers in the Fukuoka Prefecture o Japan in 1884 and was called **white whithering disease**. Originally, i was believed to be caused by acidic soil. In 1908, Takaishi found bacteria masses in dewdrops of infected rice leaves, successfully isolated a bacterium and proved its pathogenicity but did not name the organism Bokura, in 1911, also isolated a bacterium, proved its pathogenicity an named it *Bacillus oryzae* Hori & Bokura. Reitsma and Schure (1950 studied a rice disease called **'Kresek'** in Indonesia and considering i different from bacterial blight, named the pathogen as *Xanthomona kresek*. Later, Goto (1964) reported the **pale yellow leaf phase** of th disease, which occurs at a later stage of growth. Both kresek phase an pale yellow leaf phase are different phases of bacterial blight.

The disease is widely spread and occurs in the Philippines, Thailand Indonesia, China, Taiwan, Malaysia, Cambodia, Vietnam, Myanmar

Laos, Bangladesh, Mexico, S.E. Russia, Australia, Pakistan, West Africa, Latin America, the Caribbean and India. The presence of the disease has not been confirmed in the United States and not yet found in Europe. Quarantines for BB pathogen are in place in the United States and other rice growing countries where the disease is not endemic but also in places where it is present, to prevent the introduction of new virulent strains. It is also an EPPO A1 quarantine pest.

In India, the disease was first reported by Sreenivasan *et al.* (1959) from Maharashtra but the symptoms described by them were not exactly of bacterial blight. Bhapkar *et al.* (1960) reported correct symptoms of the disease but they did not isolate the bacterium. Later, Srivastava and Rao (1963) reported its epiphytotic on cv. BR 34 from Shahabad district of Bihar. During this period, a Formosian rice variety, Taichung Native 1, a high nitrogen-responsive but resistant to lodging was introduced into India to step up the rice production. This variety was highly susceptible to BB. With the introduction of this variety, BB that was unknown in all but two states of India earlier, became pandemic within a couple of years.

Although, the disease occurs in many countries, it is most destructive in Southeast Asia. The yield losses vary with host cultivar, age and nutrition of the host and environmental conditions. They are usually high in early infected crop and the occurrence of kresek phase may lead to total failure of the crop. Losses due to BB increased significantly following the widespread cultivation of nitrogen-responsive, high yielding, semi-dwarf rice cultivars derived from Taichung Native 1 and its derivatives in the 1960s. Prior to the introduction of resistant varieties and implemen-tation of strict quarantine measures in Japan, losses generally ranged from 20 - 30% but in some cases went as high as 50% (Ou, 1972). Reports from the Philippines, Indonesia and India estimate that losses due to kresek phase have reached 60 - 75% depending on weather, location and rice variety (Ou, 1985; Reddy *et al.* 1979). Ray and Sengupta (1970) also reported losses up to 50% from semi-dwarf high yielding varieties. Singh *et al.* (1977) reported that four rice cultivars with a similar leaf score of 9 showed variable loss in grain yield. Reddy *et al.* (1979a) established a relationship between severity of BB at different growth stages of host plant and rice yield and found that early epiphytotics reduced grain yield, panicle fertility and kernel weight while late epiphytotics had no measurable effect on grain yield. They also devised a scale for calculating BB associated losses. In addition to reduction in yield, the disease may also affect grain quality by interfering with maturation.

13.1.1. Symptoms

It is a typical vascular disease. The below given three different types of symptoms appear.

13.1.1.1. Leaf blight phase: These are the most common symptoms of the disease and generally appear 4 - 6 weeks after transplanting. Initially, the symptoms start as small, green, water-soaked spots at the tips and margins of fully developed leaves. The spots expand forming linear, yellow to straw-coloured stripes having wavy inner margins. The stripes appear generally on both the margins and rarely on one margin of the leaves. The stripes progress rapidly, extending lengthwise and crosswise. The drying and twisting of the leaf tips occur (Fig. 13.1a). Occasionally, linear stripes may develop anywhere on the leaf lamina or midrib with or without marginal stripes. The blightening extends to leaf sheaths, culms, killing the tillers. In humid weather, yellow- or amber-coloured drops of bacterial ooze appear on young lesions at the margins of the leaves. On drying, these form yellowish hard beads. The symptoms on panicles appear as grey to light brown lesions on glumes, resulting in poor fertility and low quality grains. The symptoms of BB often occur simultaneously with those of bacterial leaf streak and individual leaves may show symptoms of both the diseases.

13.1.1.2. Kresek (wilt) phase: It is the most destructive phase of the disease and results from systemic infection of plants in the nursery or early infection in the main field. In an infected plant, the bacterium spreads through xylem of the infected leaf to the culm and infects the crown (base of other leaves). Subsequently, all the tillers of the plants are infected resulting in quick wilting of the entire plant. The leaves of the affected plants roll completely, droop and turn yellow or grey. Ultimately, the diseased tillers wither away and in severely infected plants, stools are killed. The plants that survive, exhibit arrested growth of tillers and stunted appearance (Fig. 13.1b). Kresek phase is common in tropical countries. Temperature range of 28 - 34°C favours kresek development. Kresek phase is generally confused with stem borer attack. However, in the latter case, affected tillers/panicles can be easily pulled out and detached part of the tiller shows the signs of insect feeding at the broken end.

13.1.1.3. Pale yellow leaf phase: The youngest leaves in a clump turn pale yellow or whitish. The diseased leaves later turn yellowish brown and tillers do not grow fully.

The disease can be easily diagnosed on the basis of above given symptoms. The confirmation of the diagnosis can be done by observing the bacterial ooze in the laboratory. In the field, diagnosis of the disease can be done by placing infected leaf bits in a glass of water. If the water turns turbid after some time, presence of the disease is confirmed.

13.1.2. Causal organism

Xanthomonas oryzae pv. *oryzae* (Ishiyama 1922) Swings *et al*. 1990

Syn. *Xa. oryzae, Xa. campestris* pv. *oryzae*

The bacterium was first named as *Bacillus oryzae* Hori & Bokura. Ishiyama (1922) renamed it as *Pseudomonas oryzae* Uyeda & Ishiyama according to Migula's system of classification. In 1927, it was renamed as *Bacterium oryzae* (Uyeda & Ishiyama) Nakata according to E. F.Smith's concept and subsequently to *Xanthomonas oryzae* (Uyeda & Ishiyama) Dowson with the creation of *Xanthomonas* genus. With the introduction of pathovar concept, the name *Xa. campestris* pv. *oryzae* (Ishiyama) Dye was adopted (Dye, 1978). Swings *et al.* (1990), on the basis of phenotypic, genotypic and chemotaxonomic data, proposed *Xanthomonas oryzae* (ex Ishiyama 1922) sp. nov., nom. rev. and renamed the blight pathogen as *Xanthomonas oryzae* pv. *oryzae* (Ishiyama) Swings *et al.* (hereafter referred to as *Xoo).* The following description of *Xanthomonas oryzae* (Ishiyama 1922) sp. nov., nom. rev. is taken primarily from Swings *et al.* (1990).

Cells are straight rods (0.4 - 0.8 × 1.5 - 2.9 μm) with rounded ends, Gram-negative, non-sporulating and motile with a single polar flagellum. Cells occur singly, in pairs or sometimes in chains and filaments may occur. The bacterium is obligately aerobic and catalase positive. Colonies on solid media containing glucose are round, convex, mucoid and yellow in color due to production of xanthomonadin. Xanthomonadin, though not required for virulence, protects the bacterium form UV light and probably contributes to its survival in the field. The cells produce copious, capsular, extracellular polysaccharide (EPS). EPS plays a role in virulence and *Xoo* mutants lacking this ability are severely attenuated in virulence. The tests for indole and 2-ketogluconate formation, urease, egg yolk hydrolysis, nitrate reduction and oxidase are negative. Starch is hydrolysed and carbon sources are used only oxidatively. Growth occurs on D-xylose, D-fructose, D-galactose, sucrose, cellobiose, trehalose, sodium fumarate, sodium L-lactate, sodium L-malatate, sodium oxaloacetate and sodium succinate but not on L-arabinose, D-ribose, L-rhamnose, lactose, raffinose, inulin, esculin, methanol, ethanol, *meso*-inositol, sorbitol and dulcitol. Acid is produced from D-glucose, D-fructose, D-galactose, trehalose and cellobiose but not from D-ribose, L-rhamnose, maltose, lactose, salicin, inulin, sorbitol, *meso*-inositol, dulcitol and adonitol. Litmus milk is not acidified but H_2S is produced. Esculin, Tween 40 and 80 are hydrolysed. No growth occurs at 4 or 35°C or in the presence of 3% NaCl but poor growth occurs at 10 and 32°C. Optimum temperature for growth is between 25 and 30°C. The G + C contents of the DNA range from 64.6 to 65.0 mol%.

Xoo differs from *Xa. oryzae* pv. *oryzicola* in causing different type of symptoms on rice. *Xoo* also differs from pathovar *oryzicola* in not producing acetoin, inability to use L-alanine as sole source of carbon and unable to grow on 0.2% vitamin-free casamino acids. *Xa. oryzae* pv.

oryzicola is positive for all the above given three characteristics. *Xoo* is able to grow in the presence of 0.001% cupric nitrate while *Xa. oryzae* pv. *oryzicola* is not.

The genomes of two *Xoo* strains, namely, Japanese strain MAFF 311018 (also called T 7174) and Korean strain KACC 10331 (also called KXO 85) have been sequenced by Lee *et al.* (2005) and Ochiai *et al.* (2005), respectively. Very recently, Salzberg *et al.* (2008) have reported the sequencing of genome of a third strain, PXO 99^A of the bacterium. The genome of PXO 99^A is a single circular chromosome of 5,240,075 bp and considerably longer than the genomes of other two strains (4,941,439 bp for KACC 10331 and 4,940,217 bp for MAFF 311018). The genome of PXO 99^A contains 5083 protein coding genes, including 87 not found in the genomes of other two strains. The findings of Salzberg *et al.* (2008) provide striking evidence of genome plasticity and rapid evolution within *Xoo*. The comparisons point to sources of genomic variation and candidates for strain-specific adaptation of this pathogen that help to explain the extraordinary diversity of *Xoo* genotypes and races that have been isolated throughout the world.

Zhao *et al.* (2007) have developed a novel and highly sensitive real-time method for the identification of *Xoo* based on a TaqMan probe, which is designed to recognize the sequence of a putative siderophore receptor gene cds specific to *Xoo* and can be indentified from either a bacterial culture or naturally infected rice seeds and leaves in about 2 hr. The sensitivity of the method is 100 times higher than that of the currently used polymerase chain reaction gel electrophoresis method for diagnosis.

Pathogenic variation: A high degree of race-cultivar specificity occurs in the host-pathogen interactions. The first indication of host specialization in *Xoo* came when Asakaze, a resistant rice cultivar was unexpectedly found to be severely attacked in Kyushu Island areas of Japan. This prompted active research on this aspect. Sato *et al.* (1976) analyzed the distribution of bacterial population in Kyushu area and divided it into 2 races: one attacked all differential cultivars including Wase Aikoku 3 and the other attacked Kinmaze and Wase Aikoku 3 but not the other cultivars. In 1977, Yamamoto and associates tested 71 isolates from Indonesia on Kozaka's differential cultivars (included Rantaj-emas along with others) and found no isolates belonging to Kozaka's groups (Kozaka, 1969) I or II but 46 isolates with virulence similar to group III in Japan, i.e., virulent to Rantaj-emas. They also identified two more groups, IV and V and the last two groups showed pathogenicity similar to the two races identified by Sato *et al.* (1976). These results not only demonstrated pathogenic specialization/variation but also showed a strong interaction between rice cultivars and strains of the pathogen. At present, there are more than 30 races of the pathogen reported from different countries (Nino-Liu *et al.*, 2006). A set of races identified in the Philippines using five differential rice cultivars (Mew, 1987) has been used widely for identifying and classifying resistance to the pathogen in other cultivars.

13.1.3. Disease cycle

13.1.3.1. Perpetuation: It has been carefully studied in temperate regions, particularly in Japan (Tagami and Mizukami, 1962; Mizukami, 1966). The following sources help in the survival of the bacterium in different regions of the world.

13.1.3.1.1. Weed hosts: In Japan, natural infection is found on weeds such as *Leersia* spp. and *Zizania latifolia* and among them, *L. sayanuka* is the most important source of perpetuation. The bacterium survives in rhizosphere of this weed and multiplies in early spring when new shoots of the weed begin to emerge. From the weed, the bacterium is dispersed through irrigation water to the nursery or transplanted fields of rice. Gonzalez *et al.* (1991) reported that *L. hexandra* acts as a symptomless carrier of the bacterium. In India, Brar and Thind (1994) reported *Echinochloa colonum* as a natural host of the bacterium and also proved that seeds from infected plants transmit the bacterium to the seedlings. In the tropics, *Leptochloa* spp., *Cyperus difformis* and *C. rotundus* also act as natural hosts of the bacterium. Other hosts of the bacterium include several species of wild rice (*O. rufipogon*, *O. australiensis*) and a number of weeds such as, *Cynodon dactylon*, *Cenchrus ciliaris*, *Brachiaria mutica*, *E. crusgalli*, *Panicum maxicum* and *Paspalum scrobiculatum*.

13.1.3.1.2. Rice straw: The bacterium can survive in piles of rice straw kept out door in temperate regions (Isaka, 1962). In tropical areas like Punjab, India, the bacterium has been found to survive in rice straw kept in big heaps protected from weathering agencies but it does not play any role in the perpetuation of the bacterium (personal observation).

13.1.3.1.3. Rice stubbles: The bacterium survives in the stubbles (base of stems and roots) of the plants left behind in the field after harvest until the following spring in Japan. It is the most important source of infection in Thailand (Tabei and Eamchit, 1974).

13.1.3.1.4. Volunteer plants: Infected seeds from the previous crop shed in the field, give rise to diseased seedlings and these serve as a source of infection in double-cropped areas. In North India, off-type tall plants usually infected with bacterial blight are generally found in rice fields. As these plants mature before harvesting of the crop, they shed their seeds and these infected seeds initiate the disease in the subsequent year (Durgapal, 1985).

13.1.3.1.5. Rhizosphere of succeeding crops: Thind and Brar (1989), on the basis of two-year field trials, demonstrated that the bacterium survived in the rhizosphere of berseem, wheat and winter maize (sown in fields after harvesting bacterial blight infected rice) till the sowing of next rice crop. The survival in the rhizosphere of wheat and rapeseed has also been reported by Wakimoto (1956) and Mohiuddin *et al.* (1977). The survival in the rhizosphere of wheat and berseem in Punjab and other North Indian states, where rice is generally sown after these crops, has an important bearing because most of the rice acreage in these areas is sown after harvesting wheat and berseem.

13.1.3.1.6. Seed: The seed-borne nature of the bacterium was first reported by Fang *et al.* (1956) and according to them the bacterium was present mainly in glumes and occasionally in the endosperm. Since then, the seed-borne nature of the bacterium has remained controversial, some workers reporting positive results while others reporting negative findings. Different methods like isolation of the bacterium from seed, inoculation of seed macerate on young seedlings, bacterial streaming from seed parts and seedling symptom test have been used to prove the seed-borne nature. Among them, the seedling symptom test is considered best, being natural and reliable.

Srivastava and Rao (1963, 1964) and Singh (1972) reported positive seed-plant transmission of the bacterium. Reddy (1983) using nine-month-old seeds observed typical blight symptoms 60 days after sowing. Brar and Thind (1995) developed a technique called, **"Nitrogen-Supplemented Seedling Symptom Test"** (NSSST) using irrigation water containing 300 μg of nitrogen/ml, to prove seed-plant transmission of the bacterium. The typical symptoms of the disease appear in 10 - 12 days after sowing in NSSST. Thind and Brar (1998) reported that the bacterium survived in all the seed samples, till the sowing of next rice crop. These seed samples were collected from naturally infected crop from different places in Punjab, India and tested at monthly intervals. However, the number of seeds showing positive seed-plant transmission decreased with storage period. The seed-borne inoculum is a major source of perpetuation in Punjab and possibly in other Northwestern states of India where only one crop of rice is taken in a year.

13.1.3.2. Disease development: The bacterium enters the host through hydathodes, wounds and growth cracks formed due to the development of new roots. The new wounds are more conducive than old wounds for infection. Bacteria multiply in the intercellular spaces of the underlying epitheme, then enter and spread into the plant through the xylem. The pathogen moves vertically through the leaf through primary veins but also progresses laterally through commisural veins. Within a few days, bacterial cells and extracellular polysaccharide (EPS) produced by the bacterium fill the xylem vessel and bacteria ooze out from hydathodes, forming beads or strands of exudate on the leaf surface which acts as a source of secondary inoculum. The EPS protects the bacterial cells from desiccation.

The temperature range of 25 - 31°C is highly favourable for disease development. The kresek phase appears in 20 days when rice plants are kept at 31°C while it takes 40 or more at 21°C. Younger plants, i.e., less than 21-day-old are the most susceptible and the susceptibility decreases with the increase in age. The common practice of cutting leaf tips of seedlings before transplanting, particularly when the seedlings are quite tall, plays an important role in the development of kresek symptoms because the cut ends serve as infection courts for the pathogen. In addition, broken roots resulting from pulling the seedlings off the seed-bed serve as entry points for bacteria present in flood-irrigated fields.

The disease is favoured by high relative humidity. The epiphytotics of the disease are common in rainy or monsoon season in India. The same varieties, which show high disease severity in rainy season, either escape infection or show traces to moderate infection in the *Rabi* season. Secondary spread of the disease occurs through wind-splashed rain, irrigation water and contact of healthy leaves with diseased leaves. Wind-driven rains and persistent warm weather are highly conducive for the spread of pathogen and disease development. Severe epiphytotics often occur following typhoons, the fierce winds, wind-blown rain and hail, and all these wound rice plants and disperse bacteria.

Heavy doses of nitrogenous fertilizers influence the disease incidence and even the source of nitrogen also influences disease severity. However, increase in disease intensity with increased levels of fertilizers differs with different cultivars. In general, nitrogen induced susceptibility is expressed by susceptible cultivars while it is not so with resistant cultivars. Reddy *et al.* (1979) showed that high levels of nitrogen (> 100 kg/ha) caused reduction in yield due to increased bacterial blight incidence in susceptible variety IET 2895 but resistant variety IET 4141 was least affected by all the nitrogen levels. The increase in disease severity due to application of excess doses of nitrogen results from enhanced vegetative growth leading to high humidity.

The disease severity also increases under shade. In Punjab, India, it is a common sight to see high disease severity on plants growing under shade in transplanted rice (personal observation). The high disease severity results from high relative humidity and favorable temperature maintained for longer period under shade.

Attempts have been made to forecast the disease by different methods based on climatic factors and counting bacterial and bacterio-phage populations. The unusual changes in climate affect forecasting by all these methods. In temperate regions, monitoring of bacteriophage strains specific to *Xoo* has been used for forecasting since 1960s. The time when phage population in irrigation water exceeds 200 particles/ml corresponds to time 10 days before the initial disease development (Yoshimura *et al.*, 1960). For an accurate forecasting, the water sample should be taken in the morning at the out flow. However, the bacteriophage forecasting system is not practised extensively in tropical Asia, because rice cultivation is mostly rain fed, limiting the use of phage detection in paddy fields. In general also, disease forecasting has been of limited utility because chemical control of the disease is unavailable or impracticable.

13.1.4. Host resistance

The development of resistant cultivars carrying major resistance (*R*) genes has been the most effective approach for controlling BB. Till date, 31 genes imparting resistance to BB pathogen have been identified (Wang *et al.*, 2009). Most of these genes have been found in cultivars of *O. sativa* subspecies *indica* but some are also from *japonica* cultivars and from wild species of rice such as, *O. longistaminata, O. rufipogon, O. nivara, O. minuta* and *O. officinalis.* Some of these *R* genes like *Xa21* are effective only in adult plants while most of them do not seem to be related with plant growth stage. However, *Xa3* is typically effective only in adult plants bearing at least one race against which it is effective at all stages of growth. Some genes like *Xa21* and *Xa31* impart resistance against a wide range of *Xoo* races while others are effective against only one or a few races of the pathogen. Most *R* genes are dominant while some are recessive (*xa5, xa8, xa13, xa19, xa20, xa24 and xa28*) while still others are semi-dominant.

Near isogenic lines (NILs) have been developed by introducing single *R* genes into the background of susceptible *indica* cultivar IR 24 to serve as differential lines. These differential lines are used for the identification of races of the bacterium. Rice differentials recommended by the scientists of All India Coordinated Rice Improvement Programme of Directorate of Rice Research, Hyderabad are given in Table 13.1.1. Rice workers also include commercially grown rice varieties in their regions, as additional differentials, to have a better assessment of the pathogenic variation in the region.

Table 13.1.1. Rice differentials recommended by the scientists of All India Coordinated Rice Improvement Programme of Directorate of Rice Research, Hyderabad for field monitoring virulences in *Xanthomonas oryzae* pv. *oryzae.*

Sr. No.	Line/Differential	Gene combination
1	IRBB 1	*Xa1*
2	IRBB 3	*Xa3*
3	IRBB 4	*Xa4*
4	IRBB 5	*xa5*

5	IRBB 7	*Xa7*
6	IRBB 8	*xa8*
7	IRBB 10	*Xa10*
8	IRBB 11	*Xa11*
9	IRBB 13	*xa13*
10	IRBB 14	*Xa14*
11	IRBB 21	*Xa21*
12	IRBB 50	*Xa4+xa5*
13	IRBB 51	*Xa4+xa13*
14	IRBB 52	*Xa4+Xa21*
15	IRBB 53	*xa5+xa13*
16	IRBB 54	*xa5+Xa21*
17	IRBB 55	*xa13+Xa21*
18	IRBB 56	*Xa4+xa5+xa13*
19	IRBB 57	*Xa4+xa5+Xa21*
20	IRBB 58	*Xa4+xa13+Xa21*
21	IRBB 59	*xa5+xa13+Xa21*
22	IRBB 60	*Xa4+xa5+xa13+Xa21*
23	DV 85	*xa5+Xa7*
24	Ajaya (IET 4141/CR 98-7216)	
25	RP B10-226	
26	TN 1	*Xa14*

1 - 9 scale of Standard Evaluation System for Rice (IRRI, 1996) is used to record the disease reaction.

Resistance of rice to specific *Xoo* races is governed by major *R* genes with a qualitative effect that impart complete resistance and polygenes with a quantitative effect, which impart partial resistance. It was believed that gene *Xa21* derived from *O. longistaminata* conferred resistance to all the pathotypes/races of *Xoo* from India and Philippines. However, Saini and Goel (1998) reported that even this growth stage specific gene was ineffective in 70-day-old plants against some isolates of the bacterium from Punjab, India. Priyadarisini and Gnanamanickam (1999) also found that out of 140 isolates of *Xoo* isolated from an epiphytotic affected plants that occurred in Kerala, South India in 1998, 20 isolates were virulent on rice line IRBB 21 (carrying *Xa21* gene) but the rice line NH 56 carrying four resistance genes (*Xa4, xa5, xa13 and Xa21*) was resistant to all the isolates. Sirisha *et al.* (2004) again reported breakdown of resistance of *Xa21* against *Xoo* due to possible gain of virulence by the pathogen in the process of co-evolution. Swamy *et al.* (2006) also reported that Pusa Basmati 1 transgenic lines expressing *Xa21* gene were susceptible to all the six isolates collected from different regions of India while another line MH 2R into which three resistant genes, *xa5*, *xa13* and *Xa21* were introduced through marker-assisted selection (MAS) showed resistant reaction to all the isolates. Sanchez *et al.* (2000) has already emphasized the usefulness of MAS for pyramiding of bacterial blight resistance genes, particularly for recessive genes like *xa5* and *xa13* that are difficult to select through conventional breeding in the presence of dominant genes, such as *Xa21*. They also reported that MAS showed an accuracy of 95 and 96% for identifying homozygous resistant plants containing *xa5* and *xa13* genes, respectively. Host resistance conferred by single genes against *Xoo* is generally not stable due to considerable variation in the pathogen population. Therefore, pyramiding of BB resistance genes in rice is based on pathogen population response to these genes. Thus, the selected resistance genes are combined in a cultivar through MAS. This helps in developing varieties with broader-spectrum and longer-lasting resistance. Shanti *et al.* (2001) also found that three gene combinations, i.e., *Xa4* +*xa5*, *xa5* + *Xa21* and *Xa4* + *xa5* + *Xa21* conferred a broad-spectrum of resistance to all the strains evaluated (collected from Orissa, Madhya Pradesh and Uttar Pradesh of India), thus supporting the strategy of pyramiding appropriate resistance genes.

Xa21 is the first resistance gene cloned in the monocots. It is a member of multigene family in rice and of a larger family of genes in plants and animals involved in recognition of pathogens and innate immunity. In addition to *Xa21*, other genes like *Xa1, xa5*, *xa13* and *Xa27* have been cloned and represent gene families. The *avrXa7* and *avrXa10* corresponding to *R* genes *Xa7* and *Xa10,* respectively are the first avirulence genes of *Xoo* cloned from race 2 strain PXO 86. These avirulence genes are members of the large gene family of *avrBs3* and share an extensive sequence homology with *avrBs3* gene cloned from the pepper and tomato pathogen, *Xa vesicatoria*. This gene family is widely distributed in *Xanthomonas* species, and *Xoo* possesses 12 - 14 copies. Type III secretion system-mediated secretion and delivery of AvrXa10 into rice cells by *Xoo* have been demonstrated using a reporter protein fusion.

Kano *et al.* (2010) investigated the response of rice plant to three rare sugars, namely, D-sorbose, D-altrose and D-allose and found that only D-allose significantly inhibited the root and shoot growth of rice plants. They further revealed (on the basis of large-scale gene expression analysis using a rice microarray) that D-allose treatment caused a high upregulation of many defence-related, pathogenesis-related protein genes in rice. Hence, D-allose treatment of rice plants conferred limited resistance to rice against the pathogen, *Xoo*. These results indicate that D-allose has a growth inhibitory effect but might prove to be a candidate elicitor for reducing disease development in rice.

13.1.5. Management

The measures adopted for the management of BB vary depending on the location and disease incidence. The following measures help in reducing the disease severity.

1. Destruction of diseased plant debris including rice straw and stubbles.
2. Removal of weeds from canals and ridges.
3. Seed treatment before sowing is practised in tropical Asia. Seed can be treated with Streptocycline (100 μg/ml) + Agallol (0.1%) or Aretan/Tafasan (0.05%) followed by hot water treatment at 52 - 54°C for 30 minutes. Alternately, use the seed from disease-free crop.
4. Nursery should be raised away from diseased fields and avoid its flooding with contaminated water. It should also not be sown under shade. Raising the nursery under upland dry conditions has been found effective in Sri Lanka.
5. Keep proper plant spacing and avoid high doses of nitrogenous fertilizers.
6. Chemical control of BB began in 1950s with the prophylactic sprays of Bordeaux mixture. Since then many fungicides including copper and mercurial compounds and several antibiotics have been tried singly and in combination with limited success. In temperate regions, probenazole is applied in the irrigation water in nursery and in the transplanted rice to prevent the bacterial multiplication. The chemicals used for spraying include tecloftalam, phenazine oxide and nickel dimethyl dithiocarbamate. In the tropical monsoon climate of Asia, chemical sprays are not generally given due to the non-availability of effective commercial chemicals and difficulty in spraying due to uncertain weather. However, the following three reports suggest the possibility of managing the disease with chemical sprays. Foliar sprays of Blitox-50WP (0.25%) + Streptocycline (100 μg/ml) have been found to give significant disease control along with significant increase in grain yield (Thind and Mehra, 1992; Sinha and Sinha, 2000). Tilak *et al.* (2001) obtained significant disease control along with significant increase in grain yield with 2 sprays of Bacterimycin - 2000 (0.04%) in a multilocation trial.

7. Bacterial antagonists of *Xoo* have received special attention as biocontrol candidates and as a result of it; some success has been achieved in this respect. In India, more than 40 bacterial isolates antagonistic to the bacterium have been identified through *in vitro* and *in vivo* tests. Among them, rice-associated rhizobacteria, *Psm. fluorescens* and *Psm. putida* strain V14i significantly suppressed the disease severity when sprayed on leaves (Sivamani *et al.*, 1987). The application of *Bacillus* species, as seed treatment before sowing, root-dip prior to transplanting and foliar sprays in the field, has also given encouraging results. In at least one study, the antagonists suppressed the disease by almost 60% and increased the plant height and grain yield by two-fold (Vasudevan *et al.*, 2002).

Thind and Ahmad (1994) reported that out of 32 microorganisms, isolated from phylloplane and spermosphere of 4 rice cultivars and from phylloplane of 4 rice weeds, 9 proved antagonistic to the bacterium in *in vitro* evaluation and these antagonists also gave significant control of the disease under field conditions for two consecutive years along with an increase in grain yield. Three antagonists, namely, *Ba. subtilis*, *Erwinia herbicola* and *Penicillium oxalicum* gave significantly higher grain yield in addition to significantly better disease control in comparison to Streptocycline (100 μg/ml). *Ba. subtilis*, the most promising antagonist, gave 35.4% disease control and 10.55% increase in grain yield. Vidhyasekaran *et al.* (1996) reported that treatment of rice seeds with talc-based formulation of *Psm. fluorescens* @ 10 g/kg of seed induced systemic resistance and gave protection up to 60 days against bacterial blight pathogen while the foliar sprays imparted resistance up to 15 days. Kamala and Vidhyasekaran (1998) tested talc-based formulation of *Psm. fluorescens* str. Pf 1 as seed treatment, root-dip application and foliar sprays individually and in combination, and found that the combination treatment induced resistance and gave 60 - 70% disease control in field along with an increase in grain yield. Despite these encouraging results, biocontrol agents have not attained a widespread use for the control of BB.

8. Use of resistant cultivars is the most effective and economical method to control bacterial blight of rice. However, the development of resistant varieties and durability of the resistance is complicated by the high diversity in the pathogen population. Resistant varieties, based on pathogen population structure of different regions, have been developed and released for cultivation. Rice cultivars, namely, PR 111, PR 113, PR 114, PR 115, PR 116, PR 118 and PR 120 developed by Punjab Agricultural University, Ludhiana, Punjab have been released for cultivation in Punjab, India (Anonymous, 2009). Pusa 1460, a variety

developed by pyramiding resistance genes *Xa1* and *Xa21* in the background of Pusa Basmati 1 through marker-assisted back-cross breeding has been released for cultivation in Punjab, Haryana, J & K, Delhi and Uttarakhand of India. Transgenic rice developed from cultivar IR 72 showed high level of resistance to existing races of the pathogen in the Philippines (Mew and Vera Cruz, 2001). The growing of cultivar mixtures containing appropriate single resistance genes has also been tried for the management of the disease.

REFERENCES

Anonymous (2009). *Package of Practices for Kharif Crops*. Punjab Agricultural University, Ludhiana, pp. 344.

Bhapkar, D.G., Kulkarni, N.B. and Chavan, V.M. (1960). Bacterial blight of paddy. *Poona Agri. Coll. Mag.* 51: 36 - 46.

Brar, J.S. and Thind, B.S. (1994). A new weed host of *Xanthomonas oryzae* pv. *oryzae*, the causal agent of bacterial leaf blight of rice. *Ann. Plant Prot. Sci.* 2: 79 - 80.

Brar, J.S. and Thind, B.S. (1995). Nitrogen supplemented seedling symptom test-a new method for proving seed transmission of *Xanthomonas oryzae* pv. *oryzae*. *Oryza* 32: 42 - 45.

Durgapal, J.C. (1985). Self sown plants from bacterial blight infected rice seeds– a possible source of primary infection in North West India. *Curr. Sci.* 54: 1283 - 84.

Dye, D.W. (1978). Genus IX *Xanthomonas* Dowson 1939. In: Young, J.M., Dye, D.W., Bradbury, J.F. *et al.* A proposed nomenclature and classification for plant pathogenic bacteria. *N. Z. J. Agric. Res.* 21: 162 - 66.

Fang, C.T., Lin, C.F. and Chu, C.L. (1956). A preliminary study on the disease cycle of the bacterial blight of rice. *Acta Phytopathol. Sin.* 2: 173 - 85 (Ch. en).

Gonzalez, C.F., Xu, D.W. and Li, H.L. (1991). *Leersia hexandra*, an alternate host for *Xanthomonas campestris* pv. *oryzae* in Texas. *Plant Dis.* 75: 159 - 62.

Goto, M. (1964). 'Kresek' and pale yellow leaf systemic symptoms of bacterial leaf blight of rice caused by *Xanthomonas oryzae* (Uyeda et Ishiyama) Dowson. *Plant Dis. Reptr* 48: 858 - 61.

Isaka, M. (1962). Overwintering of bacterial leaf blight organism in damaged leaf in paddy field. *Proc. Assoc. Plant Prot.* Hokuriku, 10, 90.

Ishiyama, S. (1922). Studies on the white leaf disease of rice plants. *Rept. Agri. Expt. Sta. Tokyo* 45, 233 - 51 [Ja.].

Kamala, N. and Vidhyasekaran, P. (1998). Evaluation of new biopesticide formulation for the management of some foliar diseases of rice. In: G.S. Dhaliwal, N.S. Randhawa, R. Arora and A.K. Dhawan (eds). *Ecological Agricultural and Sustainable Development*. Vol. 2, pp. 382 - 88.

Kano, A., Gomi, K., Yamasaki-Kokudo, Y. *et al.* (2010). A rare sugar, D-allose, confers resistance to rice bacterial blight with upregulation of defence-related genes in *Oryza sativa*. *Phytopatholgy* 100: 85 – 90.

Kozaka, T. (1969). Control of rice diseases with resistant varieties. *Agric. Hortic.* 44: 208 -12.

Lee, B.M., Park, Y.J., Park, D.S. *et al.* (2005). The genome sequence of *Xanthomonas oryzae* pathovar *oryzae* KACC 10331, the bacterial blight pathogen of rice. *Nucleic Acids Res.* 33: 577 – 86.

Mew, T.W. (1987). Current status and future prospects of research on bacterial blight of rice. *Annu. Rev. Phytopathol.* 25: 359 - 82.

Mew, T.W., Alvarez, A.M., Leach, J.E. *et al.* (1993). Focus on bacterial blight of rice. *Plant Dis.* 77: 5 - 12.

Mew, T.W. and Vera Cruz, C.M. (2001). Bacterial blight of rice. In: O.C. Maloy and T.D. Murray (eds). *Encylopedia of Plant Pathology*. Vol. 1, John Wiley and Sons, Inc., New York, pp. 71 - 74.

Mizukami, T. (1966). Epidemiology of bacterial leaf blight of rice and use of phages for forecasting. *Eleventh Pacific Science Congress, Symposium on Plant Diseases in the Pacific.* 15 - 32.

Mohiuddin, M.S., Rao, Y.P., Mohan, S.K. *et al.* (1977). Survival of *Xanthomonas oryzae*, the casual incitant of bacterial blight of rice in the rhizosphere of wheat. *Sci. & Cult.* 43: 124 - 25.

Nino-Liu, D.O., Ronald, P.C. and Bogdanove, A.J. (2006). *Xanthomonas oryzae* pathovars: Model pathogens of a model crop. *Mol. Plant Pathol.* 7: 303 - 24.

Ochiai, H., Inoue, Y., Takeya, M. *et al.* (2005). Genome sequence of *Xanthomonas oryzae* pv. *oryzae* suggests contribution of large numbers of effector genes and insertion sequences to its race diversity. *Jpn. Agr. Res. Q.* 39: 275 – 87.

Ou, S.H. (1972). *Rice Diseases*. 1st Ed. Commonwealth Mycol. Int. Kew, Surrey England, 368 pp.

Ou, S.H. (1985). *Rice Diseases*. 2nd Ed. C.A.B. International, Slough, UK, 380 pp.

Priyadarisini, V.B. and Gnanamanickam, S.S. (1999). Occurrence of a subpopulation of *Xanthomonas oryzae* pv. *oryzae* with virulence to rice cv. IRBB 21 (*Xa21*) in Southern India (Abstr.). *Plant Dis.* 83: 781.

Ray, P.R. and Sengupta, T.K. (1970). A study on the extent of loss of yield in rice due to bacterial blight of rice. *Indian Phytopathol.* 23: 713 - 14.

Reddy, A.P.K., Katyal, J.C., Rouse, D.I. *et al.* (1979). Relationship between nitrogen fertilization, bacterial leaf blight severity and yield of rice. *Phytopathology* 69: 970 - 73.

Reddy, A.P.K., MacKenzie, D.R., Rouse, D.I. *et al.* (1979a). Relationship of bacterial leaf blight severity to grain yield of rice. *Phytopathology* 69: 967 - 69.

Reddy, P.R. (1983). Evidence for seed transmission of *Xanthomonas campestris* pv. *oryzae*. *Curr. Sci.* 52: 265 - 66.

Reitsma, J. and Schure, P.S.J. (1950). 'Kresek', a bacterial disease of rice. Contributions of the Central Agricultural Research Station, Bogor No. 117, 17 pp.

Saini, R.G. and Goel, R.K. (1998). Disease resistance in rice (*Oryza sativa* L.)- an appraisal of the Indian work. *Crop Improv.* 25: 141 - 54.

Salzberg, S.I., Sommer, D.D. Schatz, M.C. *et al.* (2008). Genome sequence and rapid evolution of the rice pathogen *Xanthomonas oryzae* pv. *oryzae* PXO 99[A]. *BMC Genomics* 9: 204 - 19.

Sanchez, A.C., Brar, D.S., Huang, N. *et. al.* (2000). Sequence tagged site marker-assisted selection for three bacterial blight resistance genes in rice. *Crop Sci.* 40: 792 - 97.

Sato, T., Choi, Y.S., Iwasaki, M. *et al.* (1976). Distribution of races of *Xanthomonas oryzae* in Kyushu (Abstr.). *Ann. Phytopathol. Soc. Jpn.* 17: 42.

Shanti, M.L., George, M.L.C., Vera Cruz, C.M. *et al.* (2001). Identification of resistance genes effective against rice bacterial blight pathogen in Eastern India. *Plant Dis.* 85: 506 - 512.

Singh, G.P., Srivastava, M.K., Singh, R.V. *et al.* (1977). Variation in quantitative and qualitative losses caused by bacterial blight in different rice varieties. *Indian Phytopathol.* 30: 180 - 185.

Singh, R.N. (1972). Perpetuation of the bacterial blight disease of rice in North India. *Indian Phytopathol.* 25: 148 - 50.

Sinha, B.B.P. and Sinha, R.K.P. (2000). Integrated disease management of bacterial blight of rice with special reference to neem product. *Ann. Plant Prot. Sci.* 8: 260 - 61.

Sirisha, C., Reddy, J.N., Mishra, D. *et al.* (2004). Susceptibility of IRBB 21 carrying the resistance gene *Xa21* to bacterial blight. *Rice Genet. Newslet.* 21: 75 - 77.

Sivamani, E., Anuratha, C.S. and Gnanamanickam, S.S. (1987). Toxicity of *Pseudomonas fluorescens* towards bacterial plant pathogens of banana (*Pseudomonas solanacearum*) and rice (*Xanthomonas campestris* pv. *oryzae*). *Curr. Sci.* 56: 547 - 48.

Sreenivasan, M.C., Thirumalachar, M. J. and Patel, M. K. (1959). Bacterial blight of rice. *Curr. Sci.* 28: 469 - 70.

Srivastava, D.N. (1972). Bacterial blight of rice. *Indian Phytopathol.* 25: 1 - 16.

Srivastava, D.N. and Rao, Y.P. (1963). Epidemic of bacterial blight disease of rice in North India. *Indian Phytopathol.* 16: 393 - 94.

Srivastava, D.N. and Rao, Y.P. (1964). Seed transmission and epidemiology of bacterial blight of rice in North India. *Indian Phytopathol.* 17: 77 - 78.

Swamy, P., Panchbhai, A.N., Dodiya, P. *et al.* (2006). Evaluation of bacterial blight resistance in rice lines carrying multiple resistance genes and *Xa21* transgenic lines. *Curr. Sci.* 90: 818 - 24.

Swings, J., Van den Moorter, M., Vauterin, L. *et. al.* (1990). Reclassification of the casual agents of bacterial blight (*Xanthomonas campestris* pv. *oryzae*) and bacterial leaf streak (*Xanthomonas campestris* pv. *oryzicola*) of rice as pathovars of *Xanthomonas oryzae* (ex Ishiyama 1922) sp. nov., nom. rev. *Int. J. Syst. Bacteriol.* 40: 309 - 11.

Tabei, H. and Eamchit, S. (1974). Infection source of bacterial leaf blight of rice in Thailand. *Jpn. Agric. Res. Q.* 8, 123 - 25.

Tagami, Y. and Mizukami, T. (1962). Historical review of the researches on bacterial leaf blight of rice caused by *Xanthomonas oryzae* (Uyeda *et* Ishiyama) Dowson. *Special Report of the Plant Disease and Insect Pests Forecasting Service* No. 10, 112 pp.

Thind, B.S. (2002). Bacterial blight of rice- an overview. *Plant Dis. Res.* 17: 227 - 51.

Thind, B.S. and Ahmad, M. (1994). Biological control of *Xanthomonas oryzae* pv. *oryzae*. In: Lemattre, M., Freigoun, S., Rudolph, K. *et al.* (eds). *Proc. 8th Int. Conf. Plant Path. Bact.* Versailles, France, pp. 867 - 73.

Thind, B.S. and Brar, J.S. (1989). Perpetuation of *Xanthomonas campestris* pv. *oryzae* (Ishiyama) Dye in the rhizosphere of succeeding crops. In: Z. Klement (ed.). *Proc. 7th Int. Conf. Plant Path. Bact.* Part A, Budapest, Hungary, pp. 1021 - 27.

Thind, B.S. and Brar, J.S. (1998). Perpetuation of *Xanthomonas oryzae* pv. *oryzae* under North Indian conditions. In: R. Mahadevan (ed.). *Proc. 9th Int. Conf. Plant Path. Bact.* Chennai, India, pp. 409 - 19.

Thind, B.S. and Mehra, R.K. (1992). Chemical control of bacterial blight of rice. *Plant Dis. Res.* 7: 226 - 34.

Tilak, K.V.B.R., Udaiyan, K. and Ramarethinam, S. (2001). Bioefficacy of bacterimycin - 2000, a plant immunomodulator in the control of bacterial leaf blight (BLB) in rice. *Pestology* 25: 1 - 10.

Vasudevan, P., Kavitha, S., Priyadarisini, V.B. *et al.* (2002). Biological control of rice diseases. In: S. Gnanamanickam (ed.). *Biological Control* of Crop Diseases. Marcel Dekker, New York, pp. 11 - 32.

Vidhyasekaran, P., Kamala, N. and Ramanathan, A. (1996). Induced systemic resistance against rice bacterial blight. Paper presented in *9th Int. Conf. Plant Path. Bact.* held during August 26 - 29, 1996, Chennai, India, p.98.

Wakimoto, S. (1956). Considerations of the overwintering and the infection mechanisms of *Xanthomonas oryzae. Plant Prot. Jpn.* 10: 421 - 24.

Wang, C., Wen, G., Lin, X. *et al.* (2009). Identification and fine mapping of the new bacterial blight gene, *Xa31* (t) in rice. *Eur. J. Plant Pathol.* 123: 235 – 40.

Yoshimura, S., Aoyagi, K., Morihasni, T. *et al.* (1960). *Proc. Assoc. Plant Prot.* Hokuriku, 8: 31 - 41.

Zhao, W.J., Zhu, S.F., Liao X.L. *et al.* (2007). Detection of *Xanthomonas oryzae* pv. *oryzae* in seeds using a specific TaqMan probe. *Mol. Biotechnol.* 35: 119 - 27.

13.2. BACTERIAL LEAF STREAK OF RICE

This is another important disease of rice but comparatively less destructive and less widespread than bacterial blight (hereafter referred to as BB). Reinking (1918) was the first to report the disease from the Philippines but he named it as bacterial stripe. However, the disease was erroneously referred to as BB for several years. Fang *et al.* (1957) studied the disease in southern China, distinguished it from BB and renamed it as bacterial leaf streak (hereafter referred to as BLS). He also found its causal agent different from that of BB and named it as *Xanthomonas oryzicola*.

The occurrence of the disease is restricted largely to tropical and subtropical Asia including Thailand, Indonesia, Bangladesh, Malaysia, Vietnam and India in addition to southern China and the Philippines. It is one of the major rice diseases in southern China. It is also present in regions of northern Australia and has become a significant problem in parts of West Africa. Quarantines for the causal pathogen are in place in the United States and other rice growing countries where the disease is not present, but also in places where the disease occurs, to guard against the introduction of new virulent strains. The pathogen is also listed in the 2002 Agricultural Bioterrorism Protection Act of the USA as a potential bioterrorism agent, necessitating strict biosecurity measures in addition to standard biosafety measures. The available reports suggest that the yield losses range from 0 - 20% depending upon the rice variety, climatic conditions and severity of the disease (Ou, 1985). Under conditions favorable for disease development, reduction in grain weight up to 32% has been reported. The incidence of BLS is increasing in parts of Asia, where hybrid rice is grown, such as China, because hybrid rice cultivars are susceptible to the pathogen.

13.2.1. Symptoms

Although it is mainly a foliage disease infecting the parenchymatous tissue exclusively, infection of the seed coat is common. The symptoms start as minute, water-soaked, translucent, interveinal pinheads that can be better seen when the leaf is held against light. These pinheads enlarge gradually into 1 - 10 cm long, water-soaked, dark green, translucent, discrete vein-limited streaks, which are typically yellow in colour. The width of the streaks varies from 0.5 - 1.00 mm depending on the rice cultivar but is generally more in broad-leaf varieties. At later stage, infected leaves turn grayish-white and die. Under moist conditions, yellow- to amber-coloured bacterial exudate is deposited in the form of beads in linear rows all along the streak on both the leaf surfaces. This bacterial ooze is a typical sign of the disease. Irregular, diffused, cholorotic areas appear around the lesions in highly susceptible cultivars (Fig. 13.2). In severe cases, streaks may coalesce resulting in blotchy lesions and blightening of the leaves. Sheath infection is rare but symptoms are similar to that of leaf lamina. Glume infection occurs in the form of water-soaked lesions. In resistant varieties, water-soaked pinheads turn brown without elongation or enlarge into 5 - 10 mm long reddish-brown streaks. Under conditions favourable for the spread of pathogen, the disease may affect the entire field.

When infection results from entry of the bacterium through breaks in the leaf, which might occur due to high wind velocity, symptoms may expand length-wise killing most or whole of the leaf. The symptoms of BB and BLS are quite distinct in the initial stage but they become less differentiable in the advanced stage. Moreover, the severe infection of BLS may also result in partial or total leaf blightening resembling BB symptoms. However, even in such cases, both the diseases can be correctly and conveniently diagnosed on the basis of diagnostic hints given below. BB infected leaves, when viewed against light, appear uniformly opaque. In case of BLS, numerous interveinal, translucent streaks are seen. In case of BB, bacterial exudate appears in the form of isolated and infrequently formed beads near the margins on the upper surface of the leaves. In case of BLS, numerous beads of yellow to amber colour, arranged closely in a linear fashion, appear all along the streak length on both the leaf surfaces (Shekhawat and Rao, 1972).

The symptoms of BB and BLS often occur simultaneously in rice fields and the symptoms of both the diseases may be present on the same leaves.

13.2.2. Causal organism

Xanthomonas oryzae pv. *oryzicola* (Fang *et al.* 1957) Swings *et al.* 1990

Syn. *Xa. campestris* pv. *oryzicola*

Fang *et al.* (1957) first named the bacterium as *Xa. oryzicola*. Later on, it was changed to *Xa. translucens* f. sp. *oryzae* (Fang *et al.*) Pordesimo (Pordesimo, 1958) and then to *Xa. translucens* f. sp. *oryzicola* (Fang *et al.*) Bradbury (Bradbury, 1971). With the introduction of pathovar concept, it was renamed as *Xa. campestris* pv. *oryzicola* (Fang *et al.*) Dye (Dye 1978). Swings *et al.* (1990) reclassified it as *Xa. oryzae* pv. *oryzicola*. *Xa. oryzae* pv. *oryzicola* (hereafter referred to as *Xoc*), being a pathovar of *Xa. oryzae*, shares the characteristics of *Xa. oryzae*. The description of *Xanthomonas oryzae* (Ishiyama 1922) sp. nov., nom. rev. is given under bacterial blight of rice. *Xoc* differs from *Xa. oryzae* pv. *oryzae* (hereafter referred to as *Xoo*) in its ability to produce acetoin, to utilize L-alanine as sole source of carbon, to grow on 0.2% vitamin-free casamino acids and sensitivity to 0.001% cupric nitrate. The growth of *Xoc* on solid media is fast in comparison to *Xoo*. Variation in *Xoc* has not been studied as extensively as in *Xoo*. However, the analysis of strains isolated from different locations in the Philippines indicates the diversity in the pathogen (Raymundo *et al.*, 1999). Clusters of isolates were not robust, and there was no correlation between clusters and the reactions of a set of diverse rice varieties. Like *Xoo, Xoc* also serves as a model for understanding fundamental aspects of bacterial interaction with plants (Nino-Liu *et al.*, 2006).

The genome of strain BLS 256 of *Xoc* has been sequenced (available in the Comprehensive Microbial Resource, http://www.tigr.org). The genome has overall similarity to *Xoo* genomes sequenced with respect to size, G + C content, abundance of IS elements and TAL effector genes, and the presence of inversions and rearrangements. However, the genome also has some notable unique features like a gene cluster with predicted functions in synthesis and transport of lipopolysaccharides, and bordered by two highly conserved genes and is distinct both in gene order and gene content from the corresponding locus in the *Xoo* genomes. Half of the cluster is orthologous with *Xa. citri* subsp. *citri* cluster present at this locus while half is apparently unique.

13.2.3. Disease cycle

Infected seed is the main source of perpetuation of the bacterium. The bacterium can also survive on wild rice species. In tropical areas, where only one crop of rice is taken in a year, plant debris, stubbles and soil do not play any role in the survival. In India, where 2 or 3 crops of rice are taken in a year, the bacterium can survive on the crop itself. The problem of survival is only in areas like North or Northwestern India, where only one crop of rice is taken in a year.

During germination of seed, the bacterium present in the inner surface of the glumes infects the developing plumule. The first leaf carries the bacterium to the subsequent leaves, which are also infected successively as they emerge. The bacterium penetrates the leaf mainly through stomata, multiplies in the substomatal cavity and then colonizes the intercellular spaces of parenchyma. Leaf veins act as a barrier to the lateral spread of the pathogen. The penetration can also occur through wounds but it remains restricted to the apoplast of the mesophyll tissue and does not invade the xylem. In humid weather, masses of bacterial cells ooze out in the form of bacterial exudate from stomata and act as inoculum for secondary spread of the pathogen through wind, rain, irrigation water, insects and contact. Wind-driven rain is the most important agent for the dissemination of the bacterium. Bacterial exudate on the flag leaf provides abundant inoculum for the infection of seeds in the panicle. Without strong winds and rain, secondary spread of the disease is limited and the effect of disease dwindles rapidly as the growth of new leaves compensates for damage to infected leaves.

The disease is more severe in rainy season but disappears during short periods of dry spell or bright sunshine. Relative humidity of 90% or above or dew during morning hours for 2 - 3 days is necessary for rapid infection. Temperature is more important than relative humidity for the enlargement of lesions. Optimal temperature for disease development is 26 - 30°C and below 22°C, enlargement of lesions is checked. Younger plants are more susceptible and after 60 days of age, only top young leaves remain susceptible.

Different methods, like spraying of bacterial suspension without injury, spraying the cell suspension containing carborundum powder and smearing the cell suspension on the underside of the leaves with a piece of cloth, are used for artificial inoculation of plants to evaluate the germplasm. Out of these methods, smear method gives better and consistent results even under dry conditions, provided water is sprayed twice a day on inoculated plants. The best results are obtained up to one-month age of plants.

13.2.4. Host resistance

No major genes resistant to *Xoc* have been identified in rice. In contrast to resistance to BB, available genetic resistance to BLS in rice is quantitative. Several quantitative trait loci conditioning resistance to BLS have been identified in *indica* rice. Recently, *Rxo1* gene from maize was cloned and shown to provide resistance against several strains of *Xoc*, as a transgene in rice (Zhao *et al.*, 2005). As a transgene, *Rox*1 holds a great promise for rice breeding as a simply inherited source of resistance to BLS. An avirulent gene *avrRxo1* has also been cloned and it is present in all the strains of *Xoc* tested. Hence, the indication is there that it may play a critical role in fitness including the virulence of the bacterium. The activity of *avrRxo1* is dependent on type third secretion system but its product is not TAL effector and lacks similarity to any known protein.

13.2.5. Management

Not many studies have been conducted for the management of the disease but many of the measures used for controlling BB can be expected to work for this disease also. As in case of BB, host genetic resistance is also important for the control of BLS. Brief information on this aspect is given above under host resistance. The other management measures include:

(a) Use of pathogen-free seed or treatment of seed by soaking in Streptocycline (250 μg/ml) solution for 12 hours or in hot water at 50°C for 30 minutes.

(b) As the disease is exclusively parenchymatous, it can be effectively controlled by spraying Streptocycline (100 μg/ml). The number of sprays will vary with disease severity but in general, 2 - 3 sprays at 10 - 15 days intervals should give satisfactory results.

REFERENCES

Bradbury, J.F. (1971). Nomenclature of the bacterial leaf streak pathogen of rice. *Int. J. Syst. Bacteriol.* 21 : 72.

Dye, D.W. (1978). Genus IX *Xanthomonas* Dowson 1939. In: Young, J.M., Dye, D.W., Bradbury, J.F. *et al.* A proposed nomenclature and classification for plant pathogenic bacteria. *N. Z. J. Agric. Res.* 21: 162 - 66.

Fang, C.T., Ren , H.C., Chen. T.Y. *et al.* (1957). A comparison of the rice bacterial leaf blight organism with the bacterial leaf streak organism of rice and *Leersia hexandra* Swartz. *Acta Phytopathol. Sin.* 3: 99 - 124 (Ch. en).

Nino-Liu, D.O., Ronald, P.C. and Bogdanove, A.J. (2006). *Xanthomonas oryzae* pathovars: Model pathogens of a model crop. *Mol. Plant Pathol.* 7: 303 - 24.

Ou, S.H. (1985). *Rice Diseases.* 2nd Ed. C.A.B. International, Slough, UK, 380 pp.

Pordesimo, A.N. (1958). Bacterial blight of rice. *Philippine Agri.* 42: 115 – 28.

Raymundo, A.K., Briones, A.M. Jr., Ardales, E.Y. *et al.* (1999). Analysis of DNA polymorphism and virulence in Philippine strains of *Xanthomonas oryzae* pv. *oryzicola. Plant Dis.* 83: 434 - 40.

Reinking , O.A. (1918). Philippine economic plant diseases. *Oryza sativa* L. Rice bacterial leaf stripe. *Philippine J. Sci.* Sec. A, 13: 225 – 26.

Shekhawat, G.S. and Rao, Y.P. (1972). Bacterial leaf streak of rice. In: P.N. Patel (ed.). *Plant Bacteriology.* Vol. I, *Bacterial Diseases of Plants in India.* Indian Agricultural Research Institute, New Delhi, pp. 13 - 21.

Swings, J., Van den Moorter, M., Vauterin, L. *et al.* (1990). Reclassification of the causal agents of bacterial leaf blight (*Xanthomonas campestris* pv. *oryzae)* and bacterial leaf streak (*Xanthomonas campestris* pv. *oryzicola)* of rice as pathovars of *Xanthomonas oryzae* (ex Ishiyama 1922) sp. nov., nom. rev. *Int. J. Syst. Bacteriol.* 40: 309 - 11.

Zhao, B., Lin. X., Poland, J. *et al.* (2005). A maize resistance gene functions against bacterial streak disease in rice. *Proc. Natl. Acad. Sci.* USA, 102: 15383 - 88.

13.3. YELLOW EAR ROT OF WHEAT

The disease is also known by other names such as, **yellow slime rot** and **spike blight** and in India as ***Tundu* disease**. It was first reported by Hutchinson (1917) from Punjab, India but he did not establish the

Koch's postulates. According to him the disease was present since 1908. The disease is present in the entire wheat belt of North India. In general, the losses in India range from 1 - 2% but in extreme cases, losses up to 50% have also been reported. Paruthi and Gupta (1987) reported an infection level of 60.8% from India. The disease is quite severe in Pakistan and the yield losses ranging from 69.6 - 93.5% (depending upon the cultivars) have been reported by Anwar and Inam-ul-Haq (1992). They also reported drastic reductions in other plant parameters, i.e., 15.1 - 21.7% in plant height, 20.8 - 27.7% in tillering, 10.9 - 14.5% in ear length and 28.6 - 84.3% in number of grains. Besides India and Pakistan, the occurrence of disease has been reported from China, Egypt, Australia, Iraq, Iran, Cyprus and Ethiopia.

Yellow ear rot is actually a **disease complex,** i.e., it is caused by a bacterium, *Rathayibacter tritici* and a nematode, *Anguina tritici*. *A. tritici* alone causes another disease, namely, **ear cockle of wheat**. Milne in 1925 gave the indication that the disease is caused by a bacterium and a nematode when he observed that some ear cockle galls could be obtained from certain ear heads showing yellow ear rot. However, the experimental evidence for this was provided by Fahmy and Mikhail (1925) by showing that bacterium alone could not produce the disease. Later on, Chaudhuri (1935) claimed that the bacterium alone can cause the disease but his claim has not been confirmed by the subsequent workers. Vasudeva and Hingorani (1952) confirmed the work of Fahmy and Mikhail (1925) and the latest confirmation is by Sahebani *et al.* (2006).

13.3.1. Symptoms

As *A. tritici* is also a part of the causal complex, initial symptoms of ear cockle and yellow ear rot are the same. The first symptom is the enlargement of the stem near the soil line in 20- to 25-day-old seedlings. The leaves emerging from such seedlings are twisted and wrinkled. Affected plants are dwarfed and spreading type, and many die in case of severe infection.

The symptoms of yellow ear rot become evident at the time of ear emergence and are characterized by the presence of a bright yellow slime or gum on abortive ears and the leaves in contact with such ears in the boot leaf. The affected ears fail to produce any grain. In humid weather, yellow slimy mass in the ears trickles down and may fall on the lower leaves. On drying, the slimy mass becomes hard, brittle and brown. Affected culms either die in young stage or grow until heading. In the latter case, the emerging spike is narrow and short with grains partially or completely replaced by the bacterial mass. Sometimes, the spike may fail to come out of the boot leaf. The stalk is always distorted when the bacterial mass is produced in the ears. If the conditions favourable for the multiplication of the bacteria continue, the slimy mass completely fills the ear resulting into a **complete *Tundu* ear**. However, if the conditions are not that favourable, then partial ear cockle and partial yellow ear rot symptoms develop. In the latter case, ear cockle galls carry the bacterial inoculum for initiating the disease in the next year. Bright yellow streaks or slimy mass also appear on the leaf surface of young plants.

13.3.2. Causal organism

1. *Rathayibacter tritici* (ex Hutchinson 1917) Zgurskaya *et al*. 1993

Syn. *Clavibacter tritici, Corynebacterium michiganense* pv. *tritici, Cor. tritici*

2. *Anguina tritici*

Rat. tritici: The bacterium was first named *Pseudomonas tritici* by Hutchinson (1917) and Burkholder (1948) changed it to *Corynebacterium tritici*. With the introduction of pathovar concept for the classification of plant pathogenic bacteria, it was changed to *Cor. michiganense* pv. *tritci*. Later on, Davis *et al.* (1984) reclassified it as *Clavibacter tritci*. Zgurskaya *et al.* (1993) created a new genus *Rathayibacter* and renamed it as *Rat. tritici*. The below given description of the bacterium is taken from Zgurskaya *et al.* (1993).

Cells are irregular, pleomorphic, short rods and measure 0.4 - 0.8 × 0.5 - 1.8 μm. V-forms and sometimes short cell chains are formed but a clear rod-coccus cycle is not observed. Nonmotile, nonsporulating, Gram-positive, not acid-fast and obligately aerobic. Chemoorganotrophs, having an oxidative type of metabolism and catalase test positive. The optimal growth occurs between 24 and 28°C. The cell walls contain galactose and xylose. The bacterium is capable of utilizing xylose, lactose, L-arabinose, mannitol, sorbitol, inulin, citrate, tartrate and sebacinate. Acid is produced from glucose, galactose, fructose, mannose, maltose, sucrose and xylose. Glutamic acid and DL-ornithine are utilized but not DL-valine. The bacterium is able to hydrolyze Tween 21, 40 and 85 and grow in the presence of 5% NaCl and 0.03% potassium tellurite. Voges-Proskauer test is negative but methyl red is positive. The bacterium contains a 50 kb plasmid.

It is generally believed that both the bacterium and nematode are equally important in causing the disease. However, the opinion is changing in favour of a major role played by the bacterium. Now it is believed that the disease is caused mainly by the bacterium and the nematode helps in carrying the bacterium to the ears and in its entry into the host tissue by causing wounds in addition to the survival of the bacterium in or the nematode galls.

13.3.3. Disease cycle

The primary source of inoculum for both the bacterium and the nematode are cockles or nematode galls. These galls may be present in the soil or in the seed. The galls carry the bacterium inside and on their surface. Under dry conditions, the nematode can survive inside the galls for several years while the survival of the bacterium is at least for five year.

The nematode galls present in the soil, or brought to the soil along with the contaminated seed, absorb moisture, swell and release 2nd stage larvae. These bacteria laden larvae find their way to growing point of the seedlings and feed ectoparasitically. With the natural growth of the plants, the larvae are carried to the inflorescence, where they enter the floral primordia and feed endoparasitically. The bacteria also gain entry into the developing ear heads along with the nematode. Under favourable weather conditions, the bacteria multiply profusely inside the ears producing an abundant slimy mass, which envelops the developing spikelets. Poorly drained soils, early sowing and more number of irrigations increase the disease incidence.

13.3.4. Management

As the nematode galls of *A. tritici* are responsible for the survival of the bacterium and its larvae initiate the disease, the control of this nematode will automatically eliminate the disease. The following measures are adopted for the control of *A. tritici* and ultimately the yellow ear rot.

1. Use of gall-free seed. If the seed is not gall-free, it should be made gall-free by adopting the following measures.
   (a) Soak the seed in 20% brine solution (NaCl or KCl). Vigorously stir the seed and remove the galls, which float on the surface of water. Destroy these galls by burning. After soaking in brine solution, wash the seed 2 - 3 times with water to avoid reduction in germination.
   (b) Soak the seed in water for 2 hr and heat the water to raise its temperature to 51°C and maintain it for 30 minutes. The heat treatment activates the dormant larvae and then kills them.
   (c) Galls can also be removed by sieving, fanning or winnowing. In China, nematode eliminators are used to remove the galls.
2. Crop rotation for 2 - 3 years with non-graminaceous crops.
3. Grow the disease resistant or tolerant cultivars.
4. Use of chemicals for seed/soil treatment.

SELECTED REFERENCES

Anwar, M.J. and Inam-ul-Haq, M. (1992). Effect of *Tundu* disease on growth and yield of wheat. *Pakistan J. Phytopathol.* 4: 66 - 69.

Bradbury, J.F. (1973). *Corynebacterium tritici: IMI Description of Fungi and Bacteria.* No. 38, Sheet 377. CAB International, Wallingford, UK.

Fahmy T. and Mikhail, T. (1925). The bacterial disease of wheat caused by *Pseudomonas tritci* Hutchinson. *Agri. J. Egypt* 1: 64 - 72.

Hutchinson, C.M. (1917). A bacterial disease of wheat in Punjab. Mem. Dept. Agric. India Bact. Ser.1: 169 - 75.

Sahebani, N., Kheiri, A., Rahimian, H. *et al.* (2006). A study on interaction between wheat seed gall nematode *Anguina tritici* and wheat yellow ear rot *Rathayibacter tritici. Iranian J. Agric. Sci.* 36: 1355 - 60.

Swarup, G. (1972). Tundu disease of wheat. In: P.N. Patel (ed.). *Plant Bacteriology.* Vol. I, *Bacterial Diseases of Plants in India.* Indian Agricultural Research Institute, New Delhi, pp. 33 - 37.

Vasudeva, R.S. and Hingorani, M.K. (1952). Bacterial disease of wheat caused by *Corynebacterium tritici* (Hutchinson) Bergey *et al. Phytopathology* 42: 291 - 93.

Zgurskaya, H.I., Evtushenko, L.I., Akimov, V.N. *et al.* (1993). *Rathayibacter* gen. nov., including the species *Rathayibacter rathayi* comb. nov., *Rathayibacter tritici* comb. nov., *Rathayibacter iranicus* comb. nov., and six strains from annual grasses. *Int. J. Syst. Bacteriol.* 43: 143 - 49.

13.4. BACTERIAL STRIPE AND BLACK CHAFF OF BARLEY

The disease was first reported by Jones *et al.* (1917) as bacterial blight of barely (*Hordeum vulgare* L.) from the USA. The currently used name of the disease is bacterial stripe and black chaff of barley. It is prevalent in Mexico, USA, South America, France, Belgium, Spain and India. In India, the disease was first reported by Patel and Shekhawat (1971), as bacterial blight of barely from Ludhiana, Punjab, India. Later on, Maharshi (1996) reported it from Rajasthan in 1989 where it is has become more or less endemic (Maharshi and Sharma, 2006). The disease incidence may reach nearly 100% in overhead-irrigated fields in contrast to fields irrigated with conventional flood irrigation system. However, there is not much information available on losses caused by the disease.

13.4.1. Symptoms

All the above ground plant parts are affected. Mainly two types of symptoms, i.e., stripes and black chaff appear. The stripes appear on leaf sheath, leaf lamina and stems while the black chaff appears on ears and neck. Initially, water-soaked, dark green flecks/lesions appear on leaf sheaths and leaf blades, which elongate to linear stripes parallel to the veins. The stripes also appear on the stem internodes. Soon the stripes lose their greenish-yellow and water-soaked appearance and become necrotic, brown and translucent. The length of the stripes varies from 15 - 25 cm or may cover the entire length of the leaf blade (Fig. 13.4a & b). Severe infection in susceptible varieties may cause chlorosis of the leaves. Severe infection of flag leaf and its sheath results in sealing of sheath lumen with bacterial mass, resulting in the distortion of emerging spike and its rachis. Oftenly, the emergence of spike is completely prevented. Frequently, the stripes coalesce to from yellowish-brown blotches leading to shriveling and drying of the infected leaves/sheaths. In the field, the symptoms are more pronounced at the tillering stage. In the early morning hours, copious viscous slimy mass of bacterial ooze appears on the surface of stripes, which dries up into numerous beads or crusty translucent scales under dry conditions.

Seed-borne nature of the bacterium leads to the infection of young seedlings. Severely affected plants remain weak and stunted, and oftenly the ears fail to emerge. If the ears emerge, they are dwarfed, twisted and even sterile having compressed and closed spikelets. Young infected plants at 2 - 3 leaf stage are sometimes so severely stunted and may be killed or if they grow further, the grain formation is poor accompanied by deformation of spikes. Sometimes, only the awns emerge from the ears followed by the death of plants. The symptoms of the disease can be confused with those of barley stripe caused by *Helminthosporium graminieum* but it can be conveniently diagnosed on the basis of bacterial ooze/turbidity test in the laboratory and field, respectively.

The initiation of leaf lesions from the apex and their downward extension and starting of the lesions primarily from central part of the blade where the leaf bends have been reported by different workers from different regions. It appears that the symptom development and further progress may very in different geographical areas.

13.4.2. Causal organism

Xanthomonas translucens pv. *translucens* (Jones *et al.* 1917) Vauterin *et al.* 1995

Syn. *Xa. campestris* pv. *translucens*

Jones *et al.* (1917) originally described the bacterium as *Bacterium translucens* and with the establishment of genus *Xanthomonas;* it was transferred to it as *Xa. translucens*. Later on, it was designated as *Xa. translucens* f. sp. *hordei*. With the introduction of pathovar system, it was named as *Xa. campestris* pv. *translucens*. Vauterin *et al.* (1995), while reclassifying the species of *Xanthomonas,* changed it to *Xa. translucens* pv. *translucens*.

The colonies on nutrient agar are yellow, mucoid, convex and smooth having entire margins. Cells measure 0.4 × 1.0 μm and are motile by a single polar flagellum. The bacterium produces hydrogen sulphide but does not reduce nitrate, does not produce indole, does not hydrolyse starch and does not utilize lactose; catalase test is positive while oxidase test is negative. The optimum temperature for its growth ranges from 28 - 30°C.

The following four pathogens, often grouped under the name "**translucens group"** cause bacterial stripe and streak disease on cereals

Xa. translucens pv. *translucens*

Xa. translucens pv. *cerealis*

Xa. translucens pv. *secalis*

Xa. translucens pv. *undulosa*

Jones *et al.* (1917), who first isolated the bacterium from barley, found that the bacterium was pathogenic only to barley and did not infect wheat, rye, oat, timothy, spelt emmer and einkorn (all *Triticum* spp.) Maharshi (1996) also reported that the Rajasthan isolate of the bacterium (CABIMIB 12929) was also non-pathogenic to wheat. Following the rules of bacterial nomenclature (International Society for Plant Pathology, Young *et al.*, 1996), the name *Xa. translucens* pv. *translucens* should be reserved for strains pathogenic only to barley as originally reported by Jones *et al.* (1917). As *Xa. translucens* pv. *undulosa* strains infect wheat and triticale but can also be isolated from barely and rye, hence its host range is not only broader than that of *Xa. translucens* pv. *translucens* but also covers the host range of *Xa. translucens* pv. *cerealis*. Based on amplified fragment length polymorphism and fatty acid methyl ester analysis, Bragard *et al.* (1997) showed that strains pathogenic on barley and not on wheat, clustered in a genetically different group. They further reported that the pathovars *cerealis, translucens* and *undulosa* corresponded to true biological entities.

3.4.3. Disease cycle

The bacterium survives in infected seed and it is both externally and internally seed-borne. Maharshi (1996), using paper towel and seedling system test, reported 30% seed-plant transmission at harvesting time, which decreased to 4% after 7 months, i.e., up to the sowing of next crop.

Xa. translucens specific medium is better than nutrient glucose agar for the detection of the pathogen from the seeds. No inoculum threshold is known for infected barley seeds but wheat seeds, the washings of which contained approximately 1000 cfu of the bacterium per ml per gram of seed, produced the diseased seedlings (Schaad, 1988).

Jones *et al.* (1917) reported that the bacterium from the seeds enters into lemma and leaf through stomata and further moves into intercellular spaces of thin-walled parenchymatous tissue. Sharma (2003) traced the movement of the bacterium from husk (glumes) and pericarp, but not from endosperm or embryo, to the coleoptile, first true leaf and then into the aerial parts of the seedlings. The transmission from seed to seedlings is facilitated by the presence of free water, i.e., it is more in high rainfall or under overhead-irrigation.

The survival of the bacterium in infected crop debris and soil is only for a short period and hence of no significance. The bacterium enters the host through stomata and wounds. The dissemination of the bacterium occurs through wind-splashed rain, overhead-irrigation, plant contact and insects. In Montana (USA), under sprinkler irrigation the bacterium has been found to spread from a single infection locus over an area of 28 m^2 in 39 days. Sprinkler irrigation during the daytime also promotes the penetration of the bacterium, as the stomata are open during the day.

The bacterium grows under a wide range of temperature and moisture conditions but the disease is more severe in warm and humid conditions. The bacterium multiplies rapidly in the host tissue in tissue creases and intercellular spaces. The incubation period is short and is of approximately 10 days. The bacterium multiplies in all the aerial plant parts including spikes, which contaminate the developing seeds.

13.4.4. Management

1. Use of pathogen-free seed is the best approach for the management of the disease. The seed certification programmes and seed assays are necessary for the production of pathogen-free seed.
2. Dry heat treatment of barley seeds at 72°C for 5 - 7 days has been found to eradicate the pathogen. Fourest *et al.* (1990) found that dry heat treatment of heavily infested barley seed at 71, 75 or 84°C for 11 days eliminated the pathogen while treatment at 72°C for 4 days apparently eliminated the pathogen from moderately infested seed. Reduction in seed germination was practically negligible for seed treated at 71 or 72°C for 7 days or less while the treatment at 71°C for 11 days caused an average reduction of 8%.
3. Avoid the overhead-irrigation or limit its use.
4. Use of disease resistant cultivars is the most effective method of disease management but no disease resistant cultivars ar

available. However, the available cultivators differ in their level of susceptibility to the pathogen.

REFERENCES

Bragard, C., Singer, E., Alizadeh, A. *et al*. (1997). *Xanthomonas translucens* from small grains: Diversity and phytopathological relevance. *Phytopathology* 87: 1111 - 17.

Fourest, E., Rehms, L.D., Sands, D.C. *et al*. (1990). Eradication of *Xanthomonas campestris* pv. *translucens* from barley seed with dry heat treatments. *Plant Dis.* 74: 816 - 18.

Jones, L.R., Johnson, A.G. and Reddy, C.S. (1917). Bacterial blight of barley. *J. Agric. Res.* 11: 625 - 43.

Maharshi, R.P. (1996). Bacterial stripe and black chaff disease of barley caused by *Xanthomonas campestris* pv. *translucens*: A report on the natural occurrence in Rajasthan, India. *Proc. 9th Int. Conf. Plant Path. Bact.*, Chennai, India, pp. 367 - 70.

Maharshi. R.P. and Sharma, H. (2006). Bacterial stripe and black chaff of barley (*Hordeum vulgare* L.) caused by *Xanthomonas campestris* pv. *translucens* in Rajasthan, India. *Ann. Exptl. Agri. Al. Sci.* 1: 103 - 10.

Patel, P.N. and Shekhawat, G.S. (1971). Occurrence of *Xanthomonas translucens* f. sp. *hordei* in India and its pathogenicity to rice. *Plant Dis. Reptr* 55: 365 - 68.

Schaad, N. W. (1988). Bacteria. In: Symposium on Inoculum Thresholds of Seed-borne Pathogens. *Phytopathology* 78: 872 - 75.

Sharma, H. (2003). Role of Seed Transmission and Environmental Factors on Development of Bacterial Stripe Disease of Barley (*Hordeum vulgare* L.) Caused by *Xanthomonas campestris* pv. *translucens* (Jones, Johnson & Reddy) Dye. Ph.D. Thesis. Rajasthan Agricultural University Bikaner, 68 pp.

Shekhawat, G.S. and Patel, P.N. (1972). Bacterial blight of barley. In: P.N. Patel (ed.). *Plant Bacteriology*. Vol. I, *Bacterial Diseases of Plants in India*. Indian Agriculture Research Institute, New Delhi, pp. 38 - 42.

Vauterin, L. Hoste, B., Kerster, K. *et al*. (1995). Reclassification of *Xanthomonas*. *Int. J. Syst. Bacteriol*. 45: 472 - 89.

Young, J.M., Saddler, G.S., Takikawa, Y. *et al*. (1996). Names of plant pathogenic bacteria 1864-1995. *Rev. Plant Pathol*. 75: 721 - 63.

13.5. BACTERIAL STALK ROT OF MAIZE

Four bacterial plant pathogens, namely, *Dickeya zeae* (syn. *Erwinia hrysanthemi* pv. *zeae*, *Pectobacterium chrysanthemi* pv. *zeae*), *Entero-bacter dissolvens* (syn. *Er. dissolvens*), *Pseudomonas syringae* pv. *lapsa* (syn. *Psm. lapsa*) and *Acidovorax avenae* subsp. *avenae* (syn. *Psm. alboprecipitans*) have been reported to cause stalk rot of maize (*Zea mays* L.). Among the diseases caused by the above mentioned four bacterial pathogens, stalk rot caused by *D. zeae* is the most important and widespread disease of the maize and is described below. It was first reported from United Arab Republic by Sabet (1954) and he identified the bacterium as *Er. carotovora* f. sp. *zeae*. Since then, it has been reported from many countries including the USA, where it is called **top rot**.

Hingorani *et al.* (1959) reported the disease for the first time from India. In India, it is one of the major diseases of maize and is prevalent in Himachal Pradesh, Uttar Pradesh, Delhi, Punjab, Rajasthan, Bihar, West Bengal and Andhra Pradesh. During *Kharif* 1969, an outbreak of the disease occurred in Sundar Nagar area of Mandi district of Himachal Pradesh on hybrid Ganga 5 and the disease incidence was as high as 80-85% in several fields (Thind, 1970). The work done on the disease in India was reviewed by Thind and Payak (1985).

13.5.1. Symptoms

In nature, the symptoms usually appear at flowering stage. In parts, the seriousness of the disease lies in its often sudden and dramatic appearance and the swiftness with which it can kill the plants within 2 - 4 days under favourable conditions. The first symptoms are water-soaking and rotting of basal leaves, especially the leaf sheaths. The close examination of infected plants reveals water-soaking, rotting and discolouration of basal internodes. The rind loses its healthy green colour and looks pale straw. The diseased stalks appear as they have been boiled in water. Infected plants wilt.

Infection advances rapidly covering the considerable length or the entire length of the stalk. Two most characteristic symptoms of the disease are: (1) Infected plants emit a sweet fermenting smell (2) Infected plants usually topple down from 2nd or 3rd basal internode (Fig. 13.5a). It is mainly a parenchymatous disease. In advanced infection, the pith is completely destroyed, leaving the vascular bundles in a disorganized state but intact (Fig. 13.5b).

Occasionally, developing ears, shanks and stalks get infected at a level higher than the ground level. In majority of these cases, the borer attack is found. In such cases, ears droop down and hang limply. This results due to the softening of shanks. The husk leaves, the grains, if any, as well as the cob become soft and appear water-soaked. In USA, the disease is known as top rot because the symptoms usually appear on apical portion of the plants including the leaves.

The symptoms of stalk rots caused by other bacteria are given below to distinguish them from the present one.

Psm. syringae pv. *lapsa* causes rapid rotting of stalks and leaves. Affected plants collapse but do not emit a fermenting smell. *Aci. avenae* subsp. *avenae* causes bacterial leaf blight and stalk rot of maize. The symptoms include leaf blight accompanied by shredding of leaves and rotting of stalks. Infection occurs at the top, leading to rotting and bleaching of the top. Multiple earing is also common. Infection caused by *En. dissolvens* is generally localized and may not cover the whole circumference or length of the stalk.

13.5.2. Causal organism

Dickeya zeae (Burkholder *et al.* 1953) Samson *et al.* 2005

Syn. *Er. chrysanthemi* pv. *zeae, Pe. chrysanthemi* pv. *zeae*

Sabet (1954), on the basis of similarity of stalk rot pathogen to *Er. carotovora* in cultural, morphological and biochemical characters and its ability to attack maize and other graminaceous hosts in addition to other known hosts of *Er. carotovora,* named it as *Er. carotovora* f. sp, *zeae.* However, assigning the epithet of *forma specialis* to this bacterium by Sabet (1954) was not correct, as there was no specialization or restriction in attacking the hosts by the maize pathogen. Rather it attacked more hosts than *Er. carotovora.* Graham (1964) while reviewing the taxonomy of soft rot coliform bacteria disfavoured the use of *forma specialis* for maize pathogen, as there was no specialization in pathogenicity similar to that found in rust fungi. In the 8th edition of *Bergey's Manual of Determinative Bacteriology*, it was named as *Er. chrysanthemi* Burk. et *al.* corn pathotype. Victoria *et al.* (1975) named it as *Er. chrysanthemi* pv. *zeae.* Hauben *et al.* (1998) renamed it as *Pectobacterium chrysanthemi.* Samson *et al.* (2005) have renamed it as *Dickeya zeae* sp. nov.

The cells are Gram-negative rods (0.6 - 0.8 × 1.5 - 3.0 μm) with rounded ends, motile by peritrichous flagella, facultative anaerobes and metabolism is fermentative. The bacterium produces indole and H_2S, reduces nitrates, liquefies gelatin and degrades pectin. Oxidase, urea hydrolase, phenylalanine dreaminase and urease tests are negative. Acid is produced from glucose, arabinose, cellobiose, mannose, glycerol, melibiose, mannitol, sorbitol, fructose, esculin, galactose, raffinose, rhamnose, salicin, xylose, ribose and sucrose but not from adontitol, dulcitol, trehalose, maltose, dextrin and starch. Citrate, formate, lactate and malonate salts are used as sources of carbon and energy. The G+C content of the DNA is 56.4 mol %.

The bacterium is highly sensitive to 1 μg/ml of chlorine and 1, 10, 25 and 50 μg/ml of Streptocycline. Highly motile cultures are also highly virulent while nonmotile cultures are non-infective. Phages specific to the bacterium occur in nature (Thind, 1970).

13.5.3. Disease cycle

Infected plant debris seems to be the only known source of survival of the bacterium. The bacterium can survive for 9 months in soil containing naturally and artificially infected stalks. Maize-potato-maize rotation increases the disease incidence as potato also acts as the host of bacterium. The bacterium enters the host through wounds. Maize borer (*Chilo partellus*) have been shown to transmit the bacterium (Thind and Singh, 1976). Dalmacio *et al.* (2007) have also reported the association of Asian corn borer (*Ostrinia furnacalis*) with the occurrence of the disease. The optimum temperature for disease development is 28 - 30ºC and relative humidity above 90% also favours the disease development. Heavy doses of nitrogenous fertilizers increase the incidence of the disease. Nitrogen doses of 120 and 180 kg per hectare, without any level of potash and phosphorus, significantly increased the disease incidence than 60 kg of nitrogen (Saxena and Lall, 1981). They recommended 60, 60 and 40 kg of nitrogen, phosphorus and potash, respectively per hectare to keep the disease under check. A positive correlation exists between early infection/disease incidence and loss in grain yield. Early infection leads to heavy losses or death of plants. Screening of 32 maize genotypes including inbred lines, hybrids, composites and open-pollinated varieties revealed the disease incidence ranging from 42.9 - 100%, yield losses ranging from 21.3 - 98.8% and high susceptibility of most of the hybrids and composites (Thind and Payak, 1978). The high levels of phenols in hybrid Ganga 2 plants after 60 and 75 days of planting, i.e., flowering stage in comparison to 40 and 95 days after planting were implicated with relatively more resistance at flowering.

13.5.4. Host range

The bacterium has a very wide host range. It infects many fleshy fruits and vegetables and solanaceous plants like brinjal, tomato and tobacco. The bacterium also infects sorghum, pearl millet and sugarcane.

13.5.5. Management

1. Removal and destruction of diseased plant debris are of utmost importance for the management of the disease.
2. Avoid water logging of the field.
3. Number of plants should not exceed the recommended population density (55,000/ha).
4. Use of resistant cultivars. Dalmacio *et al.* (2007) reported that transgenic hybrids showed significantly less bacterial stalk rot mortality and ear rot incidence and higher grain yield in comparison to their conventional counterparts.

5. Chlorinated water have been tried and found effective in USA (Thompson, 1965) and India (Thind and Payak, 1972; Sharma *et al.,* 1982). Thind and Payak (1972), on the basis of 3-year field trials, reported that soil drenching with chlorinated water (100 μg/ml chlorine) at 2-week intervals, starting from knee-high stage to flowering stage, reduced the disease incidence by 75.92%. One pre-planting application was also given. Lall and Saxena (1978) recommended two applications of bleaching powder @ 25 kg/ha per application. The first application should be given at flowering and the second, 10 days after the first application.

REFERENCES

Buchanan, R.E. and Gibbons, N.E. (eds) (1974). *Bergey's Manual of Determinative Bacteriology*. 8th Ed. The Williams and Wilkins Co. Baltimore, USA. 1268 pp.

Dalmacio, S.C., Lugod, T.R., Serrano, E.M. *et al.* (2007). Reduced incidence of bacterial stalk rot on transgenic insect-resistant maize in the Philippines. *Plant Dis*. 91: 346 – 51.

Graham, D.C. (1964). Taxonomy of the soft rot coliform bacteria. *Annu. Rev. Phytopathol.* 2: 13 – 42.

Hauben, L., Moore, E.R.B., Vauterin, L. *et al.* (1998). Phylogenetic position of phytopathogens within the Enterobacteriaceae. *Syst. Appl. Microbiol.* 21: 384 - 97.

Hingorani, M.K., Grant, U.J. and Singh, N.J. (1959). *Erwinia carotovora* f. sp. *zeae, a* destructive pathogen of maize in India. *Indian Phytopathol.* 12: 151 - 57.

Lall, S. and Saxena, S.C. (1978). Bacterial stalk rot of maize: Loss assessment and possibility of disease control by application of calcium hypochlorite (Abstr.). *Indian Phytopathol.* 31: 120.

Sabet, K.A. (1954). A new bacterial disease of maize in Egypt. *Empire J. Exptl. Agri.* 22: 65 - 67.

Samson, R., Legendre, J.B., Christen, R. *et al.* (2005). Transfer of *Pectobacterium chrysanthemi* (Burkholder *et al.* 1953) Brenner *et al.* 1973 and *Brenneria paradisiaca* to the genus *Dickeya* gen. nov. as *Dickeya chrysanthemi* comb. nov. and *Dickeya paradisiaca* comb. nov. and delineation of four novel species, *Dickeya dadantii* sp. nov., *Dickeya dianthicola* sp. nov., *Dickeya dieffenbachiae* sp. nov. and *Dickeya zeae* sp. nov. *Int. J. Syst. Evol. Microbiol.* 55: 1415 - 27.

Saxena, S.C. and Lall, S. (1981). Effect of fertilizer application on the incidence of bacterial stalk rot of maize. *Indian J. Mycol. Plant Pathol.* 11: 164 - 68.

Sharma, S.C., Randhawa, P.S., Thind, B.S. *et al.* (1982). Use of Klorocin for the control of bacterial stalk rot and its absorption, translocation and persistence in maize tissue. *Indian J. Mycol. Plant Pathol.* 12: 185 - 90.

Thind, B.S. (1970). Investigations on Bacterial Stalk Rot of Maize (*Erwinia carotovora* var. *zeae* Sabet). Ph.D. Thesis, Postgraduate School, Indian Agricultural Research Institute, New Delhi, 113 pp.

Thind, B.S. and Payak, M.M. (1972). Antibiotics and bleaching powder in the control of bacterial stalk rot of maize. *Hindustan Antibiot. Bull.* 15: 9 - 13.

Thind, B.S. and Payak, M.M. (1978). Evaluation of maize germplasm and estimation of losses due to maize stalk rot. *Plant Dis. Reptr* 62: 319 - 23.

Thind, B.S. and Payak, M.M. (1985). A review of bacterial stalk rot of maize in India. *Trop. Pest Manag.* 31: 311 - 16.

Thind, B.S. and Singh, N.J. (1976). Maize borer (*Chilo partellus* Swinhoe) as carrier of *Erwinia carotovora* var. *zeae*, the causal agent of bacterial stalk rot of maize. *Curr. Sci.* 45: 117 - 18.

Thompson, D.L. (1965). Control of bacterial stalk rot of corn by chlorination of water in sprinkler irrigation. *Crop Sci.* 5: 369 - 70.

Victoria, J.I., Arboleda, F. and Munoz, S. (1975). La pudricion suave del tallo de maiz (*Zea mays* L.) en Colombia. *Noticas Fitopatologicas* 4: 136 - 47.

13.6. BACTERIAL LEAF STRIPE OF MAIZE

Six bacterial plant pathogens have been reported to attack maize foliage in different parts of the world. Important among them are *Acidovorax avenae* subsp. *avenae* (syn. *Pseudomonas rubrilineans*), *Burkholderia andropogonis* (syn. *Psm. andropogonis*) and *Psm. alboprecipitans*.

Ullasa *et al.* (1967) observed for the first time a severe leaf stripe disease in some inbred lines of maize during 1965 and 1966 *Kharif* seasons at Delhi, India and identified the bacterium as *Xanthomonas rubrilineans*. Later, Dange (1972) while working on this disease found that it is caused by *Psm. rubrilineans,* the pathogen that causes red stripe of sugarcane and has already been reported on sugarcane from India by Chona and Rao (1963). The first natural occurrence of this pathogen on maize was observed by Orian (1957) in Mauritius. As the disease being described here causes symptoms mainly on the leaves, it is appropriate to call it **'Bacterial Leaf Stripe of Maize'** rather than bacterial stripe disease of maize.

Bacterial leaf spot disease of maize due to *Xanthomonas maydis,* reported by Rangaswami *et al.* (1961), has not been found subsequently and the identity of the pathogen is also in doubt (Dye, 1966).

Since 1965, bacterial leaf stripe has been regularly observed at Delhi and Pantnagar but has never caused significant losses. The disease has also been seen at Ludhiana (Punjab) and in Rajasthan and Himachal Pradesh and probably occurs in other maize growing areas of India.

13.6.1. Symptoms

Initially, water-soaked spots appear on both the leaf surfaces, which develop rapidly into long stripes. The stripes are distributed throughout the leaf lamina but are generally more concentrated towards the leaf base. The length of stripes varies from 1 – 25 cm or may be more and they may cover the whole length of the leaf. The stripes often coalesce forming elongated broad bands of necrotic tissue. Later, the stripes turn dark brown and papery. On old papery stripes, many saprophytic fungi grow.

Under humid weather, bacterial exudate appears on the stripes, which on drying forms a dark brown crust. In later stages, stripes split and disintegrate giving the leaves a shredded or tattered appearance. Occasionally, narrow, blackish, longitudinal stripes appear on the mid-ribs. Generally leaf sheaths, stalks and ears are not infected in nature.

13.6.2. Causal organism

Acidovorax avenae subsp. *avenae* (Manns 1909) Willems *et al.* 1992

Syn. *Pseudomonas rubrilineans*

It is primarily a pathogen of sugarcane. Therefore, its taxonomy and other cultural, morphological and biochemical characters are given under red stripe of sugarcane.

Maize, sugarcane and teosinte isolates have the same cultural, morphological and biochemical characters. They cross infect the hosts of each other on artificial inoculation. A phage isolated from diseased maize tissues typed all the 12 isolates of maize but did not type sugarcane isolates in spot testing. It caused lysis of teosinte isolates also but later on resistant colonies appeared in the lysed areas (Dange, 1972).

13.6.3. Disease cycle

The bacterium is internally and/or externally seed-borne. It enters the host through stomata or wounds. Secondary spread of the pathogen occurs mainly through wind-splashed rain. Young plants are more susceptible than older plants.

Although the maize, sugarcane and teosinte isolates artificially cross infect the hosts of each other, it appears that in mature they do not cross infect other hosts. In fields, the pathogen may be present on maize but absent on sugarcane crop growing adjacent to it, suggesting that maize strain does not infect sugarcane in nature.

Inbred lines, namely, CM 104, CM 105, CM 106, CM 109, CM 112, CM 201, CM 300 and Eto group were found resistant on artificial screening and most of the commercial grown hybrids and composites are not severely infected in nature (Dange, 1972).

13.6.4. Host range

In addition to maize, sugarcane and teosinte, the bacterium attacks sorghum, barley, *Pennisetum purpureum*, *Setaria verticillata* and *Sorghum halepanse* (Ullasa *et al*., 1967; Dange, 1972).

13.6.5. Management

1. Use of pathogen-free seed is the most effective method of disease control. In the absence of pathogen-free seed, there is an urgent need to devise the methods for the eradication of seed-borne inoculum.
2. As the pathogen attacks only the parenchyma of the host tissue, it can be effectively checked with Streptocycline. Thind (1978) obtained 67% control of the disease with two sprays of Streptocycline (100 μg/ml) given at an interval of 10 days.

REFERENCES

Chona, B.L. and Rao, Y.P. (1963). Association of *Pseudomonas rubrilineans* (Lee *et al.*) Stapp with the red stripe disease of sugarcane in India. *Indian Phytopathol.* 16: 392 – 93.

Dange, S.R.S. (1972). Studies on Bacterial Stripe Disease of Maize Caused by *Pseudomonas rubrilineans*. Ph.D. Thesis, Postgraduate School, Indian Agricultural Research Institute, New Delhi.

Dye, D.W. (1966). A comparative study of some atypical 'Xanthomonads'. *N. Z. J. Sci.* 9: 843 – 54.

Orian, G. (1957). Plant Pathology Division. Rep. Dep. Agric. Mauritius, 1995 pp. 90 – 93.

Rangaswami, G., Prasad, N.N. and Easwaran, K.S.S. (1961). A new bacterial leaf spot disease of maize. *Madras Agric. J.* 48: 392 – 93.

Thind, B.S. (1978). Chemical control of bacterial leaf stripe of maize caused by *Pseudomonas rubrilineans. Indian J. Plant Prot.* VI: 70 – 71.

Ullasa, B.A., Mehta, Y.R., Payak, M.M. *et al.* (1967). *Xanthomonas rubrilineans* on *Zea mays* in India. *Indian Phytopathol.* 20: 77 – 78.

13.7. STEWART'S WILT OF CORN

In 1895, F.C. Stewart first observed a bacterial wilt of sweet corn from Long Island, New York (Stewart, 1897). He described the symptoms of the disease, isolated the bacterium and proved Koch's postulates. In 1898, E.F. Smith described and named the causal bacterium, the culture of which was sent to him by Stewart, as *Pseudomonas stewartii* to honour Stewart. Prior to Stewart's report, T.J. Burrill reported a new bacterial disease of field corn from southern Illinois. However, Burrill and an entomologist, S.A. Forbes attributed this disease to dry conditions and large population of clinch bugs in spite of the presence of bacteria in the diseased tissue. Moreover, Burrill could not prove Koch's postulates in spite of isolating the bacterium from diseased plants and insects, and also described many symptoms of the disease presently not attributed to Stewart's wilt. Hence, the credit of reporting Stewart's wilt correctly for the first time goes to Stewart. The disease was also called as **bacterial wilt** or **bacterial leaf blight**.

The disease is generally more severe on sweet corn than on dent corn or popcorn, but some hybrids and inbred lines of popcorn and field corn are also highly susceptible. Gamma grass (*Tripsacum dactyloides*) and teosinte (*Zea maxicana*) are also natural host of the bacterium. The artificially infected hosts include sorghum, millet, Sudan grass and some weed species like yellow foxtail. Economic losses are caused due to reduction in yield and restrictions on import of seed, as the causal bacterium is a quarantined pathogen in several countries (EPPO A2 quarantine pest).

The disease has been a major threat to corn producers in the eastern corn belt of the United States for over 100 years. Severe outbreaks of the disease occurred in Nebraska in 1999 and 2000. However, the economic losses due to the disease have been reduced due to the development of resistant hybrids, adequate warning systems and effective control measures. Golden Cross Bantam, a single-cross sweet corn hybrid developed by Glenn Smith and released in 1932 reduced enormous disease losses. It was so widely grown that within a few years of its release, it accounted for 70 - 80% of all the sweet corn canned in USA. The disease also occurs in other countries of South and Central America, Europe and Asia. Some countries including South Africa are believed to be free from the pathogen.

13.7.1. Symptoms

There are two phases of the disease, namely, **wilt phase** and **leaf blight phase**. In both the phases, the initial symptoms appear as leaf lesions originating from corn flea beetle feeding wounds.

13.7.1.1. Wilt phase: In general, this phase is more severe as the plants may be killed outrightly. The symptoms usually occur on young seedlings but can also appear on older plants of very susceptible sweet corn cultivars. Initially, the leaf tissue surrounding flea beetle feeding wounds becomes water-soaked. Later on, pale-green to white-yellow linear streaks having irregular or wavy margins develop parallel to leaf veins. In susceptible cultivars, these lesions may extend the entire length of the leaves. With age, water-soaked and chlorotic tissues become necrotic and may result into leaf blight, if weather conditions remain favourable.

The wilt phase is associated with systemic infection. After entering the host tissue, the bacteria multiply in the xylem vessels and spread throughout the plant and occasionally infect kernels. The multiplication of bacteria and production of large quantities of extracellular polysaccharides inside the xylem vessels results in plugging of xylem vessels and reduction in flow of water and nutrients. Bacterial slime may exude in the form of droplets from the cob husks. Bacteria penetrate the seed but do not enter the embryo. The systemically infected plants become stunted, wilt and die (Fig. 13.7a). The infected plants produce dwarfed and bleached tassels, which ultimately shrivel and die. The bacterial multiplication in the stalk tissue causes browning and necrosis of vascular bundles. Yellow bacterial ooze comes out from the cut ends when systemically infected stalks are pressed. In severely infected plants, open cavities may develop in the stalks at the base.

13.7.1.2. Leaf blight phase: These symptoms usually appear after tasselling although these can appear any time during the crop season. Foliar symptoms of this phase are similar to those of wilt phase. The blighted leaves often wither and die but they are replaced by new leaves in young seedlings. On adult plants, this phase becomes more obvious because no new leaves are produced. In severe cases, the entire leaves may be blighted due to coalescence of lesions (Fig.13.7.b). Secondary fungi often grow on the dead tissue.

The leaf symptoms similar to those of Stewart's wilt are also produced by other leaf blights caused by *Acidovorax avenae* subsp. *avenae*, *Burkholderia andropogonis* and *Aspergillus* species. Therefore, it is advisable to confirm the identity of the disease by isolating the causal agent and proving its pathogenicity. Stewart's wilt lesions on plants after tasseling can also be confused with those of Goss's wilt. However, these two diseases can be distinguished by using a selective growth medium for Goss's wilt bacterium and an ELISA antibody technique for Stewart's wilt bacterium.

13.7.2. Causal organism

Pantoea stewartii subsp. *stewartii* (Smith 1898) Mergaert *et al.* 1993

Syn. *Erwinia stewartii*

Cells are Gram-negative, non-flagellate, nonsporing, facultatively anaerobic rods and measure approximately 0.4 - 0.8 $\times$ 0.9 - 2.2 μm. Capsules are formed. Colonies on nutrient glucose agar are cream-yellow to orange-yellow. The tests for methyl red, Voges-Proskauer, H_2S and indole production, gelatin and esculin hydrolysis, and nitrate reduction are negative. The bacterium also does not produce oxidase, urea hydrolase, arginine dihydrolase and ornithine dicarboxylase but catalase is produced. Acid is produced from glucose but not gas. Acid production occurs from L-arabinose, D-mannitol, D-mannose, melibiose, raffinose, D-sorbitol, trehalose and D-xylose but not from D-adonitol, cellobiose, dulcitol, glycerol, myo-inositol, maltose, lactose, L-rhamnose, salicin and α-methyl D-glucoside. The metabolism is fermentative. Pathogenicity is related to the production of extracellular polysaccharides, which require quorum-sensing regulatory proteins. Maltose, D-arabitol, quinate, citrate, betaine, D-saccharate and 3-*O*-methyl-D-glucose are not used as sole source of carbon and energy. The mol% G + C of the DNA is 54.6 - 55.1 (T_m).

13.7.3. Disease cycle

The bacterium primarily overwinters (survives) in the digestive tracts of hibernating adult corn flea beetles (*Chaetocnema pulicaria* Melsheimer, order Coleoptera, family Chrysomelidae). The beetles hibernate and pass the winter in soil of grassy areas near the corn fields. The beetles come out from soil in spring when soil temperature is around 18 - 21°C and feed on young corn plants and deposit the bacteria in wounds as faecal contamination. The bacteria enter into the leaf veins and cause infection. These beetles are also the primary mode of secondary spread of the pathogen. The beetles feed on the infected plants, acquire the bacterium and spread it to the healthy plants. An infested beetle can carry and transmit the bacterium throughout its life. Other insects like rootworms and flea beetle species can be vectors of the bacterium but their role is negligible in comparison to the corn flea beetles. Corn fields near to grassy areas and winter wheat are generally more severely affected by the disease. Menelas *et al.* (2006) reported that after 48 hr of feeding period on diseased plants, corn flea beetles required a minimum period of 3 hr to successfully transmit the bacterium and 50% transmission occurred by 7.6 ± 0.87 hr. They further concluded that the beetles are highly efficient vectors, both in acquiring and transmitting the bacterium; hence only those insecticides should be used for seed treatment and foliar sprays, which act quickly to prevent beetles from acquiring and transmitting the bacterium.

The bacterium can also survive in the infected seed but the risk of seed-plant transmission is extremely low or almost non-existent under commercial conditions. Only severely and systemically infected plants produce infected seed. As most of this seed is of poor quality, it would be removed during routine seed conditioning. Moreover, among infected seeds, the frequency of seed-plant transmission is extremely low ranging from 0.022 to 0.038% (Block *et al.*, 1998; Michener *et al.*, 2002). Besides, the seed-borne infection is also influenced by certain disease parameters.

It is more common in systemically infected plants. The ratio of infected seeds in systemically infected susceptible plants is 1 in 50,000 seeds while it is 1 in 20,000,000 seeds in resistant plants showing leaf blight symptoms. The seed-plant transmission of the bacterium is least significant in North America where disease is endemic. But more than 60 countries outside North America enforce quarantine restrictions on import of corn seed to prevent the entry of the pathogen (Pataky, 2003).

The severity of disease is influenced by many factors like host nutrition, host reaction to the pathogen, temperature and the number of over-wintering corn flea beetles. High levels of ammonium nitrogen and phosphorus tend to increase the susceptibility while high levels of calcium and potassium impart resistance. Meyer *et al.* (2010), on the basis of 741 observations made from 79 field trials conducted at 15 locations throughout Illinois and one each in Kentucky and Delaware, found that host resistance affected the incidence of systemic seedling wilt within ranges of winter temperatures used for Stewart's wilt forecasts. In case of moderate and resistant hybrids, frequency distributions of Stewart's wilt incidence and mean incidence ranging from 0.7 to 1.8% did not differ among three winter temperature ranges above -2.8°C. On the other hand, distributions of Stewart's wilt incidence on susceptible hybrids differed among each of the four ranges of winter temperature from Stevens-Boewe forecast (i.e., >0.6, -1.1 to 0.6, -2.8 to -1.1 and <-2.8), with mean incidence ranging from 0.5 to 8.5%. They also found that levels of host resistance and winter temperature influenced the probability of exceeding economic thresholds of 1 or 5% incidence of systemic Stewart's wilt. Stewart's wilt is unlikely to exceed economic thresholds when the mean winter temperature is below -4.4°C. When mean winter temperature was above -2.8°C, the probability of exceeding 1% incidence of systemic Stewart's wilt was 0.59 for susceptible sweet corn hybrids and 0.28 for moderate and resistant hybrids. With mean winter temperature below -2.8°C, the probability of exceeding 1% incidence of systemic Stewart's wilt was 0.22% for susceptible hybrids and 0.04 for moderate and resistant sweet corn hybrids. The probability of exceeding 5% incidence was less than 0.1, except when the mean winter temperature was above -2.8°C and susceptible hybrids were grown. High temperatures also favour the disease development. The temperature during winter indirectly affects the disease by influencing the survival of corn flea beetles. Stevens (1934) developed a forecasting system based on winter temperature to predict the occurrence of the disease. Later on, Boewe (1949) modified Stevens's method to predict more accurately the leaf blight phase of the disease. The simplified version of Stevens-Boewe's method uses the average of mean monthly temperatures for December, January and February. If the mean daily temperature for December, January and February is above 1°C, the flea beetles will survive and the disease is likely to be severe. If this average daily temperature is less than -3°C, flea beetles are unlikely to survive and it is unlikely that Stewart's wilt will be severe. Stevens-Boewe's forecasting system appears to be a reliable predictor of the disease, because severe epiphytotics of the disease, which occurred in the early 1930s, early 1950s and 1990s in Midwest and eastern United States, followed winters that were among the warmest of that century.

Yield losses vary with host cultivars, growth stage of the plant at the time of infection and whether the infection is systemic or non-systemic. The losses may range from 40 - 100% when infection of susceptible sweet corn hybrids occurs before 5-leaf stage. In sweet corn, the yield loss is about 0.8% for every 1% incidence of systemically infected seedlings.

13.7.4. Management

1. Planting resistant hybrids of sweet corn is the best control measure. Many sweet corn hybrids contain sufficient resistance to the wilt pathogen (Michener *et al.*, 2003; Meyer *et al.*, 2010).
2. Another equally effective control measure is the control of corn flea beetles with insecticides. The insecticides can be applied as seed treatment; furrow application before sowing or foliar application but seed treatment is most effective. Use of insecticide is more viable in sweet corn and seed corn production. Treatment of sweet corn seed of susceptible variety 'Sprint' with imidacloprid @ of 4 ml/kg caused 37 - 83% reduction in disease incidence (Kuhar *et al.,* 2002).
3. Use of pasthogen-free seed. The seed-borne infection can be detected using ELISA.

REFERENCES

Block, C.C., Hill, J.H. and McGee, D.C. (1998). Seed transmission of *Pantoea stewartii* in field and sweet corn. *Plant Dis.* 82: 775 - 80.

Boewe, G.H. (1949). Late season incidence of Stewart's disease on sweet corn and winter temperatures in Illinois, 1944-1948. *Plant Dis. Reptr* 33. 192 - 94.

Kuhar, T.P., Stivers-Young, L.J., Hoffmann, M.P. *et al.* (2002). Control of corn flea beetle and Stewart's wilt in sweet corn with imidacloprid and thiamethoxam seed treatments. *Crop Prot.* 21: 25 - 31.

Menelas, B., Block, C.C., Esker, P.D. *et al.* (2006). Quantifying the feeding periods required by corn flea beetles to acquire and transmit *Pantoea stewartii. Plant Dis.* 90: 319 - 24.

Meyer, M.D., Pataky, J.K., Joos, D.K. *et al.* (2010). Influence of host resistance on Stewart's wilt forecasts and probability of exceeding thresholds for use of seed-treatment insecticides on sweet corn. *Plant Dis.* 94: 1111 – 17.

Michener, P.M., Freeman, N.D. and Pataky, J.K. (2003). Relationship between reactions of sweet corn hybrids to Stewart's wilt and incidence of systemic infection by *Erwinia stewartii. Plant Dis.* 87: 223 – 28.

Michener, P.M., Pataky, J.K. and White, D.G. (2002). Rates of transmitting *Erwinia stewartii* from seed to seedlings of a sweet corn hybrid susceptible to Stewart's wilt. *Plant Dis.* 86: 1031 – 35.

Pataky, J.K. (2003). Stewart's wilt of corn. APSnet Feature Story, July-August 2003 [http://www.apsnet.org/feature/stewarts].

Stevens, N.E. (1934). Stewart's disease in relation to winter temperatures. *Plant Dis. Reptr* 18: 141 – 49.

Stewart, F.C. (1897). A bacterial disease of sweet corn. *N.Y. Agric. Exp. Sta. Bull.* 130: 422 - 439.

13.8. RED STRIPE OF SUGARCANE

Lyon (1922) was the first to observe red stripe on tip canes in Hawaii. Lee and Martin (1925) established the bacterial nature of the disease. Another kind of symptom/condition called **top-rot of sugarcane** was reported from Queensland by Tryon (1923). Cottrell-Dormer (1932) showed that red stripe and top-rot are two different phases of the same disease but not two different diseases as considered earlier, and are caused by the same bacterium. Since these early studies, the disease has been reported from most of the sugarcane growing countries of the world for which the readers can refer to chapter XXII of the book entitled, "*Diseases of Sugarcane*: *Major Diseases*" edited by Ricaud *et al.* (1989).

In India, the disease was first observed simultaneously by Mc Rae (1933) and Desai (1933). Since then, several reports of severe infections have appeared infecting varieties like Co 419 in Maharashtra (Albuquerque and Arakeri, 1956; Bhide *et al.*, 1956; Summanwar and Bhide, 1962), Co 449, Co 527 and Fiji B in Tamil Nadu (Rangaswami, 1960, 1960a) and Co 312 in Delhi and eastern districts of Punjab (Chona and Rao, 1963).

The disease was of great economic importance in commercial plantings of Tip canes and occasionally on other varieties in Hawaii. The losses can occur if susceptible varieties are grown under environmental conditions favourable for development and dissemination of the pathogen. The losses up to 15% or more have been reported due to top-rot phase (Martin and Wismer, 1961). The red stripe phase alone generally does not affect yield and the symptoms usually disappear with the onset of the new growth.

Patro *et al.* (2006) reported top-rot phase of the disease from Srikakulam district of Andhra Pradesh, India but the bacterium associated with the disease was identified as *Acinetobacter baumannii.*

13.8.1. Symptoms

There are two types/phases of symptoms of red stripe, namely, leaf stripe and top-rot. These may occur singly or simultaneously and under field conditions both are favoured by high relative atmospheric humidity.

13.8.1.1. Leaf stripe: In the beginning, the symptoms appear as watery green stripes on the leaf lamina. They usually appear in the middle of the leaf and near the midrib but sometimes they may concentrate near the leaf base. The stripes spread rapidly upward and downward and develop a reddish colour, which later on changes to dark red, hence the name red stripe. They are straight and uniform in width and remain confined to the parenchymatous tissue in the early stages of infection. Two or more stripes often coalesce to form broad bands of diseased leaf tissue. The width of the stripes varies from 0.5 to 4.0 mm and the length from a few centimetres to the entire length of the leaf lamina (Fig. 13.8a). However, in some varieties, the length of the stripes may extend to leaf sheath also. The stripes may also appear on the lower surface of the midrib. During the warm moist weather, the bacterial ooze comes out through the stomata on the lower leaf surface of the affected tissue, which often dries to form whitish flakes.

The disease appears mostly on the young and middle-aged leaves and may attack the partially unrolled youngest leaves. In young cane, the entire stool may be killed while in older cane, the individual stalks die. Usually the young ratoons are more susceptible to infection than plant cane of the same age.

13.8.1.2. Top-rot: As the name indicates, it is the rotting of the top portion of the plant. The losses from this phase are much higher than those from leaf stripes. The yellowing and wilting of older leaves occurs and the affected plants may show the typical reddish leaf stripes. Top-rot may result from bud or stem infection without showing leaf symptoms as well as from infection of the leaves. The reddish discolouration on the outside and reddish splashes on the inner surfaces often appear on the leaf sheaths attached to the affected internodes. The sunken areas, initially water-soaked in appearance and later turning reddish-brown, often appear on the affected internodes (Fig. 13.8b). As the rotting advances, large cavities are formed inside the internodes. In advanced stage, the leaf spindle is easily pulled out of enclosing sheaths.

A characteristic strong unpleasant odour, an important diagnostic feature of the disease, is emitted by the rotted spindles. Sometimes, the odour is so strong that it may be felt from a distance. Stalks exhibiting top-rot are stunted in growth and usually die; their tops often break and fall down. Sometimes, the uppermost healthy buds of top-rot affected stalks give rise to side shoots, and the leaves of these shoots may exhibit stripe symptoms.

A new disease called **false red stripe,** different from the above-described disease of sugarcane, has been reported from central and southern areas of Brazil (Mantovani *et al.* 2006). It does not occur in North and northeastern areas of Brazil or any other country. The causal bacterium was identified as a species of *Xanthomonas.* Artificial inoculation, by injecting and spraying of bacterial cell suspension, of 31 host species comprising crops, grasses and weeds revealed that only sorghum, maize and oats expressed disease symptoms 15 days after inoculation.

Occasionally, the leaf symptoms of red stripe and those of gumming (*Xa. axonopodis* pv. *vasculorum*) and mottled stripe (*Pseudomonas rubrisubalbicans*) can be confused with each other. Gumming can be distinguished due to its some specific symptoms like formation of gum pockets. There is no top rot in mottled stripe and in general its symptoms are paler in colour.

13.8.2. Causal organism

Acidovorax avenae subsp. *avenae* (Manns 1909) Willems *et al.* 1992

Syn. *Pseudomonas avenae* subsp. *avenae, Psm. rubrilineans*

The taxonomy of bacterium causing red stripe has undergone many changes. It was first classified as *Phytomonas rubrilineans* by Lee *et al.* (1925). It was renamed as *Pseudomonas rubrilineans* by Stapp (1928) and as *Bacterium rubrilineans* by Elliott (1930). Following the establishment of genus *Xanthomonas* by Dowson in 1939, Starr and Burkholder (1942) transferred the bacterium to this genus as *Xa. rubrilineans.* However, they pointed out that as far as lipolytic activity is concerned, red stripe pathogen does not agree with a typical *Xanthomonas.* Wernham (1948) also included it in doubtful species of *Xanthomonas.* In spite of these doubts, Breed *et al.* (1957) listed it as a species of *Xanthomonas* in the 7th edition of *Bergey's Manual of Determinative Bacteriology.*

After a detailed systematic study of bacteria causing sugarcane diseases, Hayward (1962) found that red stripe pathogen has many characteristics of genus *Pseudomonas* and hence proposed that the name, *Psm. rubrilineans* (Lee *et al.*) Stapp should replace *Xa. rubrilineans* (Lee *et al.*) Starr and Burkholder. Later on, Dye (1963) and Bradbury (1967) supported the Hayward's proposal. Dange (1972) reported that the red stripe of sugarcane and bacterial stripe disease of maize were caused by the same pathogen. He further reported that earlier reports from India identifying *Xa. rubrilineans* as the causal agent of red stripe of sugarcane (Desai, 1933; Albuquerque and Arakeri, 1956; Bhide *et al.*, 1956; Rangaswami, 1960, 1960a; Summanwar and Bhide, 1962; Chona and Rao, 1963) and of bacterial stripe disease of maize (Ullasa *et al.*, 1967) were probably erroneous. Willems *et al.* (1992), on the basis of DNA-DNA hybridization and other studies, grouped *Psm. avenae, Psm. rubrilineans* and *Psm. setariae* into *Acidovorax avenae* subsp. *avenae* subsp. nov., comb. nov. The below given description of *Psm. rubrilineans* (recent name, *Aci. avenae* subsp. *avenae*) is taken from Hayward (1962).

Cells are straight rods and motile by one polar flagellum. Poly-β-hydroxybutyrate is accumulated. Colonies are non-mucoid on 2% glucose peptone agar. No pigments are produced. Oxidase positive; H_2S produced. Nitrates are reduced to nitrites, but denitrification is negative. Tween 80 is hydrolyzed, starch hydrolysis is negative and gelatin liquefaction is weak. Capable of growth at 40°C. Acid is produced from glucose, fructose, galactose, arabinose, mannitol, glycerol and sorbitol. No acid from sucrose, raffinose, salicin, lactose, maltose, cellobiose or m-inositol.

13.8.3. Disease cycle

As the sugarcane crop is of 10 months duration or is taken as a ratoon crop, undecompsed plant debris or the infected crop may serve as primary source of inoculum. In Java, Bolle (1929) showed that the bacterium isolated from old withered leaves showing red stripes was virulent after 4 months. Cottrell-Dormer (1932) was able to isolate the pathogen from leaf stripe infected plant debris kept in a dry cardboard box for 7 months.

The bacterium is disseminated mainly through wind-blown rain. After multiplying in large numbers in parenchyma of leaves, the bacteria ooze out on the surface of leaf lesions in moist warm weather. Then with the help of wind and rain, these bacteria spread from plant to plant, from one field to another or fall down on the lower plant parts or may run down the leaves to cause the stem infection. The bacterium is rarely disseminated through cane knives and mechanical equipment.

The bacterium enters the host through stomata and wounds. The leaf wounds are caused due to the scratches made by the marginal spines of the leaves through the action of the wind. The bacterium enters through the stomata present on both the leaf surfaces but the symptoms from natural infection usually appear first on the lower leaf surfaces. Barnum and Martin (1925) found that the bacterium artificially infected all the plant parts but the youngest leaves and the youngest internodes showed more susceptibility. Vigorously growing plants were also much more susceptible than the less vigorous plants. Pal *et al.* (1996) showed that out of four methods tried for screening of sugarcane cultivars, whorl inoculation method was found to be most effective as it gave highest disease incidence and also induced both the phases of the disease.

13.8.4. Host range

In addition to sugarcane, the bacterium naturally infects maize (Orian, 1957; Ullasa *et al.*, 1967), maize and teosinte (Dange and Payak, 1973) and two grasses, namely, *Paspalum nutans* and *P. paniculatum*

(Orian, 1956). A few more hosts have been shown to be artificially infected. Bhide *et al.* (1956) reported successful inoculation of sorghum and Summanwar and Bhide (1962) of *Pennisetum typhoides* (Burm. f.) Stapf and Hubb. with red stripe bacterium. Ullasa *et al.* (1967) could only successfully inoculate sorghum and barley with *Xa. rubrilineans* isolated from maize but not several other graminaceous hosts. Dange (1972) reported that maize isolates infected *bajra*, barley, sorghum, elephant grass (*Pennisetum purpureum*) and *Setaria verticillata*. Martin and Wismer (1961), who could infect only several varieties of *Sorghum bicolor* out of many wild and cultivated grasses, stated that the common field grasses were of little or no importance as a source of inoculum for sugarcane crop.

13.8.5. Management

The use of resistant varieties is the most effective and economical measure for managing the disease. The replacement of susceptible varieties with resistant varieties has relegated red stripe to be of little economic importance. Before release, the varieties must be screened using artificial inoculation techniques. In North Queensland, top-rot phase of the disease was markedly reduced by altering the planting date. Autumn (March-April) sown crop showed much less disease incidence compared to spring (August-September) sown crop.

Quarantine measures are helpful in restricting the movement of the pathogen to disease-free areas.

REFERENCES

Albuquerque, M.J. and Arakeri, H.R. (1956). Sugarcane red stripe disease on Co 419. *Indian Sugar* 6: 323 - 24.

Barnum, C.C. and Martin, J.P. (1925). The susceptibility of roots, stalks, leaf sheath and leaf blades to red-stripe disease, and the relationship of maturity of tissues to increasing resistance to red stripe. Red-Stripe Disease Studies. Hawaii. Sugar Plant. Assoc. Exp. Stn., Pathol. Dept., pp. 35 - 48.

Bhide, V.P., Hedge, P.K. and Desai, M.K. (1956). Bacterial red-stripe disease of sugarcane caused by *Xanthomonas rubrilineans* in Bombay state. *Curr. Sci.* 25: 330.

Bolle, P.C. (1929). De roodestrepenziekte. *Arch. Suikerind.* 37: 1147 - 1218.

Bradbury, J.F. (1967). *Pseudomonas rubrilineans. C.M.I. Description of Pathogenic Fungi and Bacteria*. Set 13, No. 127.

Breed, R.S., Murray, E.G.D. and Smith, N.R. (eds) (1957). *Bergey's Manual of Determinative Bacteriology*. 7th Ed. Williams and Wilkins Co., Baltimore, 1094 pp.

Chona, B.L. and Rao, Y.P. (1963). Association of *Pseudomonas rubrilineans* (Lee *et al.*) Stapp with the red stripe disease of sugarcane in India. *Indian Phytopathol.* 16: 392 - 93.

Cottrell-Dormer, W. (1932). Red-Stripe disease of sugar-cane in Queensland. *Bull. Bur. Sugar Exp. Stn. Div. Pathol.* 3: 25 - 59.

Dange, S.R.S. (1972). Studies on Bacterial Stripe Disease of Maize Caused by *Pseudomonas rubrilineans*. Ph. D. Thesis, Postgraduate School, Indian Agricultural Research Institute, New Delhi.

Dange, S.R.S. and Payak, M.M. (1973). The taxonomic position of the pathogen of red stripe of sugar-cane. *Sugarcane Pathol. Newsl.* 10: 25 - 28.

Desai, S.V. (1933). Occurrence of the red stripe disease of sugarcane in India. (Abstr.). *Proc. 20th Indian Sci. Congress*, Patna, pp. 61 - 62.

Dye, D.W. (1963). Comparative study of the biochemical reactions of additional *Xanthomonas* spp. *N. Z. J. Sci.* 6: 483 - 86.

Elliott, Charlotte (1930). *Manual of Bacterial Plant Pathogens*. Bailliere, Tindall and Cox Publ., London, 349 pp.

Hayward, A.C. (1962). Studies on bacterial pathogens of sugar cane. II. Differentiation, taxonomy and nomenclature of the bacteria causing red stripe and mottled stripe diseases. Mauritius Sugar Ind. Res. Inst., Occas. Pap. 13: 13 - 27.

Lee, H.A. and Martin, J.P. (1925). The cause of red-stripe disease of sugar cane. Red-Stripe Disease Studies. Hawaii. Sugar Plant. Assoc. Exp. Stn., Pathol. Dept., pp. 83 - 93.

Lee, H.A., Purdy, H.A., Barnum, C.C. *et al.* (1925). A comparison of red-stripe disease with bacterial diseases of sugar cane and other grasses. Red-Stripe Disease Studies. Hawaii. Sugar Plant. Assoc. Exp. Stn., Pathol. Dept., pp. 64 - 74.

Lyon, H.L. (1922). A leaf disease of the Tip canes. *Proc. Hawaii. Sugar Plant. Assoc.* 246.

Mantovani, E.S., Marini, D.C. and Giglioi, E.A. (2006). Host range of *Xanthomonas* sp., casual agent of the false red stripe of sugarcane, among grasses. *Summa Phytopathol.* 32: 124 – 30.

Martin, J.P. and Wismer, C.A. (1961). In: J.P. Martin, E.V. Abbott and C.G. Hughes (eds). *Sugar-Cane Diseases of the World.* Vol. I, Elsevier, Amsterdam, pp. 109 - 26.

Mc Rae, W. (1933). New diseases reported during the year 1932. *Int. Bull. Pl. Prot.* 7: 79 - 80.

Orian, G. (1956). Occurrence of a disease similar to red stripe of sugar cane in Mauritius. *Proc. Int. Soc. Sugar Cane Technol.* 9: 1042 - 48.

Orian, G. (1957). Plant Pathology Division. Rep. Dep. Agric. Mauritius, 1955, pp. 90 - 93.

Pal, V., Kumar, Anil and Vir, S. (1996). Evaluation of inoculation methods for red stripe induction in sugarcane. *Proc. 9th Int. Conf. Plant Path. Bact.* Chennai, India, pp. 430 - 33.

Patro, T.S.S.K., Nageswara Rao, G.V. and Gopalakrishnan, J. (2006). Association of *Acinetobacter baumannii* with a top rot phase of sugarcane red stripe disease in India. *Indian Phytopathol.* 59: 501 - 02.

Rangaswami, G. (1960). Studies on two bacterial diseases of sugarcane. *Curr. Sci.* 29: 318 - 19.

Rangaswami, G. (1960a). Further studies on bacterial gummosis and red stripe disease of sugarcane. *J. Annamalai Univ.* 22: 135 - 50.

Ricaud, C., Egan, B.T., Gillaspie A.G. Jr. and Hughes, C.G. (eds) (1989). *Diseases of Sugarcane : Major Diseases.* Elsevier, Amsterdam, 399 pp.

Stapp, C. (1928). In Sorauer's Handb. *Pflanzenkr.* 2(5): 35, Berlin.

Starr, M.P. and Burkholder, W.H. (1942). Lipolytic activity of phytopathogenic bacteria determined by means of spirit blue agar and its taxonomic significance. *Phtopathology* 32: 598 - 604.

Summanwar, A.S. and Bhide, V.P. (1962). Bacterial red stripe disease of sugarcane (*Saccharum officinarum*) caused by *Xanthomonas rubrilineans* var. *indicus* in Maharashtra. *Indian J. Sugarcane Res. Dev.* 6: 65 - 68.

Tryon, H. (1923). Top rot of the sugar cane. *Bull. Bur. Sugar Exp. Stn., Div. Pathol.* 1.

Ullasa, B.A., Mehta, Y.R., Payak, M.M. *et al.* (1967). *Xanthomonas rubrilineans* on *Zea mays* in India. *Indian Phytopathol.* 20: 77 - 78.

Wernham, C.C. (1948). The species value of pathogenicity in the genus *Xanthomonas. Phytopathology* 38: 283 - 291.

Willems, A., Goor, M., Thielemans, S. *et al.* (1992). Transfer of several phytopathogenic *Pseudomonas* species to *Acidovorax* as *Acidovorax avenae* subsp. *avenae* subsp. nov., comb. nov., *Acidovorax avenae* subsp. *citrulli, Acidovorax avenae* subsp. *cattleyae,* and *Acidovorax konjaci. Int. J. Syst. Bacteriol.*42: 107 - 19.

13.9. RATOON STUNTING DISEASE OF SUGARCANE

Ratoon stunting disease was first observed by Steindl in Queensland, Australia in the summer of 1944 - 1945 on ratoon crops of sugarcane cultivar Q 28. He also coined the term 'ratoon stunting' as the ratoon crops were more severely infected. Due to the seed-borne nature of the pathogen, the disease continued to spread from one country to another country along with the movement of the germplasm and now it is present in most of the cane growing countries of the world. Fifty-two countries from where the disease has been reported are given in the list compiled in 1986 by the I.S.S.C.T.'s Standing Committee on Sugarcane Diseases (Ricaud *et al.,* 1989).

In India, the disease was first observed by Prof. S.J.P. Chilton in 1956 in Co S 510 from Golagokarnath, U.P. (Chona, 1956). However, the presence of disease was reportedly questioned as most of the experiments conducted subsequently failed to confirm the presence of the disease. Later on, Singh (1966) confirmed the presence of the disease in Co 290 obtained from Jaora, Madhya Pradesh.

The name ratoon stunting disease (hereafter called RSD) is a misnomer as the disease retards the growth of the plant crop as well and causes considerable losses. The cumulative losses due to RSD are generally more than the losses caused by any other disease of sugarcane. The losses caused by the disease are influenced by many factors such as susceptibility of the variety grown, weather conditions, amount of irrigation water available and whether the crop is freshly planted or a ratoon crop. Overall losses ranging from 5 - 10% have been reported by Hughes (1974). The losses up to 30% or more have also been reported in some years and in highly susceptible cultivars, these may go as high as 50%. In spite of the reduction in cane yield and sugar per unit area, the sugar content of diseased canes may be increased. This results from the reduced growth and lower water content of the canes. This is the only disease of sugarcane in which the sugar content of the canes increases.

13.9.1. Symptoms

There are two types of symptoms, internal and external symptoms.

13.9.1.1. External symptoms: The stunting and unthrifty growth of the plants are the only external symptoms of the disease, which can occur not only due to RSD but due to several other factors like nutritional deficiency, lack of moisture, poor agronomic practices, etc. The stunting and unthriftiness due to RSD are generally more pronounced when the crop grows in reduced soil moisture. The slow growth of the infected crop accompanied with production of small and thinner canes causes reduction in yield in spite of total numbers of canes remaining unchanged. However, in a draught year, the total number of stalks may also be reduced along with the variation in the length of individual canes in a stool. The stunting is not uniform in all the plants even if all the stools in a field are infected. There is a characteristic **'up and down'** appearance, which means that all the plants are not of uniform height.

The varieties differ in their response to infection and show different degrees of stunting. Ratoon crops usually show more disease severity than plant crops. The losses in clone Q 28 ranged from 12 - 37% in plant crops while these usually exceeded 60% in ratoon crops. The roots and underground parts of the diseased canes do not show any abnormality but there is general reduction in root mass.

13.9.1.2. Internal symptoms: In contrast to generalized external symptoms, there are well-defined and characteristic internal symptoms of the disease. The discolouration of individual vascular bundles in fully differentiated nodes of mature stalks is generally observed in diseased canes. The discolouration occurs in lower part of the node just below the region of leaf sheath attachment and is confined to the nodes.

In the longitudinally splitted diseased stalks, the discolouration appears as dots, commas or short straight or bent lines of 2 - 3 mm length. The discolouration varies in type and intensity with severity of disease and the variety involved. Some varieties do not show any discolouration and in those which show discolouration, it may vary in the same variety form time to time. The discolouration may be of yellow, orange, pink-red and reddish–brown shade (Fig. 13.9).

The discolouration is due to the plugging of many large xylem vessels with a coloured gummy substance. For a reliable diagnosis, all the nodes in the fully developed stalks should show discolouration. Only the undamaged stalks should be examined for the diagnosis of RSD as the damage to the internal tissues of the sugarcane due to mechanical/pest injuries leads to discolouration of tissues. The diseased stalks should be examined immediately after splitting because the symptoms become inconspicuous as the tissues darken or dry. The nodes of young canes infected with RSD also show pinkish colour below the apical meristem, which may extend downward about a centimeter. These symptoms are not as reliable as the mature node symptoms for diagnosing the disease but they are more helpful in detecting the disease at an early stage in some varieties.

The vascular discolouration also occurs in leaf scald and gumming disease of sugarcane caused by *Xanthomonas albilineans* and *Xa. axonopodis* pv. *vasculorum,* respectively. In these diseases, the intensity of discolouration is more and the discolouration also extends from nodes to internodes in comparison to RSD. Moreover in gumming disease, the formation of gum pockets and exudation of gum form cut ends of the canes are other distinguishing features.

Viswanathan (1996) found that internal/nodal symptoms were often unreliable for diagnosing the disease as they were not produced in all the susceptible varieties and they also varied in varieties in which they were produced. He also found that indirect ELISA and dot blot techniques were more suitable for the detection of the disease, the former being more sensitive than the latter. Duttamajumder (1996) reported that many workers have questioned the reliability of nodal symptoms for diagnosing the disease and at present it is taken as an indication but not the confirmation of the disease. He further reported that Co 975, a genotype universally susceptible to RSD with 100% nodal symptoms did not yield any RSD bacterium in extracts taken from most of the samples. It shows that Co 975 is not susceptible to the disease and may be a false positive case for RSD. He also tested 25 commercially grown genotypes for the presence of RSD bacterium using dark field, phase contrast and fluorescence microscopy and found that no genotype was completely free from the bacterium. Dark field microscopy was superior to phase contrast microscopy. The other techniques commonly used to detect RSD bacterium in the infected plants include tissue-blot enzyme immunological assay and fluorescent antibody staining (FAS) and among these techniques, including the above-mentioned ones, FAS is considered more effective than others.

In nature, RSD is often found in association with mosaic. Simultaneous occurrence of both the diseases has been more common in certain cultivars. The combined infection of both the diseases causes marked reduction in germination and yield.

13.9.2. Causal organism

Leifsonia xyli subsp. *xyli* (Davis *et al.* 1984) Evtushenko *et al.* 2000

Syn. *Clavibacter xyli* subsp. *xyli*

For nearly about three decades, the causal agent of RSD was believed to be a virus because of its mode of transmission and the failure of the workers to either observe or isolate the pathogen. Gillaspie *et al.* (1973), Teakle *et al.* (1973) and Maramorosh *et al.* (1973) independently reported the association of a microorganism with the disease. Davis *et al.* (1980) isolated the bacterium and proved its pathogenicity. Colonies on SC medium after 2 weeks of incubation at 28°C are circular with entire margin, non-pigmented, convex and measure 0.1 - 0.3 mm in diameter. Bacterial cells measure 0.25 - 0.35 × 1.0 - 4.0 μm and are straight or slightly curved rods, some cells swollen in the middle or at the tip. Variations observed in cell size have been attributed to different methods of sample preparation. Cells are nonmotile. Gram-positive, nonsporing, aerobic, not acid-fast, oxidase negative and catalase positive. The cell walls contain fucose and rhamnose as the major sugars and 2, 4-diaminobutyric acid, glutamic acid, glycine and alanine as the major amino acids. All the coryneform plant pathogenic bacteria that contain 2, 4-diaminobutyric acid were placed in the genus, *Clavibacter*. The guanine + cytosine content of the bacterium is 66 mol% and the whole cell fatty acid extracts contain 17 - 24% 15:0-anteiso, 5 - 14% 16:0-iso and 62 - 72% 17:0-anteiso acids. Evtushenko *et al.* (2000) transferred *Cla. xyli* subsp. *xyli* to a newly created genus *Leifsonia* as *L. xyli* subsp. *xyli*.

13.9.3. Disease cycle

The bacterium perpetuates in the diseased canes and infected stubbles and the latter give rise to infected ratoon crop. There is no evidence of its survival in the true seed or soil.

As the bacterium can be readily inoculated mechanically to the healthy plants through the sap of diseased canes, cutting knives/cutter-planter machines and harvesting equipment play an important role in the dissemination of the bacterium. Hughes and Steindl (1955) found that in one case, the cutter machine spread the bacterium from one diseased cane of clone Q 28 to 60 consecutive plants and thereafter to some scattered plants in Queensland. In Louisiana, the dissemination by harvester equipment increased the infected plants in the plant crop from 16 to 47% and then in the subsequent ratoon crops from 50 to 80 - 90% (Steib *et al.*, 1957). The exchange of germplasm from one country to another country has also played a major role in the spread of disease. The field rats, dogs and foxes play a limited role in spreading the disease.

There seems to be no possibility of the spread of the pathogen through wind-splashed rain as the susceptible varieties grown adjacent to the infected cane fields remained free from infection. In the varietal evaluation trials also, the rows of healthy canes planted adjacent to the rows of inoculated canes, remained free from infection for three years when proper sanitation measures were observed (Hughes and Steindl, 1955).

The bacterium penetrates through wounds and then it enters the xylem vessels. After penetration, it colonizes the xylem vessels of roots, stalks and leaves. It seems that the bacterium generally multiplies and spreads in dysfunctional vessels. The enzymatic activity of the bacterium dissolves pit membranes to facilitate its movement through xylem vessels. The extent and nature of anatomical barriers, which restrict the movement of the bacterium in the host, are considered mainly responsible for the susceptible/resistant reaction of the cultivars. The susceptibility of the host is influenced more by the accessibility of vascular tissue to the pathogen than the ability of the pathogen to multiply within the host.

13.9.4. Host range

No natural hosts other than sugarcane have been reported. Gillaspie and Teakle (1989) have listed 14 hosts, which were artificially infected by the bacterium and from these hosts the bacterium was successfully reinoculated to the sugarcane. These hosts included maize, sorghum, sweet Sudangrass and many other grasses commonly found in sugarcane fields.

13.9.5. Management

1. The use of pathogen-free seed and prevention of the spread of the pathogen from the diseased plants to the healthy plants are the most effective control measures. The selection of apparently healthy, thick canes for seed greatly reduces the disease incidence. However, the selection of healthy canes is hampered due to the absence of clearly defined external symptoms of the disease.

2. In the absence of pathogen-free seed, the diseased canes/setts should be treated with heat therapy to eliminate the seed-borne inoculum. Four methods that employ heat therapy for treating sugarcane setts are given below.

Hot water treatment: This method is used globally. In India, the hot water treatment is normally carried out at 50°C for ½ hr. The same is done at 50°C for 3 hr in Queensland (Australia), Florida and Hawaii (USA). The seed setts of mature stalks are usually less affected and give better germination after hot water treatment than those from immature or senescing stalks.

Hot air treatment: In India, it is used to a lesser extent than hot water treatment. It is carried out at 54°C for 8 hr. In this treatment, the bud damage is less than that of hot water treatment. It is more suitable for soft, succulent and immature canes, which cannot withstand hot water treatment. After the treatment, dip the seed cane in water at ambient temperatures to reduce the loss in germination.

Moist-hot–air treatment: The method was developed in India to check the loss of moisture in hot air treatment. Within 1 hr, the required temperature of 54°C is obtained and it is maintained for 7 hr in a sealed chamber (Shukla *et al.*, 1974).

Aerated-steam method: It is the most recent method in which an aerated-steam, a mixture of air and steam is used. In India, the recommended temperature is 54°C for 4 hr while in USA the treatment is done at 51°C for 4 hr. The smooth and uniform distribution of heat in the treatment chamber makes this method more effective. It is also economical than the other methods. In addition to RSD, the treatment is also effective for grassy shoot, leaf scald and red rot of sugarcane.

The major drawbacks with heat therapy are reduction in germination, high cost of equipment and failure to eradicate the pathogen completely. The percentage of disease-free setts never exceeds more than 90 - 95, especially in hot water treatment. In the remaining 5 - 10% setts, the number of bacteria goes so low, which is difficult to detect with the commonly used methods. Hence, there is an urgent need to develop the more sensitive and accurate methods to detect this low population of the bacterium.

3. Sanitation including disinfection of cutting knives, cultivators and harvester blades is very helpful in preventing the spread of the pathogen. The commonly used disinfectants include 5 - 15% Lysol (neutralized cresylic acid), 1% dettol, 50% ethanol or 10% formalin. A minimum contact of 5 min of the disinfectant with the cutting surface is essential for proper disinfection.

4. Strict quarantine measures should be followed to check the spread of the pathogen to the disease-free areas.

5. Give proper irrigation to the crop to minimize the losses.
6. Use of resistant varieties is the most economical method of disease control. Though most commercial sugarcane clones are susceptible to the pathogen, substantial resistance has been found in clones such as CP 29 - 116, Q 50 and H 60 - 6909.

REFERENCES

Chona, B.L. (1956). Presidential address for the pathology section. *Proc. Int. Soc. Sugar Cane Technol., 9th Congress,* pp. 975 - 980.

Davis, M.J., Gillaspie, A.G. Jr., Harris, R.W. *et al.* (1980). Ratoon stunting disease of sugarcane: Isolation of the causal bacterium. *Science* 240: 1365 - 67.

Duttamajumder, S.K. (1996). Ratoon stunting disease in subtropical India: Prevalence and control. *Proc. 9th Int. Conf. Plant Path. Bac.* Chennai, India, pp. 333 - 41.

Evtushenko, L.I., Dorofeeva, L.V., Subbotin, S.A. *et al.* (2000). *Leifsonia poae* gen. nov., sp. nov., isolated from nematode galls on *Poa annua* and reclassification of *Corynebacterium aquaticum* Leifson 1962 as *Leifsonia aquatica* (ex Leifson 1962) gen. nov., nom. rev., comb. nov. and *Clavibacter xyli* Davis *et al.* 1984 with two subspecies as *Leifsonia xyli* (Davis *et al.* 1984) gen. nov., comb. nov. *Int. J. Syst. Evol. Microbiol.* 50: 371 - 80.

Gillaspie, A.G. Jr., Davis, R.E. and Worley, J.F. (1973). Diagnosis of ratoon stunting disease based on the presence of a specific microorganism. *Plant Dis. Reptr* 57: 987 - 90.

Gillaspie, A.G. Jr. and Teakle, D.S. (1989). Ratoon stunting disease. In: C. Ricaud, B.T. Egan, A.G. Gillaspie, Jr. and Hughes, C.G. (eds). *Diseases of Sugar-cane: Major Diseases.* Elsevier, Amsterdam, pp. 59 - 80.

Hughes, C.G. (1974). The economic importance of ratoon stunting disease. *Proc. Int. Soc. Sugar Cane Technol.* 15: 213 - 17.

Hughes, C.G. and Steindl, D.R.L. (1955). Ratoon stunting disease of sugarcane. *Queensl. Bur. Sugar Exp. Stn. Tech. Comm.* No. 2, 54 pp.

Maramorosh, K., Plavsic-Banjac, B., Bird, J. *et al.* (1973). Electron microscopy of ratoon stunted sugarcane: Microorganisms in xylem. *Phytopathol. Z.* 77: 270 - 73.

Ricaud, C., Egan B.T., Gillaspie, A.G. Jr. and Hughes, C.G. (eds) (1989). *Diseases of Sugarcane: Major Diseases.* Elsevier, Amsterdam, 399 pp.

Shukla, U.S., Ram R.S. and Tripathi, R.C. (1974). Effect of moist hot air treatments on the control of GSD and RSD. *Annu. Rep. Indian Inst. Sugarcane Res.* Lucknow, pp. 72 - 74.

Singh, K. (1966). Ratoon stunting disease of sugarcane in India. *Indian Sugar* 16: 335 - 37.

Steib, R.J., Forbes, I.L. and Chilton, S.J.P. (1957). A report on further studies on the ratoon stunting disease of sugarcane in Louisiana. *Sugar J.* 19: 35 - 37.

Teakle, D.S., Smith, P.M. and Steindl, D.R.L. (1973). Association of small coryneform bacterium with the ratoon stunting disease of sugarcane. *Aust. J. Agric. Res.,* 24: 869 - 74.

ꞏwanathan, R. (1996). Serological techniques for the detection of a bacterium causing ratoon stunting disease (RSD) in sugarcane. *Proc. 9th Int. Conf. Plant Path. Bac.* Chennai, India, pp. 162 - 68.

13.10. GUMMOSIS OF SUGARCANE

Gummosis also known as **gumming or Cobb's disease** was the first disease of sugarcane reported in the literature. Dranert (1869) reported this disease for the first time from Brazil but its bacterial nature was first demonstrated by Cobb (1893) while working in Australia. Immediately its occurrence was reported from Mauritius in 1894, from Fiji in 1895 and later on, from Reunion, Caribbean, Argentina, Belize, Columbia, Dominican Republic, French Guyana, Jamaica, Ghana, Madagascar, Malawi, Mozambique, Mexico, Panama, Puerto Rico, South Africa, Swaziland and Zimbabwe (Ricaud and Autrey, 1989; Janse, 2006). In India, Rangaswami (1960) reported it from Tamil Nadu on Co 449 and Co 527. So far, there is no report of its occurrence form North India.

The disease caused heavy losses till 1930's where highly susceptible noble canes where grown. However, the development of resistant inter-specific hybrids brought the disease under control and it was eradicated from Australia, Brazil, Fiji, and the West Indies (Ricaud and Autrey, 1989).

At present it is a minor disease in most countries, except in those where climatic conditions are highly favorable for the development and spread of the disease.

The losses from the disease occur due to reduction in yield and sugar content of the canes. Besides the loss in yield, the presence of gum (polysaccharides) in the cane juice interferes with the milling process leading to low sugar recovery. The race of the pathogen involved and the degree of tolerance of the varieties affected, influence the extent of loss. In the same variety, the loss from systemic infection of the stalks will be more than from stalks showing foliar infection only. The losses as high as 30 - 40% in cane yield and 9 - 70% in sugar content have been reported under epiphytotic conditions in Australia.

13.10.1. Symptoms

Two types of symptoms, namely, **leaf-streaking phase** and **systemic phase** are produced.

In the first phase, yellow to orange, 3 - 5 mm wide streaks appear on the leaves and progress along the affected veins. With age, their colour changes to ashy grey and the affected tissue becomes necrotic. Usually, they originate from the leaf margin and extend towards the base but they can arise from anywhere of leaves when the infection occurs through the wounds and develop in either direction. The streaks may cover the entire length of lamina and may even extend to leaf sheath, in the later case leading to the systemic infection of the stalks (Fig. 13.10). The streaks of gummosis differ in colour and patterns of appearance form those caused by leaf scald and chlorotic streak.

The severity of leaf infection is influenced by the age of leaf, environmental conditions and varietal reaction of the cane. The intensity of the streaks is maximum on the mature leaves. Under highly favourable conditions of disease development, many streaks may appear on a leaf and several leaves of a stalk may show symptoms.

The characteristic symptoms of the disease are observed, when transversely cut and longitudinally splitted stalks are examined. Longitudinally splitted canes may show reddish discolouration of vascular bundles, which is more prominent at the nodes (Fig. 13.10). The breakdown of the internal stalk tissue results in formation of pockets/ cavities, particularly below the apical meristem. These pockets contain gum consisting of bacterial cells and extracellular polysaccharide.

The bacteria ooze out in the form of yellow shiny droplets from infected bundles of the transversally cut surface of stumps or stalks having systemic infection. The bacterial ooze can be seen more clearly by performing a **sweating test**. In this test, cut pieces of affected stalks kept in sealed polythene bags or moist chambers are incubated at 30 - 35°C. Within a few hours, yellowish bacterial ooze can be seen on the cut ends. This test is more helpful in detecting the bacterial ooze from doubtful cases.

Systematic infection may also cause malformation of the stalk making it flat on one side and bulging on the opposite side. Transverse splits appearing as **'knife cuts'** can be seen near the apex of affected stalk. Systemically infected plants remain stunted. Ratoon crop suffers less than the plant crop.

Systemic infection is usually, but not always associated with the chlorosis of leaves. The leaves of mature stalks may show chlorosis but it is usually more common in young ratoon shoots. In highly susceptible varieties, the death of plants may occasionally occur after the chlorotic sympto ms while in tolerant varieties, the affected stalks may completely recover from these symptoms, especially under highly favourable growth conditions.

A part or whole of the lamina may show chlorosis and from one to all leaves of a stalk may be affected. In case of partial chlorosis, white patches appear on lamina; starting from the base of the oldest leaf and moving to the tip of the youngest leaf. The chlorotic symptoms of gummosis and leaf scald are indistinguishable from each other. As the pathogen is absent in the affected tissue in both the diseases, the involvement of a toxin is suspected.

Narrow red stripes, indistinguishable from those of red stripe disease, may also appear on the leaves of systemically infected stalks. A film of dried bacterial exudate is often seen on these stripes. The bacterium infects both xylem vessels and mesophyll cells.

13.10.2. Causal organism

Xanthomonas axonopodis pv. *vasculorum* (ex Cobb 1893) Vauterin *et al.* 1995

Syn. *Xa. vasculorum, Xa. campestris* pv. *vasculorum*

Cells are Gram-negative, straight rods (0.4 - 0.5 × 1.0 - 1.5 μm); occur predominantly singly, motile by a single polar flagellum, nonsporing, non-capsular and not acid-fast. Xanthomonadins are produced. Hydrolysis of starch and esculin is positive. Growth on nutrient agar is poor. Amino acids are required for growth. Acid is produced within 21 days from glucose, sucrose and trehalose in Dye's medium C but not from arabinose, mannose, galactose, cellobiose and fructose. Catalase test and H_2S production from peptone are positive but nitrate reduction and lysine dicarboxylase tests are negative. The PCR-RFLP of the 16S - 23S rDNA spacer region has been used to differentiate *Xanthomonas* species attacking sugarcane. The method also allows differentiation of Type A from Type B of *Xa. axonopodis* pv. *vasculorum* (Destefano *et al.*, 2003).

13.10.3. Disease cycle

The bacterium perpetuates in diseased canes and infected stubbles. The diseased setts disseminate the bacterium to short as well as long distances. The diseased setts and contaminated setts give rise to primary infection of the crop. The secondary spread occurs through the cutting knives, harvesting implements, wind-blown rain and cyclones; the spread of the bacteria is very fast through the last two means. The cutting knives and implements are important agents for spreading the bacterium at harvest time.

During humid weather and periods of heavy dew, bacteria ooze out from wounds of infected stalks and leaves and may be also through stomata of leaves and leaf sheaths. Then they are splashed away/carried with the help of wind/rain and enter the host through wounds and stomatal openings. The bacteria may also enter the stalks via leaf scars and wounds caused due to farm operations. The wounds on the leaves are caused due to the shearing action of the wind and the serrated margins of leaves. For successful infection of the host, high humidity or free moisture on host surface is essential. From vascular infection, leaf streaks develop within 1 - 3 weeks under favourable weather conditions.

In Mauritius, the severity of epiphytotics was positively correlated with the number and intensity of cyclones in the growing season. Epiphytotics were more severe when cyclones immediately preceded the early beginning of a dry and cool maturing season (Ricaud, 1969).

Ricaud and Autrey (1989) have listed 6 natural hosts of the bacterium, in addition to sugarcane, reported by different workers. These are maize, three palms, namely, princess palm (*Dictyosperma album*), royal palm (*Roystonia regia*) and betel nut palm (*Areca catechu*), broom bamboo (*Thysanolaena maxima*) and Guatemala grass (*Tripsacum laxum*). The infection was rarely observed on maize, *A. catechu* and Guatemala grass.

T. maxima often shows a systemic infection but the bacterium is quite different from the bacterium infecting the sugarcane. Isolates identical to the races infecting sugarcane obtained from *T. maxima* cause only foliar symptoms on this host. It is evident from the above given account that sugarcane is the only main host of gumming pathogen. The other hosts are only temporary and accidental hosts and do not harbour the pathogen for long periods in the absence of sugarcane.

A more or less similar disease occurs on maize and Guatemala grass. Strains of *Xanthomonas* causing this disease differ in several aspects from the gumming pathogen and probably they belong to a group of strains reclassified as *Xa. vasicola* pv. *vasculorum*.

13.10.4. Management

1. Use of resistant varieties is the most effective method of disease management. In Australia and Mauritius, legislative measures were adopted for the replacement of susceptible varieties with resistant varieties. The varieties having high degree of resistance to foliar infection should be recommended in areas where environmental conditions favour frequent epiphytotics of the disease.
2. The other management measures like sanitation have a limited utility due to the rapid spread of the disease. However, they can be helpful in areas where conditions for rapid aerial spread occur very infrequently. The solution of an iodophore (250 μg/ml available iodine) can be used to disinfect cutting knives and blades of harvesters.
3. Use of pathogen-free seed. The seed should not be taken form fields showing systemic infection.

4. Rouging of diseased plants particularly, those showing systemic infection (chlorosis) will reduce the disease incidence.
5. Do not grow the collateral host like maize, sorghum and pearl millet near the sugarcane fields.

REFERENCES

Cobb, N.A. (1893). Plant diseases and their remedies-Diseases of the sugarcane. *Agric. Gaz. N. S. W. for 1893,* 4: 777 - 98.

Destefano, S.A.L., Almeida, I.M.G., Rodrigues, N.J. *et al.* (2003). Differentiation of *Xanthomonas* species pathogenic to sugarcane by PCR-RFLP analysis. *Eur. J. Plant Pathol.* 109: 283 - 88.

Dranert, F.M. (1869) Bericht uber die Krankheit des Zuckerrohres. *Z. Parasitenkd. Bd.* 1: 13 - 17.

Janse, J.D. (2006). *Phytobacteriology: Principles and Practice.* CABI Publishing, Wallingford, UK, 360 pp.

Rangaswami, G. (1960). Studies on two bacterial diseases of sugarcane. *Curr. Sci.* 29: 318 – 19.

Ricaud, C. (1969). Investigation on the systemic infection of gumming disease. *Proc. Int. Soc. Sugar Cane Technol.* 13: 1159 - 69.

Ricaud, C. and Autrey, L.J.C. (1989). Gumming disease. In: C. Ricaud, B.T. Egan, A.G. Gillaspie Jr. and C.G. Hughes (eds). *Diseases of Sugarcane: Major Diseases.* Elsevier, Amsterdam, pp. 21 - 31.

13.11. LEAF SCALD OF SUGARCANE

Leaf scald of sugarcane was first reported by Wilbrink (1920) from Java and later on by North (1926) from Australia and Fiji. Since then, it has been reported from many countries and Ricaud and Ryan (1989) have listed 44 countries from where it has been reported. Later on, Rott and Davis (2000) reported that the disease occurs in at least 66 countries. In India, the disease was first detected by Egan in 1960 from I.A.R.I., New Delhi during his visit to India (Egan, 1979) but the first detailed report of the disease was published by Satyanarayana (1974) from Andhra Pradesh. Since then, the incidence of the disease has increased many folds and now it has become a major disease of sugarcane in the country, occurring in Punjab, Uttar Pradesh and Bihar. Among the susceptible cultivars, prominent ones include CoJ 64, CoS 767 and CoLK 8001.

Leaf scald was a serious disease of noble canes, which were widely grown in the early years of 20th century. North (1926) reported substantial losses due to the disease from Australia and in severe outbreaks; the losses were more than 25%. Egan (1971) reported that outbreaks of the disease in early 1900s caused heavy losses in highly susceptible varieties in Queensland. The replacement of susceptible noble canes with varieties containing *Saccharum spontaneum* genes, resistant to the disease, has considerably reduced the incidence of the disease. It is still a major constraint in the successful cultivation of sugarcane in many countries like Mauritius, Fiji, Cuba, Brazil, Taiwan, Indonesia and the United States.

Apart from reducing cane yield; the disease also lowers the quality of the juice by reducing Brix, Pol and Purity. The slow growth of affected stalks, reduction in number of millable canes per stool and poor ratooning reduce the yield. The acute stage causes death of stools and reduction in sugar yield. The factors, which influence the growth of sugarcane plants, also influence the disease severity. The onset of adverse conditions soon after the maturity of the canes, give rise to large number of side shoots. The number and size of side shoots are directly related to deterioration of juice quality.

13.11.1. Symptoms

There are two types of symptoms, namely, the chronic phase and the acute phase. The latent infection is also an important characteristic feature of the disease.

13.11.1.1. Chronic phase: These are the most commonly observed symptoms. The most characteristic symptom of the disease is the appearance of **'white pencil-line'** streaks, which are narrow, about 1-2 mm wide and sharply defined on the leaf lamina. The streaks develop parallel to the midrib and often extend to the base of leaf lamina or the leaf sheath. The specific epithet of the causal pathogen has been derived from the description of this streak.

Initially, the streaks are usually well-defined but they become diffused and broader with the passage of time. However, the close examination of these broader and diffused stripes will show the white pencil-line at the centre. As the disease progresses, the stripes start drying causing withering away of the leaves. The drying always starts from the tip and extends downward, giving a scalded appearance to the leaves. The streaks also appear on leaves in gummosis or chlorotic streak of sugarcane but these streaks are devoid of central white pencil-line. The partial or complete chlorosis of the leaf blade also occurs. The white pencil-line or its reddish necrotic patches can be seen on close examination of these chlorotic blades.

Generally the side shoots, varying in number, develop on the mature infected stalks. Some of the side shoots die when they are still young while the surviving shoots generally exhibit typical white pencil-line or chlorosis of leaves (Fig. 13.11).

The necrosis of the vascular bundles in the form of bright to dark red streaks can be seen on the longitudinally splitted canes. These streaks are generally more prominent at the nodes than in the internodes and also appear in the side shoots. Inside the severely infected canes, longitudinal lysigenous cavities may be formed in the nodal and internodal areas, particularly near the shoot apex. These cavities are relatively dry compared to those of gumming disease as they contain less amount of bacterial slime or gum.

13.11.1.2. Acute phase: The sudden wilting and death of mature stalks, often without any previous symptoms, marks the onset of the acute phase. Occasionally, the small shoots having typical streaks appear at the base of stalks. The other symptoms like proliferation of buds, bunchy top appearance and reductions of the root systems were observed by Agnihotri (1986). Acute phase generally occurs after a stress resulting from a prolonged dry weather or dry weather after a rainy season. To confirm the occurrence of acute phase, one should look for canes showing chronic phase symptoms among the affected plants or wait till the new tillers come out bearing the white pencil-line symptoms.

In India, the symptoms of acute phase have not been observed. However, gradual wilting of plants showing chronic-phase symptoms, especially during the summer months and after the first monsoon shower is commonly observed in some susceptible varieties like CoLK 8001, CoJ 81 and Co 62399 (Agnihotri and Duttamajumder, 1992).

13.11.1.3. Latent infection: The disease is characterized by a latent phase/infection in which the infected plants do not show any symptoms. The plants suffering from chronic phase recover from the disease under certain conditions. The recovery is more common in tolerant varieties or when the conditions are suitable for plant growth. The latency may involve a few shoots or stools or an entire field. The symptoms may appear on such plants after ratooning and on the plants growing from setts taken from the infected stools. The latent infection can also lead to the development of acute phase. Various techniques like fluorescent antibody technique, ELISA and use of indicator plants are used to detect the latent infection. Among them, fluorescent antibody technique is more effective

13.11.2. Causal organism

Xanthomonas albilineans (Ashby 1929) Dowson 1943

Syn. *Phytomonas albilineans*

In 1929, Ashby named the bacterium as *Bacterium albilineans.* Later on, it was transferred to genus *Phytomonas* as *Ph. albilineans*. With the establishment of genus *Xanthomonas*, it was transferred to it as *Xa. albilineans.* Wilbrink's agar, now known as sucrose peptone agar (SPA) was earlier used for isolation and culturing of this bacterium. Glutamic acid and methionine are required for growth. The optimum temperature for growth on SPA is 25°C and the maximum is 37°C. The isolation of the bacterium can be made by using the macerated infected tissue or juice of infected plants after diluting it 1000-fold. On SPA, the colonies appear after incubation of 5 - 7 days, and they are honey-yellow to Naples-yellow, minute, glistening and viscid having an entire margin.

A selective medium (XAS medium) have been devised by suppleme-nting Wilbrink's medium with 5 g of KBr, 100 mg of cycloheximide, 2 mg of benomyl, 25 mg of cephalexin, 30 mg of novobiocin and 50 mg of kasugamycin per litre. The isolation on this medium gives more than 98% frequency from symptomatic plants. The growth rate, morphology and pigmentation of colonies on the medium serve as useful differential characteristics to identify the bacterium.

Cells are straight rods (0.25 - 0.3 × 0.6 - 1.0 μm), predominantly single, Gram-negative and motile by a single polar flagellum. Xantho-monadins are produced. The bacterium neither liquefies gelatin nor reduces nitrate and does not produce H_2S, ammonia and indole. Esculin is hydrolysed but not starch. Acid is produced within 21 days on Dye's medium C from glucose, sucrose and mannose. Ammonium salts, nitrates or asparagine are not used as a sole source of nitrogen.

Variation in genetic, physiological and serological characteristics, and pathogenicity of the bacterium, indicating the existence of different pathotypes in the species, has been found but so far there is no evidence of existence of races. According to Silva *et al.* (2007), there is evidence for occurrence of pathogenic variation in the bacterium as the same cultivar gives a susceptible reaction to the pathogen in one region but a different reaction in another region. Rep-PCR analysis of 50 isolates, collected by them from different regions of Sao Paulo state of Brazil, revealed genetic diversity among these isolates and they divided them into three groups. Pathogenicity test of these isolates revealed the variation in aggressiv-eness among the isolates, suggesting a possible relationship between occurrence of bacterial variants and reaction of sugarcane cultivars in Sao Paulo State. It can be inferred from these findings that pathogenic races of the bacterium occur in nature and there is an urgent need to develop suitable differentials for the identification of these races.

13.11.3. Disease cycle

The main sources of perpetuation of the bacterium are infected stubbles and infected setts used for sowing. Major dissemination of the bacterium occurs at the time of harvesting and planting and it occurs mainly through the infected cuttings and implements used for harvesting and cutting of canes. Infected cuttings are an important mode of spread over long distances. Duttamajumder (1996) showed that setts obtained from the upper half of the infected canes were much more efficient (48%) in producing the disease than those from lower half (21%)

The cane knives and base cutters of mechanical harvesters also play a role in the secondary spread of the bacterium within a crop. The harvesting implements carry the inoculum after cutting an infected cane and transmit the same up to the third stool, after which the transmission becomes irregular. Fluff (seed)-plant transmission of the bacterium has also been proved. Aerial dissemination of the pathogen has been reported to occur in several geographical areas of the world, and was associated with outbreaks of the disease (Comstock, 2001; Daugrois *et al.*, 2003). The role of phyllosphere colonization by the bacterium in the epidemiological cycle of the disease in Guadeloupe was also shown by Daugrois *et al.* (2003). Role of insects such as grasshoppers, leafhoppers, leaf beetles and cane borers, and rodents in transmission is suspected but not yet proved.

Xa. albilineans is a systemic and xylem-invading pathogen and colonizes only the xylem of the vascular strands of the affected parts. However, occasionally the bacterium may break the xylem vessels and destroy the surrounding cells, producing cavities. Generally, this situation is manifested by the appearance of red areas on the white pencil-line. The bacterium produces gum, a xanthan-like polysaccharide composed of glucose, mannose and glucuronic acid. The gum occludes phloem and xylem vessels and partially enters mesophyll cells (Solas *et al.*, 2003). The chlorosis-inducing isolates of the bacterium produce toxic compounds, of which the **albicidin** being the most important plays a key role in pathogenicity. Albicidin inhibits DNA replication of protoplastids, thereby blocking plastid development. There is no correlation between toxin production and pathogenicity of *Xa. albilineans*. Hence, it is apparent that some other virulence factors play a key role concurrently with albicidin in pathogenicity (Champoiseau *et al.*, 2006)

The disease is favoured by wet weather, particularly in combination with cyclones. Usually, the appearance of disease after the first rain shower indicates the involvement of wind-blown rain in disease development. Stress caused by drought and water logging usually increases the disease severity.

13.11.4. Host range

In addition to sugarcane, the bacterium causes natural infection on *Paspalum dilatatum, P. congugatum, Brachiaria piligera* and *Imperata cylindrica* var. *major* (blady grass).

The possibility of survival of the bacterium in blady grass for sometime can make this host as a source of inoculum for the sugarcane crop. *Zea mays, Rottboellia cochinchinensis, I. cylindrica, Panicum maximum, Pennisetum purpureum, Paspalum* sp. and *Sorghum* sp. also act as natural hosts of the bacterium. It seems that *Xa. albilineans* has a narrow host range, restricted to sugarcane and some other members of the family Graminae.

13.11.5. Management

An integrated approach consisting of the following measures can prove very effective for the management of the disease.

1. Resistant varieties: The use of resistant varieties is the most effective, practical and economical method of control. Before release, the resistant varieties must be tested by artificial inoculation and both plant and ratoon crops should be screened. For assessing the reaction of varieties, take into account only the systemic infection on the non-inoculated leaves and ignore the initial symptoms on the inoculated plants.

2. Pathogen-free seed: As the bacterium is transmitted mainly through the diseased setts, the use of pathogen-free seed is of utmost importance for the management of the disease. In areas free from severe epiphytotics, pathogen-free seed can be obtained by regular inspection of seed plots, rouging of diseased plants and then protecting the crop by adopting phytosanitary measures. Duttamajumder (1996) showed that a directed-selection for an apparently healthy cane progeny yielded symptomless plants in the fourth progeny with no trace of leaf scald bacterium on isolation. From 25% of the apparently healthy canes in the first year, it increased to about 90% at the end of the 4th year.

3. Heat therapy: Heat therapy have been tried to eradicate the bacterium from the diseased planting material. Hot air treatment at 54°C for 8 hr is not effective. Long hot water (50°C for 3 hr) and short hot water (52°C for 20 min) treatments are effective but they considerably reduce the germination of buds. Nowadays, moist-hot-air treatment (54°C for 4 hr) is mostly used. Agnihotri (1984) obtained an effective control of the disease with this method. It is superior to all other heat therapy methods used in India and it also controls grassy shoot and ratoon stunting diseases of sugarcane.

4. Chemotherapy: Dip the single bud sets after desiccating for 12 hr, in 100 μg/ml solution of streptomycin for 1 day. However, its higher concentrations are injurious to plants.

5. Phytosanitary measures: The disinfection of cane knives, mechanical harvesters and planters reduces the spread of the bacterium. Ten percent commercial formalin, 1% dettol or 5% Lysol can be used for the disinfection of implements. While cutting setts, the knives should be disinfected at frequent intervals. The destruction of infected debris, stubbles and of weeds like *I. cylindrica* helps in reducing the inoculum load. *Sorghum* sp. and *Z. mays,* which act as collateral hosts of the bacterium, should not be grown near the sugarcane crop.

6. Quarantine measures: Strict quarantines should be enforced to check the spread of the pathogen to the disease-free areas. However, the latent infection of the bacterium is a big hurdle in the effective implementation of the quarantines. All the quarantine stations should be equipped with adequate facilities for the detection of the bacterium in the seed material.

7. Cultural practices: High doses of nitrogenous fertilizers, water stress and water logging increase the incidence of the disease. The correction or removal of these factors will reduce the incidence of the disease. The early ploughing of heavily infected fields considerably reduces the source of inoculum.

REFERENCES

Agnihotri, V.P. (1984). Control of leaf scald of sugarcane, a bacterial disorder, by heat therapy and streptomycin. *Indian Sugar Crops J.* 10: 13 - 14.

Agnihotri, V.P. (1986). Some new or interesting symptoms of leaf scald of sugarcane. *Indian Sugar Crops J.* 11: 21.

Agnihotri, V.P. and Duttamajumder, S.K. (1992). Leaf scald disease of sugarcane. In: A.N. Mukhopadhyay, J. Kumar, H.S. Chaube and U.S. Singh (eds). *Plant Diseases of International Importance.* Vol. IV, *Diseases of Sugar, Forests and Plantation Crops.* Prentice Hall, Englewood Cliffs, New Jersey, pp. 27 - 40.

Champoiseau, P., Daugrois, J.H., Pieretti, I. *et al.* (2006). High variation in pathogenicity of genetically closely related strains of *Xanthomonas albilineans,* the sugarcane leaf scald pathogen, in Guadeloupe. *Phytopathology* 96: 1081 - 91.

Comstock, J.C. (2001). Foliar symptoms of sugarcane leaf scald. *Sugar J.* 64: 23 - 32.

Daugrois, J.H., Dumont, V., Champoiseau, P. *et al.* (2003). Aerial contamination of sugarcane in Guadeloupe by two strains of *Xanthomonas albilineans. Eur. J. Plant Pathol.* 109: 445 - 58.

Duttamajumder, S.K. (1996). Dynamics of field spread of *Xanthomonas albilineans* (Ashby) Dowson causing leaf scald disease in sugarcane. *Proc. 9th Int. Conf. Plant Path. Bact.*, Chennai, India, pp. 324 - 30.

Egan, B.T. (1971). The decline of leaf scald as a major disease in northern Queensland. *Proc. Queensland Soc. Sugarcane Technol.* 38: 157 - 61.

Egan, B.T. (1979). Leaf scald disease in India: letters to the editor. *Sugarcane Pathol. News.* 22: 53.

North, D.S. (1926). Leaf scald, a bacterial disease of sugarcane. Colonial Sugar Refining Co. Ltd. Sydney Agric. Rep. No. 8, 80 pp.

Ricaud, C. and Ryan, C.C. (1989). Leaf scald. In : C. Ricaud, B.T. Egan, A.G. Gillaspie, Jr. and C.G. Hughes (eds). *Diseases of Sugarcane: Major Diseases.* Elsevier, Amsterdam, pp. 39 - 58.

Rott, P. and Davis, M.J. (2000). Leaf scald. In: P. Rott, R.A. Bailey, J.C. Comstock, B.J. Croft and A.S. Saumtally (eds). *A Guide to Sugarcane Diseases.* La Librairie du Cirad, Montpellier, France, pp. 38 - 44.

Satyanarayana, Y. (1974). Leaf scald of sugarcane: A new disease in Andhra Pradesh. *Indian Sugar* 24: 23 – 26.

Silva, M. de S. e, Bedendo, I.P. and Casagrande, M.V. (2007). Molecular and pathogenic characterization of isolates of *Xanthomonas albilineans* (Ashby) Dowson, causal agent of sugarcane leaf scald. *Summa Phytopathol.* 33: 341 – 47.

Solas, M. T., Pinon, D., Acevedo, R. *et al.* (2003). Ultrastructural changes and production of xanthan-like polysaccharide associated with scald of sugarcane leaves caused by *Xanthomonas albilineans. Eur. J. Plant Pathol.* 109: 351 - 59.

Wilbrink, G. (1920). De Gomziekte van het Suikerriet, hare Oorzaak en hare Bestrijding. Arch. Suikerind, Ned. Indie, 28: 1399 - 1525

13.12. BACTERIAL WILT OF SOLANACEOUS PLANTS

Bacterial wilt caused by *Ralstonia solanacearum* is a highly destructive disease of many solanaceous and some non-solanaceous crops. It is considered the world's single most destructive bacterial plant disease. It is also considered the second most important disease of potato worldwide after late blight. It is also known by several other names like **brown rot of potato, southern bacterial wilt of tomato** (due to its prevalence in southern states of USA), **Moko disease of banana** and **Granville wilt of tobacco.** The bacterium infects many solanaceous plants, namely, potato, tomato, eggplant, tobacco and pepper, and some non-solanaceous hosts like banana, peanut and Heliconia; geranium and other ornamental plants and weeds. More than 200 plant species belonging to 50 families are attacked by the bacterium. The disease was first described by E.F. Smith (1896) who also described the bacterium and named it as *Bacillus solanacearum.*

The disease occurs in tropical, sub tropical and temperate regions of the world. In the past, the disease has caused serious losses to different crops in different countries. The damage to the potato crop occurs on account of premature wilting of the standing crop and rotting of potato tubers both in field and stores. In Granville county of North Carolina, USA, tobacco crop suffered losses ranging from 25 - 100% and in 1908 alone; the loss was above $ 100,000 for the entire state of North Carolina. During 1935 - 45, the disease caused an annual loss of two million dollars to the tobacco crop (Kelman, 1953). Around 1890, the disease almost eliminated the moko plantain; a relative of banana, in Trinidad and later on, it destroyed one million plants on the pacific coast of Costa Rica. Tomato and brinjal cultivation was abandoned due to the disease in some parts of the Madhya Pradesh, India. In India losses ranging from 5 - 6% for tobacco, 50% for potato and 62 - 65% for brinjal have been reported (Rao, 1972). According to Champoiseau and Allen (2009), the bacterium is responsible for an estimated loss of 1 billion US dollars worldwide every year, mostly to poor small-scale farmers in the tropical high lands

In India, the disease was first reported by Cappel in 1892 from Pune district of the Maharashtra as bangle blight or bungdi disease of potato; although it's bacterial nature was established in 1909 by Coleman. Hutchinson in 1913 reported it as Rangpur wilt of tobacco. It is the first bacterial disease recorded in India. The disease has not been noticed in the Northwestern high hills (excluding Kumaon hills), where the mean monthly minimum temperatures during December - February remain below 0°C and in the Northwestern and North central plains having mean monthly maximum summer temperatures above 40°C coupled with water deficit in soil continuously for a minimum period of 45 days. In India, race 1 is more predominant and it has a wide genetic diversity. The strains of bacterium prevalent in India are also more virulent than those found in other countries particularly the US and South America.

13.12.1. Symptoms

As the bacterium infects many plant species, the symptoms differ on different hosts although some of the symptoms are common on all the hosts.

Potato: Foliage symptoms include rapid wilting of leaves and stems, usually first visible at the warmest time of the day. Initially plants recover from wilt during the nighttime. The leaves may have bronze tint and petioles may develop epinasty. Eventually, plants fail to recover, become yellow and brown, necrotic and die. As the disease progresses, a streaky brown discolouration appears on the stems above the soil line and also on the branches of the plants. Brown discolouration of the vascular system of the stems occurs, which can be seen by cutting the infected stems longitudinally or transversely. A white, slimy bacterial mass oozes from vascular bundles of the cut or broken stems. The bacterial ooze in the form of threads can be seen when cut potato stems are suspended in water. Under cool growing conditions, wilting and other foliage symptoms may not occur.

On tubers, external symptoms may or may not appear. However, in damp soils, the bacterial ooze often emerges from the eyes and stolen-end attachment of infected tubers. Soil may adhere to bacterial mass exuded from the tubers. Infected cut tubers reveal a browning and eventual necrosis of the vascular ring and immediately surrounding tissues. A few minutes after cutting, a creamy fluid exudate usually appears spontaneously on the vascular ring of the cut surface (Fig. 13.12a). In case of ring rot of potato, tuber has to be squeezed in order to press out a mass of yellowish macerated vascular tissue and bacterial slime. Plants with foliar symptoms of wilt may produce healthy and diseased tubers, while those showing no symptoms of the disease may sometimes produce diseased tubers. Water-soaked lesions, 0.5 – 1.0 cm in diam and 5 mm deep develop at the site of infected lenticels. The tuber skin remains intact but becomes shriveled. A tuber cut through the lesion exposes a dirty white bacterial mass below the skin.

Tomato: The youngest leaves show the first sign of the disease as they become flaccid in appearance, usually at the warmest time of the day. Under less favourable conditions, the disease develops slowly. Stunting may occur (Fig. 13.12b) and large number of adventitious roots develops on the stem. The vascular tissues of the stem show a brown discoloration (Fig. 13.12c). Drops of white or yellowish bacterial ooze come out from the transverse sections of the stems.

Tobacco: On young seedlings, the wilt develops rapidly. The leaves of infected plants become light green and gradually turn yellow. The midrib and veins become flaccid and large leaves droop in an umbrella-like fashion. Necrotic areas may frequently occur between the veins. One of the main symptoms is the unilateral wilting. One-sided wilting of the entire plants and of half leaves is a common feature of the disease. In severe cases, the leaves wilt without changing colour and remain attached to the stem. When stems are cut open, the vascular tissue shows a brown discolouration.

Banana: Most characteristic symptom is the vascular discolouration in the inner leaves and fruit stalk (peduncle). On young and fast-growing plants, the youngest leaves turn pale-green or yellow and collapse. Lower leaves become yellow, droop and dry out. Sometimes the first symptom is in the fruit, which appears distorted and pulp exhibits a dark brown discolouration. Premature yellowing and cracking occurs in infected banana fruits. Brown discolouration of vascular tissue is seen in cross sections of the stem. Extensive breakdown of pith occurs in the lower parts of the stem and taproot, and the stems become hollow. Young suckers may be blackened, stunted or twisted. The pseudostems show brown vascular discolouration.

Geranium: First symptoms are wilting followed by yellowing of leaves, often sectorial. Plants develop a systemic wilt but initially recover during cooler temperatures or at night. The lower leaves tend to wilt first and become chlorotic. Brown discolouration of the vascular tissue is often visible internally. In the later stages, leaves become brown and necrotic; plants never recover from wilt, desiccate and die.

Eggplant: Most characteristic symptom is leaf drop, which occurs before there is much wilting.

Peanut: Foliage loses its green colour, wilts suddenly and turns dull black. Leaf yellowing is quite conspicuous.

Pepper: Plants rarely show true wilting. Leaves turn yellow and roll or curl before dropping. Vascular browning has been reported to be a characteristic symptom of the disease.

However, Shekhawat *et al.* (2000) found that vascular browning in the host depends on the ability of the infecting bacterial strains to produce brown pigment. The strains, which produce brown pigment in the cultures, produce vascular browning. The strains, which do not produce vascular browning, have been reported from Portugal, Kenya and Australia, besides India.

Apart from the symptoms, the streaming test is a valuable diagnostic tool for quick identification of the disease in the field. It is performed by placing cut stem pieces of infected plants in clear water. After a few minutes, threads of a viscous white slime can be seen streaming from the cut end of the stem. These threads are of bacterial ooze exuding from the infected xylem vascular bundles.

13.12.2. Causal organism

Ralstonia solanacearum (Smith 1896) Yabuuchi *et al.* 1996

Syn. *Burkholderia solanacearum, Pseudomonas solanacearum*

The bacterium can be easily isolated from the symptomatic material on YPGA, a non-selective medium or on Kelman's tetrazolium chloride medium. For isolating from symptomless material, SMSA medium modified by Elphinstone *et al.* (1996) is useful and has been used successfully in Europe.

The bacterium is Gram-negative, rod-shaped, non-fluorescent, nonsporing, non-encapsulated, strictly aerobic with a polar tuft of flagella. The bacterium is very sensitive to desiccation and is inhibited in culture by 2.0% sodium chloride. On solid agar media, individual bacterial colonies are usually visible after 36 - 48 hr of growth at 28°C, and colonies of the normal or virulent type are white or cream-coloured, irregularly shaped, highly fluidal and opaque. The colonies of the mutant or non-virulent type, which appear occasionally, are uniformly round, smaller, butyrous or dry. The colonies of virulent strains on tetrazolium chloride medium are irregularly shaped, fluidal and white having pink centres while those of non-virulent strains are round, smaller and dark red. This medium is also used for the differentiation of virulent and weakly virulent/avirulent strains (Kelman, 1954). Polyhydroxy butyrate is accumulated. Gelatin hydrolysis is negative or weak, nitrate reduced by nearly all strains and many produce gas. Levan is not produced from sucrose; starch and esculin are not hydrolysed, ammonia is produced and only glucose is metabolised oxidatively. The tests for oxidase and catalase are positive while those for arginine dihydrolase and lipase are negative. For most strains, the optimum growth temperature is between 28 and 32° C. The bacterium can utilize glucose, sucrose and n-propanol but not D-arabinose, L-rhamnose, D-fucose and D-raffinose.

The bacterium can be stored for many years at room temperature in sterilized tap, distilled or deionized water. It can also survive for a long time at - 80°C in liquid culture broth amended to 40% glycerol. Repeated transfers on agar media result in rapid loss of virulence and cultures stored at 4°C also lose viability.

Schaad *et al.* (2001) developed Real-time PCR for the detection and identification of *Ral. solanacearum* race 3 biovar 2. This technique was further refined by developing a Real-time Bio-PCR by Ozakman and Schaad (2003). Real-time Bio-PCR was more sensitive (detected 20 cells/ml) than Real-time PCR and it detected *Ral. solanacearum* race 3 biovar 2 from latently infected potato tubers. Serological field kits, which incorporate highly specific monoclonal antibodies, are also commercially available (immunostrips-AgDia, CSL Pocket Diagnostics) for rapid on-site diagnosis. However, a confirmed diagnosis should always be obtained through laboratory tests on a pure culture.

Most recently, Priou *et al.* (2010) developed two sensitive, specific and user-friendly serological methods, namely, double-antibody sandwich (DAS)-ELISA and indirect ELISA on nitrocellulose membrane after enrichment of the plant extract in a semi-specific broth for detection of *Ral. solanacearum* (biovar 2A) in stems of symptomless plants 20 days before harvest of potato crop. Detection probalities were higher with DAS-ELISA. Being a reliable and user-friendly technique, DAS-ELISA can easily be used by national plant protection services and seed certification programmes in developing countries.

The bacterium undergoes spontaneous conversion from fluidal wild-type to afluidal, avirulent or weakly virulent state when stored in water or cultured in media. Such afluidal variants can be easily recognized by their characteristic colony morphology on tetrazolium agar medium. This conversion was considered unidirectional till the report of Shekhawat *et al.* (1992), which showed that the reversion back to fluidal virulent forms also occurred. The average of reversion rate was 3.45%. This reversion to virulent fluidal form may be playing a role in maintaining the virulence of the bacterium. The exposure of the bacterium to Terramycin and tetracycline increased the number of fluidal revertants by 5- to 16-fold. The incubation at higher temperatures of 35 and 37°C, in comparison to optimum temperature of 30°C, also increased the reversion rate by 4.0- to 6.4-fold (Gadevar *et al.,* 1999).

Genome of 3 strains of *Ral. solanacearum* has been sequenced. The first genome sequenced was that of strain GMI 1000, which has a wide host range (Salanoubat *et al.,* 2002). It is organized in two replicons, a chromosome of 3.7 Mb and a mega plasmid of 2.1 Mb, with a very similar G + C% (about 67%). The strains Mo IK2 (a strain of race 2, isolated from a banana tree) and 1609 (a race 3 strain, isolated from potato) have genomes, which are comparable to that of GMI 1000 (about 6 Mb, G + C % estimated at 67%).

The bacterium belongs to RNA homology group II (non-fluorescent) within the pseudomonads and within this group it is readily distinguished from other members by its failure to grow at 40°C. On the basis of RFLP and other genetic fingerprinting studies (Hayward, 2000), the species is divided into Division I called the Asiaticum division (biovars 3, 4 and 5 originating in Asia) and Division II called the Americanum division (biovars 1, 2A and 2T, originating in South America).

The species *Ral. solanacearum* is a complex taxonomic unit with a great degree of both phenotypic and genetic diversity. Thousands of genetically distinct strains of *Ral. solanacearum* have been characterized worldwide, revealing significant variation within the group. Therefore, *Ral. solanacearum* is considered a **'species complex.'** Unlike other phytopathogenic bacteria, race system of *Ral. solanacearum* is not based on gene-for-gene interactions, i.e., different cultivars carrying different *R* gene(s). Instead, the races are based on their pathogenicity on different host species. Although the biovar and race systems are widely accepted for the classification of *Ral. solanacearum,* there is no correlation between biovar and race; each race transects the biovars and each biovar contains various races. The only positive correlation between the biovar and race systems exists for biovar 2 and race 3. The differentiation of biovars is given in Tables 1A and 1B.

Table 1A. Tests for differentiation of *Ral. solanacearum* biovars

Utilization of	Biovar				
	1	2	3	4	5
Maltose	-	+	+	-	+
Lactose	-	+	+	-	+
Cellobiose	-	+	+	-	+
Mannitol	-	-	+	+	+
Sorbitol	-	-	+	+	-
Dulcitol	-	-	+	+	-

+ = Utilized, - = Not utilized   After EPPO Bulletin 34

Table 1B. Tests for differentiation of subphenotypes of biovar 2 of *Ral. solanacearum*

Test	Biovar 2A RFLP group26	Biovar 2A RFLP group 27	Biovar 2T or 2N
Utilization of trehalose	-	+	+
Utilization of inositol	+	-	+
Utilization of D-ribose	-	-	+
Pectolytic activity	Low	Low	High

+ = Utilized, - = Not utilized

Till today, 5 races of the bacterium have been identified on the basis of their ability to attack different host species. Race 1 occurs in tropical areas all over the world and is endemic in southern US. It has a wide host range and attacks tobacco, many other solanaceous crops and many other hosts and floriculture crops. It infects hundreds of different plant species in 50 families.

Race 2 occurs mainly in tropical areas of South America, in the Caribbean and Brazil, and infects bananas and *Heliconia* species, causing so-called Moko disease. In the Philippines, it causes bugtok disease on plantains.

Race 3 (biovar 2) is the most important race, occurs in subtropical and temperate regions and at higher altitudes in the tropics. It is of special concern to potato industry in the US and Canada because being a cold-temperature tolerant race it has caused serious disease problems in potato crop in other temperate countries of the world. It is also an extremely destructive pathogen of potato in the highland tropics of Africa, Asia and Latin America. This race was imported into the US from Guatemala in geranium cuttings on several occasions in 1999. In the United States, it was listed in 2002 as a Select Agent plant pathogen, and is now subjected to the strictest biosecurity regulations. Apart from its status as Select Agent, race 3 is considered a serious quarantine pest by the Animal and Plant Health Inspection Service of the US Department of Agriculture and by the European Plant Protection Organization (Champoiseau and Allen, 2009). Government regulations in the US include a zero tolerance for this race. It attacks potato, tomato, occasionally *Pelargonium zonale,* aubergine, eggplant and capsicum, some solanaceous weeds like black nightshade (*Solanum nigrum*) and bittersweet or climbing nightshade (*S. dulcamara*). Race 3 also infects a number of non-solanaceous weeds, which often do not show symptoms. This is the most predominant race and is mostly comprised of biovar 2A or RFLP group 26 with a worldwide distribution, RFLP group 27 found in Chile and Colombia or biovar 2T, sometimes also called 2N, found in tropical areas in South America. Race 3 is not found in Canada. It is on the NAPPO list of A1 quarantine pests as well as Canada's list of quarantine pests. Milling *et al.* (2009) compared different traits of tropical, temperate (race 3 biovar 2) and warm-temperate strains of the bacterium in *in vitro* and *in vivo* and found that interaction with plants is required for expression of the temperate epidemiological trait of race 3 biovar 2. Race 4 affects ginger in most parts of Asia. Race 5 (biovar 5) infects mulberries in China.

Nowadays, the identification of *Ral. solanacearum* into races and biovars is considered unsatisfactory because it is not predictive and some groups (e.g. race 1) contain very large variation (Champoiseau and Allen, 2009). Recently, a phylogenetically meaningful classification scheme was developed based on DNA sequence analysis (Fegan and Prior, 2005; Prior and Fegan, 2005). In this classification, twenty-three sequevars and four phylotypes have been recognized within the species complex and these broadly reflect the ancestral relationships and geographical origins of the strains.

Molecular studies have shown *Ral. solanacearum* as a species complex consisting of highly diverse strains (Poussier *et al.* 2000; Fegan & Prior, 2005). Molecular analyses using the nucleotide sequences of genes *egl* (encoding endoglucanase, a conserved virulence factor), *mut S* (a DNA mismatch repair enzyme), *hrp B* (a regulator of type III secretion) and the ITS region between the 16S and 23S ribosomal RNA genes have generated a phylogenetically meaningful subdivision of the species complex into four phylotypes, which group strains by ancestral geographic origin. Phylotype I strains come from Asia, phylotype II strains from the Americas, phylotype III strains from Africa and phylotype IV strains from Indonesia, which is the probable origin of the group (Poussier *et al.*, 2000; Fegan & Prior, 2005). These four groups are distinguished by a multiplex PCR reaction. Within phylotypes, strains are further clustered into sequevars based on nucleotide sequences of the *egl* and *mut S* genes (Poussier *et al.*, 2000; Fegan & Prior, 2005).

13.12.3. Disease cycle

Ral. solanacearum is primarily soil-borne and also water-borne pathogen. Primary inoculum comes from soil but there is no conclusive evidence that it is a persistent soil inhabitant. Top 30 cm soil contains the high population of the bacterium. The bacterium also survives in infected tubers including symptomless tubers, infected stock, reservoir plants and weeds. Although air-borne spread has never been reported, there is some evidence of epiphytic survival of the bacteria on the plant surfaces under conditions of high relative humidity. Depending upon biological and temperature conditions, the bacterium can survive from days to years in contaminated water, wet soils or deep soil layers forming reservoir of inoculum from which it can disperse. In European countries, bittersweet or climbing nightshade (*S. dulcamara*), a solanaceous weed growing along waterways, is considered to be a source of primary inoculum. *Ral. solanacearum* in contaminated water in waterways, infects the roots of this weed. The source of bacteria in waterways may come from waste generated by the processing of infected potatoes. Growers may pull *Ralstonia*-contaminated water from the waterways for irrigation, thereby inoculating their potato and tomato crops. Recent outbreaks of brown rot of potato in the United Kingdom were traced to *Ral. solanacearum*-infected *S. dulcamara* plants, which release the pathogen from roots into river waters that were subsequently used for potato irrigation. Recently, Hong *et al.* (2008) have also shown the survival of the bacterium in irrigation ponds in Gadsden county of North Florida (U.S.A.) and in common aquatic weeds growing in these ponds, including *Hydrocotyle ranunculoides* (dollar weed) and *Polygonum pennsylvanicum* (Pennsylvania smart weed). Both weeds were latently infected and showed no symptoms of wilt. There was a positive correlation between ambient temperature and *Ral. solanacearum* population in irrigation water, as shown by previous research workers. In Australia, England, Kenya and Sweden, race 3 biovar 2 was not detected in previously diseased fields after 2 years, suggesting that long-term survival in temperate regions is reduced.

The bacterium spreads through infected tubers, banana rhizomes and transplants, contaminated equipment, irrigation water, insects and in banana through pruning knives. Its movement from plant to plant occurs through soil. Race 2 is transmitted by insects and has a high potential for natural spread. Race 3 spreads more easily with surface water when infected *S. dulcamara* grows with its roots floating in water. The long distance spread of the bacterium occurs through infected (latently) seed potatoes and other vegetative propagating plant materials. Among the hosts of the bacterium producing true seed, only groundnut seed has been shown to be naturally infected.

The bacterium enters the host through wounds caused by agricultural operations and pruning knives, and through the rifts caused by emergence of lateral roots. The feeding injuries caused by root and spiral nematodes also aid in the penetration of the pathogen. Besides, certain gall forming insects, cut worms, white grubs and root invading parasitic fungi also provide avenues for the entry of bacterium. The secondary spread of the bacterium is not of much significance, as the bacterium does not become air-borne. It occurs to a limited extent through irrigation water, insects and contaminated equipment. The bacterial cells released in the soil after the death of diseased roots, invade the nearby roots.

Soil and air temperature, and soil moisture play an important role in disease development. Temperature and soil moisture has a synergistic effect on disease development. High temperature and high soil moisture alone cannot induce the disease. Light intensity, day length and mineral nutrition also affect disease development. Race 3 has comparatively low optimum temperature of 27°C while the range for maximum disease development is between 24 and 35°C. The virulence of the bacterium decreases when temperature exceeds 35°C or falls below 12°C. The optimum temperature for disease development for races 1, 2, 4 and 5 is 35°C. Favourable temperature favours rapid multiplication of the bacterium in the host, its release and infection of neighbouring plants. High soil moisture (50 - 100% water holding capacity) favours rapid disease development. It increases survival of the bacterium in soil, enhances multiplication and disease development and spread through soil. The disease is found in diverse soil types and in both acidic (pH 5.0 – 5.5) and alkaline (pH 7.0 – 8.5) soils.

The bacterium migrates first in the large xylem vessels, and in succulent stems it multiplies in the intercellular spaces of the cortex and the pith causing lysigenous cavities. In potato tubers, the pathogen may produce an internal decay emanating from the vascular bundles and causing a breakdown in the bud tissue of the eyes.

The physiology of the host-parasite interaction has been excellently reviewed by Buddenhagen and Kelman (1964). Various biochemical and physiological changes occur in the host during pathogenesis, which are manifested in the form of internal and external symptoms on the host.

After entering the xylem vessels, the bacteria multiply and travel rapidly to the upper parts of the plants. This results in the production of large amounts of exopolysaccharide (EPS) slime. The presence of this slime along with large number of bacterial cells in the xylem vessels leads to reduction in sap flow resulting in wilting of the plants. The wilting can also occur due to rupturing of vessel elements on account of excessive hydrostatic pressure. The main virulence factor of *Ral. solanacearum* is EPS I, a long (>10^6 Da) polymer. Many diverse isolates of the bacterium produce EPS I (or a very similar polymer) and 85% of the EPS I exists as released, cell-free slime while 15% remains in a cell surface-bound capsular form. EPS I accounts for >90% of its total exopolysaccharide produced in the plants. Mutants of *Ral. solanacearum* specifically blocked in synthesis of only EPS I rarely cause wilting or killing of plants, even if large number of bacterial cells are directly injected into the stem. EPS I also promotes stem colonization, because EPS I-deficient mutants colonize stems much more slowly and to a much lesser extent in comparison to wild-types.

The bacterium produces ~ 10 major exoproteins that are readily detectable in addition to many others that are less abundant and/or conditionally produced. The bacterium uses two different systems to export polypeptides from its cytoplasm to the outside environment. These systems consist of a Type II secretion system encoded by *eep* (enhanced expression of pheromone) genes, which exports most major exoproteins, including all the cell wall degrading exoenzymes, and a Type III secretion system encoded by *hrp* (hypersensitive reaction and pathogenicity) genes, which appears to deliver toxic proteins directly into plant cells. Mutants lacking either system are severely reduced in colonization ability and/or multiplication in plants. These exoproteins are absolutely essential for the disease development.

Ral. solanacearum produces pectin methylesterase (PME) that removes methoxy groups from pectin, converting it into methanol and polygalacturonic acid units. An endopolygalacturonase, PehA (also called PglA) and two exopolygalacturonase, PehB (PglB) and PehC (PglC) attack the polygalacturonic acid units. In pectin-rich regions of plants and also in soil, these four enzymes breakdown pectin into oligomers and also provide monogalacturonic acid for the growth of the bacterium. PehB-deficient mutants show only slight reduction in virulence, while PehA-deficient mutants take at least two times more time than the wild-type strains to wilt and kill the plants. This reduction in virulence is even more pronounced when soil infestation is used for inoculation in comparison to stem or petiole inoculation. The role of PME in wilt development is not clear.

Ral. solanacearum produces two extracellular glucanases, Egl and CbhA *in vitro* and these enzymes attack β-1-4 glycosidic linkages of cellulose. Egl releases cellobiose, cellotriose and cellotetraose from soluble cellulose and CbhA releases cellobiose from the reducing ends of crystalline cellulose. Mutants deficient in Egl are slower to cause the disease than wild-type strains. It seems that Egl facilitates invasion of roots and/or penetration of xylem vessels by degrading cellulosic glucans in the cell walls.

Schell (2000) has reviewed the information on control of virulence and pathogenicity genes of the bacterium by an elaborate sensory network. Many of key extracytoplasmic virulence and pathogenicity factors are transcriptionally controlled by an extensive network of distinct, interacting signal transduction pathways. A five-gene Phc (*ph*enotype *c*onversion) system, which is the core of this sensory network, regulates exopolysaccharide, cell wall-degrading exoenzymes, and other factors in response to self-produced signal molecule that monitors the pathogen's growth status and environment. Four additional environmentally responsive two-component systems work independently, and with the Phc system to fine-tune virulence gene expression. Another critical system is Prh (*pro*line rich *h*omoeodomain), which transduces plant cell-derived signals through a six-gene cascade to activate deployment of the Type III secretion pathway encoded by the *hrp* genes.

13.12.4. Management

The management of the disease is very difficult due to the soil-borne nature and wide host range of the pathogen, high pathogenic variability and limited host resistance. However, the following measures are helpful in reducing the disease incidence and severity.

1. Crop rotation of 5 - 7 years with nonhost crops is recommended.
2. Altering soil pH based on the cropping period can be effective. For potatoes, lowering the pH to 4 - 5 in the summer and raising it to 6 in autumn helped to eradicate the pathogen (Smith *et al.*, 1997).
3. Use of resistant varieties. Development of resistant varieties saved the tobacco industry in the United States from near extinction around the middle of 20th century. However, host plant resistance has not been very effective for the control of bacterial wilt on potato and tomato, since host resistance often varies among locations due to strain variation. In potato, resistance often fails at warmer temperatures. Grafting susceptible tomato cultivars onto resistant tomato or other solanaceous rootstock is effective on the commercial scale in Japan, Bangladesh and the Philippines, but grafting has not been tested for use against race 3 biovar 2. Lin *et al.* (2008) obtained a significant control of tomato wilt by integrating the use of a resistant tomato rootstock (Hawaii 1996) with soil amendment consisting of urea and slaked lime.
4. Destruction and removal of weed hosts like *S. dulcamara* with herbicides.
5. Many attempts to manage the disease through soil fumigation and antibiotics have not given encouraging results. However, dipping of tomato seedlings in Streptocycline solution (1 g in 30 liters of water) proved effective and it gave better results than soil drench and soil drench + foliar sprays. Recently, Ji *et al.* (2007) obtained very effective control of tomato wilt and a significant increase in yield with the combined application of thymol (applied as preplant fumigation through drip irrigation @ 73 kg/ha) and acibenzolar-S-methyl (foliar spray @ 25 mg/l). Hot air treatment of ginger rhizomes for 30 min at 50°C is also successful.
6. *Psm. fluorescens* isolates and vascular arbuscular mycorrhizae, *Glomus mosseae* have given encouraging results for the control of the disease in the preliminary trials. Recently, Messiha *et al.* (2007) has obtained a significant control of brown rot of potato with *Stenotrophomonas maltophilia* in the Nile Delta area of Egypt. Sood and associates have suggested the integrated management of tomato wilt utilizing various cultural (soil amendments, soil solarization, adjusting time of transplanting and biofumigation) and biological methods (host resistance, antagonistic rhizobacteria and use of VAM) in Himachal

Pardesh, India where race 1 biovar 3 of the bacterium is prevalent. Ambadar and Sood (2010) obtained 43 and 63% reduction in terminal wilt incidence of tomato through soil solarization for 10 and 8 weeks, respectively under Palampur (H.P., India) conditions.

Biological soil disinfestation (BSD- induction of anaerobic soil conditions by increasing microbial respiration through incorporation of fresh organic amendments and by reducing resupply of oxygen by covering with airtight plastic sheets) has also been tried to manage the disease. Recently, Messiha *et al.* (2007a) reported that BSD reduced the soil population of *Ral. solanacearum* race 3 biovar 2 by 92.5 to 99.9% in the field and therefore, it has the potential to become an important element in a sustainable and effective management strategy to control brown rot of potato especially in areas where the disease is endemic.

Shekhawat *et al.* (2000), after reviewing the exhaustive work done on bacterial wilt of potato in India; have suggested the following measures for its management.

1. Use of pathogen-free whole tubers for sowing.
2. Application of stable bleaching powder @ 12 Kg/ha mixed with fertilizers in furrows at the time of sowing.
3. Application of urea @ 1000 kg/ha.
4. Bacterization of seed tubers with *Bacillus subtilis*.
5. Use of resistant varieties.
6. Two to three year crop rotation with wheat, lupin, maize, paddy and bean.
7. Pre-treatment of seed tubers with 250 μg/ml of Streptocycline solution followed by 2 foliar sprays of 100 μg/ml concentration.
8. Adjustment of date of planting and minimum tillage in the standing crop. Post-emergence earthing up in potatoes causes the maximum injury ranging from 10 - 50%.
9. One or two deep ploughings of infested soil during May-June in plains or plateau and in October-November in hills.

The following measures are adopted to manage the geranium wilt. Some of these can be applicable also to other greenhouse grown crops.

1. Sanitation of greenhouses including weed control.
2. Geranium plants should be irrigated by trickle-tubes or drip-irrigation. No water flow should occur between plants.
3. Water quality should be ensured through treatments like UV, chemical or heat treatment.

4. Knives, tools and gloves used by workers should be sterilized between plants with quaternary ammonium or sodium hypochlorite.
5. At the end of season, greenhouses and irrigation systems should be cleaned with nitric acid, quaternary ammonium or sodium hypochlorite.
6. To ensure the pathogen-free stock, the plant material is culture-indexed over 1 - 3 years under temperature favourable for symptom expression.

REFERENCES

Ambadar, V.K. and Sood, A.K. (2010). Effect of solarization on tomato wilt incidence and population dynamics of *Pseudomonas fluorescens* and *Ralstonia solanacearum*. *J. Mycol. Plant Pathol.* 40: 120 – 23.

Buddenhagen, I.W. and Kelman, A. (1964). Biological and Physiological aspects of bacterial wilt caused by *Pseudomonas solanacearum. Annu. Rev. Phytopathol.* 2: 203 - 30.

Champoiseau, P.G. and Allen, C. (2009). *Ralstonia solanacearum* race 3 biovar 2: Tropical losses, temperate anxieties. APSnet Feature Story, [http://www.apsnet.org/online/feature/ralstonia/].

Elphinstone, J.G., Hennessey, J., Wilson, J.K., *et al.* (1996). Sensitivity of different methods for the detection of *Pseudomonas solanacearum* in potato tuber extracts. *Bulletin OEPP / EPPO Bulletin,* 26: 663 - 78.

Fegan, M. and Prior, P. (2005). How complex is the "*Ralstonia solanacearum* species complex"? In: C. Allen, P. Prior and A.C. Hayward (eds). *Bacterial Wilt: The Disease and the Ralstonia solanacearum Species Complex.* APS Press, St. Paul, MN, USA, pp. 449 - 61.

Gadewar, A.V., Shekhawat, G.S., Chakrabarti, S.K. *et al.* (1999). Temperature induced non-directed heritable changes in virulence of an unstable avirulent strain of *Ralstonia solanacearum. J. Indian Potato Assoc.* 26: 11 - 18.

Hayward, A.C. (2000). *Ralstonia solancearum. Encyclopedia of Microbiology.* 2nd Ed. Vol. 4, Academic Press, London, pp. 32 - 42.

Hong, J.C., Momol, M.T., Jones, J.B. *et al.* (2008). Detection of *Ralstonia solanacearum* in irrigation ponds and aquatic weeds associated with ponds in North Florida. *Plant Dis.* 92: 1674 – 82.

Ji, P., Momol, M.T., Rich, J.R., *et al.* (2007) Development of an integrated approach for managing bacterial wilt and root-knot on tomato under field conditions. *Plant Dis.* 91: 1321 - 26.

Kelman, A. (1953). The Bacterial wilt caused by *Pseudomonas solanacearum. N.C. Agr. Exp. Stn. Tech. Bull.* 99: 194 pp.

Kelman, A. (1954). The relationship of pathogenicity of *Pseudomonas solanacearum* to colony appearance in a tetrazolium medium. *Phytopathology* 44: 693 – 95.

Lin, C.-H., Hsu, S.T., Tzeng, K.C. *et al.* (2008). Application of a preliminary screen to select locally adapted resistant rootstock and soil amendment for integrated management of tomato bacterial wilt in Taiwan. *Plant Dis.* 92: 909 - 16.

Messiha, N., Diepeningen, A., Farag, N. *et al.* (2007). *Stenotrophomonas maltophilia*: A new potential biocontrol agent of *Ralstonia solanacearum*, casual agent of potato brown rot. *Eur. J. Plant Pathol.* 118: 211 - 25.

Messiha, N., Diepeningen, A., Wenneker, M. *et al.* (2007a) Biological soil disinfestation (BSD), a new control method for potato brown rot, caused by *Ralstonia solanacearum* race 3 biovar 2. *Eur. J. Plant Pathol.* 117: 403 - 15.

Milling, A., Meng, F., Denny, T.P. *et al.* (2009). Interactions with hosts at cool temperatures, not cold tolerance, explain the unique epidemiololgy of *Ralstonia solanacearum* race 3 biovar 2. *Phytopathology* 99: 1127 – 34.

OEPP/EPPO (2004). EPPO diagnostic protocols for regulated pests: *Ralstonia solanacearum. Bulletin OEPP/EPPO Bulletin* 34: 173 - 78.

Ozakman, M. and Schaad, N. W. (2003). A real-time Bio-PCR assay for detection of *Ralstonia solanacearum* race 3, biovar 2 in asymptomatic potato tubers. *Can. J. Plant Pathol.* 25: 232 – 39.

Poussier, S., Prior, P., Luisetti, J. *et al.* (2000). Partial sequencing of the *hrpB* and endoglucanase genes confirms and expands the known diversity within the *Ralstonia solanacearum* species complex. *Syst. Appl. Microbiol.* 23: 479 - 86.

Prior, P. and Fegan, M. (2005). Recent developments in the phylogeny and classification of *Ralstonia solanacearum. Acta Hort.* 695: 127 - 36.

Priou, S., Gutarra, L., Aley, P. *et al.* (2010). Detection of *Ralstonia solanacearum* (biovar 2A) in stems of symptomless plants before harvest of the potato crop using post-enrichment DAS-ELISA. *Plant Pathol.* 59: 59 -67.

Rao, M.V.B. (1972). Bacterial wilt due to *Pseudomonas solanacearum.* In: P.N. Patel (ed.). *Plant Bacteriology.* Vol. I, *Bacterial Diseases of Plants in India.* Indian Agricultural Research Institute, New Delhi, pp. 94 - 121.

Salanoubat, M., Genin, S., Artiguenave, F. *et al.* (2002). Genome sequence of the plant pathogen *Ralstonia solanacearum. Nature* 415: 497 – 502.

Schaad, N.W., Gaush, P., Postnikova, E., *et al.* (2001). On-site one hr PCR diagnosis of bacterial diseases. APS Congress, 2001.

Schell, M.A. (2000). Control of virulence and pathogenicity genes of *Ralstonia solanacearum* by an elaborate sensory network. *Annu. Rev. Phytopathol.* 38: 263 - 92.

Shekhawat, G.S., Gadewar, A.V. and Chakrabarti S.K. (1992). Spontaneous phenotypic reversion from afluidal to fluidal state in strains of *Pseudomonas solanacearum. ACIAR Bacterial Wilt Newsletter,* No. 8, 5 - 6.

Shekhawat, G.S., Gadewar, A.V. and Chakravarti, S.K. (2000). Potato bacterial wilt in India. Technical Bulletin No. 38, Central Potato Research Institute, Shimla, H. P., India.

Smith, E.F. (1896). A bacterial disease of the tomato, eggplant and Irish potato (*Bacillus solanacearum* nov. sp.). U.S. Dept. Agr. Div. Veg. Phys. and Path. Bull. 12: 1 - 28.

Smith, I.M., McNamara, D.G., Scott, P.R. *et al.* (1997). *Ralstonia solanacearum.* In: *Quarantine Pests for Europe.* 2nd Ed. EPPO/CABI.

Yabuuchi, E., Kosako, Y., Yano, I. *et al.* (1996). Validation of the publication of new names and new comtinations previously effectely published outside the IJSB. *Int. J. Syst. Bacteriol.* 46: 625 - 26. Effective publication: Yabuuchi, E., Kosako, Y., Yano, I. *et al.* (1995). Transfer of two *Burkholderia* and an *Alcaligenes* species to *Ralstonia* gen. nov.: Proposal of *Ralstonia pickettii* (Ralston, Palleroni and Doudoroff 1973) comb. nov., *Ralstonia solanacearum* (Smith 1896) comb. nov. and *Ralstonia eutropha* (Davis 1969) comb. nov. *Microbiol. Immunol.* 39: 897 - 904.

13.13. SOFT ROT OF FLESHY VEGETABLES AND FRUITS

Fleshy tissues of fruits and vegetables often undergo rotting in field, transit and in storage. The bacteria are often associated with these rotting tissues. In majority of these cases, the bacteria involved are not plant pathogens. They are either saprophytes, i.e., they do not attack the living cells, or secondary parasites, which grow in tissues already damaged or weakened by pathogens or adverse environmental conditions. However, some bacteria are real plant pathogens that attack living plant tissues and cause soft rots. *Pectobacterium carotovorum* subsp. *carotovorum* (syn. *Erwinia carotovora* subsp. *carotovora*) and *Pe. atrosepticum* (syn. *Er. carotovora* pv. *atroseptica, Pe. carotovorum* subsp. *atrosepticum*) are the main soft rotting bacteria responsible for enormous losses due to soft rots. Soft rot of fleshy vegetables and fruits caused by *Pe. carotovorum* subsp. *carotovorum* and blackleg of potato caused by *Pe. atrosepticum* are described in this chapter. The other soft rotting bacteria of less economic importance include:

1. *Pseudomonas* spp.: *Psm. fluorescens, Psm. marginalis, Psm. cichorii* and *Psm. viridiflava* are found associated with soft rot of fleshy fruits and vegetables, pink eye disease of potato, slippery skin disease of onion and sour skin of onion. These bacteria cause soft rots at lower temperatures than *Pectobacterium* and *Bacillus* spp.

2. *Ba. subtilis* and *Ba. megaterium* cause rotting of potato tubers and tobacco leaves. They cause rotting at higher temperatures compared to other soft rotting bacteria.

3. *Clostridium* spp. cause rotting of potato tubers and tobacco leaves, and wet wood symptoms in poplar. These bacteria cause rotting under unaerobic conditions.

The disease is worldwide in occurrence but is most prevalent in warm and humid climates. It occurs most commonly on fleshy vegetables like potato tubers, onion bulbs, roots of carrot, radish and turnip, cabbage and cauliflower heads, fleshy fruits such as tomato, eggplant, cucumber, watermelon, strawberry, peach and papaya, celery petioles and succulent leaves of lettuce and spinach. The losses may occur in the field, transit and storage/market. The cumulative losses due to this disease are more than any other bacterial disease. L.R. Jones was the first person to report the soft rot of carrots from Vermont in 1901.

13.13.1. Symptoms

As the name indicates, it is a characteristic soft decay of fleshy organs of fruits and vegetables.

Soft rot symptoms on potato tubers start as small, water-soaked, tannish lesions on the surface, which enlarge rapidly in diameter and in depth. The affected inner tissue becomes soft, mushy and watery while its surface appears slightly depressed and wrinkled. Affected tissues become cream-coloured and slimy, leading to disintegration and disorganization of plant cells. In certain cases, such as potato tubers and tomato fruits, the outer surface remains intact while the inner entire contents change to a thin paste. The slimy mass consisting of bacterial cells and cell debris oozes out on the outer surface through cracks. The slimy mass turns tan, grey or dark brown due to exposure to air. Within 3 to 4 days, a whole tuber or fruit may rot completely under favourable weather conditions (Fig. 13.13a). Onion bulbs, potato tubers and cruciferous plants, when infected by soft rot bacteria, often emit an offensive sulphurous odour while in most other plants, the infected tissues remain almost odourless until the secondary invaders grow on the rotten tissues and produce offensive smell (Agrios, 2004).

Soft rot bacteria also invade swollen potato lenticels, which commonly swell when tubers are soaked in water or exposed to wet soil. The infected areas around the lenticels become slightly raised and darker in colour than the potato skin. The lesions are initially ¼ inch in diameter and ¼ - ½ inch deep but under favourable conditions; these lesions develop into soft rot (Fig. 13.13b). When low temperature prevails, the lenticel infection dries out, leaving a shallow spot with a chalky-white deposit under the normal skin colour.

The aerial symptoms manifest as weak and chlorotic plants with margins of the leaflets curled upward. Irregular, brownish to black (but not as black as in blackleg of potato) lesions appear on the stems. These lesions enlarge causing a soft mushy rot of the stems. These symptoms are referred to as **aerial stem rot** or **bacterial stem rot**.

Succulent leaves are rarely infected in field but their infection in storage or in plastic containers leads to their rapid rotting and disintegration.

13.13.2. Causal organism

Pectobacterium carotovorum subsp. *carotovorum* (Jones 1901) Waldee 1945 emend. Hauben *et al.* 1999

Syn. *Erwinia carotovora* subsp. *carotovora*

Waldee (1945) created a new genus *Pectobacterium* to include soft rotting bacteria and suggested that all the soft rot causing species of the then genus *Erwinia* should be transferred to this genus. Graham (1964) also supported Waldee's proposal. However, Waldee's proposal did not get widespread acceptance but the name of *Pectobacterium* appeared in Approved Lists of Bacterial Names in 1980. Hauben *et al.* (1998) emended the description of genus *Pectobacterium* and its type species, *Pe. carotovorum* and also proposed five new subspecies combinations including *Pe. carotovorum* subsp. *carotovorum*. The below given description of this subspecies is taken from Hauben *et al.* (1998) and Gardan *et al.* (2003).

Cells are Gram-negative rods with rounded ends and measure 0.5 - 1.0 × 1.0 - 3.0 μm. They occur mostly singly or in pairs but chains are also formed and are usually motile by means of peritrichous flagella. The bacterium produces H_2S from cysteine, liquefies gelatin, reduces nitrate, produces catalase and degrades pectin. The tests for indole production, oxidase and starch hydrolysis are negative. Acid is produced from α-methylglucoside, glucanate, inulin and sorbitol. The bacterium assimilates 1-*O*-methyl α-galactopyranoside, 1-*O*-methyl β-galactopyranoside and L-glutamate but does not assimilate palatinose, *meso*-tartrate or L-alanine. It can grow in the presence of 7% NaCl and can utilize citrate. The bacterium does not utilize hypoxanthin as nitrogen source. The bacterium is able to grow at 36°C. It does not produce reducing substances from sucrose. The G + C content of the DNA of the strains of the subspecies ranges from 50.5 - 53.1 mol%. The selective medium of Beraha (1968) is helpful in the isolation of soft rot bacteria while the differential medium of Friedman (1964) is useful in separating virulent and avirulent colonies.

13.13.3. Disease cycle

The bacterium perpetuates in infected plant refuse, fleshy organs in storage, in ponds and streams used for irrigation water, in infested soil, in contaminated stores/warehouses and in pupae of several species of maggot fly. The contamination of potato tubers mostly occurs during harvest, handling and washing. Plant tissues weakened, invaded or killed by pathogens or by mechanical means are more susceptible to soft rot bacteria.

The penetration of bacteria occurs usually through wounds caused by cultural operations and insects. The bacteria can also penetrate through lenticels of freshly harvested tubers. The dissemination of bacteria occurs through wind-splashed rain, irrigation water and insects. Several species of maggot fly are known to carry the bacteria.

Aerial stem rot infection usually results from bacteria present in the infested soil and brought by irrigation water, wind-blown rain and insects. Stem infections occur through wounds and leaf scar avenues. In the field, the disease first appears on plants grown from previously infected seed pieces or healthy seed pieces contaminated from infested soil. The further spread and inoculation of bacteria into fleshy organs are facilitated greatly by insects (maggot flies). These bacteria can live in all growth stages of the insect. Adult insects lay eggs on decaying tissues. When the eggs hatch, larvae feed on rotting tissue and their bodies become contaminated externally and internally with bacteria. These larvae inoculate the bacteria into healthy plant parts when they feed on them. The pupae and adult flies developing from internally contaminated larvae also become internally infested. These adult flies lay eggs smeared with soft rot bacteria. The larvae coming out from these eggs get contaminated and when they feed on healthy tissue, not only cause wounds but also introduce the pathogen. By causing wounds, the maggots also break natural barrier of host resistance, thereby making even the resistant plants or plant parts susceptible.

Soft rot bacteria are typical wound parasites. After entering wounds, the bacteria feed and multiply on cell sap released by wounded cells. During multiplication, the bacteria produce a considerable amount of pectolytic enzymes and a lesser amount of cellulolytic enzymes. The pectolytic enzymes break down the pectic substances of middle lamella and cell walls. The pectic substances are present mainly in the middle lamella as calcium pectate. The pectolytic enzymes produced move ahead of bacteria and cause maceration of the host tissue by breaking the pectic substances of middle lamella and cell walls. Due to high osmotic pressure of the macerated tissue, water from the cells comes into the intercellular spaces leading to plasmolysis, collapse and death of cells. Bacteria move into intercellular spaces of macerated tissue, multiply, produce pectolytic enzymes, which move ahead of bacteria and prepare that tissue for invasion. The invasion of host cells occurs at later stages when cell walls are dissolved partially. The infected plant parts become soft and are converted into a slimy mass containing numerous bacteria present in the cell sap. The slimy mass containing bacteria exudes through cracks or wounds formed on the outer surface of infected plant parts and spreads the bacteria to other fleshy organs present in the field or in stores.

A fairly high relative humidity is essential for the progress of the disease. When a decaying fleshy organ is placed in a dry atmosphere, the rotted organ dehydrates rapidly and further progress of the disease may be checked completely. The progress of the disease is best above 24°C. Oxygen depletion in potato tubers also favours soft rot. A film of water on potato tubers in wet soil or in stores depletes the oxygen from tubers.

13.13.4. Management

The combination of sanitary measures and cultural practices can result in a very effective management of soft rot.

1. Remove all the debris from the warehouses and disinfect the floor and walls of the warehouses with copper sulphate (1 lb in 5 gal of water) or formaldehyde (1 pint in 10 gal of water).
2. All efforts should be made to avoid wounding of plants and their storage organs.
3. Do not wash potatoes and other products before shipping or storage. If absolutely necessary, use chlorinated water for washing. The product must be dried before shipping or storage.
4. Temperature (4°C) and humidity in stores should be kept low.
5. Use pathogen-free potato seed and if possible use whole tubers instead of seed pieces.
6. Harvest potato only after the vines are completely dead to ensure skin maturity.
7. Insecticides should be sprayed in the fields and in warehouses to kill the insects responsible for the spread of bacteria.
8. Chemical sprays are generally not recommended except for the control of soft rot of tomatoes, for which Bordeaux mixture (0.8%) should be sprayed repeatedly. Currently, chlorine compounds are used to control postharvest soft rot of fleshy fruits and vegetables. Mahovic *et al.* (2007) found that the application of chlorine dioxide gas (exposure to 99 mg ClO_2 for a 2- or 24-hr period) protected the harvested tomato fruits from soft rot.
9. Rotate the susceptible crops with cereals and other nonhost crops.
10. *Pantoea agglomerans* (*Er. herbicola* 252) has shown promising results in controlling both soft rot and blackleg diseases. Potato seed pieces/tubers are treated with antagonist before sowing.

13.14. BLACKLEG OF POTATO

Blackleg of potato has many similarities with soft rot disease. However, it affects mainly potatoes in contrast to the wide host range of soft rot bacterium. It is worldwide in occurrence. In the first decade of the 20th century, the disease was independently studied/described by Appel in Germany; van Hall in the Netherlands and Harrison in Canada. It used to cause considerable economic losses in all potato growing regions but in recent years, its incidence has greatly decreased in countries where micropropagated plantlets are used instead of seed tubers reducing the number of field generations.

13.14.1. Symptoms

Blackleg usually begins from a contaminated seed piece or tuber. Under favourable weather conditions, entire seed piece or tuber and sprouts emerging from it may rot in the soil, resulting in a poor stand. The non-emergence of plants, also called **blanking**, is a serious economic consequence of the disease. Under less favourable conditions, plants emerging from infected tubers show blackleg symptoms.

The symptoms usually start on the stems at the point of attachment to the tubers and extend upward. In the early growing season, a light brown to inky decay originating from the seed piece extends to the stem portions that are entirely below ground and the plant shows yellowing and stunting, and exhibits a stiffened and upright habit. However, the typical phase of the disease, i.e., blackleg usually develops after the plants have emerged from the soil. If there is sufficient moisture in the soil, the shoots arising from the infected tubers show watery decay of the cortical tissue and the shoots wilt. If the soil moisture is reduced, the base of the shoot develops a blackened, shriveled cortex, **blackleg phase** (Fig. 13.14) and the growth of the shoot is stunted. The leaves turn yellow, margins of the leaflets roll upward and later, entire plant may wilt and often die. Stem decay and associated blackening is not always uniform throughout the entire length of the stem. Sometimes, the blackening of the tissue occurs near the soil level and then at a node or stem injury distant from the previous one while the intervening tissue showing no external symptoms.

The tubers of the infected plants also become infected by the movement of pathogen through stolons. The dark-coloured decay of the tubers starts at the stolon end, extends inward causing decay of pith tissue and turning the tuber into a hollow shell (Fig. 13.14). These decaying tubers are often invaded by the secondary soil microorganisms, which modify the original disease symptoms. In improperly stored tubers, the pathogen causes decay at the sites of tuber lenticels and wounds. Under dry conditions, lenticel infections result in formation of restricted, small, sunken lesions around the lenticels. These restricted lesions are ..erred to as **"hard rot"** in the literature. However, under moist conditions and in association with weakly pathogenic or saprophytic microorganisms, lenticel infections can lead to decay of entire tubers. The flesh of the affected tubers first turns creamy, then greyish and finally black. Irregular cavities having black walls may extend through the centre of the potato tubers. Near mid-season, inky-black, slightly sunken lesions are formed at the stem end of the tubers.

13.14.2. Causal organism

Pectobacterium atrosepticum (van Hall 1902) Gardan *et al.* 2003

Syn. *Erwinia carotovora* pv. *atroseptica, Pe. carotovorum* subsp. *atrosepticum*

Van Hall in 1902, first named the bacterium as *Er. atroseptica.* Later, Dye in 1969 renamed it as *Er. carotovora* subsp. *atroseptica* and in 1978 as *Er. carotovora* pv. *atroseptica.* Hauben *et al.* (1999), on the basis of 16S rDNA sequences reclassified it as *Pe. carotovorum* subsp. *atrosepticum*. Gardan *et al.* (2003) raised it to the species level and named it as *Pe. atrosepticum.* The below given description of the bacterium is based on Hauben *et al.* (1998) and Gardan *et al.* (2003).

Cells are Gram-negative rods having round ends and measure 0.5 - 1.0 × 1.0 - 3.0 μm. They occur mostly singly or in pairs but chains occur as well and are usually motile by means of peritrichous flagella. Acid is produced from gluconate, α-methyl-glucoside, lactose, cellobiose, raffinose and melibiose but not from inulin, sorbitol and D(+)-arabitol. The bacterium assimilates platinose, lactulose, L-alanine, 1-*O*-methyl β-galactopyranoside and D-glucuronate but does not assimilate *meso*-tartrate, or D(-)–tartrate. The bacterium is unable to grow at 36°C and produces reducing compounds from sucrose. The G + C content of the DNA ranges from 51.3 to 53.1 mol%. Bell *et al.* (2004) sequenced the genome of the bacterium and demonstrated the presence of novel pathogenicity determinants in it.

De Boer *et al.* (1987) recognised four different serological groups on the basis of antigenicity to the lipopolysaccharide. Serogroup I accounts for about 95 strains of this bacterium associated with blackleg in Canada, the United States and several countries of Western Europe. Monoclonal antibodies specific to this group are available commercially and have been used to index seed tubers for the presence of the bacterium. Fragments of genomic DNA to be used as primers in the polymerase chain reaction have been identified for the detection of the bacterium.

Production of pectolytic enzymes, namely, pectate lyase, polygalacturonase and pectin methyl esterase, is the primary but not the sole pathogenicity factor. In addition, cellulase, xylanase and protease activity is present but their role in tissue maceration has not been studied in detail.

13.14.3. Disease cycle

The bacterium mainly survives in or on the contaminated potato tubers in the lenticels and in the damaged periderm tissue (Perombelon and Kelman, 1980). It does not survive well in soil for more than one year; however, the survival in infected debris is possible. The dissemination of the bacterium is mainly through contaminated potato tubers. The other means of dispersal are wind-blown rain, seed planting equipment, cutting knives and dipterous insects. The bacterium enters mainly through wounds and bruises caused on the potato tubers.

Temperature and moisture are the critical factors in the initiation and development of the disease. The extent of seed lot contamination and seed handling techniques also influence the severity of the disease. Most of the blackleg outbreaks have resulted from contaminated seed lots. Toth *et al.* (2003) showed that the disease development was directly related to seed tuber contamination and the threshold level of seed tuber contamination remains an important factor in predicting both disease development and progeny tuber contamination. High soil moisture and high temperature favour seed-piece decay and pre-emergence blackleg. Cool, wet soils at planting and high temperature after emergence of plants are favourable for blackleg phase. Even in the absence of disease, population of the bacterium may increase considerably in the root zone of the plants and may contaminate the progeny tubers. Therefore, visual evaluation of blackleg incidence is not a proper assessment of the level of contamination on progeny tubers.

13.14.4. Management

1. As the bacterium does not survive well in soil, the rotation with nonhost crops is not very helpful in the management of the disease.
2. Use of pathogen-free seed and seed handling techniques, which cause minimum injuries to the seed, are very effective.
3. Disposal and burial of infected plant refuse and cull piles and disinfection of tools used in cutting and planting of potato tubers reduce the disease severity (De Boer, 2001).
4. Proper wound healing and drying of potato tubers before storage and storage under proper conditions reduce the spread of the pathogen during storage.

REFERENCES

Agrios, G.N. (2004). *Plant Pathology.* 5th Ed. Elsevier Academic Press, San Diego, 922 pp.

Bell, K.S., Sebaihia, M., Pritchard, L. *et al.* (2004). Genome sequence of the enterobacterial phytopathogen *Erwinia carotovora* subsp. *atroseptica* and characterization of virulence factors. *Proc. Natl. Acad. Sci.* USA, 101: 11105 - 10.

Beraha, L. (1968). A rapid method for the preparation of a semi-soild agar medium for detection of pectolytic enzyme activity in *Erwinia carotovora. Plant. Dis. Reptr* 52: 167.

De Boer, S.H. (2001). Black leg/soft rot of potato. In: O.C. Maloy and T.D. Murray (eds). *Encyclopedia of Plant Pathology.* Vol. 1, John Wiley and Sons, Inc., New York, pp. 146 - 48.

De Boer, S.H., Verdonck, L., Vruggink, H. *et al.* (1987). Serological and biochemical variation among potato strains of *Erwinia carotovora* subsp. *atroseptica* and their taxonomic relationship to other *E. carotovora* strains. *J. Appl. Bacteriol.* 63: 487 - 95.

Friedman, B.A. (1964). Carbon source and tetrazolium agar to distinguish virulence in colonies of *Erwinia carotovora. Phytopathology* 54: 494 - 95.

Gardan, L., Gouy, C., Christen, R. *et al.* (2003). Elevation of three subspecies of *Pectobacterium carotovorum* to species level : *Pectobacterium atrosepticum* sp. nov., *Pectobacterium betavasculorum* sp. nov. and *Pectobacterium wasabiae* sp. nov. *Int. J. Syst. Evol. Microbiol.* 53: 381 - 91.

Graham, D.C. (1964). Taxonomy of the soft rot coliform bacteria. *Annu. Rev. Phytopathol.* 2: 13 – 42.

Hauben, L., Moore, E.R.B., Vauterin, L. *et al.* (1999). Validation of publication of new names and new combinations previously effectively published outside the IJSB. *Int. J. Syst. Bacteriol.* 49: 1. Effective Publication: Hauben, L., Moore, E.R.B., Vauterin, L. *et al.* (1998). Phylogenetic position of phytopathogens within the enterobacteriaceae. *Syst. Appl. Microbiol.* 21: 384 - 97.

Mahovic, M.J., Tenney, J.D. and Bartz. J.A. (2007). Applications of chlorine dioxide gas for control of bacterial soft rot of tomatoes. *Plant Dis.* 91: 1316 - 20.

Perombelon, M.C.M. and Kelman, A. (1980). Ecology of the soft rot erwinias. *Annu. Rev. Phytopathol.* 18: 361 - 87.

Toth I.K., Sullivan, L., Brierley, J.L. *et al.* (2003). Relationship between potato seed tuber contamination by *Erwinia carotovora* ssp. *atroseptica,* black leg disease development and progeny tuber contamination. *Plant Pathol.* 52: 119 - 26.

Waldee, E.L. (1945). Comparative studies of some peritrichous phytopathogenic bacteria. *Iowa St. Coll. J. Sci.* 19: 435 - 84.

13.15. COMMON SCAB OF POTATO

Common Scab of potato was first discovered in 1825 in the United States but its causal agent was identified as *Oospora scabiei* by Thaxter in 1891. In 1914, Gussow transferred the casual agent to Actinomycetes. Later on, Waksman and Henrici in 1943 transferred it to *Streptomyces scabies.* In 1953, potato scab was found in Maine in soils having pH as low as 4.3 and it was named as **"acid scab"**. Common scab of potato occurs in potato growing regions throughout the world. It is more prevalent in neutral or slightly alkaline soils and the severity increases during the dry years. Generally, the disease does not cause reduction in yield but significant losses result from reduction in the market value of potato tubers. Severe infection may cause reduction in yield also and deep-pitted lesions increase losses resulting from wastage due to peeling. Also deep-pitted lesions, which are common in diseased tubers, affect processing quality of potatoes where the normal steam-peeling processes do not remove lesioned material. The disease is rated among the top five destructive diseases of potato by potato seed producers in the United States. It is also considered one of the most economically important diseases affecting potato around the world. According to Wilson (2004), the disease annually cost the Tasmanian (Australia) potato industry in excess of A$ 3.66 million or approximately 4% of the industry value.

13.15.1. Symptoms

The symptoms appear on the underground plant parts, mainly on the tubers. The symptoms are usually noticed late in the growing season or at harvest. Corky or scabby lesions of three types, namely, surface (shallow), raised (erumpent) and pitted (deep) appear on the surface of potato tubers. Sometimes surface lesions are also referred to as russeting **(russet scab)**, particularly on white skinned potato varieties, as the general appearance resembles the skin of a russet-type tuber. The lesions start as small, superficial, roughened, circular spots, sometimes raised above and often slightly below the plane of the healthy skin. The spots enlarge; coalesce with each other forming large scabby lesions. The lesions become corky due to abnormal proliferation of cells of the periderm of tubers. The lesions vary in shape and size and are slightly darker in colour than the healthy skin (Fig. 13.15). In **deep** or **pitted scab**, the lesions extend deep into the tuber and on an average may be 3 - 4 mm deep. These lesions are darker in colour than the surface (shallow) lesions. The association of insects is suggested in the formation of pitted lesions (Walker, 1969). The type of lesion formed is considered to be the outcome of the interaction of host response, aggressiveness of the pathogen strain, time of infection and environmental conditions. Another scab, called acid scab is caused by *St. acidiscabies.* The symptoms of acid scab are similar, if not identical, to those of common scab. However, the acid scab occurs in soils having pH below 5.2, as well at higher pH levels while the common scab is found in soils having pH range of 5.2 to 8.0.

Netted scab, caused by *St. reticulsiscabiei* is another disease of potato, which differs from common scab in the type of symptoms produced, causal agent (Bouchek-Mechiche *et al.* 2000), host range, varietal reaction and optimum soil temperature required for disease development (Bouchek-Mechiche *et al.* 2000a). Netted scab is reported only from European countries. The symptoms consist of superficial brown lesions on the tuber's skin and on the roots, resulting in yield reductions. Several potato cultivars are immune to the pathogen. However, the most popular cultivar for commercial cropping in continental Europe, such as Bintje, is susceptible. Sugarbeet is another host but carrot and radish are not infected. The differences in climatic requirements, i.e., soil temperature and moisture for the development of common and netted scab suggest that netted scab may become a problem in areas or in seasons unfavourable for common scab, if susceptible cultivars are grown.

13.15.2. Causal organisms

The following species have been reported, till today, as causal agents of potato scabs:

Streptomyces acidiscabies Lambert & Loria 1989

St. europaeiscabiei Bouchek-Mechiche *et al.* 2000

St. luridiscabiei Park *et al.* 2003

St. niveiscabiei Park *et al.* 2003

St. puniciscabiei Park *et al.* 2003

St. reticuliscabiei Bouchek-Mechiche *et al.* 2000

St. scabiei (syn. *St. scabies*) (ex Thaxter 1892) Lambert & Loria 1989

St. steliiscabiei Bouchek-Mechiche *et al.* 2000

St. turgidiscabies Miyajima *et al.* 1998

St. acidiscabies and *St. reticulsiscabiei* cause acid scab and netted scab, respectively while the remaining species cause common scab. Among these species, *St. scabiei* is the most frequently isolated species from common scab lesions and is worldwide in distribution. Unlike most other plant pathogenic procaryotes, *Streptomyces* species consist of slender, branched, prostrate mycelium containing few or no septa. The diameter of the hyphae is about 1 micrometer.

St. scabiei: The formation of grey-coloured spores in spiral chains is a characteristic feature of the species. The spores are smooth, cylindrical and measure 0.5×0.9 - $1.0\ \mu m$. About 20 or more spores are formed in a chain. Brown melanoid pigment is formed in spores on tyrosine nutrient agar and peptone iron agar. The septa formation on sporogenous hyphae starts from the tip and proceeds towards the base. As the spores mature, the septa constrict, eventually breaking the cells at the tip. On germination, the spores produce one or two germ tubes, which grow into mycelium.

The organism is tyrosinase positive and gives Gram-positive reaction. It utilizes L-arabinose, D-fructose, D-glucose, D-mannitol, rhamnose, sucrose, D-xylose and raffinose but not xanthine as a sole source of carbon. It is sensitive to 20 μg/ml streptomycin and 0.5 μg/ml crystal violet. The lowest pH for its growth is 5.0 and it's G + C content of the DNA is 70% mol (Goto, 1992).

St. acidiscabies: The spores are formed as flexuous chains consisting about 20 or more spores in a chain. The spores are smooth, white but reddish on certain media having high pH and measure 0.4 × 0.5 - 1.1 μm. The organism utilizes L-arabinose, D-fructose, D-glucose, D-mannose, D-rhamnose, D-xylose and sucrose but not raffinose as a sole source of carbon. The lowest pH for its growth is 4.0. It is tolerant to 20 μg/ml streptomycin and 0.5 μg/ml crystal violet. The G + C content of the DNA is 71%mol. The species is distinguished from other pathogenic *Streptomyces* species by the production of red pigment, white to reddish spore mass and resistance to high acidic reactions.

Despite the morphological differences and genetic diversity in scab-causing species, it seems that these species share pathogenicity genes and mechanism(s) of pathogenicity. These pathogenicity genes are clustered in a large block called pathogenicity island. The pathogenicity island described in *St. turgidiscabies* is a single 425 kb restriction endonuclease fragment which contains genes for synthesis of the pathogenicity determinant thaxtomin *(txtA, txtB, txtC,* and *nos)* near one end and genes for pathogenicity factors tomatinase (*tomA*) and a necrosis inducing protein (*nec1*) near the other end. These three loci are also found in a similar large contiguous chromosomal region of *St. acidiscabies* and *St. scabiei* (Wanner, 2006).

The pathogenicity island genes have been found to be transferred horizontally from pathogenic species to saprophytic species resulting in the emergence of new pathogenic forms. The independent emergence of *St. acidiscabies* and *St. turgidiscabies* in the recent years in the USA and Japan is the result of this phenomenon (Bouchek-Mechiche *et al.,* 2006).

The nitrated dipeptide phytotoxin, thaxtomin, inhibits cellulose biosynthesis in expanding plant tissues, stimulates Ca^{2+} spiking and causes cell death. The thaxtomin production by *Streptomyces* species which cause common, acid and pitted scabs, the quantitative relationship between toxin production and virulence and the lack of its production by saprophytic species support the hypothesis that this toxin is a pathogenicity determinant on potato tubers (Loria *et al.,* 2006). Thaxtomin is not associated with the development of netted scab lesions, suggesting that *St. reticulsiscabiei* might possess some other pathogenicity determinants. Thaxtomin production is also not found associated with species, which cause russet scab.

Most recently, Lerat *et al.* (2010) reported that addition of both cellobiose (0.5%) and suberin (0.1%) in a starch-containing minimal medium doubled bacterial growth and triggered thaxtomin A production, which correlated with the upregulation (up to 342-fold) of genes involved in thaxtomins synthesis. The addition of either suberin or cellobiose alone did not affect these parameters. Suberin appeared to stimulate the onset of secondary metabolism, which is a prerequisite to the production of molecules such as thaxtomin A, while cellobiose induced the biosynthesis of this secondary metabolite.

Recently, Flores-Gonzalez *et al.* (2008) developed a PCR-based method to directly detect pathogenic *Streptomyces* causing common scab of potato using primers designed to amplify a fragment of the *txtA* and *txtB* genes, which are pathogenicity determinants in the main pathogenic *Streptomyces* species. They detected pathogenic *Streptomyces* from tuber lesions in 70 samples out of 84 naturally infected samples comprising 19 potato cultivars collected during 2000 to 2004 from the Netherland, the UK, France, Germany and Spain by PCR. The results of pathogen detection by PCR were confirmed by isolation. All the pathogenic isolates showed the basic general phenotypic traits of the *St. scabiei* phenetic cluster. RFLP analysis of amplified rRNA sequences, together with carbon sources utilization and repetitive BOX profiles, allowed most isolates to be assigned to *St. europaeiscabiei,* which emerged as the main cause of potato common scab in Western Europe. Another PCR assay (SYBR green quantitative real-time) developed by Qu *et al.* (2008) using primers designed from the *txtAB* (genes encoding thaxtomin synthetase) operon allows rapid, accurate and cost effective quantification of pathogenic *Streptomyces* strains in potato tubers and in soil. The assay was specific for pathogenic *Streptomyces* strains and the detection limit was 10 fg of the target DNA, or one genome equivalent. The amount of pathogenic *Streptomyces* DNA in total DNA extracts from 1 g asymptomatic and symptomatic tubers quantified ranged from 10^1 to 10^6 pg.

13.15.3. Disease cycle

St. scabiei can survive indefinitely, as a saprophyte, in most soils except the most acidic ones. Its population in soil in the absence of hosts can be reduced but cannot be eliminated completely as it reproduces to some extent on organic matter present in the soil. In contrast *St. acidiscabies* is a poor soil inhabitant and does not survive as well as *St. scabiei. St. scabiei* can also survive on infected potato tubers and in infected plant debris.

The pathogen is spread through infected potato seed tubers, soil water and soil carried by wind and implements. The penetration into the host occurs through lenticels, wounds, stomata and directly through the cuticle in freshly harvested and young tubers. Young tubers are more susceptible to infection than older ones. None of the potato cultivars is immune although a very few show highly resistant reaction. The other hosts include radish, carrot, turnip, beet and sweet potato.

The infection by *St. scabiei* usually starts with the onset of tuberization. The optimum temperature for tuber infection is 20 - 22°C while the infection range varies from 10 to 31°C. After penetration, the pathogen grows between or through a few layers of peridermal cells, these cells die due to the pathogenic action and the bacterium then feeds saprophytically on them producing a metabolite. The metabolite promotes rapid division of living cells surrounding the lesion. This results in the formation of several layers of corky (suberized) cells that isolate the pathogen and surrounding tuber cells. After the first layer of suberized cells is penetrated by the bacterium, the second layer is formed and it goes on till several such layers are formed in the growing season. As these layers are pushed outward and sloughed off, the bacterium grows and multiplies in additional dead cells, thereby leading to formation of large scab lesions. The depth of lesion is influenced by the host variety, time of infection, soil pH, soil moisture and invasion of the scab lesion by secondary invaders including insects. Insect attack usually results in breakdown of the corky layers and allows the pathogen to invade the tuber deeply. The following factors influence the disease severity.

13.15.3.1. Soil pH: Soil pH from 5.2 to 8.0 favours the development of common scab whereas the acid scab can also occur up to pH 4.0.

13.15.3.2. Soil moisture: Maintaining soil moisture near field capacity (above - 0.4 bars) during 2 - 6 weeks period following tuber initiation will reduce the incidence of the disease. Bacteria that grow at high soil moisture outgrow *St. scabiei* on the tuber surface.

13.15.3.3. Soil type and nutrition: Light-textured soils and those containing high organic matter are conducive to disease development.

The disease cycle of *St. acidiscabies* is similar to that of *St. scabiei* except that the former can grow in the more acidic soils, i.e., up to pH 4.0.

13.15.4. Management

1. Use of resistant varieties is the most effective strategy for the management of scab. There is a good correlation between resistance to common and acid scab among potato cultivars. Wilson *et al.* (2009) have suggested a somatic cell selection approach using phytotoxin, thaxtomin A as a positive selection agent. The best disease-resistant variant, among the 13 disease-resistant variants obtained from cv. Iwa, showed on an average 85 - 86% lower disease score (or 91 - 92% less estimated tuber

surface cover) than the parent cultivar. Wilson *et al.* (2010) further advocated the use of somatic cell selection with taxtomin A to obtain stable and extremely resistant variants of potato cv. Russet Burbank to common scab.

2. Maintaining soil moisture levels near field capacity by light and frequent irrigation during 2 to 6 weeks period following tuberization will inhibit/reduce infection.
3. Severity of common scab is significantly reduced in soils by maintaining pH at 5.2 or below (by using fertilizers like ammonium sulphate which produce acidic reaction) but this measure is associated with other drawbacks. Firstly, the number of crops, which can be grown in soils having such a low pH, is greatly reduced. Secondly, plant nutrients are best available in soils having pH levels near 6.5.
4. Rotation with small grain crops, corn and alfalfa reduces the soil borne inoculum. In Punjab, India, potato rice rotation almost eliminated the disease, which at one time was very severe.
5. Use of disease-free seed and treatment of infected seed will prevent the introduction of the pathogen into disease-free fields. It also reduces the disease severity in infested fields.
6. Biological control with *Streptomyces* phages has shown promising results.
7. Green manuring reduces soil-borne inoculum by encouraging antagonistic microorganisms.
8. Tegg *et al.* (2008) have suggested a novel approach for controlling the disease by foliar application of 2, 4 - D @ 0.90 mM (0.20 g/litre). 2, 4 - D is not inhibitory to the growth of *St. scabiei* but it reduces the toxicity of phytotoxin, thaxtomin A, essential for induction of common scab. The application of 2, 4 - D can be incorporated into the integrated management strategy of the disease.

EFERENCES

ouchek-Mechiche, K., Gardan, L., Andrivon, D. *et al.* (2006). *Streptomyces turgidiscabies* and *Streptomyces reticulsiscabiei:* One genomic species, two pathogenic groups. *Int. J. Syst. Evol. Microbiol.* 56: 2771 - 76.

ouchek-Mechiche, K. Gardan, L., Normand, P. *et al.* (2000). DNA relatedness among strains of *Streptomyces* pathogenic to potato in France: Description of three new species, *S. europaeiscabiei* sp. nov. and *S. steliiscabiei* sp. nov. associated with common scab, and *S. reticulsiscabiei* sp. nov. associated with netted scab. *Int. J. Syst. Evol. Microbiol.* 50: 91 - 99.

Bouchek-Mechiche, K., Pasco, C., Andrivon, D. *et al.* (2000a). Differences in host range, pathogenicity to potato cultivars and response to soil temperature among *Streptomyces* species causing common and netted scab in France. *Plant Pathol.* 49: 3 - 10.

Flores-Gonzalez, R., Velasco, I. and Montes, F. (2008). Detection and characterization of *Streptomyces* causing potato common scab in Western Europe. *Plant Pathol.* 57: 162 - 69.

Goto, M. (1992). *Fundamentals of Bacterial Plant Pathology.* Academic Press, Inc. San Diego, 342 pp.

Lerat, S., Simao-Beaunoir, A.-M., Wu, R. *et al.* (2010). Involvement of the plant polymer suberin and the disaccharide cellobiose in triggering thaxtomin A biosynthesis, a phytotoxin produced by the pathogenic agent *Streptomyces scabies. Phytopathology* 100: 91 – 96.

Loria, R., Kers, J. and Joshi, M. (2006). Evolution of plant pathogenicity in *Streptomyces. Annu. Rev. Phytopathol.* 44: 469 - 87.

Qu, X., Wanner, L.A. and Christ, B.J. (2008). Using the *txtAB* operon to quantify pathogenic *Streptomyces* in potato tubers and soil. *Phytopathology* 98: 405 - 12.

Tegg, R.S., Gill, W.M., Thompson, H.K. *et al.* (2008). Auxin-induced resistance to common scab disease of potato linked to inhibition of thaxtomin A toxicity. *Plant Dis.* 92: 1321 - 28.

Walker, J.C. (1969). *Plant Pathology.* 3rd Ed. McGraw-Hill Book Co., New York, 819 pp.

Wanner, L.A. (2006). A survey of genetic variation in *Streptomyces* isolates causing potato common scab in the United States. *Phytopathology* 96: 1363 - 71.

Wilson, C.R. (2004). A summary of common scab disease of potato research from Australia. In: S. Naito, N. Kondo, S. Akino *et al.* (eds). *Proc. Int. Potato Scab Symp.* Sapporo, Japan, Hokkaido Univ., pp. 198 - 214.

Wilson, C.R., Luckman, G.A., Tegg, R.S. *et al.* (2009). Enhanced resistance to common scab of potato through somatic cell selection in cv. Iwa with the phytotoxin thaxtomin A. *Plant Pathol.* 59: 137 - 44.

Wilson, C.R., Tegg, R.S., Wilson, A.J. *et al.* (2010). Stable and extreme resistance to common scab of potato obtained through somatic cell selection. *Phytopathology* 100: 460 – 67.

13.16. RING ROT OF POTATO

The disease was first observed in Germany in 1906, when Spieckermann described it in 1913. In early 1930s, the disease spread to Canada and the United States and by the 1940s, it was found in most major potato growing regions of the North America. The rapid sprea occurred mainly through the infected tubers. The disease caused heavy losses in Canada and the United States and in some instances the losses were even up to 100%. After 1940s, yield losses due to the disease have been considerably reduced by the adoption of zero tolerance in seed certification standards in most seed producing areas, including the United States, Canada and the European Union. Many countries, including the United States have attempted unsuccessfully, the eradication of the disease. The main reason for the persistence of the disease is the unnoticed presence of the pathogen in apparently healthy tubers due to lack of any symptoms. The disease still poses a continuous threat to potato seed industry. Most recently, the occurrence of the disease has been reported from Turkey (Altundag *et al.,* 2009).

13.16.1. Symptoms

In continuous cool weather, the symptoms are not observed before harvest. In warm weather, the symptoms usually appear after mid-season. Severely infected tubers may not sprout or shoots may be killed before emergence from soil. The shoots, which emerge, appear stunted, bear less number of leaves and finally die.

When the disease develops late in the season, often only one or two stems in a hill show symptoms. Affected stems show stunting, and wilt during the day and recover at night. The wilting progresses slowly in the humid weather. The wilting and yellowing usually start from the lower leaves and progress upward. The interveinal areas of the leaflets turn yellowish; leaf margins become necrotic and roll upward (Fig. 13.16a). Brown stripes appear on the stems, which may extend to leaf petioles. In advanced stages of disease development, longitudinally splitted stems show brown discolouration of vascular tissue. When an infected stem is cut across at the base, a milky white to creamy bacterial exudate can be seen at the cut end by squeezing the stem. The deterioration of roots, particularly young roots, also occurs. The extent and type of symptoms can vary with the environment and host cultivar. However, in some cases, there are no above ground symptoms.

The infection of tubers occurs through stolons. In tubers, sometimes the symptoms may not appear till harvest and appear later in storage. In most cases, there are no symptoms on the surface of tubers except in severely infected tubers. When an infected tuber is cut across the stem end, a creamy-yellow to light brown discolouration of the vascular ring, about 1/4 inch below the skin, is seen along with bacterial ooze that may be increased by squeezing the tuber. As the disease progresses, a creamy-yellow or light brown crumbly or cheesy rot develops in the vascular ring and on squeezing the tuber, a soft pulpy exudate comes out from the diseased areas (Agrios, 2004). This is the most **characteristic symptom** of the disease and the name; **"ring rot"** (Fig. 13.16b) has been derived from this diagnostic symptom. As the disease advances further, bacterial exudate becomes more copious and breakdown of the tissue of vascular ring occurs forming macroscopic cavities. Due to dehydration, the affected areas may become crumbly and powdery, or soft, mushy and foul-smelling, if invaded by soft-rot bacteria, which invariably invade. In the advanced stage, the rot involves pith of the tuber, leaving a hollow but a firm shell of the tuber. The rot also extends to the cortex of tubers leading to formation of canker lesions on the surface of tubers and cracking of skin (Fig. 13.16c). In stores, the disease continuously progresses in tubers, occasionally forming cracks on the surface and finally causing an extensive rot. Symptoms of ring rot in the vascular tissue of infected tubers are not always as evident/prominent as described above, appearing as only a broken, sporadically appearing dark line, or as a continuous yellowish discolouration. Ring rot has some overlapping symptoms with brown rot.

Diagnosis: Infected tubers may grow for several generations in the field without producing any visible symptoms, hence making the diagnosis of the disease difficult. Tomato and eggplant have been used as indicator plants for the diagnosis of the disease, as foliar symptoms on these hosts appear within two to three weeks in comparison to four to six weeks on potato plants. Since late 1980s, serological techniques like latex agglutination test, enzyme-linked immunosorbent assay (ELISA) and immunofluorescent antibody staining (IFAS) are routinely used for the diagnosis of ring rot. Nowadays, polymerase chain reaction (PCR) methods developed for detecting specific DNA sequences found in the chromosome or plasmid of *Clavibacter michiganensis* subsp. *sepedonicus* are being used for this purpose (Lee *et al.*, 2001; van Beckhoven *et al.*, 2002). Recently, Smith *et al.* (2008) integrated an internal reaction control into a TaqMan PCR assay for the detection of this bacterium. The reaction control, cloned into plasmid pCmsC4, consisted of a sequence unrelated to *Cla. michiganensis* subsp. *sepedonicus* flanked by primer sequences used in the TaqMan PCR, thus eliminating the need for multiplexing. The use of the reaction control plasmid permitted the validation of negative results, and thus facilitated the use of TaqMan real-time PCR in the routine testing of diagnostic samples for *Cla. michiganensis* subsp. *sepedonicus*. This method takes care of the drawbacks like occurrence of false negative results due to reagent or thermocyler failure and amplicon carry-over into samples or reagents leading to false positive results, associated with some of the earlier described methods. Another real-time PCR assay developed by Gudmestad *et al.* (2009) uses the cellulase A (CelA) gene sequence as the basis for primer design. CelA primers were specific to *Cla. michiganensis* subsp. *sepedonicus* and did not react with any other coryneform bacteria or bacteria pathogenic to potato. The assay was more sensitive than Cms50/72A PCR primers, immunofluorescence antibody and enzyme-linked immunosorbent assays for the detection of the bacterium from infected and symptomless potato tubers. The assay proved to be a very robust detection tool with the added advantage of detecting only virulent strain of the bacterium.

13.16.2. Causal organism

Clavibacter michiganensis subsp. *sepedonicus* (Spieckermann and Kotthoff 1914) Davis *et al.* 1984

Syn. *Corynebacterium michiganense* subsp. *sepedonicum, Cor. sepedonicum*

For many years, the bacterium was classified in the genus, *Corynebacterium*. However, due to considerable differences in cell wall and DNA base composition, it was placed in a newly created genus, *Clavibacter*. The below given description of the bacterium has been taken mainly from Davis *et al.* (1984).

As the bacterium grows slowly in culture media, its isolation is difficult from the infected plant material. It is rarely isolated when the samples contain less than 10^5 cells/cm^2 or a gram of host tissue. The colonies on nutrient agar appear within three to six days and are white to slightly yellow. Cells are Gram-positive, club-shaped rods that vary from small coccoids through rod shapes and may appear as snapping, V-shaped cells. Cells are nonmotile, nonsporing, obligate aerobes and measure 0.4 - 0.6 × 0.8 - 1.2 μm. Optimal temperature for growth is 18 - 21°C while the maximum is 30°C. Starch hydrolysis is positive but neither galatin liquefaction nor production of pectinolytic enzymes occurs. Acid is produced from mannitol but not from lactose, trehalose, inositol and mannitol. Pathogen strains vary in virulence and production of extracellular slime. Both mucoid and non-mucoid colony types have been isolated from potato.

In nature, the bacterium infects only potato, although it can colonize roots of sugar beet (Rowe *et al.*, 1995).

13.16.3. Disease cycle

Infected tubers are the main source of perpetuation of the bacterium. The bacterium also survives in the infected plant refuse and volunteer plants growing from infected tubers left in the soil from the previous season. The survival also occurs for 10 - 24 months in dried bacterial slime spread from infected tubers and stems on crates, baskets, storage bins, sacks and warehouses and also on the surface of cultivating, harvesting and grading equipment. The survival is longest under cool and dry conditions.

The dissemination of the bacterium occurs through cutting knives, planters, particularly picker-type planters, chewing insects like Colorado potato beetles and flea beetles, and irrigation water. A knife that cuts an infected potato tuber can spread the bacterium to the subsequent 20 - 100 healthy seed pieces. Secondary spread occurs by the bacteria released from the wounds caused by chewing insects and cultural practices. Secondary spread is of much less significance in comparison to primary spread from infected tubers. The penetration of the bacterium into the host occurs only through wounds. The bacterium can remain in the host for a long time without producing visible symptoms. A symptomless tuber carrying the bacterium internally may produce symptomless plants and symptomless tubers but actually all the three stages are infected. This is particularly true when the plants grow under continuous cool conditions. Favourable soil temperature for disease development is 19 - 28°C. In plants growing from infected tubers, the bacteria move slowly through the vascular tissue and reach all the lower portions of the growing plant. Bacteria also move out of the vessels into the surrounding parenchyma tissues, where they multiply and cause their breakdown forming cavities. From the parenchymatous tissue, the bacteria again move to the newly formed xylem vessels. Late in the season, the bacteria migrate from the stem to the newly formed tubers via stolons and these infected tubers serve as a source of inoculum for the next crop.

13.16.4. Management

The management of the disease is primarily through the exclusion and eradication of the pathogen. The following measures are effective in controlling or minimizing the disease.

1. Use of certified pathogen-free seed. In the U.S. and Canada, certified seed has a zero tolerance for ring rot.
2. Destruction of infected plant debris and volunteer plants.
3. Follow crop rotation with nonhost plants for 3 years.
4. If ring rot is present on a farm, a through clean up must be done. Dispose of all infected potato tubers away from potato production areas. Clean all surfaces of warehouses and farming equipment by removing all dirt, debris, trash and plant refuse and then wash with a strong detergent in hot water applied by a high pressure washer. After cleaning, disinfect all warehouses and equipment with a disinfectant (1000 - 2000 μg/ml chlorine solution, 5% Lysol solution or 2 pounds of copper sulphate dissolved in 10 gallons of water). For proper disinfection, the disinfectant must be present for at least 10 minutes, preferably 20 - 30 minutes, on any surface being treated. Disinfect also crates, bins, bags etc.
5. While cutting seed tubers, frequently disinfect the cutting knives with a disinfectant or use rotary knives, which continuously run

through a disinfectant solution, which should be frequently renewed.

6. Use assist-or cup-feed planters instead of picker-type planters.
7. Control insects such as Colorado potato beetles, plant bugs, aphids and leafhoppers, which may transmit the bacterium.
8. As the highly resistant cultivars also support sizable populations of the bacterium and act as potentially silent carriers of the disease, the use of resistant cultivars is not recommended (Ishimaru, 2001).

REFERENCES

Agrios, G.N. (2004). *Plant Pathology*. 5th Ed. Elsevier Academic Press, San Diego, 922 pp.

Altundag, S., Karahan, A., Kilinc, A.O. *et al.* (2009). First report of *Clavibacter michiganensis* subsp. *sepedonicus* causing bacterial ring rot of potato in Turkey. *Plant Pathol.* 58: 794.

Davis, M.J., Gillaspie, A.G. Jr., Vidaver, A.K. *et al.* (1984). *Clavibacter*: A new genus containing some phytopathogenic coryneform bacteria, including *Clavibacter xyli* subsp. *xyli* sp. nov., subsp. nov. and *Clavibacter xyli* subsp. *cynodontis* subsp. nov., pathogens that cause ratoon stunting disease of sugarcane and Bermuda-grass stunting disease. *Int. J. Syst. Bacteriol.* 34: 107-17.

Gudmestad, N.C., Mallik, I., Pasche, J.S. *et al.* (2009). A real-time PCR assay for the detection of *Clavibacter michiganensis* subsp. *sepedonicus* based on the cellulase A gene sequence. *Plant Dis.* 93: 649 - 59.

Ishimaru, C. (2001). Ring rot of potato. In: O.C. Maloy and T.D. Murray (eds). *Encyclopedia of Plant Pathology*. Vol. 2, John Wiley and Sons, Inc. New York, pp. 863 - 64.

Lee, I.M., Lukaesko, L.A. and Maroon, C.J.M. (2001). Comparison of dig-labeled PCR, nested PCR and ELISA for the detection of *Clavibacter michiganensis* subsp. *sepedonicus* in field-grown potatoes. *Plant Dis.* 85: 261 - 66.

Rowe, R.C., Miller, S.A. and Riedel, R.M. (1995). Bacterial ring rot of potatoes. Extension Fact Sheet, Plant Pathology. Ohio State Univ. USA.

Smith, D.S., De Boer, S.H. and Gourley, J. (2008). An internal reaction control for routine detection of *Clavibacter michiganensis* subsp. *sepedonicus* using a real-time TaqMan PCR-based assay. *Plant Dis.* 92: 684 - 93.

Van Beckhoven, J.R.C.M., Stead, D.E. and van der Wolf, J.M. (2002). Detection of *Clavibacter michiganensis* subsp. *sepedonicus* by AmpliDet RNA, a new technology based on real-time monitoring of NASBA amplicons with a molecular beacon. *J. Appl. Microbiol.* 93: 840 - 49.

13.17. BLACK ROT OF CRUCIFERS

Black rot (bacterial wilt) is one of the most destructive diseases of cruciferous crops. The disease was first reported by Garman in 1891 from Lexington Kentucky, USA on cabbage and Pammel first described the pathogen from rutabaga in Iowa in 1895. The pathogen was one of the first bacterium shown to be seed-borne by Harding and associates in 1904. According to Williams (1980), it is the most important worldwide disease of crucifers, attacking all cultivated brassicas, radishes and numerous cruciferous weeds. It affects many cultivated and wild crucifers. Cabbage, cauliflower and kale are the most susceptible host while broccoli, mustards, rape, rutabaga and turnip are susceptible hosts. Radish, *gobhi sarson* (canola) and Chinese *sarson* are also infected. In warm and wet climate, the losses due to black rot may exceed 50% on some crops. It is particularly more severe in developing countries, in the tropical and subtropical regions where small-scale and uneducated farmers often use their self-produced seed to raise the transplants (Schaad, 2001).

The disease is worldwide in occurrence where cabbage is grown. In India, it was first reported by Patwardhan in 1928. According to Patel and Kulkarni (1953), the pathogen was not only introduced in India but it was repeatedly brought with the imported seed. Several epiphytotics of the disease on cabbage and cauliflower have been reported from West Bengal, Rajasthan and Himachal Pradesh. In 1969, National Seeds Corporation of India suffered an estimated loss of Rs. 5,00,000.00 from a 10 hectare seed production plot of cauliflower laid in Himachal Pradesh, due to combined infection of black rot and stump rot (*Erwinia* sp.) (Trivedi and Patel, 1972).

13.17.1. Symptoms

The most characteristic symptom of the disease is the formation of V-shaped chlorotic lesions, base of V towards midrib, on the margins of leaves developing from hydathodal infections. Lesions enlarge and spread towards midrib of the leaf and the affected areas later turn brown and dry. In the chlorotic areas, veins and veinlets turn black and this network of black venation can be seen more clearly when the infected leaf is held against light (Fig. 13.17a). The bacteria spread through veins to midrib of the leaf from where it moves to stem through petiole. From the stem, the bacteria move upward and downward to other leaves and roots. The chlorotic lesions may appear anywhere on the leaf lamina from the systemic infection moving from stem to midrib of the leaves. Infected leaves show stunted growth and may fall off prematurely. The cross sections of infected stems and petioles show black rings due to the infection of water-conducting vessels. The chlorotic lesions also appear from invasions occurring through wounds caused by insects or other means. Under ideal conditions, the symptoms appear 10 - 14 days after entry of the bacterium.

Cabbage and cauliflower heads are also invaded and become disloured. Fleshy tissue of radish, turnip, knolkhol and rutabaga infected by black rot show vascular discolourtion and later internal break down. Such fleshy tissues are often invaded by soft rot bacteria and undergo rotting, emitting a foul smell.

On seedlings, developing from infected seeds or those developing from healthy seeds but infected by bacteria present in plant refuse or soil, the symptoms appear as black lesions on the margins of the cotyledons. The lesions enlarge and coalesce and the cotyledons turn yellow to black. The infected cotyledons shrivel, hang down and drop prematurely but before this the bacteria move into stem and young leaves. Affected seedlings wilt, collapse and die (Fig. 13.17b).

In addition to *Xanthomonas campestris* pv. *campestris,* four other pathovars of *Xa. campestris,* namely, *armoraciae, aberrans, barbareae* and *incanae* have been reported to cause non-vascular leaf spots (some also cause sunken lesions on stem and petioles) on many cruciferous and some non-cruciferous plants. McCulloch (1929) was the first to report *Xa. campestris* pv. *armoraciae* (syn. *Bacterium campestris* var. *armoraciae*) as the causal agent of non-vascular leaf spots on horseradish (*Armoracia rusticana*) but weekly pathogenic on cabbage and cauliflower (causing only leaf spots but no lesions on stems and petioles) and nonpathogenic on kale, mustarad and radish. White (1930) reported for the first time *Xa. campestris* pv. *raphani* (syn. *Bac. vesicatorium* var. *raphani*) as the cause of bacterial leaf spot of radish, turnip, cabbage, cauliflower, kale, Brussels sprout and mustard but nonpathogenic to horseradish. Although these pathovars generally cause non-vascular infections, vascular infection by *Xa. campestris* pv. *aberrans* (Vicente *et al.,* 2006) and by *Xa. campestris* pv. *armoraciae* and *Xa. campestris* pv. *incanae* (Vicente *et al.,* 2001) have also been reported.

13.17.2. Causal organism

Xanthomonas campestris pv. *campestris* (Pammel 1895) Dowson 1939

Syn. *Xa. campestris*

Cells are small, Gram-negative, nonsporing rods having a single polar flagellum. It is catalase and hydrogen sulphide positive but negative for oxidase, nitrate reduction and indole production. The colonies on nutrient agar and yeast dextrose carbonate agar are yellow, convex, mucoid and shining. It produces a yellow extracellular polysaccharide called xanthan on media containing glucose. The bacterium hydrolyzes starch and this character is used for its easy recognition on Schaad's selective medium.

It prefers neutral or alkaline medium and its growth is inhibited or suppressed by acids. The optimum growth occurs at 30ºC while lethal temperature is around 50ºC (Onsando, 1992).

Acid is produced within 21 days in Dye's medium C from arabinose, mannose, galactose, trehalose, cellobiose and fructose. Complete genome sequencing has been done for two strains, namely, ATCC 33913 (CFBP 5241) and 8004 (CFBP 6650) of the bacterium.

The bacterium is composed of genetic and serological heterogeneous groups of strains. The variability in *Xa. campestris* pv. *campestris* continuously endangers cultivars with a narrow genetic base for resistance.

Vicente *et al.* (2001), on the basis of reaction of the pathogen to cultivars of *Brassica oleracea, B. rapa, B. napus, B. carinata* and *B. juncea,* identified six races of the pathogen. Races 1 and 4 are predominant on *B. oleracea* crops. Based on 4 avirulence genes of *Xa. campestris* pv. *campestris* and four matching *Brassica* resistance genes, they also proposed a gene for gene hypothesis to explain the interaction of races and differential cultivars. Fargier and Manceau (2007) described three new races, hence making the total number of races to nine. However, they added no new avirulence gene in bacterial genotypes and no resistance gene in plant genotypes to those proposed in the gene-for-gene model of Vicente *et al.* (2001).

13.17.3. Disease cycle

The bacterium survives in infected seed (internally and externally), plant refuse and numerous weeds. Infected cruciferous weeds growing nearby the crop are an important source of carryover and infection of the subsequent crops. Most of these weeds do not show typical disease symptoms and many show no symptom at all under most favourable conditions for disease development. If the weed population around the seed production fields becomes infected, it becomes extremely difficult to produce the pathogen-free seed. The local spread of the pathogen is mainly through wind-splashed rain while the distant spread is through infected seed and diseased nursery plants. As low as 0.03% infected seed can result in the development of an epiphytotic. The other means of dissemination include wind-blown detached leaves, dust particles, insects, agricultural equipment and workers. The bacterium penetrates mainly through hydathodes and wounds caused by insects and cultural operations. The stomatal penetration occurs in cotyledons or very young leaves. In developed leaves, the penetration through stomata occurs in highly susceptible crops like cauliflower, in water-soaked tissue or when inoculum is applied with a suitable detergent.

The optimum temperature for disease development is 25 - 28°C. The bacteria can multiply at 16 - 18°C but the progress of disease is very slow and that too with no visible symptoms. Free moisture in the form of rain or dew on plant surface is required for infection and disease development. Under favourable conditions, the symptoms may appear 8 - 10 days after infection. Generally, the epiphytotics occur due to favourable weather conditions but occasionally these can be traced to an infected seed lot or infected nursery bed.

After penetration through stomata and hydathodes, the bacteria multiply and progress intercellularly until they reach and invade the spiral vessels. The bacteria multiply in the vascular tissue and produce a mucilaginous extracellular polysaccharide, xanthan. The multiplication of bacterial cells and accumulation of xanthan cause plugging of xylem vessels impeding the normal flow of water and nutrients. The bacteria then spread in the vessels throughout the plant reaching the seed via xylem of funiculus. In the infected seeds, the bacteria remain protected below the seed coat. External seed infestation occurs either through local lesions on siliques or during threshing. Infected xylem vessels disintegrate at places and the bacteria move to the surrounding parenchyma cells, killing and disintegrating these cells and forming cavities. In wet weather, the bacteria often ooze to the surface of leaves through hydathodes and wounds from where they are disseminated by rain splashes, wind and insects. Unfavourable weather conditions and the absence of disseminating agents may result into disappearance of the visible symptoms. In these plants, the bacteria progress slowly within the vascular tissue and the visible symptoms appear on the return of favourable weather conditions.

Schaad and Kendirck (1975) developed a laboratory technique capable of detecting low number of *Xa. campestris* pv. *campestris* cells in crucifer seeds. The method was capable of detecting a single diseased seed per 1000 inoculated seeds and was superior to several other methods tested. Schaad (1978) also demonstrated the use of immunofluorescence for rapid identification of the bacterium in crucifer seeds. The indirect immunofluorescence test (using antiserum to ribosomes) was better than the direct test for quick identification of the bacterium. Zaccardelli *et al.* (2007) also developed a method for detection and identification of the bacterium from crucifer seeds and from naturally infected and inoculated plants. The method is based on PCR amplification of fragments from the pathogenicity-associated gene *hrcC* (a PCR product of 519 bp). The method was very specific for *Xa. campestris* pv. *campestris* because no amplicons were obtained from 39 pure cultures of phytopathogenic bacteria including 4 pathovars of *Xa. campestris*, 2 pathovars of *Xa. axonopodis* and species of *Pseudomonas* and *Pectobacterium*. Moreover,

PCR reactions were also negative for 50 unidentified environmental isolates purified from the surface of crucifers.

13.17.4. Management

Management of the disease is based mainly on the eradicative and preventive measures given below:

1. Use certified pathogen-free seed
2. Hot water treatment (50ºC for 30 min) of all crucifer seeds is essential. However, Carisse and associates have recommended treatment of cabbage, broccoli and Brussel's sprouts seed at 50ºC for 25 min and those of cauliflower, turnip, rutabaga and kale for 15 min. Immediately after the treatment, wash the seed in cold water and allow it to dry. Proper hot water treatment also helps to eliminate seed-borne pathogens of black leg, anthracnose, downy mildew, Alternaria leaf spot and Fusarium yellows. Another recommendation is soaking of seed in 0.5% cupric acetate dissolved in 0.005 N acetic acid at 40ºC for 20 min (Schaad *et al.*, 1980) or in 10% household bleach at 50ºC for 20 min.
3. Nursery beds should be at least ¼ mile away from the transplanted fields.
4. Three-year rotation should be practised both for nursery beds and transplanted fields.
5. Avoid sprinkler irrigation or use it in moderation.
6. Agricultural equipment and tools should be disinfected with bleach after use.
7. Destroy the plant refuse by burning or by deeply burying in the soil immediately after harvest. High soil moisture hastens the decomposition of plant refuse.
8. Control the insects, which spread the pathogen and destroy the cruciferous weeds, which serve as source of inoculum for the bacterium.
9. Maintain balanced soil fertility in both seedbed and main field.
10. Spray the crop with copper fungicides at 10-day intervals.
11. Use of biocontrol agents. Wulff *et al.* (2002) have shown the possibility of managing the disease with an endophytic strain of *Bacillus subtilis* (strain BB). The control was better in dry and short rainy seasons (in comparison to main rainy season) and in clay loam soil (in contrast to sandy loam soil).
12. Grow resistant varieties. Two cabbage cultivars, namely, Early Fuji and Badger are resistant while cultivar, White rock of cauliflower is tolerant to the pathogen.

REFERENCES

arisse, O., Wellman-Desbiens, E., Toussaint, V. *et al.* Preventing black rot. Horticultural Research and Development Centre, Quebec, Canada.

argier, E. and Manceau, C. (2007). Pathogenicity assays restrict the species *Xanthomonas campestris* into three pathovars and reveal nine races within *X. campestris* pv. *campestris*. *Plant Pathol.* 56: 805 - 18.

AcCulloch, L. (1929). A bacterial leaf spot of horseradish caused by *Bacterium campestris* var. *armoraciae,* n. var. *J. Agric. Res.* 38: 269 - 87.

)nsando, J.M. (1992). Black rot of crucifers In: H.S. Chaube, U.S. Singh, A.N. Mukhopadhyay and J. Kumar (eds). *Plant Diseases of International Importance.* Vol. II, *Diseases of Vegetables and Oilseed Crops.* Englewood Cliffs, New Jersey, pp. 243 - 52.

atel, M.K. and Kulkarni, Y.S. (1953). A review of bacterial plant disease investigation of India. *Indian Phytopathol.* 6: 131- 40.

chaad, N.W. (1978). Use of direct and indirect immunofluorescence tests for identification of *Xanthomonas campestris. Phytopathology* 68: 249 – 52.

chaad, N.W. (2001). Black rot of crucifers. In: O.C. Maloy and T.D. Murray (eds). *Encyclopedia of Plant Pathology*. Vol. 1, John Wiley and Sons, Inc. New York, pp. 145 - 46.

chaad, N.W., Gabrielson, R.L. and Mulanax, M.W. (1980). Hot acidified cupric acetate soaks for eradication of *Xanthomonas campestris* from crucifer seeds. *Appl. Erviron. Microbiol.* 39: 803 - 07.

chaad, N.W. and Kendrick, R. (1975). A qualitative method for detecting *Xanthomonas campestris* in crucifer seed. *Phytopathology* 65: 1034.

rivedi, B.M. and Patel, P.N. (1972). Black rot of crucifers. In: P.N. Patel (ed.). *Plant Bacteriology*. Vol. I, *Bacterial Diseases of Plants in India.* Indian Agricultural Research Institute, New Delhi, pp. 122 - 33.

icente, J.G., Conway, J., Roberts, S.J. *et al.* (2001). Identification and origin of *Xanthomonas campestris* pv. *campestris* races and related pathovars. *Phytopathology* 91: 492 - 99.

icente, J.G., Everett, B. and Roberts, S.J. (2006). Identification of isolates that cause a leaf spot disease of brassicas as *Xanthomonas campestris* pv. *raphani* and pathogenic and genetic comparison with related pathovars. *Phytopathology* 96: 735 - 45.

/hite, H.E. (1930). Bacterial spot of radish and turnip. *Phytopathology* 20: 653 - 62.

/illiams, P.H. (1980). Black rot: A continuing threat to world crucifers. *Plant Dis.* 64: 736 - 42.

/ulff, E.G., Mguni, C.M., Mortensen, C.N. *et al.* (2002). Biological control of black rot (*Xanthomonas campestris* pv. *campestris*) of brassicas with an antagonistic strain of *Bacillus subtilis* in Zimbabwe. *Eur. J. Plant Pathol.* 108: 317 - 25.

Zaccardelli, M., Campanile, F., Spasiano, A. *et al.* (2007). Detection and identifi ation of the crucifer pathogen, *Xanthomonas campestris* pv. *campestris*, b PCR amplification of the conserved Hrp/type III secretion system gene *hrc*(*Eur. J. Plant Pathol.* 118: 299 - 306.

BACTERIAL DISEASES OF TOMATO

Many bacterial diseases occur on tomato (*Solanum lycopersicum* L Among them, the important ones are bacterial spot, which also occurs o pepper (*Capsicum annuum* L.), bacterial speck, bacterial canker an bacterial wilt. The first three diseases are described below while th bacterial wilt is covered under the description of bacterial wilt solanaceous plants.

13.18. BACTERIAL SPOT OF TOMATO AND PEPPER

Bacterial spot of tomato and pepper is a serious disease of both th crops. Localized epiphytotics occur in areas where warm and mois conditions prevail during the crop season. It is a major problem in tropic and subtropical regions. Especially under overhead-irrigation, the diseas can cause enormous losses both on tomato and pepper. The disease i much less damaging in greenhouses. Crop losses occur due to defoliatio resulting in yield reductions and severe spotting of fruits leading t reduction in their market value or making the fruits unsalable (Jone 2001).

Bacterial leaf spot of chillies was first observed in the USA in 191 and of tomatoes in 1917. The first detailed account of bacterial spot tomato was published by Doidge in 1920 and she named the caus bacterium as *Bacterium vesicatorium*. The specific name was derived i part from the development of vesicles in host cells during pathogenesi Gardner and Kendric in 1923 showed that the same bacterium als caused a leaf spot of pepper. Approximately at the same time, Higgin described bacterial spot of pepper in detail. Since then, the disease ha been reported from many countries including Canada, Central and Sout America, China, Australia, Austria, Bulgaria, France, Germany, Nige Nigeria, Philippines, Poland, Romania, Russia, Senegal, Spain, Suda Taiwan, Tanzania, Thailand, Turkey, Zambia, Zimbabwe, Caribbea Israel, Hungary, New Zealand, Italy, Yugoslavia and Southeast Asi *Xanthomonas gardneri* (group D strains) is predominant in Ontari

Canada. Due to its efficient overwintering under southern Ontario conditions, it is considered to be an aggressive form. The bacterium was responsible for the epiphytotic of the disease in Ontario processing tomato fields in 2000. The occurrence of this species has also been reported from Brazil. In India, bacterial spot was first observed on pepper from Pune in 1948 by Patel *et al.* (1950). It is now present in most of the tomato and pepper growing areas of the country.

13.18.1. Symptoms

The symptoms appear on all the above ground plant parts of tomato and pepper. On tomato leaves, small (up to 3 mm in diam), irregular and dark brown to black, greasy lesions appear. After a few days, the lesions are often surrounded by yellow haloes. The centres of the lesions dry out and frequently tear. Lesions coalesce to form big irregular dead areas, which give a scorched appearance. On mature plants, the lesions are more on fully-expanded and older leaves. The leaves having numerous lesions turn yellow and may defoliate (Fig. 13.18a). Defoliation begins from lower leaves and it is generally more on pepper plants. The shot-hole symptoms caused on tomato leaves are usually associated with *Xa. perforans.* On stem and petioles, large black lesions are formed. Infection of flower parts, particularly on pedicel, generally results in severe blossom drop/blossom abortion resulting in yield loss and split sets of fruit.

On green tomato fruits, small, water-soaked, slightly raised spots appear. The spots increase in size up to 4 - 6 mm in diam and become dark brown, scabby and slightly sunken. Initially, the spots may be surrounded by greenish-white haloes, which eventually disappear (Fig. 13.18b). The symptoms do not appear on ripe fruits.

On pepper, the spots on leaves, petioles, stem and fruits are distinctly raised. Under conditions of high rainfall and high temperature, the lesions on pepper may take the form of blight.

The symptoms of bacterial spot and bacterial speck on tomato are similar, except on fruits, and are often confused. Symptoms of bacterial speck on green fruits appear as black, slightly raised stipplings, which cause lesions less than 1.5 mm in diameter.

13.18.2. Causal organisms

Xanthomonas euvesicatoria Jones *et al.* 2004

Xanthomonas gardneri (ex Sutic 1957) Jones *et al.* 2004

Xanthomonas perforans Jones *et al.* 2004

Xanthomonas vesicatoria (ex Doidge 1920) Vauterin *et al.* 1995

Syn. *Xa. campestris* pv. *vesicatoria*

The taxonomy of the bacteria causing bacterial spot of tomato and pepper has undergone many changes. In 1921, Ethel Doidge identified it as *Bacterium vesicatorium.* It was reclassified as *Pseudomonas vesicatoria* in 1925, as *Phytomonas vesicatoria* in 1930 and as *Xanthomonas vesicatoria* in 1939. With the introduction of pathovar system in 1978, it was changed to *Xa. campestris* pv. *vesicatoria.* Vauterin *et al.* (1995) subdivided the bacterium into two species, placing the strongly amylolytic strains in *Xa. vesicatoria* and non-amylolytic strains in *Xa. axonopodis* pv. *vesicatoria.* Jones *et al.* (2000) found that xanthomonads pathogenic to tomato, pepper or both consist of at least four phenotypic groups, namely, A, B, C and D. Later on, Jones *et al.* (2004) renamed group A strains as *Xa. euvesicatoria,* group B strains as *Xa. vesicatoria*, group C strains as *Xa. perforans* sp. nov. and group D strains as *Xa. gardneri* on the basis of DNA:DNA hybridization data. Group A strains reclassified here as *Xa. euvesicatoria* represent the strains originally identified by Doidge in 1921 as *Bac. vesicatorium* referring to them as feebly amylolytic. The specific epithet "*gardneri*" was proposed by Sutic in 1957, who identified a bacterial disease of tomato from Yugoslavia. However, this bacterium has not been a major plant pathogen except in Ontario, Canada since its discovery. Strains of *Xa. euvesicatoria* and *Xa. vesicatoria* are widely distributed compared to those of other two species (Jones *et al.,* 1998).

Xa. euvesicatoria and *Xa. vesicatoria* differ from each other in utilization of cis-aconitic acid, reaction to monoclonal antibodies, amylolytic and pectolytic activity, whole cell protein profiles, rep-PCR genomic fingerprinting and reaction on differential hosts (races).

The bacteria are Gram-negative rods, measure 0.7 - 1.0 × 2.0 - 2.4 μm, motile by a single polar flagellum and strictly aerobic. They grow slowly on nutrient agar and produce colonies, which are circular, shinning, wet, entire and Naples yellow. Xanthomonadin, a water-insoluble yellow pigment is produced.

Most recently, Stall *et al.* (2009) have given an excellent account of races of bacterial spot pathogens that infect pepper and tomato, resistance genes and their corresponding avirulence genes involved in imparting resistance to both the hosts and occurrence of hypersensitive and quantitative forms of resistance. They also emphasized the need to intensify the research on quantitative or multigenic resistance and to transfer this resistance to commercial cultivars of tomato and pepper.

Races: Pathogenic variation in bacterial spot pathogens is clearly established. Some strains are pathogenic only to tomato or pepper while others infect both the hosts. Both hypersensitive and quantitative forms of resistance to bacterial spot pathogens occur in tomato and pepper. Five resistance genes involved in hypersensitive reaction (HR) in pepper and four in tomato have been identified so far. The evolution of resistance in tomato and pepper to bacterial spot pathogens has been slow in development. The genes in tomato and pepper that impart resistance against these pathogens are listed in Table 13.18.1. The origin of these genes and their corresponding avirulence genes present in the bacteria are also given.

Table 13.18.1. Resistances in tomato and pepper and bacterial effectors that interact with them (source: Stall *et al.*, 2009)

Resistance gene	Plant		Bacterium	
	Source	Species	Effector	Location
Pepper				
Bs1	PI 163192	*Capsicum annuum*	AvrBs1	Plasmid
Bs2	PI 260435	*C. chacoense*	AvrBs2	Chromosome
Bs3	PI 271322	*C. annuum*	AvrBs3	Plasmid
Bs4	PI 235047	*C. pubescens*	AvrBs4	Plasmid
bs5	PI 163192 or	*C. annuum*	ND	ND
	PI 271322	*C. annuum*	ND	ND
bs6	PI 163192 or	*C. annuum*	ND	ND
	PI 271322	*C. annuum*	ND	ND
BsT	Commercial pepper	*C. annuum*	AvrBsT	Plasmid
Tomato				
rx1, rx2, rx3	Hawaii 7988	*Solanum lycopersicum*	AvrRxv	Chromosome
Xv3	PI 128216 & Hawaii 7981	*S. pimpinellifolium* & *S. lycopersicum*	AvrXv3	Chromosome
Xv4	716	*S. pennellii*	AvrXv4	Chromosome
Bs4	Commercial tomato	*S. lycopersicum*	AvrBs4	Plasmid

ND = Not determined.

The occurrence of pathogenic races in bacterial spot pathogens is known. Some of these races attack both the hosts. There are 11 races (0 to 10), which attack pepper. Differential reaction of these races to 4 resistance genes is given in Table 13.18.2. The resistance of plants of accession PI 163192 of pepper was the result of an HR to the bacterial spot pathogen. This was the first report of an HR occurring in resistant host plants to a bacterial plant pathogen. Initially two races attacking pepper were indentified. The bacterium pathogenic on plants (accession PI 163192) having resistance gene *Bs1* was designated as race 1, while the prevalent bacteria in the field that caused bacterial leaf spot were designated as race 2. Later on, an avirulence gene cloned from race 2 caused race 1 to be hypersensitive in pepper leaves containing the *Bs1* gene. This work strongly supported the gene-for-gene hypothesis of disease resistance established by Flor in 1955 as a resistance gene *Bs1* in the host (cv. Early Calwonder 10R) and an avirulence gene (*avrBs1*) in the bacterium was necessary for resistance to bacterial leaf spot. Subsequently *Bs2, Bs3* and *Bs4* resistance genes were identified. Each of these single dominant genes, i.e., *Bs1* to *Bs4* can be overcome individually by specific races of the pathogens in the field-grown plants. To improve the level of resistance in the host, three genes, namely, *Bs1, Bs2* and *Bs3* from *C. annuum* were transferred to a single plant of Early Calwonder (ECW) and designated as ECW 123. Each gene was inherited independently of the others. The combination of all the three resistance genes in a plant provided a level of disease less than on the universally susceptible control to strains of race 6, which is pathogenic on plants with all the three genes. The combination of *Bs1* and *Bs2* provided some control against race 6 while the combination of *Bs1* and *Bs3* was not effective. None of the combinations provided a satisfactory disease control.

Table 13.18.2. Differential reaction of races of xanthomonads on pepper resistance genes (source: Stall *et al.*, 2009)

Pathogen race	Resistance gene[a]			
	Bs1	*Bs2*	*Bs3*	*Bs4*
0	HR[b]	HR	HR	HR
1	Sus	HR	HR	HR
2	HR	HR	Sus	Sus
3	Sus	HR	Sus	HR
4	Sus	Sus	HR	HR
5	HR	Sus	Sus	Sus
6	Sus	Sus	Sus	HR
7	Sus	HR	HR	Sus
8	Sus	HR	Sus	Sus
9	Sus	Sus	HR	Sus
10	Sus	Sus	Sus	Sus

[a]Sources of resistance genes are listed in Table 13.18.1.

[b]HR refers to hypersensitive reaction which is involved in resistance, Sus refers to susceptibe reaction.

A Hawaiian tomato accession (H 7998) showed a highly resistant reaction to the bacterial spot pathogen. The resistant plants exhibited an HR but the resistance was not inherited as a single dominant gene as in most other hypersensitive resistances. Three genes were involved in this case and these genes were designated as *rx1, rx2* and *rx3*. The inoculation of pathogen strains from Brazil on H 7998 plants produced two types of reactions. Some strains produced the susceptible reaction and these strains were designated as race T 2, and the strains giving HR on H 7998 were lesignated as race T1 (Table 13.18.3). T2 strains were abundant in 3razil, Asia, Argentina and midwestern states of United States but were ever found in Florida. As T 2 strains had many characteristics different rom those of T1 strains, they were identified as *Xa. vesicatoria.* Race T1 emained the sole pathogen of bacterial spot of tomato in Florida for at east 30 years. In 1991, H 7998 showed susceptible reaction to bacterial pot pathogen. This bacterium showed pectolytic and amylolytic activity nd was similar to previously discovered T 2 strains. However, unlike the ' 2 strains, it caused HR in many PI accessions of *S. pimpinellifolium.* 'herefore, this strain was designated as race T 3. A near-isogenic line ontaining *Xv3* was developed by backcrossing plants with the resistance ene from PI 128216 into FL 7060 seven times. This line was designated s FL 216 and is useful in determining races of the tomato bacterial spot athogens along with H 7998. In 1991, T3 strains comprised only a small ercentage of bacterial spot pathogen's population in Florida. It gradually liminated the T1 race in Florida and in 2006 no culture of T1 was solated. The elimination of T1 strains by T3 strains was thought to be he result of an antagonism of T 3 strains to T1 strains caused by acteriocin production by the T 3 strains (Tudor-Nelson *et al.,* 2003). At east three bacteriocins are produced by T3 strains and at least one of hem is antagonistic to all the T1 strains tested. The *avr* gene associated vith T3 hypersensitivity was cloned and designated as *avrXv3.* Tomato trains with this *avr* gene also initiate hypersensitivity in pepper plants.)ue to genetic differences of race T3 strains from the other strains of anthomonads that cause bacterial spot of tomato, T3 strains were dentified as *Xa. perforans.*

The strains of *Xa. perforans* were found in Florida which proved athogenic on plants having gene *Xv3.* These strains were designated as ace T4. The cultivation of grape tomatoes in Florida apparently provided he selection pressure to cause the development of T4 strains or their ncrease over T3 strains. In a survey of races on tomato in 2006 in 'lorida, approximately 77% of the strains collected were of the T4 race nd the rest were of T3 race. The *avr* gene in race T4 that incites HR in lants having *Xv4* gene was cloned and designated as *avrXv4.* Both the '3 and T4 races contain *avrXv4.* Probably, race T4 originated by nutation of *avrXv3* in race T3 because race T3 contains both *avrXv3* and *vrXv4* genes, and T3 and T4 are genetically similar.

The durability of a resistance gene based on HR in a plant depends n the stability of the avr gene in the pathogen. Many mechanisms perate in bacterial spot pathogens that result in instability of avr genes. Iany avr genes of bacterial spot pathogens are contained in self-ransmissible plasmids. The loss of such a plasmid can result in the loss f an *avr* gene. Another mechanism of inactivation of *avr* genes is ddition of an insertion element into the *avr* gene coding sequence. Total genomic sequencing of one strain of the bacterial spot pathogen revealed 58 insertion elements in the chromosome and another eight on four plasmids of the bacterium. One of these elements was *IS476,* which is associated with inactivation of *avrBs1* in race 2 of pepper strains. The frequency of these changes affects race composition among pathogen populations and determines the durability of the corresponding plant resistance.

Table 13.18.3. Differential reactions of races of xanthomonads on tomato resistance genes (source: Stall *et al.*, 2009)

Pathogen race[a]	Resistance gene[b]		
	rx1, rx2, rx3	*Xv3*	*Xv4*
T1	HR[c]	Sus	Sus
T2	Sus	Sus	Sus
T3	Sus	HR	HR
T4	Sus	Sus	HR

[a]Letter T before each race indicates the reaction of races from tomato plants.

[b]Sources of resistance genes are listed in Table 13.18.1.

[c]HR refers to hypersensitive reaction which is involved in resistance, Sus refers to susceptible reaction.

13.18.3. Disease cycle

The bacteria survive in diseased plant debris, volunteer plants and as epiphytes on weed species. Epiphytic survival on weeds seems to be of minimal importance as a primary source of inoculum. These bacteria have also been found to persist as latent infections on host leaves and furthermore to survive there under unfavourable conditions. This is important in distribution of these pathogens on transplants that may not have visible symptoms. The role of plant debris in the survival of bacteria is more important in temperate regions where the crop is raised in summer months and the debris does not decompose readily due to low temperature. The bacteria also survive on seed as contaminants and the seed contamination occurs during the seed extraction process.

The local or short distance spread of the bacteria occurs through wind-splashed rain, rain splashes and overhead-irrigation. The dissemination to distant places and new areas occurs through infested seed and diseased transplants.

The bacteria penetrate into the host through stomata and wounds caused by wind-blown sand, insects, mechanical means and breaking of hairs. The entry of the bacteria in the fruits occurs through cracks in the epidermis and through fruit hairs, which are present only on immature fruits.

The disease development is favoured by high relative humidity and overhead-irrigation. The optimum temperature for disease development is 24 - 30°C. The bacteria may occur in low populations as epiphytes on the surface of host plants. At the onset of favourable conditions, these low populations increase rapidly and cause infection. Marcuzzo *et al.* (2009) made the following observations on the epidemiological aspects of bacterial spot of tomato in Cacador, Brazil. The epiphytotic onset of the disease coincided with physiological maturation of the first fruit clusters. Bacterial leaf spot continued to develop even in dry period, possibly leaf wetness from dewdrops favoured disease development. The bacterial population on leaf surface varied during the first 10 weeks after planting depending on weather but it remained stable after the appearance of the first symptoms of the disease.

13.18.4. Host range

In addition to tomato and pepper, the bacterial spot pathogens infect chilli pepper (*C. frutescens*), currant tomato (*L. pimpinellifolium*), Aztec tobacco (*Nicotiana rustica), Datura tatula,* nightshade (S. *douglassii*), black nightshade (*S. nigrum*), buffalo bur (*S. rostratum*) and bittersweet or climbing nightshade (*S. dulcamara*). The survival of pathogens on these hosts may be an important source of primary inoculum in specific localities.

13.18.5. Management

For an effective management of the disease it is desirable to integrate the following control measures.

1. Use of pathogen-free seed. If the pathogen-free seed is not available, it must be treated by any of the following treatments.
   a. Hot water treatment at 50°C for 25 minutes after prewarming the seed at 38°C in water for 10 minutes. Immediately after the treatment, put the seed in cold water for 5 minutes (Leboeuf *et al.,* 2005).
   b. Acetic acid (0.8%) steep for 24 hr or 5% HCl for 5 – 10 hr at 21°C.
2. Use of disease-free transplants and this is achieved by growing the nursery where the crop production does not occur.
3. Rotation with nonhost plants for 2 - 3 years.
4. Destroy all the volunteer plants.
5. Avoid cull piles near greenhouse and field locations.
6. Use of copper compounds and streptomycin has created a serious problem of development of pathogen strains resistant to these

chemicals. In order to control copper resistant strains, a combination of copper and mancozeb is used and these chemicals should be applied as preventive sprays after transplanting.

7. Acibenzolar-S-methyl (ASM, trade names Actigard 50 WG and CGA 245704) is a plant activator that induces systemic acquired resistance in many crops against a number of pathogens. Louws *et al.* (2001) showed that it (35 g a.i. ha^{-1}) can be used as a viable alternative to copper based bactericides for field management of the disease, and also for speck of tomato, where copper resistant populations of these pathogens predominate. Obradovic *et al.* (2004) obtained better control of the disease with the combination of phage and ASM than phage, ASM or copper-mancozeb alone. Phage treatment, integrated with other practices, is currently widely used in greenhouse and production fields in Florida as a part of standard integrated management programme for the disease. Based on the results of greenhouse and field trials, Omniytics, Inc., in Salt Lake city, UT is the first to obtain registration from Environmental Protection Agency (EPA) of USA for using host specific phages on tomatoes in greenhouses and production fields in Florida as a part of standard integrated management programme for the control of the disease. Obradovic *et al.* (2005) also suggested the integrated use of acibezolar-S-methyl and phages as an alternative management strategy for the disease. Ji *et al.* (2006) also obtained significant control of bacterial spot and speck of tomato with combined application of *Pseudomonas syringae* str. Cit 7 and *Psm. fluorescens* str. 89B - 61.
8. Integration of foliar sprays of ammonium lignosulfonate (2 or 4%) and potassium phosphate (25 mM) in the standard bactericide management programmes has also been suggested for improving the disease management efficiency (Abbasi *et al.*, 2002).

13.19. BACTERIAL CANKER OF TOMATO

Bacterial canker of tomato was first observed by Erwin F. Smith in 1909 from Grand Rapids, Michigan (USA) and he named it as Grand Rapids disease. However, Smith soon renamed the disease as bacterial canker of tomato. By 1920s, the occurrence of the disease was reported from Central and South America, Africa, Asia, Australia and Europe. Now the disease is worldwide in occurrence and occurs wherever tomatoes are grown. Although usually sporadic in its occurrence, it is so destructive in nature that both field and greenhouse crops, especially the latter, suffer heavily from the disease. It is a quarantined pathogen and Canada, USA, European Union (EPPO A2 quarantine pest) and some other countries enforce zero tolerance for import and export of plant material infected with the pathogen. Janse (2006) has listed 60 countries where the disease occurs. Yield losses ranging from 20 – 70% have been reported from France and USA.

13.19.1. Symptoms

Tomato plants are susceptible at all the growth stages. A wide variety of symptoms, resulting from superficial or localized and systemic infections, are produced. Wilting is the most common symptom in spite of different names of the disease.

Generally there are no symptoms on the seedlings but under conditions favourable to the disease, infected seedlings show weak and stunted growth or may be killed. Wilting is first seen on the older leaves and the leaves may curl downward before wilting. Wilting of leaflets often occurs on one side of the petiole but soon the leaflets on the other side also wilt and wilting progresses involving the whole plant. The progress of the wilt is more rapid in younger plants. A light brown discolouration of vascular tissues is visible when wilted stems, branches and petioles are split opened vertically. In stems, discolouration is more prominent at nodes and just above the soil line. In the advanced stages of disease development, cankers develop on stems, petioles and midribs in the form of brownish streaks. These streaks split open exposing necrotic tissue in the pith and cortex. The pith turns granular to mealy and cavities are formed. In severely infected plants, yellow ooze may exude from the cut end of a stem when it is squeezed. The seed-borne inoculum or invasion of vascular system by the pathogen entering through wounds result in systemic infection leading to wilting and canker formation.

Superficial/localized infections resulting from the entry of epiphytic bacteria through stomata, hydathodes and broken trichomes (hairs) produce the following symptoms. Bacteria entering through hydathodes cause marginal necrosis of leaves and these brown necrotic areas (**leaf "firing"**) may extend until the entire leaflet is shriveled. Round to irregular spots also appear on the leaves. Similar spots may also be noted on stem and petioles (Fig. 13.19a).

On green fruits, first small, white, superficial spots appear. These spots enlarge up to 1 – 3 mm in diam and develop a raised dark brown centre surrounded by a distinct white halo (Fig. 13.19b). Due to their resemblance to bird's eye, they are called **bird's-eye spots.** However, their occurrence is erratic and they do not always develop in disease outbreaks. Their occurrence reduces the market value of the fruits and sometimes making them unsalable (Braun, 2001). Systemic infection results in darkening of the vascular tissues within the fruit. The bacteria can multiply in the vascular bundles within the fruits up to the seed. This leads to the formation of yellowish strands from the stem to the seeds and also the internal infection of the seed. The fruit may show a black peppering at the vascular bundles under the calyx scar.

13.19.2. Causal organism

Clavibacter michiganensis subsp. *michiganensis* (Smith 1910) Davis *et al.* 1984

Syn. *Corynebacterium michiganense* subsp. *michiganense, Cor. michiganense* pv. *michiganense*

E.F. Smith was the first to isolate and characterize the bacterium and he named it as *Aplanobacter michiganense.* Its name was changed several times before it was named *Corynebacterium michiganense,* a name that remained in use for several decades. Davis *et al.* (1984) suggested the removal of all the plant pathogenic bacteria from genus *Corynebacterium* and placed this bacterium in a new genus *Clavibacter* as *Cla. michiganensis.* The latest name of the bacterium is *Cla. michiganensis* subsp. *michiganensis.*

The bacterium grows well on nutrient agar containing 5% glucose and potato dextrose agar producing yellow mucoid colonies. Cells are Gram-positive, nonmotile, nonsporing, irregularly-shaped rods. Gelatin is weakly liquefied and H_2S is produced. Acid is produced from mannose fructose and sucrose but not from dulcitol, sorbitol, inulin and α–methyl-D-glucoside. Acetate, citrate, fumarate, malate and succinate are used as a sole source of carbon. The genes involved in pathogenicity have been identified and characterized. The following characteristics of this bacterium differentiate it from other subspecies of *Cla. michiganensis*: tolerance to sodium chloride up to 6%, H_2S production, weak liquefaction of gelatin and utilization of mannose, cellobiose and succinate but not of mannitol and propionate as the sole source of carbon.

13.19.3. Disease cycle

Infected seed is an important source of perpetuation of the bacterium and it survives in the inner most layers of the seed coat and as an external contaminant. The infection of the seed results from systemic infection of the plant while external contamination occurs during extraction of seed. One infected seed in 10,000 may be enough for the outbreak of the disease under favourable conditions. The pathogen can survive in the infected plant debris in soil up to 2 or 3 years. However, the survival in plant debris is related to the extent of decomposition of the debris in the soil. Woody stems buried deep in soil usually remain intact and help in survival of the bacterium. The bacterium also survives on volunteer tomato plants and weed hosts.

The penetration of the host occurs through wounds caused by cultural operations, broken trichomes and natural openings like stomata and hydathodes. On tomato fruits, the wounds caused by wind-blown sand, breaking of hairs and insect punctures also act as avenues for the entry of the pathogen. Secondary spread of the bacterium occurs mainly through wind-splashed rain and in greenhouses through sprinkler irrigation. Secondary spread also occurs through contaminated equipment and cultural operations like pruning, staking and training of vines. The bacteria growing epiphytically on healthy and diseased plants serve as a major source of inoculum for the secondary spread.

Seed-borne inoculum plays an important role in the outbreak of the disease. The contamination of seeds with only a few bacterial cells that too below the detection levels; give rise to high number of infected seedlings. The seed lots having a very few infested seeds pose a serious problem in greenhouse crops. No doubt the certified seed reduces the chances of infection, but is not a guarantee for a disease-free seed. When contaminated seeds germinate, the bacteria enter the cotyledons through wounds, probably through broken trichomes. The bacteria move systemically through the xylem from which it enters the phloem, pith and cortex. The optimum conditions for disease development include high moisture/relative humidity and temperature of 24 - 32°C. There is usually a long latent period of several weeks in mature plants. In addition to tomato, the bacterium attacks *Solanum douglassii* (nightshade), *S. nigrum* (black nightshade) and *S. triflorum*.

The detection of pathogen in seed lots is required for seed certification. Polymerase chain reaction (PCR) is considered the most sensitive method for the detection of plant pathogenic bacteria. Classical PR primers derived from the *pat-1* gene in plasmid pCM2 of *Cla. michiganensis* subsp. *michiganensis* were developed for its detection but the primers reacted with 75% of the strains tested. Bach *et al.* (2003) developed a real-time TaqMan PCR assay based on the primers specific to the internal transcribed spacer sequence between 16S and 23S rDNA, a common region in all the five subspecies of *Cla. michiganensis,* and a subspecies specific TaqMan probe for the detection of this bacterium. However, both the above-mentioned PCR assays do not distinguish between the living and dead cells of the bacterium and false positive results obtained with dead cells limit the use of PCR in risk analysis of seed lots. Most recently, a quantitative real-time PCR assay devised by binding the bacterial DNA with ethidium monoazide has taken care of this problem as the method allows the selective quantification of only viable cells (Lou *et al.,* 2008). The primers and TaqMan probe, based on 16S - 23S rDNA spacer sequences, were highly specific for *Cla. michiganensis* subsp. *michiganensis* at the subspecies level. The detection limit of the direct real-time PCR was 10^3 cfu/ml in samples and with an apparent sensitivity of 2 cfu of target cells in PCR reaction solution.

13.19.4. Management

The management of bacterial canker of tomato poses a significant challenge due to the lack of effective chemicals and resistant varieties and due to the systemic nature of the pathogen. The following multifaceted approach, which mainly prevents or reduces the primary source of inoculum, will reduce the incidence and severity of the disease.

1. Use only certified, pathogen-free seed, obtained from canker free plants for planting. Seed should be extracted only by the standard acetic acid extraction method or by the fermentation process. The seed extracted by these methods is free from seed coat contamination but the embryonic infection of the seed is not completely controlled. Hence, the seed must be treated with hot water or disinfected with acid or chlorine. In hot water treatment, the seed is prewarmed in water at 38°C for 10 minutes and then kept in hot water at 50°C for 25 minutes. Immediately after the treatment, put the seed in cold water for 5 minutes. The disinfection of seed with acid is done by dipping the seed in 0.8% solution of acetic acid for 24 hr at 21°C and with chlorine by dipping the seed in 1% sodium hypochorite (20% bleach) for 20 – 40 minutes. For proper seed disinfection, presoaking of seed in water, ranging from 1 – 18 hr, is essential.
2. Plant only certified disease-free transplants grown under a vigorous inspection programme.
3. Ensure the destruction of infected plant debris by its thorough decomposition. Soil solarization by covering the soil with transparent polythene sheets for six weeks proved effective in reducing the disease in plastic tunnels (Antoniou *et al.*, 1995).
4. Follow the crop rotation for at least three years.
5. Destroy the volunteer tomato plants and weeds belonging to the Solanaceae family.
6. Sterilize seedbeds and soil in the greenhouse to destroy bacteria. Steam sterilization is preferred over methyl bromide.
7. Pruning knives and other tools used in the greenhouse should be disinfected regularly. Quaternary ammonium compounds should be preferred over sodium hypochlorite or laundry bleach as a greenhouse disinfectant (Zitter, 1985).
8. Fixed copper sprays can help in protecting the crop, particularly in superficial or localized infections. The combination of copper and mancozeb is better than copper alone. Preventive sprays

which delay the disease development/outbreak until after the main fruit set will minimize the yield loss because fruit infection, responsible for major yield loss, occurs only on green immature fruits (less than 3 cm in diam).

The application of acibenzolar-S-methyl (benzo-[1, 2, 3]-thiadiazole-7-carbothioic acid-S-methyl ester, ASM; Bion 50 WG) activates defence response of the host against the bacterium. Baysal *et al.* (2003) reported reduction in disease severity up to 76.3% along with reduction in bacterial growth up to 68.2% in ASM treated plants. The enhanced resistance was due to the significant increase in peroxidase and chitinase activity.

13.20. BACTERIAL SPECK OF TOMATO

Bacterial speck of tomato caused by *Pseudomonas syringae* pv. *tomato* was first described by Okabe in 1933 from Taiwan. The economic importance of the disease was realized in late 1970s when it caused serious losses to winter tomato crop of 1977 - 78 in southern Florida and in 1978 in southern transplant fields and tomato production areas of northern United States. Cool and moist conditions favour the development of the disease. Tomato is the only economically important host of the pathogen. The bacterium has also been isolated from a wide range of crop and weed species such as *Arabidopsis thaliana*, *Stellaria media*, *Lamium amplexicaule* and *Gnaphalium* spp.

13.20.1. Symptoms

The symptoms appear on all the above ground plant parts. Small, dark brown specks, 1 – 2 mm in diam, often surrounded by discrete yellow haloes, appear on the leaves. Speck lesions coalesce to form blighted areas leading to considerable leaf distortion. On field plants, the lesions are more concentrated at the leaf margins (Fig. 13.20a). Leaf lesions of bacterial speck are difficult to distinguish from those of bacterial spot and young lesions of early blight. Unlike bacterial leaf spot, the leaf lesions of bacterial speck are not greasy but in bacterial speck, the leaf curling is more severe than bacterial spot.

In severe infection, irregular, dark brown lesions also appear on stem and petioles, and blossoms also get blighted. Lesions on stem and petioles also cannot be distinguished from those caused by bacterial spot. Infection of 3- to 5-leaf stage seedlings results in stunting of seedlings and considerable yield losses.

On immature green fruits (less than 3 cm in diam), dark brown to black, slightly raised lesions but smaller (1.5 mm in diam) than those of bacterial spot appear, which may be surrounded by dark green haloes. The lesions are very superficial and do not crack or become scabby like those of bacterial spot. Ripe fruits are not susceptible to infection but the black lesions formed on unripe fruits exist even after ripening (Fig. 13.20b).

13.20.2. Causal organism

Pseudomonas syringae pv. *tomato* (Okabe 1933) Young *et al.,* 1978
Syn. *Psm. tomato*

The bacterial cells are straight, Gram-negative rods and measure 0.7 – 1.0 × 1.8 – 2.8 μm. It produces cream-coloured colonies with a diffusible green fluorescent pigment on King's medium B. It is strictly aerobic, motile having 1 – 2 polar flagella, produces levan on nutrient agar containing 5% sucrose but oxidase and arginine dihydrolyase tests are negative. It differs from a closely related tomato pathogen, *Psm. syringae* pv. *syringae* by its fatty acid profile, ability to cause pits in pectate gel at pH 4.6 and inability to utilize erythritol or DL-lactate as a sole source of carbon (Cuppels, 2001). The G + C content of DNA is 58 – 60 mol%. The bacterium elicits a hypersensitive reaction on tobacco but does not cause rotting of potato tubers. Two races of the bacterium, namely, race 0 and race 1 are known. Recently, Yan *et al.* (2008) have proposed the division of *Psm. syringae* pv. *tomato* into two pathovars.

The bacterium has been the focus of intensive research and hence extensively used to study the molecular mechanisms of host responses to infection because *Psm. syringae* pv. *tomato* strain DC 3000 is the pathogen of a model plant, *Arabidopsis thaliana.* A major part of information about the plant immune system has been obtained by studying the interaction of *Psm. syringae* pv. *tomato* DC 3000 with its two hosts, *A. thaliana* and tomato. The molecular basis of pathogenesis in *Psm syringae* pv. *tomato* is based on the coordinated expression of multiple pathogenicity and virulence factors such as toxins, extracellular proteins and polysaccharides; and translocation of proteins into plant cells by the type III (hrp) secretion system. The contribution of individual virulence factors to parasitism and disease development varies significantly between strains (Preston, 2000). The complete genome sequence of *Psm. syringae* pv. *tomato* strain DC 3000 determined by Feil *et al.* (2005) revealed that the genome (6.5 mbp) is composed of a circular chromosome and two plasmids

13.20.3. Disease cycle

The bacterium survives in diseased plant debris, on roots and foliage of nonhost plants including weeds. The bacterium can survive in soil for more than one year. The bacterium can also survive in the crevices and cavities of seed coat. The survival in/on seed can be for a long time (over 20 years). The dispersal of the bacterium occurs through wind-driven rain, overhead-irrigation and farm equipment. It may move considerable distances in the water droplets or aerosols generated by rainstorms. The pathogen also spreads through contaminated seed and diseased transplants.

The penetration of the bacterium into the host is mainly through wounds, stomata and hydathodes. Recent studies have shown that stomata can play an active role in limiting bacterial invasion as part of plant innate immune system (Melotto *et al.*, 2008). As a counter defence, *Psm. syringae* pv. *tomato* strain DC 3000 uses the virulence factor coronatine to actively open stomata. Hail and windstorms are often associated with increased disease severity.

The disease is favoured by cool (21°C) and moist (high R.H. and prolonged period of free moisture) weather. Resident population of the bacterium present on tomato plants can cause symptoms in 3 – 5 days under favourable weather conditions and under field conditions the symptoms can develop within 6 – 10 days. Fruits that develop during cool and moist weather can be severely infected. Fruits are most susceptible to infection from anthesis till they reach a diameter of 3 cm. Grandular and non-grandular leaf hairs present on ovaries are gradually lost after anthesis, leaving openings in young fruit's epidermis. The bacterial cells associated with leaf hairs enter through these openings and cause fruit infection.

Psm. syringae pv. *tomato* produces a toxin called coronatine, which acts as a virulence factor and is responsible for the yellow haloes surrounding leaf lesions and the stunting of young seedlings. Bacterial populations that do not produce this toxin, grow poorly on host plants and cause little damage. A fragment of gene cluster controlling coronatine production has been used as a DNA probe for disease diagnosis and epidemiologically studies of the pathogen. Zaccardelli *et al.* (2005) developed a rapid method for the detection and identification of the bacterium from tomato plants and seeds based on the PCR amplification of *Psm. syringae* pv. *tomato* strain DC 3000 *hrpZpst* gene. The results confirmed the high specificity observed using pure cultures. Cuppels *et al.* (2006) developed PCR-based lesion assays using crude DNA extracts and primer sets (COR 1/2 for bacterial speck and BSX 1/2 for bacterial spot) for the detection of bacterial speck and bacterial spot pathogens.

13.20.4. Management

1. Use of resistant varieties is the most effective control of the disease. A bacterial speck resistance gene, *Pto* (single dominant) discovered in an Ontario breeding line ONT 7710 of *S. pimpinellifolium* confers resistance to race 0 but not to race 1 of

the pathogen. This was the first plant resistance gene to be isolated and characterized (Martin *et al.*, 1993).

2. Use pathogen-free seed or treat the seed with hot water (prewarm at 38°C for 10 minutes followed by 50°C for 25 minutes) or 0.8% acetic acid or 1% sodium hypochlorite (20% bleach).
3. Use of disease-free transplants.
4. Destruction of diseased plant debris, weeds and volunteer plants.
5. Use of drip irrigation to reduce leaf wetness.
6. Crop rotation for three years.
7. Protective sprays of fixed copper compounds are effective. As copper resistance is widespread among the *Psm. syringae* pv. *tomato* populations in main tomato growing areas, the copper compounds should be combined with mancozeb or chlorothalonil. The sprays should be given at regular intervals starting from anthesis until the first-formed fruits attain the size of 3 cm in diameter.
8. Treatment of tomato seeds, artificially inoculated with *Psm. syringae* pv. *tomato,* with a combination of mild chemo-thermal treatment, inoculation of seed with *Azospirillum brasilense,* a plant growth-promoting bacterium and later, a single foliar application of a copper bactericide, nearly eliminated the disease even when the plants were grown under mist for 6 weeks. However, these treatments were ineffective when tried individually (Bashan and de-Bashan, 2002).
9. For the use of acibenzolar-S-methyl (ASM) in disease management, refer to bacterial spot of tomato and pepper.

REFERENCES

Abbasi, P.A., Soltani, N., Cuppels, D.A. *et al.* (2002). Reduction of bacterial spot disease severity on tomato and pepper plants with foliar applications of ammonium lignosulfonate and potassium phosphate. *Plant Dis.* 86: 1232 - 36.

Antoniou, P.P., Tjamos, E.C., Andreou, M.T. *et al.* (1995). Effectiveness, modes of action and commercial application of soil solarization for control of *Clavibacter michiganensis* subsp. *michiganensis* of tomatoes. *Acta Hort.* 382: 119 – 28.

Bashan, Y. and de-Bashan, L.E. (2002). Reduction of bacterial speck (*Pseudomonas syringae* pv. *tomato*) of tomato by combined treatments of plant growth-promoting bacterium, *Azospirillum brasilense*, streptomycin sulphate, and chemo-thermal seed treatment. *Eur. J. Plant Pathol.* 108: 821– 29.

Baysal, O., Soylu, E.M. and Soylu, S. (2003). Induction of defence-related enzymes and resistance by the plant activator acibenzolar-S-methyl in tomato seedlings against bacterial canker caused by *Clavibacter michiganensis* subsp. *michiganensis. Plant Pathol.* 52: 747 - 53.

Braun, E. (2001). Bacterial canker of tomato. In: O.C. Maloy and T.D. Murray (eds). *Encylopedia of Plant Pathology*. John Wiley and Sons, Inc., New York, pp. 76 – 77.

Cuppels, D.A. (2001). Bacterial speck of tomato. In: O.C. Maloy and T.D. Murray (eds). *Encyclopedia of Plant Pathology*. Vol. 1, John Wiley and Sons, Inc. New York, pp. 82 - 83.

Cuppels, D.A., Louws, F.J. and Ainsworth, T. (2006). Development and evaluation of PCR-based diagnostic assays for the bacterial speck and bacterial spot pathogens of tomato. *Plant Dis.* 90: 451 - 58.

Davis, M.J., Gillaspie, A.G. Jr., Vidaver, A.K. *et al.* (1984). *Clavibacter*: A new genus containing some phytopathogenic coryneform bacteria, including *Clavibacter xyli* subsp. *xyli* sp. nov., subsp. nov. and *Clavibacter xyli* subsp. *cynodontis* subsp. nov., pathogens that cause ratoon stunting disease of sugarcane and Bermudagrass stunting disease. *Int. J. Syst. Bacteriol.* 34: 107 – 17.

Feil, H., Feil, W.S., Chain, P. *et al.* (2005). Comparison of the complete genome sequences of *Pseudomonas syringae* pv. *syringae* B728a and pv. *tomato* DC3000. *Proc. Natl. Acad. Sci.* USA, 102: 11064 - 69.

Janse, J.D. (2006). *Phytobacteriology: Principles and Practice.* CABI Publishing, Wallingford, UK, 360 pp.

Ji, P., Campbell, H.L., Kloepper, J.W. *et al.* (2006). Integrated biological control of bacterial speck and spot of tomato under field conditions using foliar biological control agents and plant growth-promoting rhizobacteria. *Biol. Control* 36: 358 - 67.

Jones, J.B. (2001). Bacterial spot of tomato. In: O.C. Maloy and T.D. Murray (eds). *Encyclopedia of Plant Pathology*. Vol. 1, John Wiley and Sons, Inc. New York, pp. 83 - 84.

Jones, J.B., Bouzar, H., Stall, R.E. *et al.* (2000). Systematic analysis of xanthomonads (*Xanthomonas* spp.) associated with pepper and tomato lesions. *Int. J. Syst. Evol. Microbiol.* 50: 1211 - 19.

Jones, J.B., Lacy, G.H., Bouzar, H. *et al.* (2004). Reclassification of the xanthomonads associated with bacterial spot disease of tomato and pepper. *Syst. Appl. Microbiol.* 27: 755 - 62.

Jones, J.B., Stall, R.E. and Bouzar, H. (1998). Diversity among xanthomonads pathogenic on pepper and tomato. *Annu. Rev. Phytopathol.* 36: 41 - 58.

Leboeuf, J., Cuppels, D., Dick, J. *et al.* (2005). Bacterial diseases of tomato: Bacterial spot, bacterial speck, bacterial canker. Ministry of Agri., Food and Rural Affairs, Ontario, Canada [http://www.omafra.gov.on.ca/english/crops/facts/05-069.htm]

Louws, F.J., Wilson, M., Campbell, H.L. *et al.* (2001). Field control of bacterial spot and bacterial speck of tomato using a plant activator. *Plant Dis.* 85: 481 - 88.

Luo, L.X., Walters, C., Bolkan, H. *et al.* (2008). Quantification of viable cells of *Clavibacter michiganensis* subsp. *michiganensis* using a DNA binding dye and a real-time PCR assay. *Plant Pathol.* 57: 332 – 37.

Marcuzzo, L.L., Becker, W.F. and Fernandes, J.M.C. (2009). Some epidemiological aspects of bacterial spot (*Xanthomonas* spp.) of tomato in Cacador/SC, Brazil. *Summa Phytopathol.* 35: 132 – 35.

Martin, G.B., Brommonschenkel, S.H., Chunwongse, J. *et. al.* (1993). Map-based cloning of a protein kinase gene conferring disease resistance in tomato. *Science* 262: 1432 - 36.

Melotto, M., Underwood, W. and He, S.Y. (2008). Role of stomata in plant innate immunity and foliar bacterial diseases. *Annu. Rev. Phytopathol.* 46: 101 - 22.

Obradovic, A., Jones, J.B., Momol, M.T. *et al.* (2004). Management of tomato bacterial spot in the field by foliar application of bacteriophages and SAR inducers. *Plant Dis.* 88: 736 - 40.

Obradovic, A., Jones, J.B., Momol, M.T. *et al.* (2005). Integration of biological control agents and systemic acquired resistance inducers against bacterial spot on tomato. *Plant Dis.* 89: 712 – 16.

Patel, M.K., Kulkarni, Y.S. and Dhande, G.W. (1950). Bacterial spot of chillies. *Indian Phytopathol.* 3: 95 - 97.

Preston, G.M. (2000). *Pseudomonas syringae* pv. *tomato*: The right pathogen, of the right plant, at the right time. *Mol. Plant Pathol.* 1: 263 - 75.

Stall, R.E., Jones, J.B. and Minsavage, G.V. (2009). Durability of resistance in tomato and pepper to xanthomonads causing bacterial spot. *Annu. Rev. Phytopathol.* 47: 265 – 84.

Tudor-Nelson, S., Minsavage, G.V., Stall, R.E. *et al.* (2003) Bacteriocin-like substances from tomato race 3 strains of *Xanthomonas campestris* pv. *vesicatoria*. *Phytopathology* 93: 1415 – 21.

Vauterin, L., Hoste, B. Kersters, K. *et al.* (1995). Reclassification of *Xanthomonas*. *Int. J. Syst. Bacteriol.* 45: 472 - 89.

Yan, S., Liu, H., Mohr, T.J. *et al.* (2008). Role of recombination in the evolution of Model plant pathogen *Pseudomonas syringae* pv. *tomato* DC3000, a very atypical tomato strain. *Appl. Environ. Microbiol.* 74: 3171 – 81.

Zaccardelli, M., Spasiano, A., Bazzi, C. *et al.* (2005). Identification and *in planta* detection of *Pseudomonas syringae* pv. *tomato* using PCR amplification of *$hrpZ_{pst}$*. *Eur. J. Plant Pathol.* 111: 85 - 90.

Zitter, T.A. (1985). Bacterial diseases of tomato. Department of Plant Pathology, Cornell University, Ithaca, New York.

13.21. BACTERIAL BLIGHT OF PEA

The disease was first reported by Sackett (1916) from Colorado, USA. It is a serious disease of pea (*Pisum sativum*) in cool and wet weather. In 2004, under favourable weather conditions, the disease caused close to 90% plant mortality in some fields in Castilla y Leon region, the main pea growing area of Spain. Although the epiphytotics of the disease in

Australia are infrequent and may be 1 in 10 years, the field pea crop suffered a loss of 10% due to an epiphytotic in 1992. In Wimmera region of Australia, the disease incidence exceeding 30% has also been reported. The yield losses as high as 70% have been reported on an early sown pea cv. Sugar Snap from Horsham, Northwest Victoria, Australia. The disease also occurs in Armenia and India.

13.21.1. Symptoms

The symptoms appear on all the above ground plant parts. The initial symptoms are discrete, dark green, water-soaked spots on leaves, pods and stems. In cool and wet weather, the spots enlarge and coalesce to cover large areas. In warm and dry weather, lesion development is checked resulting in smaller lesions and water-soaking also becomes less obvious. With age, lesions become darker and finally necrotic (Fig. 13.21 a&b). Older lesions on pods are brown and papery. In severely infected pods, the seeds may be covered with bacterial slime and pathogen may penetrate the seeds through the funicle and micropyle. When the infection takes place at the seedling stage, the entire crop may be destroyed.

13.21.2. Causal organism

Pseudomonas syringae pv. *pisi* (Sackett 1916) Young *et al.* 1978

Syn. *Psm. pisi*

The below given description of the bacterium is taken mainly from Martin Sanz *et al.* (2005).

Bacterial cells are Gram-negative, nonsporing rods, measuring 0.7 × 2.0 - 3.0 μm and motile by one to five polar flagella. Most strains produce a yellowish-green diffusible pigment on King's medium B, which fluoresces blue under ultra violet light. Moderate growth occurs in 24 hr on agar slants, which is filiform, glistening and greyish-white. The tests for nitrate reduction, starch hydrolysis, indole production and H_2S production are negative. An alkaline reaction is produced in litmus milk and soft curd formed is cleared. Gelatin is liquefied but no lipolytic activity. Acid but no gas is produced from glucose, sucrose and galactose. The bacterium is chemoorganotroph and obligate aerobe. Optimum growth occurs at 26 - 28°C and at pH 6.5 - 7.5. The bacterium has the typical characteristics of a LOPAT group 1[a] green fluorescent pseudomonad. The bacterium can be detected from the infected seedlings with the help of ELISA. An avirulence gene designated as *avrPpiB* from race 3 has been cloned and sequenced.

There are seven races of the pathogen, which have been identified on the basis of their reactions to eight differentials. However, there is a possibility of new races not yet identified (Martin Sanz *et al.,* 2005). Race 2 is the most frequently distributed worldwide followed by races 6 and 4.

Gene-for-gene relationship between pea cultivars and pathogenic races is given in Table 13.21.1.

Table 13.21.1. Gene-for-gene relationship between pea cultivars and races of *Pseudomonas syringae* pv. *pisi*

Resistance gene (R)							Avirulence gene						
							1	2	3	4	5	6	7
							1	.	.	.	.	.	.
							.	2	.	.	2	.	2
							3	.	3	.	.	.	3
							4	.	.	4	4	.	4
							.	.	.	.	(5)	.	.
							(6)	.	.	.	(6)	.	.
Kelvendon Wonder	.	.	.	.	.	.	+	+	+	+	+	+	+
Early Onward	.	2	.	.	.	.	+	-	+	+	-	+	-
Belinda	.	.	3	.	.	.	-	+	-	+	+	+	-
Hurst's Greenshaft	.	.	.	4	.	(6)	-	+	+	-	-	+	-
Partridge	.	.	3	4	.	.	-	+	-	-	-	+	-
Sleaford Triumph	.	2	.	4	(5)	.	-	-	+	-	-	+	-
Vinco	1	2	3	.	(5)	.	-	-	-	+	-	+	-
Fortune	.	2	3	4	.	.	-	-	-	-	-	+	-

+ = susceptible response, - = hypersensitive reaction, the genes in brackets are partially confirmed, . = gene absent

Source = Bevan *et al.* (1995)

13.21.3. Disease cycle

Seed-borne inoculum is the main source of perpetuation of the bacterium. The bacterium is internally and externally seed-borne and can survive for at least three years. Infected plant debris, particularly on soil surface and volunteer plants are other sources of inoculum. The survival time of bacterium is reduced by burying the debris. In southern Australia it was reduced from 2 years to 11 months when the stubbles were buried 10 cm deep in the soil.

The penetration of the bacterium occurs through stomata but mainly through wounds caused by hail, frost, storms, sand blasting and cultural operations. Ice-nucleating activity of the epiphytic cells of the bacterium increases frost damage. The local spread of the bacterium occurs through wind-splashed rain and overhead-irrigation. The long distance spread is through the infected seed. Cool and wet or humid weather favours the disease while warm and dry weather reduces the severity of the disease.

Epiphytic phase is the main characteristic of the life cycle of the bacterium. The bacterium survives, multiplies and spreads epiphytically in the crop. The inoculation of race 2 on susceptible cultivars resulted in the reduction of its population during the first three days but then it increased 1000 times more than the initial count without producing any symptoms. When it was inoculated on resistant cultivars, the bacteria did not multiply and the population levels decreased slightly. It indicates that the development of the resident phase is race specific (Grondeau *et al.*, 1996).

Soil water content and level of seed-borne inoculum greatly influence the transmission of the bacterium from seed to the seedlings. High inoculum level and high soil moisture content increase the number of infected seedlings and the number of lesions per plant (Roberts *et al.*, 1996).

The identification of seven races of *Psm. syringae* pv. *pisi* based on a set of eight different pea cultivars allowed a gene-for-gene relationship to be proposed based on six matching pairs of resistance genes in the host and avirulence genes in the pathogen (Taylor *et al.*, 1989; Bevan *et al.*, 1995). On the basis of segregation among F_2 populations of crosses between differential cultivars sequentially inoculated with races of *Psm. syringae* pv. *pisi,* Bevan *et al.* (1995) got evidence for four and possibly six putative resistance (*R*)/avirulence (*A*) gene pairs. *R1*, *R2* and *R3* are dominant resistance alleles at single loci; *R4* is a dominant allele at a single locus, which exhibits variable expression possibly dependent on genetic background. There is evidence that *R3* and *R4* are at linked loci. Homology tests provided evidence that the alleles *R2*, *R3* and *R4* were present in more than one cultivar. Two other alleles, *R5* and *R6*, were postulated to explain the observed segregation in certain crosses. They further inferred that out of races 2, 3 and 4, each carries a different single avirulence gene, race 6 carries no apparent avirulence gene and race 7 carries at least *A2*, *A3* and *A4*. Race 1 carries *A1*, *A3*, *A4* and possibly *A6* and race 5 carries *A2*, *A4* and possibly *A5* and *A6*.

Stem inoculation technique is generally used to screen pea germplasm assuming that the responses would be same in other plant parts. However, Elvira-Recuenco (2000) showed that a resistant response in the stem was not always associated with the resistance in the leaf and/or pod. This was found for both race-specific and non race-specific resistance. Elvira-Recuenco *et al.* (2003) found that race-specific resistance genes conferred resistance in the stem in a consistent way but the leaf and pod response varied depending on the matching resistance and avirulence gene combination. *R2* generally conferred resistance in all the plant parts. *R3* and *R4* singly did not confer complete resistance in leaf and pod but *R3* in combination with *R2* or *R4* enhanced leaf and pod resistance. Race-nonspecific resistance conferred stem resistance to all the races, leaf and pod resistance to races 2, 5 and 7 and variable reactions in leaves and pods to races 1, 3, 4 and 6.

In addition to *Pisum sativum*, the bacterium infects *P. sativum* var. *arvense, Lathyrus odoratus, L. latifolius, Vicia bengalensis, Lablab niger* and *Vigna* sp. (Bradbury, 1998).

13.21.4. Management

1. The cheapest and most effective method is the use of resistant cultivars. Many cultivars contain race-specific resistance genes for all the known races except the race 6.
2. The next best approach is the use of the pathogen-free seed. The following measures help in producing the pathogen-free seed.
    a. Production of seed in arid areas.
    b. Avoidance of sprinkler irrigation.
    c. Field inspection from mid-flowering to late pod-fill stage. Removal of statutory certification controls in UK in 1993 greatly increased the incidence of seed-borne infection (Reeves *et al.*, 1996). Seed test is available to detect seed-borne infection in the seed. It can detect the seed-borne infection level of less than 0.05%.
3. The following measures reduce the disease incidence by checking the spread of the pathogen.
    a. Use clean farm equipment.
    b. Avoid sprinkler irrigation.
    c. Destroy infected plant debris and volunteer plants.
4. Crop rotation for two years.
5. In Australia, the disease is successfully managed by adopting an integrated disease management approach consisting of crop rotation, use of pathogen-free seed, avoidance of planting in areas prone to frequent frosts or extreme wet weather, crop hygiene and avoiding early sowing. Seed treatment and application of foliar bactericides have limited use in controlling the disease (Hollaway *et al.*, 2007).

REFERENCES

Bevan, J.R., Taylor, J.D., Crute, I.R. *et al.* (1995). Genetics of specific resistance in pea (*Pisum sativum*) cultivars to seven races of *Pseudomonas syringae* pv. *pisi. Plant Pathol.* 44: 98 – 108.

Bradbury, J.F. (1998). *Pseudomonas pisi. IMI Description of Fungi and Bacteria.* Sheet 126, CAB International, Wallingford, UK.

Elvira-Recuenco, M. (2000). Sustainable Control of Pea Bacterial Blight: Approaches for Durable Genetic Resistance and Biocontrol by Endophytic Bacteria. Ph.D. Thesis, Wageningin University, The Netherlands (cited by Martin Sanz *et al.,* 2005).

Elvira-Recuenco, M., Bevan, J.R. and Taylor, J.D. (2003). Differential responses to pea bacterial blight in stems, leaves and pods under glasshouse and field conditions. *Eur. J. Plant Pathol.* 109: 555 - 64.

Grondeau, C., Mabiala, A., Ait-Oumeziane, R. *et al.* (1996). Epiphytic life is the main characteristic of the life cycle of *Pseudomonas syringae* pv. *pisi,* pea bacterial blight agent. *Eur. J. Plant Pathol.* 102: 353 - 63.

Hollaway, G.J., Bretag, T.W. and Price, T.V. (2007). The epidemiology and management of bacterial blight (*Pseudomonas syringae* pv. *pisi*) of field pea (*Pisum sativum*) in Australia: A review. *Aus. J. Agric. Res.* 58: 1086 - 99.

Martin Sanz, A., Garcia Vaquero, C.A. and Caminero Saidana, C. (2005). Pea bacterial blight: A problem in Castilla y Leon. Grain Legumes No. 41: 8 - 9.

Reeves, J.C., Hutchins, J.D. and Simpkins, S.A. (1996). The incidence of races of *Pseudomonas syringae* pv. *pisi* in UK pea (*Pisum sativum*) seed stocks, 1987-1994. *Plant Varieties & Seeds* 9: 1 - 8.

Roberts, S.J., Ridout, M.S., Peach, L., *et al.* (1996). Transmission of pea bacterial blight (*Pseudomonas syringae* pv. *pisi*) from seed to seedling: Effects of inoculum dose, inoculation method, temperature and soil moisture. *J. Appl. Bacteriol.* 81: 65 - 72.

Sackett, W.G. (1916). A bacterial stem blight of field and garden peas. *Bull. Colo. Agr. Expt. Sta.* 218: 1 – 43.

Taylor, J.D., Bevan, J.R., Crute, I.R., *et al.* (1989). Genetic relationship between races of *Pseudomonas syringae* pv. *pisi* and cultivars of *Pisum sativum. Plant Pathol.* 38: 364 - 75.

13.22. BACTERIAL WILT OF CUCURBITS

Bacterial wilt of cucurbits affects many cucurbitaceous species. Cucumbers and muskmelons are more susceptible to the disease. Squash, pumpkins, gourds and wild cucurbits are also infected while the water-melon is resistant or rarely attacked. The disease is present throughout the United States and in Canada east of the Rocky Mountains. It also occurs in Europe, Japan and South Africa.

13.22.1. Symptoms

The earliest symptom of the disease is drooping of one or more leaves followed by drooping and wilting of all the leaves of the infected vine. Wilt progresses through lateral shoots involving the entire plant. Wilted leaves shrivel and dry up and infected stems turn pale and become soft. Before shriveling and drying, cucumber and muskmelon leaves may turn dull green while those of squash and pumpkin become dull greyish-green and later turn bronze. The entire plant may wilt within two weeks after infection. Later on, wilted plants dry up and die (Fig. 13.22a). Squash and pumpkin plants do not show rapid wilting but may be dwarfed producing excessive blossoms and branches.

A diagnostic feature of the disease is the white bacterial ooze, which comes out from the cut ends when infected stems are cut and pressed between the fingers. This sticky bacterial mass can be drawn into fine threads by touching the cut ends of the stem with a finger and gently pulling the finger away from the cut end (Fig. 13.22b). A positive test is the confirmation of the disease but a negative test does not indicate that the disease is absent. Another diagnostic test is to put a freshly cut infected stem in a glass of water and watch it. If bacterial wilt is present, the water will start turning milky at the cut end of the stem and ultimately making the whole water milky.

The infection of the squash fruits usually occurs through infected vines. However, occasionally it may be caused by beetles feeding on the blossoms and the rind of fruits. In the stored squash fruits, slimy rot progresses internally without any external symptoms. With the further progress of the internal rot, dark spots or blotches appear on the outer surface, which enlarge and coalesce. Later on, soft rotting microorganisms invade these fruits and completely destroy them (Agrios, 2004).

13.22.2. Causal organism

Erwinia tracheiphila (Smith 1895) Bergey *et al.* 1923.

The following description of the bacterium is taken primarily from Hauben *et al.* (1998). Cells are facultatively anaerobic rods, Gram-negative and motile by peritrichous flagella. Growth factors are required for growth and metabolism is fermentative. The bacterium grows very poorly on nutrient agar but moderately well on yeast extract glucose chalk agar and glucose nutrient agar. H_2S is produced but gelatin is not liquefied. The tests for oxidase, urease, starch hydrolase, arginine dihydrolase, caseine hydrolysis, phenylalanine dreaminase and nitrate reduction are also negative. Acid is produced from sucrose, galactose, fructose, rhamnose, glucose, glycerol and N-acetylglucosamine but not from arabinose, cellobiose, dulcitol, mannitol, mannose, melibiose, lactose, maltose, raffinose, salicin, sorbitol, trehalose and xylose.

The bacterium utilizes arabinose, maltose, mannose, ribose, salicin, sorbitol and sucrose but not adonitol, dulcitol, ethanol, dextrin, cellobiose, xylose, lactose, starch and xylitol as carbon sources. Strains grow on arginine, glucosamine, leucine, phenylalanine, citrulline and tyrosine but not on betaine, choline, hydroxypyroline, thymine, tryptamine and xanthin as nitrogen sources. The strains are sensitive to ampicillin, amoxycillin, carbenicillin, cephalexine and kanamycin but not to fusidinic acid, methicillin and sulfafurazol. The G + C content of the DNA of members of the species ranges from 50.0 – 52.0 mol%.

13.22.3. Disease cycle

As the bacterium dies within 1 - 2 months in the dead plants, its survival in the diseased plant debris is overruled. The bacterium overwinters (survives) in the digestive tract of hibernating adults of two cucumber beetles, namely, striped cucumber beetle (*Acalymma vittata*) and spotted cucumber beetle (*Diabrotica undecimpunctata*). The former is more common and in North Dakota it is the main spreading agent of the bacterium.

In the spring, infected beetles feed on the leaves causing deep wounds and deposit the bacteria in the wounds along with their faeces. The bacteria enter the xylem vessels exposed by wounds, multiply in the xylem vessels and spread to other parts of the plant. All the beetles do not carry the bacterium and not all the wounds acts as avenue of infection. Only deep feeding wounds in which xylem vessels are exposed serve avenues of infection. After entering the xylem vessels, bacteria multiply rapidly producing polysaccharides and gums. The accumulation of bacterial cells, polysaccharides and gums, along with formation of tyloses results in the blockade of vascular system hindering the normal flow of water and nutrients. The plugging of xylem vessels is so extensive that it allows less than one-fifth of normal water flow, hence making it the primary cause of wilting. The rate of multiplication and the period for which the bacterium multiplies differ in susceptible and resistant cucumber cultivars. The generation time in susceptible and resistant cucumbers is 5.4 and 11.2 hr, respectively. The bacterial population starts decreasing on 4th day in the resistant cucumber, whereas in the susceptible cucumber the peak population reaches on the sixth day (Watterson *et al.*, 1972).

The beetles spread the bacteria through their contaminated mouthparts when they feed on healthy plants after feeding on infected plants. Infected wild cucurbits can also serve as source of primary inoculum and from these plants the spread occurs through beetles. Cucumber beetles prefer wilted plants for feeding resulting in an increase in number of contaminated beetles. The spread of the pathogen is fast under normal moisture conditions because the activity of beetles decreases in prolonged cool and rainy weather. In extremely dry weather, the bacteria cannot penetrate exposed vascular system due to the lack of film of water. Each contaminated beetle can inoculate several healthy plants after one feeding on an infected plant.

13.22.4. Management

1. There is no cure of the cucurbit wilt. Therefore, control is aimed at prevention of the disease by controlling cucumber beetles.
2. Control of cucumber beetles with insecticide should be started as soon as the first beetle is seen in the crop and sprays continued till the end of season at weekly intervals. Use any of the fast killing insecticides like methoxychlor, diazinon and carbryl (Sevin). The carbryl should be sprayed late in the day to prevent honeybee mortality. Organic control options for cucumber beetles include foliar application of kaolin clay (Surround WP), insecticidal soap containing neem oil and rotenone 1% dust.
3. Cover the plants with row covers, cones or other barriers to prevent feeding on vines by cucumber beetles. Row covers should be placed over the plants early in the season and their edges sealed to prevent the entry of the beetles. Row covers should be removed to allow pollination by insects.
4. Place yellow sticky cups after every 20 feet along the rows to trap the beetles. Replace these traps when they become dusty and non-sticky.
5. Remove and destroy the diseased plants to prevent further spread of the disease.
6. In fall, remove and destroy the diseased plant debris to reduce sites for beetle hibernation and to minimize the survival of other cucurbit pathogens.
7. The cultivars resistant to bacterial wilt have little or no bitterness in their fruits. Cucurbitacin B and C are the compounds responsible for bitterness in the fruits and these compounds also attract cucumber beetles. Therefore, the cultivars having no bitterness in fruits are most likely to be least susceptible to bacterial wilt.
8. To prevent squash rot in storage, harvest the fruits only from healthy vines and store them in clean and fumigated stores.
9. Do not plant cucurbits near weedy woods or brush where the cucumber beetles are likely to overwinter.

REFERENCES

Agrios, G.N. (2004). *Plant Pathology.* 5th Ed. Elsevier Academic Press, San Diego, 922 pp.

Hauben, L., Moore, E.R.B., Vauterin, L. *et al.* (1998). Phylogenetic position of phytopathogens within the *Enterobacteriaceae. Syst. Appl. Microbiol.* 21: 384 - 97.

Watterson, J.C., Williams, P.H. and Durbin, R.D. (1972). Multiplication and movement of *Erwinia tracheiphila* in resistant and susceptible cucurbits. *Plant Dis. Reptr* 56: 949 - 53.

13.23. ANGULAR LEAF SPOT OF CUCURBITS

Angular leaf spot of cucurbits, also called angular leaf spot of cucumber, is a widespread disease. The disease affects many cucurbit species but it is more severe on cucumber, honeydew melon and Zucchini squash. The disease also occurs on muskmelon, watermelon, cantaloupe, vegetable marrow, squashes and pumpkins. Yield loss occurs mainly due to less number of fruits and less weight per fruit due to reduced assimilation of infected leaves. Losses in processing cucumbers may go as high as 50% or more in wet seasons when control measures are not undertaken. Angular leaf spot of cucumber was first reported from the United States in 1913 and the causal agent was identified in 1915 by Smith and Bryan. Since then, the disease has spread to many countries. Janse (2006) has given a comprehensive list of countries where the disease occurs and the list includes Algeria, Argentina, Australia, Brazil, Canada, China, Columbia, Egypt, widespread in Europe, Gabon, India, Iran, Israel, Japan, Jordan, Kazakhstan, Kenya, Korea, Laos, Mexico, New Zealand, Philippines, Singapore, South Africa, Zimbabwe, Tajikistan, Thailand, Turkey, USA, Uzbekistan and Venezuela. In temperate regions of the world, the disease can cause up to 37 and 40% reduction in fruit number and weight, respectively besides rendering some fruits as unmarketable culls. During wet conditions, the disease renders over 80% of leaves infected by the tenth day of the first infection. In India, the disease has been found in Kashmir valley where its incidence and intensity ranges between 23.35 to 74.45% and 10.50 to 26.02%, respectively (Bhat *et al.*, 2007)

13.23.1. Symptoms

In cucumber, the symptoms appear on leaves, petioles, stems and fruits. On cotyledons, soft, transparent, round to irregular lesions are formed. Initially small, round to irregular, water-soaked spots appear on cucumber leaves. The spots enlarge until their spread is limited by large veins; hence the spots become angular in shape. The spots on the upper leaf surface turn whitish-grey to brown while on the lower surface they appear gummy and shiny. In wet weather, bacterial ooze comes out from spots in the form of droplets on the lower leaf surface. In dry weather, bacterial ooze dries to form a thin, papery, white layer. As the infected areas die and dry, they shrink and tear away from the healthy leaf tissue forming large irregular holes in the leaves (Agrios, 2004). Young fully expanded leaves are more susceptible than older leaves (Fig. 13.23a). On stem and petioles, light yellow to light brown lesions appear, which exude abundant bacterial ooze. The severe infection of stem may lead to blightening and killing of the whole plant.

On squash leaves, brown spots of varying sizes appear. The spots are surrounded by yellow haloes and the leaf area next to yellow halo on the lower leaf surface appears water-soaked following wet weather.

On watermelon leaves, the symptoms start as small, circular, dark lesions surrounded by yellow haloes. As the spots enlarge, they become angular having white centres and involve considerable area of the leaf.

Water-soaked, circular, usually superficial spots, much smaller than those on leaves, appear on the cucumber fruits. After the death of infected tissue, the centres of the lesions become chalky white and may show star-like cracks. Soft rot bacteria and fungi enter through these cracks and cause rotting of the fruit accompanied by foul small (Fig. 13.23b). Fruit infection can also occur in the transit or storage. The infection of young fruits can cause deformity and bending of fruits. The infection of young fruits can also result into their dropping from the vines. During a wet weather, droplets of milky bacterial ooze frequently appear on the infection sites. The running down of the slime along the fruits causes secondary infection of the fruit (Anonymous, 2002).

13.23.2. Causal organism

Pseudomonas syringae pv. *lachrymans* (Smith & Bryan 1915) Young *et al.* 1978

Syn. *Psm. lachrymans*

Cells are short rods, nonsporing, Gram-negative, motile by more than one polar flagellum. Poly-β-hydroxybutyrate granules are not accumulated, arginine dihydroxylase and oxidase are not produced, nitrates are reduced, protocatechuate is cleaved and levan is formed on nutrient agar containing 5% sucrose. Gelatin is liquefied, arbutin and esculin are hydrolysed, β-glucosidase test is positive and pectate gel pitting occurs at pH 4.8. Growth occurs on inositol, mannitol, sorbitol, erythritol, L-tartrate, trigonelline and quinate but not on D-tartrate, L-lactate, anthranilate, homoserine and betaine (Goto, 1992).

Significant variability in virulence, aggressiveness and genome structure has been reported among the numerous strains of the bacterium that have been identified (Olczak-Woltman *et al.*, 2007). Earlier Krivchenko and Medvedeva (1985) reported 16 races of the bacterium and occurrence of resistance against these races in cucumber cultivars.

13.23.3. Disease cycle

The main sources of perpetuation of the bacterium are infected and/or infested seeds. In infected seeds, the bacterium resides in the seed coat and can survive for more than 2 years. Bhat (2009) reported that the bacterium survived up to 21 months in the infected seed but the inoculum load of the seeds and the number of diseased seedlings decreased with the storage period. The inoculum load and the number of diseased seedlings immediately after the crop harvest were 5.3×10^3 cfu/g of seed and 19.44%, respectively while the respective figures after 21 month's storage were 2.03×10^3 cfu/g of seed and 7.48%. The bacterium also survives in the infected plant debris. Bhat (2009) also reported that in Kashmir valley, the bacterium survived for 8 months in plant debris lying on the soil surface and buried in the soil but the survival was better on the soil surface in comparison to that buried in soil. The dispersal of the pathogen occurs through wind-splashed rain, sprinkler irrigation, insects (spotted cucumber beetle, southern corn rootworm), hands of workers and farm machinery. Wind-blown soil, infected plant residue and irrigation water contaminated with the bacterium also act as dispersing agents. The penetration of the bacterium occurs through stomata and wounds.

The optimum conditions for disease development are high relative humidity and a temperature range of 23 - 28°C. High relative humidity is more important than temperature for the outbreak of disease and a severe damage occurs when relative humidity is above 90%. Lesions formed when relative humidity is below 85%, remain small and do not develop into typical angular spots. Excessive nitrogen fertilizers favour the disease development.

Olczak-Woltman *et al.* (2009) studied the inheritance of resistance to the disease based on disease severity (number and size of necrotic and chlorotic lesions on the infected leaves) and the presence or absence of a chlorotic halo around the necrotic spots, and found that the disease severity appears to be controlled by multiple genes and the heritability of the resistance was estimated to be 53%. They further reported that the presence or absence of the chlorotic halo was to be governed by a single gene, with the presence of the halo (the susceptible phenotype) being a dominant character, and also identified a RAPD marker linked to the gene conferring the chlorotic halo. Genetic distance between the marker, OP-AO07, a polymorphic 420 bp amplicon in the DNA of the susceptible plants, and the locus encoding the chlorotic halo was estimated to be 13 cm.

13.23.4. Management

1. Use pathogen-free and certified seed. Never use the seed from an infected crop. In case of a non-certified seed, treat the seed in hot water (at 50°C for 20 min) containing calcium propionate @ 4.4

oz/gal of water or acidic cupric acetate @ 6.7 oz/gal of water. Hot water treatment alone is not effective but this treatment also does not completely eliminate the pathogen.

2. Follow crop rotation with nonhost crops for 3 years.
3. Destroy the infected plant residue by burning or by burying.
4. Limit the use of sprinkler irrigation
5. Avoid excessive use of nitrogenous fertilizers.
6. Avoid cultural operations and handling of plants when they are wet.
7. Minimize wounds to the plants, particularly to the cucumber fruits. Use smooth and flexible containers for storing fruits after harvest.
8. Control cucumber beetles and other insects by following a regular spray schedule.
9. Give protected sprays of cooper hydroxide (0.2%) or mancozeb (0.2%) at weekly intervals.
10. Root application of *Trichoderma asperellum* (T-203) caused accumulation of phytoalexins, induced systemic resistance against the bacterium in cucumber plants and caused 80% reduction in disease severity (Yedidia *et al.*, 2003).

REFERENCES

Agrios, G.N. (2004). *Plant Pathology*. 5th Ed. Elsevier Academic Press, San Diego, 922 pp.

Anonymous (2002). Angular leaf spot of cucurbits. Report on Plant Diseases No. 919. Department of Crop Sciences, Univ. of Illinois, Urbana, 3 pp.

Bhat, N.A. (2009). Survival of *Pseudomonas syringae* pv. *lachrymans*- incitant of angular leaf spot of cucumber under temperate conditions of Kashmir valley. *Indian Phytopathol.* 62: 429 - 34.

Bhat, N.A., Masoodi, S.D., Ahmad, M. *et al.* (2007). Occurrence and severity of angular leaf spot of cucumber in Kashmir valley. *Ann. Plant Prot. Sci.* 15: 410 – 13.

Goto, M. (1992). *Fundamentals of Bacterial Plant Pathology*. Academic Press, Inc., San Diego, 342 pp.

Janse, J.D. (2006). *Phytobacteriology: Principles and Practice.* CABI Publishing, Wallingford, UK, 360 pp.

Krivchenko, V.I. and Medvedeva, N.I. (1985). Intraspecific differentiation of the bacterial pathogen of cucumber *Pseudomonas lachrymans*. Sbornik Nauchnykh Trudov po Prikladnoi Botanike, Genetike i Selektsii 92: 92 – 97.

Olczak-Woltman, H., Bartoszewski, G., Madry, W. *et al.* (2009). Inheritance of resistance to angular leaf spot (*Pseudomonas syringae* pv. *lachrymans*) in cucumber and identification of molecular markers linked to resistance. *Plant Pathol.* 58: 145 - 51.

Olczak-Woltman, H., Masny, A., Bartoszewski, G. *et al.* (2007). Genetic diversity of *Pseudomonas syringae* pv. *lachrymans* strains isolated from cucumber leaves collected in Poland. *Plant Pathol.* 56: 373 - 82.

Yedidia, I., Shoresh, M., Kerem, Z. *et al.* (2003). Concomitant induction of systemic resistance to *Pseudomonas syringae* pv. *lachrymans* in cucumber by *Trichoderma asperellum* (T-203) and accumulation of phytoalexins. *Appl. Environ. Microbiol.* 69: 7343 – 53.

13.24. BACTERIAL BLIGHT OF COTTON

The disease is also known by other names such as **angular leaf spot and black arm of cotton**. It was first reported from Alabama, USA in 1891 by Atkinson, who named it as black rust. During 1900 - 1905, E.F. Smith proved Koch's postulates, described the causal organism and also established a relationship between black arm and angular leaf spot phases of the disease, which were earlier considered as different diseases. The disease is worldwide in occurrence and causes heavy losses in some countries. In India, the disease was first observed in 1918 in Madras and it remained a minor disease till the introduction of tetrapoloid cottons. It is argued that in India susceptible cultivars phased out under the selection pressure of the pathogen, and the pathogen and the cultivars reached a balance. On the basis of this co-evolution, Knight and Hutchinson (1950) concluded that the disease originated in India. However, Patel and Kulkarni (1948) are of the view that the disease was introduced in the middle of 19th century through exotic material.

In 1940s, a campaign was launched in India to replace the rain fed indigenous diploid cottons with high yielding but susceptible tetraploid, *Gossypium hirsutum* and to some extent *G. barbadense* cottons. After the launch of the campaign, the severe epiphytotics of the disease occurred during 1948 - 52 and the disease became established in 1960s. Irrigation and high nitrogen applied to achieve the full potential of the introduced cotton also increased the disease severity. In India, the disease occurs in all the cotton growing areas and the losses range from 5 - 25% (Verma and Singh, 1972).

13.24.1. Symptoms

There are four different types or phases of symptoms of the disease. It is not necessary that all the four types of symptoms should appear under all the situations. Their occurrence and predominance varies under different environmental conditions.

13.24.1.1. Seedling blight: Water-soaked circular spots appear on lower surface of cotyledons. The spots spread forming irregular patches causing drying and withering of the seedlings. Many plants die. This phase is more common in countries like USA where the temperature is low at the time of sowing of cotton.

13.24.1.2. Leaf spots: Water-soaked spots first appear on the lower surface of leaves. These spots enlarge, forming dark brown to black angular dead areas and become visible also on the upper surface of leaves. As the spread of these spots is bounded by veins and veinlets, they are angular in shape. The spots enlarge forming irregular large patches (Fig. 13.24a). The infection extends along the veins turning the veins black and infected veins give a **finger-like appearance**. This phase is called **'Vein blight'** or **'black vein'** (Fig. 13.24b). From veins, the infection reaches the petioles through midrib and infects the stem. Petiole infection leads to severe defoliation of plants. Under humid conditions, yellowish bacterial exudate is formed on leaf lesions.

13.24.1.3. Black arm: Linear, sunken, dark brown to black lesions appear on the stem and branches of the plants and eventually girdle them. In severely affected stems, there is deep cracking and gummosis, which results in breaking off the stem by wind. This phase is more common in Sudan and Uganda.

13.24.1.4. Boll rot: Water-soaked, circular spots appear on bolls, which later on, turn brown and become sunken (Fig. 13.24c). Severely infected bolls drop, but in case of mildly infected bolls, the infection progresses internally. Secondary organisms also enter the infected bolls and cause extensive boll rotting. The infected bolls may fail to open.

Sometimes, atypical symptoms of the disease appear, which consist of dull green, flaccid areas extending from periphery of cotyledons and leaves, and elongated, water-soaked, dark brown areas on leaf lamina. Recently, a new type of symptomatology named **"bacterial leaf blight"** have been described by Malavolta *et al.* (2008) from Sao Paulo state of Brazil. The new type of symptomatology is characterized by leaf blight, generally exhibiting chlorotic haloes and sometimes showing V-shaped lesions beginning at the border of the leaves. The bacterium causing the new kind of symptoms was identical to the type strain of *Xanthomonas axonopodis* pv. *malvacearum*.

Earlier it was considered that the bacterium invades only the parenchyma of the host. Since 1932, information has become available that the bacterium invades the vascular system also. From infected cotyledons, the bacterium moves through the vascular system of the petioles and reaches the stem. In the stem, it moves upward within the vascular system from where it breaks outward causing cortical rotting of young stem tissues and petioles, and infection of the leaf lamina. The movement of the bacterium in the susceptible cultivars indicates the probable systemic or at least spreading nature of the disease. The isolation of the bacterium from embryos and seeds also shows the systemic nature of the pathogen.

13.24.2. Causal organism

Xanthomonas citri subsp. *malvacearum* (hereafter referred to as *Xcm*) (Smith 1901) Schaad *et al.* 2007, subsp. nov., nom. rev.

Syn. *Xa. campestris* pv. *malvacearum, Xa. axonopodis* pv. *malvacearum*

Cells are rod-shaped measuring 0.9 × 1.2 μm; occur singly or in pairs but rarely in chains, monotrichous, nonsporing, not acid-fast, encapsulated, chemoorganotrophic, facultativly aerobic and Gram-negative. The metabolism is respiratory, never fermentative, catalase test is positive but oxidase negative. Asparagine is not utilized as a sole source of carbon and nitrogen. The bacterium may utilize methionine, glutamic acid and nicotinic acid in various combinations as minimal growth requirements. Triphenyl tetrzolium chloride @ 0.1 to 0.2% in nutrient agar inhibits its growth.

The bacterium liquefies gelatin, hydrolyses starch, digests casein but not fat. Produces H_2S but not indole, nitrate is not reduced; MR and VP tests are negative and can tolerate 3% sodium chloride. Sucrose, maltose, dextrose, levulose, xylose, glycerol, galactose, lactose and raffinose are utilized with acid production but without gas. Arabinose, rhamnose, salicin, mannitol and dulcitol are not utilized. The optimum temperature for growth is 31 - 32°C while the minimum and maximum are at 6 and 42° C, respectively. Plasmids varying in number and size are present in the cells (Das and Verma, 2003). Schaad *et al.* (2005), on the basis of DNA-DNA relatedness determined by sequencing the 16S – 23S intergenic spacer regions and AFLP analysis, reclassified the bacterium as *Xa. smithii* subsp. *smithii* sp. nov. comb. nov. nom. nov. but later amended the classification and renamed it as *Xa. citri* subsp. *malvacearum* (Schaad *et al.*, 2006, 2007). Schaad *et al.* (2006) also reported the following characteristics of *Xcm*. The colonies of *Xcm* appear on YDC and FS agars after 40 – 44 and 56 – 60 hr of incubation, respectively at 28 – 30°C. The bacterium produces an alkaline reaction in litmus milk without hydrolysis, liquefies gelatin but does not hydrolyze pectate. *Xcm* does not utilize melizitose. *Xcm* can be differentiated from *Xa. campestris* pv. *campestris* and most other *Xanthomonas* pathovars, subspecies and species by DNA – DNA reassociation assays, rep-PCR profiles, ITS sequencing, serological methods, SDS-PAGE patterns of membrane proteins and phenotypic characters (see references of Schaad *et al.*, 2005, 2006, 2007 under citrus canker).

Races: Pathogenic variation in the bacterium is well established and till now 32 races have been reported from different parts of the world (Table 13.24.1). In 1946, Knight was probably the first to suggest the existence of biological specialization in the bacterium, when Stoneville-20, a line of upland cotton resistant to the pathogen was severely infected in the United States. Bird and Hadley in 1958 identified two races of the pathogen on the basis of their reaction to 3 lines of upland cotton. Brinkerhoff (1963), on the basis of screening of isolates collected from all over the world on 8 differential lines including *G. hirsutum* and *G. barbadense*, identified 10 new races. Later on, Hunter *et al.* (1968) developed a set of differentials (8 lines of upland cotton having different bacterial blight resistance genes) and identified 15 races of *Xcm.* Total number of races was raised to 17 when 2 more races, i.e., I_1 and I_2 were described from India.

Table 13.24.1. Nomenclature of races of *Xanthomonas citri* subsp. *malvacearum* (adapted from Verma and Raj, 1992)

Part A: Primary differentials for race number

Disease reaction on differentials: susceptible = +, resistant = –							Race number	
I	II	III	IV	V	VI	VII	USA	India
+	–	–	–	–	–	–	13	1
+	+	–	–	–	–	–	1, 11	2
+	–	+	–	–	–	–	--	3
+	–	–	+	–	–	–	--	4
+	–	–	–	+	–	–	--	5
+	–	–	–	–	+	–	--	6
+	+	+	–	–	–	–	2, 12	7
+	+	–	+	–	–	–	--	8
+	+	–	–	+	–	–	3	9
+	+	–	–	–	+	–	4	10
+	–	+	+	–	–	–	--	11
+	–	+	–	+	–	–	--	12
+	–	+	–	–	+	–	9	13
+	–	–	+	+	–	–	--	14
+	–	–	+	–	+	–	--	15
+	–	–	–	+	+	–	19	16
+	+	+	+	–	–	–	--	17

+	+	+	–	+	–	–	15	18
+	+	+	–	–	+	–	17	19
+	+	–	+	+	–	–	6	20
+	+	–	+	–	+	–	--	21
+	+	–	–	+	+	–	5	22
+	–	+	+	+	–	–	--	23
+	–	+	+	–	+	–	--	24
+	–	+	–	+	+	–	--	25
+	–	–	+	+	+	–	--	26
+	+	+	+	+	–	–	8	27
+	+	+	+	–	+	–	16	28
+	+	+	–	+	+	–	14	29
+	+	–	+	+	+	–	7	30
+	–	+	+	+	+	–	--	31
+	+	+	+	+	+	–	10, 18	32

I, Acala-44 (no genes); II, Stoneville 2B-S9 (polygenes); III, Stoneville-20 (B_7 + polygenes); IV, Mebane B-1 (B_2 + polygenes); V, 1-10.B (B_{In} + polygenes); VI, 20-3 (B_N + polygenes); VII, 101-102.B (B_2B_3 + unknown).

Table 13.24.1. (continued)

Part B: Additional differentials for biotypes of race 32

Disease reaction on differentials: susceptible = +, resistant = –				
VII	VIII	IX	X	Biotype number
–	–	–	–	32.0
+	+	–	–	32.1 (probably race 18 of USA)
+	–	+	–	32.2
+	+	+	–	32.3 (Follin's isolate can be placed here)

VII, 101-102.B (B_2B_3 + unknown); VIII, Empire B_4 (B_4 + polygenes); IX, Tamcot CAMD-E ($B_2B_3B_7$ + polygenes); X, a line yet to be developed.

Note: Even when the line X is attacked by a new race of *Xa. citri* subsp. *malvacearum*, the old race number will remain the same and valid. Additional lines can always be added to designate new races/biotypes. Further, additional information can always be given; e.g., race 32.3 can also attack the gene combination of TxBonham ($B_2B_3B_4B_7B_{Sm}$).

Out of 2 sets of differentials developed by Nayudu (1964) and Hunter *et al.* (1968) for the identification of races of *Xcm*, the set developed by Hunter *et al.* (1968) is more suitable and is widely used. It contains isogenic lines with different bacterial blight resistance genes, which give stable reaction under different environmental conditions. Further, the race identification is based scientifically on the number of genes attacked, and on which, till now, all the isolates have given variable reactions.

Till now 26 races have been described from India and these are *Xcm* R-1, 2, 4, 5, 7 - 10, 13, 14, 16 - 25 and 27 - 32. Race 32 was most widely distributed and was present at more than 83% of the locations tested. Out of 133 isolates tested, 59.25% belonged to this race. It was the most virulent race as it attacked five major bacterial blight resistance genes, namely, B_7, B_4, B_2, B_{In} and B_N. The other prevalent races were *Xcm* R-27, 28 and 29, which attacked 3 genes. Less virulent races, *Xcm* R-1, 2, 9 and 10, were restricted in their distribution (Verma and Raj, 1992).

Xcm R-18H described from Pakistan is capable of attacking 5 genes. A race from western and central Africa has infected the line 101-102.B, i.e., the combined genes B_2 + B_3, as well as other combined gene resistance. A race called post-Barkat race (PBR) is capable of attacking all single and developed combinations of major *B* genes for resistance in Sudan. As some of the presently known races are highly virulent and capable of attacking combined gene resistance, there is an urgent need to develop another set of differentials to properly identify these races.

More than one race may be present on the same plant, on the same leaf or even in the same lesion. Thirteen races were isolated from 4 different lesions formed on the same leaf of Acala-44, which is the most susceptible line of differentials, having no blight resistance gene. New races possessing more virulent genotypes could be generated in nature from mixed infections of pathogenic races (in a susceptible cultivar) in a synergistic reaction (Verma *et al.*, 1979).

Delannoy *et al.* (2005) reviewed the genetic basis of cotton resistance to the pathogen, with reference to resistance genes, resistance gene analogs and bacterial avirulence genes along with the physiological mechanism involved in the hypersensitive reaction and suggested that this host-pathogen interaction represented the most complex resistance gene/*avr* gene system yet known.

13.24.3. Disease cycle

Infected seed is the main source of perpetuation of the bacterium and it may carry the pathogen both externally (as a slimy mass in the fuzz) as well as internally. The other sources of perpetuation are infected self-sown seeds and infected plant debris. The bacterium does not survive in the plant debris till the next season of crop when it is buried deep in soil and the soil remains wet. However, it can survive for 7 months in trash on the soil surface. The bacterium also does not survive when wheat crop is taken between two cotton crops. However, the plant debris serves as an important source of infection where cotton is grown as a continuous crop.

It seems that under northern Indian conditions, infected seed including the self-sown seed is an important source of primary inoculum.

The bacterium enters the host mainly through stomata. Insect punctures also serve as entry points for boll infection. The disease is favoured by high humidity and moderate temperature. Primary infection occurs at relatively low temperature of 30°C at which cotton germinates and grows but secondary infection occurs at relatively higher temperature, i.e., 35°C. The presence of moisture is essential for secondary spread of the pathogen during the first 48 hr of infection. Relative humidity of 85% or above is essential for abundant infection. Dry and hot weather retard the normal development of the disease.

Wind-blown rain is the main disseminating agent of the bacterium. Surface irrigation and overhead-irrigation also spread the bacterium. In some areas, wind-blown detached leaves and wind-blown dust may serve as means of dissemination. The transmission of the bacterium by two cotton pests, namely, *Dysdercus koenigii* and *Earias* spp. is known but their role in dissemination and disease development is not clear. The bacterium survives in the excreta of larvae of *Earias* spp. for 6 days.

Jatropha curcas, Lochnera pusilla, Thurbaria thespesioides, Eriodendron anfractosum and *Thespesis* spp. act as collateral host of the pathogen but their role in the disease epidemiology is not clearly defined.

13.24.4. Management

The eradication of seed-borne inoculum and destruction of infected plant debris and self-sown seeds are of utmost importance for the management of the disease.

1. Treat the acid or machine delinted seeds with carboxin, oxycarboxin or Agrimycin - 100 @ of 2 g/kg of seed.
2. Hot water treatment of seed at 56°C for 10 min is also effective.
3. Destroy the diseased plant debris and self-sown seedlings. Plant debris should be buried deep in the soil followed by flooding of the field for 24 - 48 hr.
4. Protective sprays of phylloplane bacteria, namely, two species of *Flavobacterium* and one species each of *Pseudomonas* and *Aeromonas* are effective. This schedule should be included in the integrated management of the disease.
5. To check the secondary spread of the disease, spray the crop with Blitox (500 gm) + Agrimycin – 100 (20 gm) or Streptocycline (3 gm) per acre per spray. Give the first spray on the 4- to 6-week-old crop and subsequent sprays at 10 - 15 days intervals. Three or more sprays should be given depending upon the severity of

the disease. These chemicals possessed some curative effect als (Verma and Raj, 1992).

6. Use of resistant varieties.

REFERENCES

Brinkerhoff, L.A. (1963). Variability of *Xanthomonas malvacearum*: The cotto bacterial blight pathogen. Oklahoma State Univ. Tech. Bull., 98, 96 pp.

Das, I.K. and Verma, J.P. (2003). Plasmids in bacteria associated with bacteri leaf blight of cotton. *Indian Phytopathol.* 56: 180 - 82

Delannoy, E., Lyon, B.R., Marmey, P. *et al.* (2005). Resistance of cotton toward *Xanthomonas campestris* pv. *malvacearum. Annu. Rev. Phytopathol.* 43: 63 82.

Hunter, R.E., Brinkerhoff, L.A. and Bird, L.S. (1968). The development of a set upland cotton lines for differentiating races of *Xanthomonas malvacearu Phytopathology* 58: 830 – 32.

Knight, R.L. and Hutchinson, J.B. (1950). The evolution of black arm resistanc in cotton. *J. Genet.* 50: 36 - 58.

Malavolta, V.A. Jr., Destefano, S.A.L., Beriam, L.O.S. *et al.* (2008). Cotton le blight, a new symptomatology caused by *Xanthomonas axonopodis* p *malvacearum. Summa Phytopathol.* 34: 168 – 71.

Nayudu, M.V. (1964). Variation in *Xanthomonas malvacearum*. *Indian Cot Grow. Rev.* 18: 350 – 55.

Patel, M.K. and Kulkarni, Y.S. (1948). *Xanthomonas malvacearum* (Erw. Smith) Dowson on exotic cottons in India. *Curr. Sci.* 17: 243 - 44.

Verma, J.P., Chowdhury, H.D. and Singh, R.P. (1979). Interactions betwee different races of *Xanthomonas malvacearum* in leaves of *Gossypiu hirsutum. Z. Pflankrankh. Pflschutz.* 86: 460 -

Verma, J.P. and Raj, S. (1992). Blight and wilt of cotton. In: A.N. Mukhopadhya J. Kumar, H.S. Chaube and U.S. Singh (eds). *Plant Diseases of Internation Importance.* Vol. IV, *Diseases of Sugar, Forests and Plantation Crop* Prentice Hall, Englewood Cliffs, New Jersey, pp. 138 - 60.

Verma, J.P. and Singh, R.P. (1972). Bacterial blight of Cotton. In: P. N. Pat (ed.). *Plant Bacteriology.* Vol. I, *Bacterial Diseases of Plants in India.* India Agricultural Research Institute, New Delhi, pp. 159 - 68.

BACTERIAL DISEASES OF BEAN

Common bean or common dry bean (*Phaseolus vulgaris* L.) suffer from five important bacterial diseases, namely, common blight, fuscou blight, halo blight, bacterial brown spot and bacterial wilt. The first fou diseases are described below.

13.25. COMMON AND FUSCOUS BLIGHTS OF BEAN

Common bacterial blight or more frequently referred to, as common blight is an important disease of common bean. The disease causes significant yield losses in temperate and tropical areas and is one of the major constraints in dry bean production. Common blight was first reported in 1892 by Beach from New York and by Halsted from New Jersey. In 1897, E.F. Smith described the causal bacterium and named it as *Bacillus phaseoli*. Since its first report, the disease has been reported from many states of the USA and Canada, France, Hungary, Japan, Australia, Bulgaria, Madagascar, Norway, the Philippines, Spain, Turkey, New Zealand, Paraguay, Uruguay, Venezuela, Guatemala, Nicaragua, Honduras, Mexico, Argentina, Brazil, Colombia, Uganda, Burundi, Zaire, Zambia, Rwanda and South Africa. It is not established in a number of bean growing countries of the European and Mediterranean Plant Protection Organisation (EPPO) region. Common blight pathogen is an A2 quarantine pest for EPPO. In India, Patel and Jindal first observed the disease in 1971 at Pune, Maharashtra, where it has been occurring for a long time. Although the disease is named common blight, it is not necessarily the most common one in any given situation.

The disease occurs in temperate, subtropical and tropical regions but the maximum losses are caused in tropical areas because the most severe outbreaks of the disease occur in warm humid weather. The losses result from the reduction in photosynthetic area of plants and loss in quality and number of pods and seeds. In 1918, 75% of the fields in New York state of USA were affected with common blight and the disease caused heavy losses. In the subsequent years, losses ranged from 20 - 50%. In 1953, the disease was widespread in Nebraska (USA) and it caused an estimated loss of over 1 million US dollars. In 1976 in USA, common blight was economically the most important disease among the bacterial diseases of beans and it caused an estimated loss of 4 million US dollars. The losses due to this disease in field bean crop in Ontario (Canada) were over 1252 tons in 1970 and 218 tons in 1972. A model for the assessment of yield losses from bacterial blight has been developed in Ontario, using aerial infrared photographic surveys to assess disease levels. In 1967, yield losses up to 38% in dry bean in Ontario (Canada), and from 10 - 20% in navy bean in Michigan (USA) during 1976 were reported (Gilbertson and Maxwell, 1992). Yield losses ranging from 17 - 22% were also reported from Colombia. In addition to loss in yield, the disease causes reduction in quality of seed through discolouration of infected seeds.

In 1930, Burkholder described two bacterial diseases of bean, namely, fuscous blight and bacterial brown spot. The causal bacterium of the fuscous blight, isolated by him from bean plants collected from Switzerland, was very close to the pathogen of common blight and it also produced symptoms identical to common blight with some minor differences. Fuscous blight was first found in Switzerland in 1924. Subsequently it has been reported from Australia, Yugoslavia, Canada, New Zealand, South America, Germany, Hungary and many other countries.

13.25.1. Symptoms

The Symptoms of common and fuscous blights are similar and generally cannot be distinguished under field conditions. These diseases can be identified by the isolation and identification of their causal agents or using molecular tools.

Both the diseases are primarily foliar diseases but symptoms also appear on other above ground plant parts. The most characteristic and diagnostic symptoms appear on trifoliate leaves. Small, water-soaked lesions may develop on cotyledons and/or primary leaves and stems of seedlings raised from infected seeds.

On leaves, the symptoms start as small, water-soaked spots on the lower surface. The spots develop on the leaf margin or between the veins and there can be numerous spots on a leaf or leaflet. As these spots enlarge, their centers become necrotic and dry. The spots are surrounded by narrow bands of lemon yellow tissue, which are diagnostic features of the disease. The yellow halo may be small or absent if the temperature goes above 27°C. As the disease progresses, the spots enlarge and on susceptible varieties continue to enlarge till they merge with each other forming large dead areas of various shapes. Dark discoloured lesions also develop along the main vein, indicating systemic infection. The dry centres of the spots may tear and fall out giving the leaves a tattered and shredded appearance. Badly infected leaves fall and severely infected plants defoliate (Fig. 13.25a).

Water-soaked, occasionally sunken lesions appear on the stem, which enlarge longitudinally and turn brown. The lesions frequently break open at the surface exuding bacterial ooze. The occurrence of these lesions is more near the vicinity of the first or cotyledonary node. The lesions may girdle the stem at the cotyledonary node and this phase is known as **joint -rot**. Generally, the plants break off at this point due to the weight of the plant or action of the wind. On petioles and pulvinus, initially the lesions are water-soaked but may turn brown later on (Patel and Jindal, 1972).

On pods, small, circular, water-soaked spots appear which enlarge to form sunken, brick-red lesions. Pod lesions ultimately turn brown and vary in size depending upon the age of the pod when infection occurs. In humid weather, a slimy bacterial exudate may be formed on pod lesions, which eventually dries giving the lesions a glazed appearance (Fig. 13.25b). In severely infected pods, seeds may be shriveled, discoloured or reduced in size. The butter yellow discolouration of seeds is more prominent on white-seeded varieties in comparison to seeds of coloured varieties.

Though the most symptoms of common and fuscous blights are similar, the intensity of seed discolouration and brown discolouration at the stem nodes is more in the latter disease. In fuscous blight, the lesions on the stem are also more inclined to split the stem (Walker, 1969).

Common blight is often found in association with fuscous blight. Since both the diseases occur simultaneously in many cases, it becomes difficult to assess the amount of yield loss caused by them individually.

13.25.2. Causal organisms

Common blight: *Xanthomonas phaseoli* (*Xp*) (ex Smith 1897) Gabriel *et al.* 1989, sp. nov., nom. rev.

Syn. *Xa. campestris* pv. *phaseoli, Xa. axonopodis* pv. *phaseoli*

Fuscous blight: *Xa. fuscans* subsp. *fuscans* (*Xff*) Schaad *et al.* 2007, subsp. nov.

Syn. *Xa. campestris* pv. *phaseoli* var. *fuscans, Xa. axonopodis* pv. *phaseoli* var. *fuscans*

According to Young *et al.* (2004), *Xa. axonopodis* pv. *phaseoli* var. *fuscans* is not a valid name because it does not confirm either to the code (Rule 28 b) or the standards (Standard 18.1).

On sucrose peptone agar (SPA) and yeast extract glucose calcium carbonate agar (YGCA), the colonies of *Xp* are yellow, convex, slimy and mucoid. On media containing tyrosine or on SPA, *Xff* produces a brown diffusible pigment, melanin. *Xff* strains also produce a dark internal pigment and thus, can be distinguished from *Xp* on MXP medium. Both the bacteria are strict aerobes, non-sporing and Gram-negative rods. Xanthomonadins, having absorption peaks between 443 and 446 nm, are produced. The metabolism of glucose is strictly oxidative. Maximum temperature for growth is 38°C. The other properties include hydrolysis of starch, casein and gelatin, lipolysis of Tween 80, proteolysis of milk, lack of pectinase activity in culture and inability to reduce nitrates, to produce lysine dicarboxylase and to induce hypersensitive reaction in tobacco. *Xp* and *Xff* are sufficiently distinct genetically to be classified into different groups based on 16S – 23S intergenic spacer regions, restriction fragment length polymorphism, analysis of genomic and plasmid DNA, pulsed-field gel electrophoresis, DNA-DNA hybridization and amplified DNA polymorphism.

Lopes *et al.* (2008) developed a semi-selective medium to delect *Xa. axonopodis* pv. *phaseoli* from bean seeds. The recovery frequency of bacterial colonies in relation to nutrient agar varied between 30 and 112%, depending upon the isolate.

Pathogenic variation exists in both the pathogens but evidence for existence of races, identifiable on common bean differential genotypes, has yet to be found. *Xff* strains are generally more virulent than *Xp* strains and strains of the same pathogen from different locations also vary in virulence. Mutlu *et al.* (2008), on the basis of evaluation of common bean genotypes against 69 strains of *Xp* and 15 strains of *Xff*, found that *Xff* strains were more virulent than *Xp* strains having the same origin and pathogenic variation was more within *Xp* strains than within *Xff* strains. Opio and Male–Kigiwa (1995) reported eight races of *Xp* based on their reaction in tepary bean (*P. acutifolius*).

A DNA probe has been developed from plasmid DNA of *Xp* for the rapid detection and identification of *Xp* and *Xff* from bean plants and host debris. A 3.4 kilo base EcoRI fragment of *Xp* plasmid DNA, having a sequence that is repeated throughout the genomes of both the pathogens, was more specific than total plasmid DNA for this purpose (Gilbertson *et al.*, 1989).

13.25.3. Disease cycle

The most important source of perpetuation of the bacteria is infected or contaminated seed. Common blight bacterium can survive in the infected seed for 30 years. Internally infected and externally contaminated but apparently healthy seeds are a more important inoculum source than severely infected seeds. As low as 0.5% infected seeds are sufficient to cause epiphytotics. The minimum number of bacteria per seed required to colonize seedlings is 10^{3-4}. Infected plant residue is another source of survival, particularly when it is not buried in soil and remains undecomposed and also in areas where more than one crop of beans is taken from the same plot in a year. Resident bacteria on weed hosts and nonhost plants also act as a source of survival.

The penetration of bacteria into the host occurs through stomata, hydathodes and wounds caused by insects, hailstorms and cultural operations. The dissemination of the bacteria occurs through the infected seed, wind-splashed rain, sprinkler irrigation and implements. Seed is an important mean for short and long distance spread of the bacteria. The occurrence of devastating blight epiphytotics after wind-driven hailstorms is the result of extensive spread of the bacteria by rain splashes and their entry through wounds caused by hailstorms. Overhead-irrigation is another effective agent for the secondary spread of the bacteria. *Xp* can also be disseminated on the bodies of insects, such as *Diaprepes abbreviatus* and *Cerotoma ruficornis*, which may also provide avenues for the pathogen entry by creating wounds while feeding (Kaiser and Vakili, 1978). White flies (*Bemisia tabaci*), grasshoppers (*Melanoplus* sp.), Mexican bean beetles (*Epilachna varivestris*) and leafminers have also been reported to transmit the bacterium.

When infected seeds germinate, the bacteria multiply and proceed as vascular pathogen throughout the large xylem vessels causing blocking of the vascular system and eventually wilting of the plants. In most cases, the vascular infection results from seed-borne inoculum but it can also occur from penetration of bacteria through wounds. Stem cankers and leaf lesions may result from the systemic infection. Alternatively, the bacteria may be splashed from the infected plant debris onto the seedlings where they cause lesions on cotyledons and primary and trifoliate leaves under favourable conditions. Secondary spread of the bacteria from these lesions occurs through disseminating agents.

After entry into the host tissue through stomata and wounds, the bacteria invade the intercellular spaces of the parenchyma tissue, dissolving the middle lamella slightly in advance. The affected cells collapse and are penetrated by the bacteria leading to their disintegration and formation of lysigenous cavities. The bacteria then invade the xylem vessels in the stem and advance systemically in the larger xylem elements. The bacteria multiply in the xylem cells and when large cell masses are produced, they either rupture or dissolve the cell walls and reenter the parenchyma tissues. From the parenchyma tissue, the cells come out on the stem surface through stomata in the form of ooze and after dissemination, enter the same stem and other plant parts. Hence, the bacteria invade both parenchyma and vascular tissue.

The infection or contamination of the seed occurs in different ways. Seed infection occurs via vascular system where bacteria move through the pod pedicel into the funiculus and also through the raphe into the seed coat. Local pod infection can also result in contamination and/or infection of seed. The bacteria move through the pod wall and contaminate the surface of developing seeds or penetrate beneath the seed coat.

The disease development is favoured by warm humid weather. The optimum temperature for disease development is 24 to 32°C. Under favourable weather conditions, bacteria multiply on the host surface and reach a threshold level before the symptoms develop. This threshold population is around 5×10^6 bacteria per 20 cm^2 of leaf tissue. This requirement of a minimum bacterial population for symptom development is probably the reason why the symptoms are usually observed during the reproductive stage of plant development. When the weather conditions are unfavourable for disease development, the bacteria colonize the plants without causing symptoms and become widespread on the plants.

Age of host plant, air temperature and host nutrition influence common blight development. The plant is susceptible at all growth stages but young leaves are more susceptible than old leaves. Polysaccharide produced by *Xp* acts as a pathogenicity factor and disease severity is directly correlated with its amount produced by the pathogen. It possibly interferes with movement of water and/or nutrients leading to production of lesions and wilting. Silva *et al.* (2009) observed that bacterization of bean seeds with a fluorescent *Pseudomonas* isolate (DFs 842) for 5 hr at 10°C and challenge inoculation with *Xa. axonopodis* pv. *phaseoli* caused metabolic alterations in bean plants due to the increase in total soluble protein content and polyphenol oxidase activity. The increase in polyphenol oxidase and decrease in peroxidase activity induced resistance in bean plants against *Xa. axonopodis* pv. *phaseoli*.

The epidemiology of fuscous blight is similar to that of common blight. Very recently, Darsonval *et al.* (2009a) have shown the involvement of type III secretion system of *Xff* in its multiplication in the phyllosphere and systemic colonization of bean plants, leading to its transmission to seeds. *hrp* regulatory genes of the bacterium play a major role in host colonization processes. Strains with mutations in *hrp* regulatory genes, *hrpG* and *hrpX*, were impaired in their phyllospheric growth. Unlike the wild-type strain, strains with mutations in *hrp* genes were also not transmitted to seeds by vascular pathway. However, transmission of the bacterium to seeds by floral structures occurred with both wild and mutant strains. Darsonval *et al.* (2009) also evaluated the role of bacterial attachment and biofilm formation in host colonization processes for *Xff* and found that the mutations in five adhesion genes, namely, *pilA*, *fhab*, *xadA1*, *xadA2* and *yapH*, identified in *Xff*, altered the ability of all the mutants to adhere to polypropylene or seed. *PilA* was involved in adhesin and transmission to seed, and mutation of *PilA* led to lower pathogenicity on bean. *YapH* was required for adhesion to seed, leaves and abiotic surfaces but not for in planta transmission to seed or aggressiveness on leaves. However, none of the known adhesins was required for transmission to seed through floral structures. All mutants tested, except in *yapH*, were altered in their vascular transmission to seed.

In addition to common bean, the host range of *Xp* includes field bean, snap bean, lima bean, scarlet runner bean, hyacinth bean, soybean, *mung* bean, *urad* bean, *moth* bean, tepary bean, white flowering lupine and fenugreek (Partridge, 2003). *Xff* infects field bean, snap bean, civet bean and scarlet runner bean.

13.25.4. Management

The management of common and fuscous blights is difficult due to the lack of resistant cultivars, effective chemicals and seed-borne nature of the pathogens. However, the satisfactory control of both the diseases can be achieved by adopting the following integrated control measures.

1. Use of certified pathogen-free seed for sowing. Treat the seed with streptomycin/Streptocycline because the seed dust may contain the bacteria. Hot water treatment at 52°C for 20 min or hot air treatment (23 – 32 hr in 60°C dry air at 45 – 55% RH) has been reported effective. The latter treatment does not appear to affect seed viability (Gondreau and Samson, 1994).
2. Use of resistant varieties is the best method for the management of these diseases. However, the breeding for disease resistance is complicated as different genes control resistance in pods and leaves. Great Northern varieties, namely, Beryl, Harris, Ivory and Marquis are resistant to common blight. In comparison to common bean, scarlet runner bean has a higher level of resistance while the tepary bean has the highest level of resistance to common blight.
3. Destruction of infected plant debris by deep ploughing and by burying it in the soil.
4. Follow crop rotation for 3 years with nonhost crops. In Romania, rotation with maize reduced 85% disease incidence.
5. Escape the disease by choosing the suitable planting date.
6. Destruction of volunteer bean plants early in the following year and of weed hosts.
7. Avoid or limit the use of sprinkler irrigation.
8. Do not use the run off contaminated water for irrigation.
9. Avoid cultural operations when plants are wet.
10. Preventive sprays of copper fungicides (Kocide etc.) and antibiotics are helpful in reducing the disease severity as they reduce the epiphytic population of the bacteria. These sprays should be started early in the season and repeated at 7 - 10 days intervals. The use of antibiotics is not encouraged as it can result in the development of antibiotic resistant strains of the bacterium.

Application of Bion (0.05%) and BioZell-2000B (50% etheric oil of thyme, 20% oil of maize, 20% oil of anise, 10% oil of sesame) caused 68 and 50% reduction in common bean blight incidence along with 50 and 45% reduction in bacterial population, respectively (Abo-Elyouser, 2006).

13.26. HALO BLIGHT OF BEAN

Halo blight of bean was first described by Burkholder (1926) from the United States. It has many similarities with common blight of bean and causes serious economic losses under favourable weather conditions. The disease is worldwide in distribution, wherever beans are grown under cool and wet climatic conditions. The occurrence of the disease has not been reported from India. Under favourable weather conditions, epiphytotics resulting from a few infected seeds may cause substantial yield losses.

13.26.1. Symptoms

Most symptoms of halo blight are similar to those of common blight. The symptoms of halo blight appear on all the above ground plant parts. Water-soaked, round to irregular lesions are produced on cotyledons of seedlings raised from infected and/or infested seeds. Yellowish-brown to reddish-brown, water-soaked lesions appear on primary leaves. Severely infected plants show stunting and die.

Small (about 1.5 mm in diam), angular, greasy, water-soaked spots first appear on under surface of trifoliate leaves. As these spots enlarge and turn reddish-brown, characteristic light greenish-yellow haloes, 10 mm or more in diam, often but not always, appear around the spots, giving the leaves a yellowish appearance. The haloes, more prominent on young expanding leaves, appear due to the production of a toxin, phaseolotoxin by the bacterium and are a diagnostic symptom of the disease (Fig. 13.26a). At temperature above 27°C, haloes formed are very small or may be absent. In 2004, Oguiza and associates reported the bacterial strains that do not have the genetic information for toxin production. In severe infections, the infection may become systemic causing yellowing and death of new foliage. On stem, greasy, reddish-brown, elongated streaks appear. On pods, symptoms begin as small, water-soaked spots or streaks along the pod sutures. The spots, later on, turn reddish-brown and become sunken in the centre (Fig. 13.26b). Under humid conditions, a light cream- or silver-coloured bacterial exudate may appear on the lesions. Seeds in infected pods are usually symptomless. However, in severely infected pods, seeds may be shriveled, discoloured or reduced in size.

Pod symptoms of halo and common blight are similar. Therefore, the identification of these diseases should be based on leaf symptoms and identification of the causal bacteria. Leaf symptoms of halo blight and bacterial brown spot may be confused due to the variation in margin/width of halo influenced by the temperature. In case of brown spot, the spots are brown, necrotic having a narrow yellow halo.

13.26.2. Causal organism

Pseudomonas syringae pv. *phaseolicola* (Burkholder 1926) Young *et al.* 1978

Syn. *Psm. phaseolicola*, *Psm. savastanoi* pv. *phaseolicola*

The below given description of the bacterium is taken mainly from Burkholder (1926), who classified it as *Phytomonas medicaginis* var. *phaseolicola* nov. var.

Cells are rods with rounded ends, occur singly or in pairs and measure 0.5 - 1.25 × 1.5 - 3.0 μm. Gram-negative, nonsporing, not acid-fast, strict aerobes and motile with a polar flagellum. Colonies on nutrient agar are white to colourless or at times creamy, concentrically ringed with edges undulate. The growth of the colonies is slow to moderate. The bacterium does not reduce nitrate, does not liquefy gelatin, does not produce H_2S and indole, and does not digest starch. Litmus milk is turned alkaline without curd formation and peptonization. Pectate gel pitting occurs at pH 4.0. Weak acid production occurs in dextrose and saccharose but no acid from lactose and maltose and also no gas production from these sugars. Growth occurs on quinate and trigonelline but not on mannitol, inositol, sorbitol, erythritol, L-tartrate, D-tartrate, L-lactate, anthranilate, betaine and homoserine. Its ability to use rhamnose and raffinose as a sole source of carbon and sodium nitrate as a sole source of nitrogen varies with different strains. Maximum growth and production of phaseolotoxin occurs at 21 - 22°C. The thermal death point is around 49°C.

Taylor *et al.* (1996) reported nine races of the bacterium and among them, race 6 being the predominant.

13.26.3. Disease cycle

Infected and/or infested seed is the most important source of perpetuation of the bacterium. The bacterium can survive for more than 4 years in infected seed. One contaminated seed in 16,000 is sufficient to cause severe epiphytotic under favourable weather conditions (Dillard and Legard, 1991). Infection of bean seeds occurs through lesions formed on pods and bacteria penetrate the seed coats without causing any visible symptoms. Infestation of seeds occurs during threshing resulting from adherence of dust of diseased tissue to seed surface and to wounds. As a source of primary inoculum, both infected and infested seeds are equally important (Goto, 1992). Infected plant debris also acts as a source of survival of the bacterium. The survival in crop residue is maximum when it is on the soil surface. The dissemination of the bacterium occurs through wind-splashed rain, overhead-irrigation, windstorms and farm machinery. Severe outbreaks of disease often occur after heavy rainstorms.

The disease is favoured by cool to moderate temperature (16 - 23°C) and humid moist conditions. Temperatures above 27°C inhibit the development of yellow haloes and systemic chlorosis. When weather becomes warmer and drier, plants recover from the disease and new growth is healthy.

In addition to common bean, the host range of the bacterium includes field and snap beans, especially cranberry, red kidney and yellow-eyed beans, lima bean and scarlet runner bean. Artificially, the bacterium infects tepary bean, adzuki bean, *mung* bean, soybean, *Macroptilium atropurpureum, Phaseolus polyanthus, P. polystachyus* and *Pueraria lobata.*

13.26.4. Management

Management practices for the disease are the same as those for common and fuscous blights of bean.

13.27. BACTERIAL BROWN SPOT OF BEAN

Burkholder in 1930 described two bacterial diseases of bean and one of these was bacterial brown spot. The disease is favoured by cool to moderate temperature and high moisture. Severe outbreaks of the disease occur when temperature is below 27°C and humidity above 95% prevails for 24 hr or longer periods.

13.27.1. Symptoms

The disease is more severe on young foliage and symptoms are rarely seen on older leaves. Initially, small, water-soaked spots appear on the lower surface of trifoliate leaves. A narrow, greenish-yellow diffuse border about 1.5 mm in diam may appear around some of the lesions. The spots enlarge ranging from 3 - 8 mm in diam and on maturity they give a typical "**brown spot**" appearance (Schwartz, 2007). Dead centres of the spots may fall out producing shot-hole symptoms (Fig. 13.27a). On the stem, systemic, sunken, longitudinal lesions appear. Sunken, brown spots appear on the pods. If the infection occurs during early development of pods, infected pods may be twisted and kinked at the site of infection. Pale brown haloes, about 6 mm wide, occasionally appear around the spots on the pods (Fig. 13.27b).

Leaf symptoms of bacterial brown spot may be confused with those of halo blight due to the variation in margin/width of halo influenced by temperature.

13.27.2. Causal organism

Pseudomonas syringae pv. *syringae* van Hall 1902

Syn. *Psm. syringae* pv. *japonica*

13.27.3. Disease cycle

The bacterium survives in infected plant debris and infected seed but the former is more important. Infected bean stems and pod pieces play an important role in the survival of bacterium. Disease epiphytotics are often traced to survival of the bacterium in infected crop residue. The bacterium survives on hairy vetch and other weeds but these are not considered an important source of survival. Non-certified seed lots also carry the bacterium from one crop season to the other. These seed lots are also a potential means of dissemination of bacterium to short and long distance places including new areas. The dissemination of the bacterium also occurs through wind-splashed rain, overhead-irrigation and farm equipment. Severe infections usually occur after major rainstorms.

Different strains of *Psm. syringae* pv. *syringae* causing diseases on various plant species grow as epiphytes on various plant species. *Psm. syringae* pv. *syringae,* the causal agent of bacterial brown spot also grows epiphytically on snap and dry beans without causing disease symptoms. The symptoms of bacterial brown spot on beans generally occur after the development of large epiphytic populations of the bacteria. Severe infections usually develop after major rainstorms but the absence of symptoms before rainstorms does not mean that the bacteria were not present.

13.27.4. Management

Management practices given for common and fuscous blights are applicable to this disease also.

REFERENCES

Abo-Elyouser, K.A.M. (2006). Induction of systemic acquired resistance against common blight of bean (*Phaseolus vulgaris*) caused by *Xanthomonas campestris* pv. *phaseoli. Egypt. J. Phytopathol.* 34: 41 - 50.

Burkholder, W.H. (1926). A new bacterial disease of the bean. *Phytopathology* 16: 915 - 28.

Darsonval, A., Darrasse, A., Durrand, K. *et al.* (2009). Adhesion and fitness in the bean phyllosphere and transmission to seed of *Xanthomonas fuscans* subsp. *fuscans. Mol. Plant-Microbe Interact.* 22: 747 – 57.

Darsonval, A., Darrasse, A., Meyer, D. *et al.* (2009a). The type III secretion system of *Xanthomonas fuscans* subsp. *fuscans* is involved in the phyllosphere colonization process and in transmission to seeds of susceptible beans. *Appl. Environ. Microbiol.* 74: 2669 - 78.

Dillard, H.R. and Legard, D.E. (1991). Bacterial diseases of beans. Fact sheet, Cornell University, Ithaca, New York, Online [http://vegetablemdonline. ppath/cornell.edu.factsheets/Beans_Bacterial.html].

Gilbertson, R.L. and Maxwell D.P. (1992). Common blight of beans. In: H.S. Chaube, U.S. Singh, A.N. Mukhopadhyay and J. Kumar (eds). *Plant Diseases of International Importance.* Vol. II, *Diseases of Vegetables* and *Oilseed Crops.* Prentice Hall, Englewood Cliffs, New Jersey, pp. 18 - 39.

Gilbertson, R.L., Maxwell, D.P., Hagedorn, D.J. *et al.* (1989). Development and application of a plasmid DNA probe for detection of bacteria causing common bacterial blight of bean. *Phytopathology* 79: 518 - 25.

Gondreau, C. and Samson, R. (1994). A review of thermotherapy to free plant material from pathogens, especially seeds from bacteria. Critical Reviews in Plant Sciences 13: 57 – 75.

Goto, M. (1992). *Fundamentals of Bacterial Plant Pathology.* Academic Press, Inc., San Diego, 342 pp.

Kaiser, W.J. and Vakili, N.G. (1978). Insect transmission of pathogenic xanthomonads to bean and cowpea in Puerto Rico. *Phytopathology* 68: 1057 – 63.

Lopes, L.P., Alves, P.F.R., Zandona, C. *et al.* (2008). A semi-selective medium to detect *Xanthomonas axonopodis* pv. *phaseoli* in bean seeds and its eradication through seed treatment with tolyfluanid. *Summa Phytopathol.* 34: 287 - 88.

Mutlu, N., Vidaver, A.K., Coyne, D.P. *et al.* (2008). Differential pathogenicity of *Xanthomonas campestris* pv. *phaseoli* and *X. fuscans* subsp. *fuscans* strains on bean genotypes with common blight resistance. *Plant Dis.* 92: 546 - 54.

Opio, F. and Male-Kigiwa, S. (1995). The status of bean breeding in Uganda. Breeding for disease resistance with emphasis on durability. *Proceedings Regional Workshop for Eastern, Central and Southern Africa,* October 2 - 6, 1994, Njoro, Kenya, pp. 110 – 13.

Partridge, J.E. (2003.) Common blight of bean. University of Nebraska, Lincoln. Online [http://nu-distance.unl.edu/homer/disease/agron/drybeans/BeCmnBlt.html].

Patel, P.N. and Jindal, J.K. (1972). Common and fuscous blights of bean. In: P.N. Patel (ed.). *Plant Bacteriology.* Vol. I, *Bacterial Diseases of Plants in India.* Indian Agricultural Research Institute, New Delhi, pp. 44 - 58.

Schwartz, H.F. (2007). Bacterial diseases of beans. Colorado State University, Online [http://www.ext.colostate.edu/PUBS/crops/02913.html].

Silva, E.G. da, Moura, A.B., Bacarin, M.A. *et al.* (2009). Metabolic alterations on bean plants originated from microbiolization of seeds with *Pseudomonas* sp and inoculated with *Xanthomonas axonopodis* pv. *phaseoli. Summa Phytopathol.* 35: 98 - 104.

Taylor, J.D., Teverson, D.M., Allen, D.J. *et al.* (1996). Identification and origin of races of *Pseudomonas syringae* pv. *phaseolicola* from Africa and other bean growing areas. *Plant Pathol.* 45: 469 – 78.

Walker, J.C. (1969). *Plant Pathology.* 3rd Ed. McGraw Hill Book Co., New York 819 pp.

Young, J.M., Bull, C.T., De Boer, S.H. *et al.* (2004). Names of plant pathogenic bacteria, 1864 - 2004. Online [http://www.isppweb.org/names_bacterial_revised.asp].

BACTERIAL DISEASES OF SOYBEAN

Soybean (*Glycine max* L.) suffers from four important bacterial diseases, namely, pustule (*Xanthomonas axonopodis* pv. *glycines*), blight (*Pseudomonas savastanoi* pv. *glycinea*), wilt (*Curtobacterium flaccumfaciens* pv. *flaccumfaciens*) and wild fire (*Psm. syringae* pv. *tabaci*).

13.28. SOYBEAN PUSTULE

Soybean pustule was first described by Nakano in 1919 from Japan. Hedges in 1922 reported it from USA and named the causal bacterium as *Bacterium phaseoli* var. *sojense.* The disease occurs in most soybean-growing areas of the world where warm weather and frequent rains occur during the growing season. Janse (2006) has listed 54 countries, including USA, where the disease occurs. Severe infection causes premature defoliation that decreases yield by reducing size and number of seeds. In a study reporting 11% loss in yield, 63% of it was due to the reduction in seed size.

In India, the disease was first reported by Uppal *et al.* in 1938 from Jalgaon, Maharashtra. Now it is present in most parts of the country (Patel and Jindal, 1972).

13.28.1. Symptoms

The symptoms are confined mostly to the leaves and are most prevalent during the later growing season of the crop. The symptoms start as small, pale, yellowish-green spots (lesions) having dark reddish-brown centres, formed mostly on the upper leaf surface. As the disease progresses, the central part of each lesion develops into a minute, raised, light-coloured pustule usually on the lower surface of the leaf. Sometimes, leaf spots are formed without developing pustules. The lesions merge with each other forming large, irregular areas of dead leaf tissue surrounded by yellowish margins. Dead areas may fall out due to the action of wind giving the leaves a ragged appearance. Severe infection often results in partial defoliation. On the pods, small, reddish brown, slightly raised spots may develop on the susceptible cultivars.

The symptoms of soybean pustule can be confused with those of soybean rust. Mature rust pustules have a round opening on the upper leaf surface for the release of spores. In soybean pustule, if any opening is present, it is typically a linear crack across the surface of pustule. These features can only be seen with magnification of 20X. The absence of water-soaking of bacterial pustules and their raised centres distinguish this disease from bacterial blight of soybean. Moreover, in bacterial blight, the spots are more angular in shape (Giesler and Weissling, 2006).

13.28.2. Causal organism

Xanthomonas axonopodis pv. *glycines* (Nakano 1919) Vauterin *et al.* 1995

Syn. *Xa. glycines, Xa. campestris* pv. *glycines*

The bacterium being a pathovar of *Xa. axonopodis* species, has the same morphological, cultural and biochemical characters as those of *Xa. axonopodis.* It can be distinguished from *Xa. axonopodis* and its other pathovars on the basis of host range. The description of *Xa. axonopodis* is given under bacterial leaf spot of green gram. Some additional characteristics of the bacterium are given below. Colonies on leaf infusion agar are small, circular, smooth having entire margin; initially pale yellow but become deep yellow with age. The bacterium grows slowly in culture. The optimum growth occurs at 30 - 33°C. The maximum and minimum temperatures at which growth can take place are 38 and 10°C, respectively.

Virulence of *Xa. axonopodis* pv. *glycines* is associated with different genetic determinants, including a pathogenicity island that includes nine *hrp*, nine *hrc* and eight *hpa* genes that are regulated by HrpG and HrpX. This cluster includes genes encoding a type III secretion system that is essential for pathogenesis and for induction of hypersensitive reaction (HR) on some plant species. The HrpG seems to be essential for induction of a nonspecific HR on different nonhost plants. Mutation of HrpG shows that it plays critical roles in HR on tobacco. Recently, another gene that encodes a pectate lyase in *Xa. axonopodis* pv. *glycines* have also been shown to be necessary for induction of HR on tobacco and pepper but not on tomato, sesame and cucumber (Kaewnum *et al.,* 2006). The above given information shows that *Xa. axonopodis* pv. *glycines* contains several genes that are required for disease development and induction of HR.

More recently, Athinuwat *et al.* (2009) identified three races of the bacterium on pustule disease resistant and susceptible soybean cultivars based on virulence phenotype. For race 3, an *avrBs3* homolog, *avrXg1* was identified that conferred resistance expressed as an HR on resistant cultivar Williams 82. Mutations in two predicted functional domains of *avrXg1* led to an increase in virulence on Williams 82 and an increase in bacterial population on susceptible cultivars. Expression of *avrXg1* in race 1 that is predicted to confer an unspecific HR, led to virulence on susceptibe cultivars Spencer and PI 520733. Expression of *avrXg1* in race 2 that is predicted of carrying *avrBs3*-like genes, resulted in gained virulence and fitness of pathogen on both resistant and susceptible cultivars. These results demonstrate multifunctions of *avrXg1* depending on pathogen and plant genetic backgrounds.

13.28.3. Disease cycle

The bacterium survives in the infected crop residue, infected seed and in the rhizosphere of wheat roots. The bacterium can survive in the seed for 2½ years. The chances of survival in infected crop residue are more when the residue remains on the soil surface. The entry of the bacterium occurs through stomata and wounds. After entry into the host, the bacterium moves into intercellular spaces and multiplies. In response to infection, leaf cells increase in size and also divide at a faster rate. Increase in host growth and in bacterial cell mass in the localized area causes bulging of epidermal cells on both leaf surfaces. These raised areas rupture and become pustules.

The dissemination of the bacterium occurs through wind-splashed rain, overhead-irrigation, and cultural operations during wet foliage. Disease outbreaks occur after several days of rainstorms or hailstorms. The disease is favoured by warm and wet weather. The optimum temperature for disease development is 30 - 33°C.

Besides soybean, the bacterium infects common bean (*Phaseolus vulgaris*), horse gram (*Macrotyloma uniflorum* = *Dolichos biflorus*) cowpea (*Vigna unguiculata*), lima bean (*Phaseolus lunatus*) redvine, buckwheat-vine (*Brunnichia cirrhosa*) and many other *Phaseolus* and *Vigna* spp.

13.28.4. Management

1. Use of resistant varieties is the most effective method for the management of this disease. The disease has been relegated to minor importance in the USA mainly through host resistance. The *Rxp* gene, originally present in cultivar CNS, a variety immune to soybean pustule, has been used to develop resistant varieties. *Rxp* gene confers resistance by requiring an increased number of bacterial cells for disease development. The *avrRxv* gene inhibits disease development by inducing a hypersensitive reaction. *Xa. axonopodis* pv. *glycines* does not require some other genes, required by other phytopathogenic bacteria, for elicitation of hypersensitive reaction.
2. Crop rotation for 2 - 3 years.
3. Destruction of infected crop residue by deep ploughing.
4. Avoid or limit the use of overhead irrigation.
5. Avoid cultural operations when foliage is wet.
6. Give preventive sprays of copper fungicides.
7. Michalski *et al.* (1983) have developed computer-based models for the management of soybean disease.

13.29. BACTERIAL BLIGHT OF SOYBEAN

Bacterial blight of soybean (*Glycine max* L.) was first noticed by Heald in Nebraska (USA) in 1905, who described the disease and suggested its bacterial nature. Johnson and Coerper in 1917 isolated the bacterium, proved its pathogenicity and suggested it to be a member of genus *Pseudomonas*. It is the most common bacterial disease of soybean in areas where cool and wet weather prevails. It is widespread in soybean-growing regions of the United States, Canada, Bulgaria, Mongolia, South Africa, Brazil, Austria, Czechoslovakia and Germany. The disease is also common in coastal and wet zones of New South Wales, Queensland and Australia (Lim, 1992).

The yield losses ranging from 18 - 22% have been reported under favourable weather conditions. The bacterium also considerably inhibits germination of infected seeds and inhibition up to 68% has been reported (Nicholson and Sinclair, 1971).

13.29.1. Symptoms

The disease is most prevalent in early season in contrast to soybean pustule, which appears late. Generally, the lesions first appear on cotyledons as brown spots on margins, which enlarge and turn dark brown as the tissue collapses. Young plants may show stunting and may die if the infection reaches the growing point.

Symptoms are most conspicuous on leaves although they also appear on stems, petioles and pods. Young leaves are most susceptible. Symptoms on the leaves start as small, angular, yellow to brown lesions. The lesions are usually water-soaked in the centre and surrounded by yellowish-green haloes. Older lesions turn reddish-brown to black as the tissue dies. Angular lesions enlarge in cool and wet weather and may coalesce to produce large, irregular dead areas. Eventually the dried lesions fall out giving the leaves a ragged appearance. Often, the leaves are badly shredded after strong winds and stormy rains. Infected young leaves become stunted, distorted and chlorotic. Severe infection results in defoliation of lower leaves. Large, black lesions may also develop on stem and petioles.

The symptoms on seed appear rarely. However, when lesions appear on pods, developing seeds may show some kind of symptoms. Small, water-soaked lesions appear on pods, which enlarge and merge giving rise to dark brown to black areas. The slimy bacterial ooze may cover the developing seeds. During storage, infected/infested seeds may become shriveled, discoloured and develop sunken or raised lesions. Infected seeds lack vigour and germinate poorly.

Chlorosis, also called **systemic toxemia**, develops throughout the infected plants. It results from the production of a toxin, coronatine, by the bacterium in the older infected leaves and translocated throughout the plant. The symptoms of bacterial blight differ from those of soybean pustule that in the former the lesions are water-soaked and flat in contrast to raised centres of lesions of the latter.

13.29.2. Causal organism

Pseudomonas savastanoi pv. *glycinea* (Coerper 1919) Gardan *et al.* 1992

Syn. *Psm. syringae* pv. *glycinea*

Colonies on nutrient agar are white, circular, smooth, glistening and raised. Cells are rod-shaped (0.7 - 1.2 × 1.5 - 3.0 μm) but occasionally long chains or filaments are formed. The bacterium is Gram-negative, catalase positive, strictly aerobic, motile by polar monotrichous or lophotrichous flagella, does not liquefy gelatin and produces acid but not gas from sucrose. The other characteristics of the bacterium include formation of levan, pectate gel pitting at pH 4.0 and growth on inositol, trigonelline and quinate. The optimum temperature for growth is 25 - 30°C. Almost all the strains produce a hypersensitive reaction on tobacco. The G + C content of the DNA is 59 - 61 moles% (Bd).

The isolation of the bacterium on King's agar B or other iron-deficient media is facilitated by the production of diffusible fluorescent green pigment. However, selective and diagnostic media like M-71 and BANQ are more reliable for its isolation. The characterization of the bacterium is done through the use of LOPAT tests. Majority of the isolates produce bacteriocins. Several avirulence (*avr*) genes have been identified and characterized. The bacterium can be differentiated from *Psm. syringae* pv. *tabaci* by its inability to utilize betaine, sorbitol and erythritol.

In 1966, seven races of the bacterium were identified from 104 isolates collected from five states of USA, namely, Illinois, Iowa, Minnesota, Missouri and North Carolina. These races were identified on the basis of their reactions (resistant, intermediate and susceptible) on seven soybean cultivars, namely, Lindarin, Acme, Flambeau, Harosoy, Merit, Norchief and Chippewa. In 1980, eighth race was found in Australia. Till now, nine races of the bacterium have been identified.

13.29.3. Disease cycle

The bacterium survives in infected/infested seed and the first evidence of its seed-plant transmission was provided by Kendrik and Gardner (1921). Infection/infestation of seed occurs through the pods during the growing season or at the time of threshing and cleaning. The pathogen also survives in infected crop residue but the survival in crop residue is variable and is influenced by several factors. The survival is better under cold and dry conditions compared to warm and wet conditions. In Brazil, the bacterium overwinters on volunteer bean plants in winter wheat fields.

The bacterium grows epiphytically on leaves, buds and stems of soybean plants without causing any symptoms. The epiphytic growth is influenced by the reaction of the host to the bacterium. On the leaves of susceptible host cultivars, the population increased 1000-fold while on resistant cultivars, it remained either unchanged or declined. This epiphytic phase may serve as a continuous source of inoculum.

It is typically an early season disease favoured by cool and wet weather, which prevails during the early crop season. The optimum temperature for disease development is 21 - 27°C. Warm and dry weather after the mid-season of crop prevents the disease development. Disease epiphytotics are a common seen after windy rainstorms coupled with cool weather.

Primary infection usually occurs on the cotyledons and these lesions serve as a major source of secondary inoculum. The spread of the bacterium occurs through wind-splashed rain, overhead-irrigation and cultural operations especially during wet foliage and rubbing of wet leaves with each other. After entering through stomata and wounds, bacteria multiply in the intercellular spaces of mesophyll cells. Typical water-soaked lesions develop 5 - 7 days after infection. During multiplication in the host, pathogen produces toxin called coronatine, which is translocated throughout the plant, mainly through the phloem. Toxin interferes with photosynthesis and chlorophyll production leading to systemic chlorosis and stunting of plants. Fully expanded trifoliate leaves do not show systemic chlorosis while it is maximum on young trifoliate leaves. Infected plants recover after sometime and perform normal photosynthetic activity and chlorophyll production but stunting remains permanently. Toxin produced by the bacterium produces chlorotic lesions also on tobacco, bean and tomato, indicating its nonhost-specificity. In addition to soybean, the bacterium infects common bean, lima bean, tepary bean and cowpea.

13.29.4. Management

1. Use of resistant varieties is the best method to manage the disease.
2. Use of pathogen-free seed.
3. Follow crop rotation with nonhost crops for 2 - 3 years.
4. Destroy the infected crop refuse by burying through deep ploughing.

5. Avoid or limit the cultural operations when the foliage is wet.
6. Avoid or limit the overhead-irrigation.
7. Give the preventive sprays of copper fungicides.

REFERENCES

Athinuwat, D., Prathuangwong, S., Cursino, L. *et al.* (2009). *Xanthomonas axonopodis* pv. *glycines* soybean cultivar virulence specificity is determined by *avrBs3* homolog *avrXg1*. *Phytopathology* 99: 996 – 1004.

Coerper, F.M. (1919). Bacterial blight of soybean. *J. Agr. Res.* 18: 179 - 94.

Giesler, L.J. and Weissling, T.J. (2006). Bacterial diseases of soybean. University of Nebraska, Lincoln, [http://www.ianrpubs.unl.edu/epublic/pages/publicationD.jsp?publicationId=349].

Janse, J.D. (2006). *Phytobacteriology: Principles and Practice.* CABI Publishing, Wallingford, UK, 360 pp.

Kaewnum, S., Prathuangwong, S. and Burr, T.J. (2006). A pectate lyase homolog, *xagP,* in *Xanthomonas axonopodis* pv. *glycines* is associated with hypersensitive response induction on tobacco. *Phytopathology* 96: 1230 – 36.

Kendrick, J.B. and Gardner, M.W. (1921). Seed transmission of soybean bacterial blight. *Phytopathology* 11: 339.

Lim, S.M. (1992). Bacterial blight of soybean. In: H.S. Chaube, U.S. Singh, A.N. Mukhopadhyay and J. Kumar (eds). *Plant Diseases of International Importance.* Vol. II, *Diseases of Vegetables and Oil Seed Crops.* Prentice Hall, Englewood Cliffs, New Jersey, pp. 305 - 13.

Michalski, R.S., Davis, J.H., Bisht, V.S. *et al.* (1983). A computer-based advisory system for diagnosing soybean diseases in Illinois. *Plant Dis.* 67: 459 – 63.

Nicholson, J.F. and Sinclair, J.B. (1971). Amsoy soybean seed germination inhibited by *Pseudomonas glycinea. Phytopathology* 61: 1390 – 93.

Patel, P.N. and Jindal, J.K. (1972). Bacterial pustule of soybean. In: P.N. Patel (ed.). *Plant Bacteriology.* Vol. I, *Bacterial Diseases of Plants in India.* Indian Agricultural Research Institute, New Delhi, pp. 59 - 65.

13.30. BACTERIAL BLIGHT OF COWPEA

Bacterial blight of cowpea (*Vigna unguiculata* (L.) Walp.) was originally described as canker of cowpea by Burkholder (1944) from USA. In India, it was first reported from Pune, Maharashtra by Patel and Diwan in 1950 and subsequently from many areas like, Tamil Nadu, Delhi, Punjab, Haryana, Uttar Pradesh and Karnataka. The disease also occurs in China, Brazil, Nigeria, Tanzania, Thailand, Venezuela, Zimbabwe, Uganda, Senegal, Sudan and South Africa.

The high mortality of the seedlings is responsible for major loss due to the disease. Plant mortality as high as 60% has been reported in California Black-Eye variety in Texas, USA. In India, 15 - 20% seedling mortality was observed in certified seed production plots of Pusa Dofasli cultivar of the National Seeds Corporation of India in 1971 (Patel *et al.,* 1972).

13.30.1. Symptoms

In seedlings developing from infected seeds, first symptoms appear on cotyledons or primary leaves. The infected cotyledons turn red and shrivel. On primary leaves, necrotic spots generally appear at the margin from where the bacterium moves to stem and then systemically proceeds to other plant parts. The severe infection damages cotyledons, primary leaves and growing point resulting in the death of seedlings. In the surviving seedlings, the cankers usually develop at the points where cotyledons and primary leaves were attached. These cankers enlarge and turn brown, and ultimately the cankered stems break at this point due to the weight of plants and the action of fast winds.

The infection of trifoliate leaves results in the formation of light yellow, irregular to round spots (4 - 10 mm) having necrotic brown centers and deep yellow borders. These spots enlarge and may involve the whole leaf. Small veins within blighted areas turn reddish brown, become shriveled and extend into the green leaf area. In severe infection, infection moves along the veins turning the area around the veins yellow (Fig. 13.30). Later on, these leaves turn straw coloured and defoliate prematurely.

On infected pods, dark green, water-soaked streaks appear along the sutures. Severely infected pods turn yellow, shrivel and dry. The seeds produced in the severely infected pods are usually small, shriveled and discoloured.

13.30.2. Causal organism

Xanthomonas axonopodis pv. *vignicola* (Burkholder 1944) Vauterin *et al.* 1995

Syn. *Xa. vignicola, Xa. campestris* pv. *vignicola*

Burkholder (1944) described the bacterium for the first time and named it as *Xanthomonas vignicola* sp. nov. With the introduction of pathovar system, it was changed to *Xa. campestris* pv. *vignicola* and later on, reclassified as *Xa. axonopodis* pv. *vignicola*. The below given description of the bacterium is taken primarily from Burkholder (1944).

Bacterial cells are Gram-negative, non-sporing rods ($0.5 - 0.9 \times 1.0 - 2.8\ \mu m$) with rounded ends and motile with a single polar flagellum. Colonies on potato dextrose agar are large, mucoid and pale yellow. The optimum temperature for growth is between 27 - 30°C, maximum at 37°C and minimum between 6 and 9°C. The tests for starch hydrolysis, gelatin liquefaction, hydrogen sulphide production and lipolytic activity are positive while those for indole and ammonia production and nitrate reduction are negative. Acid is produced from glucose, dextrose, lactose, maltose, mannose, galactose, sucrose and raffinose but no growth occurs in levulose, L-arabinose, xylose, rhamnose, glycerol and salicin. Also no growth occurs in sodium salts of lactic, formic, succinic, tartaric and hippuric acids. Bacterial growth in citric and malic acid salts produces an alkaline reaction. Tyrosine and asparagine are not utilized as a combined source of carbon and nitrogen. The bacterium is aerobic and liquefies pectate medium. Khatri-Chhetri *et al.* (2003) reported that 55 strains of the bacterium obtained from 11 countries differed in metabolization of different carbon sources using the Biolog GN microplate system; the strains belonging to different origins varied greatly but the variation was not related to the virulence of the strains.

Wydra *et al.* (2004) developed the first semi-selective medium called cefazoline cellobiose-methionine medium for the isolation of the bacterium. Colonies of the bacterium on this medium were white, round, raised and 0.2 - 1.8 mm in diam after 4 days of incubation. The medium was helpful for isolation of the bacterium from cowpea plant and soil samples.

13.30.3. Disease cycle

Infected seed is an important source of perpetuation of the bacterium. Volunteer plants also act as a source of inoculum. Seed may be externally or internally infected or may be both. Heavily infested seeds usually give rise to diseased seedlings even under dry conditions and primary infection through seed readily causes stem canker and results in high plant mortality. Khan (1996) observed 73.50% disease incidence in a crop raised from naturally infected seed. Soni (1982) reported that inoculation of cowpea plants at all stages, starting from seeds before sowing to flat green pod stage, led the pathogen to become seed-borne. Methods are available to detect the bacterium in the infected seeds (Soni and Thind, 1987, 1991). The bacterium did not survive for more than 3 weeks in diseased plant debris under field conditions (Khan, 1996), indicating that diseased plant debris does not play any role in its survival. Infected crop residue and some weeds may act as source of primary inoculum when two cowpea crops are sown consecutively (Sikirou and Wydra, 2004).

Secondary spread of the bacterium occurs mostly through wind-splashed rain. The other means of secondary spread are overhead-irrigation and ground irrigation. When the planting is done in furrows running along the slope, irrigation/rain water spreads the bacterium while moving from the higher level to the lower level. Long distance spread or spread to the new areas occurs through seed. Zandjanakou-Tachin *et al.* (2007) reported the presence of the bacterium in the faeces, intestines and on the legs and mandibles of *Zonocerus variegatus* (grasshopper) and indicated the possibility of the transmission of the bacterium through this insect.

The disease is severe in humid and warm weather. Optimum temperature for disease development is 28 - 34°C. The disease is more severe in low-lying areas where water frequently stagnates even temporarily.

13.30.4. Host range

Different host species have been reported to be infected artificially but there is not much agreement among different workers on this aspect. It appears that cowpea is the main economic host under natural conditions. However, Sikirou and Wydra (2004) have reported *Sphenostylis stenocarpa* (African yam bean) as potential host of the bacterium, for the first time on artificial inoculation. Most recently, Gena *et al.* (2009) have reported successful inoculation and symptom production on French bean (*P. vulgaris*) and hyacinth bean (*Dolichos lablab*).

The bacterium multiplies epiphytically on 5 weed hosts, namely, *Parthenium hysterophorus, Tephrosia purpurea, Tylophora indica, Legascea mollies* and *Leucas aspera* and among them, the maximum population of the bacterium was found on *P. hysterophorus*. Singh and Patel (1977) reported that the resistance was governed by a single dominant gene in cowpea lines, namely, P 309, P 426 and P 910 and by a single recessive gene in Iron. In case of P 910 and Iron, the segregating populations showed some gradations in susceptible and resistant reactions indicating that in addition to major genes, some modifying factors(s) may be involved.

13.30.5. Management

1. As the infected seed is only the main source of inoculum of the bacterium, the production of pathogen-free seed or the elimination of the seed-borne inoculum will control the disease. Different methods of seed treatment have been worked out by different workers. Jindal *et al.* (1989) recommended hot water (50°C for 30 min) or solar heat (soaking of seed in tap water from 8.00 A.M. to 12.00 noon and then spreading the seed in a thin layer exposed directly to sun rays for 5 hr on a bright sunny day in summer) or Streptocycline (100 μg/ml) + Agallol (0.2%) or Streptocycline (100 μg/ml) + captan (0.2%) treatment. Jindal and Thind (1990) found that *Erwinia herbicola* (recent name, *Pantoea agglomerans*) and *Penicillium oxalicum* were as effective as the above-mentioned four treatments in controlling

seed-borne infection of the bacterium. These biocontrol agents also gave higher seed germination than hot water and chemical treatments.

2. Secondary spread of the pathogen should be checked by applying chemicals or biocontrol agents. The sprays of copper and copper + maneb mixtures have been found effective in controlling the disease along with a significant increase in grain yield (Gitaitis *et al.*, 1986). Jindal and Thind (1990a) have advocated the use of seed treatment with Streptocycline (100 μg/ml) + Agallol (2000 μg/ml) followed by three sprays of Streptocycline (100 μg/ml) + Bavistin (500 μg/ml) or two sprays of Streptocycline (100 μg/ml) + Bavistin (500 μg/ml) followed by third spray of Streptocycline (100 μg/ml) + Blitox 50-WP (2000 μg/ml). The combination of seed treatment and foliar sprays provided significantly better disease control and also gave a higher grain yield in comparison to seed treatment or spraying schedules alone. Later on, Jindal and Thind (1993) reported that foliar applications of *Bacillus* sp., *Er. herbicola* and *P. oxalicum* gave significantly better control than Streptocycline (200 μg/ml), a potent and widely used bactericide.
3. Grow disease resistant varieties. Two cultivars namely, Tuv 12349 and Tuv 15549 are resistant to the disease in the field and in greenhouse in Nigeria (Okechukwu and Ekpo, 2004).

REFERENCES

Burkholder, W.H. (1944). *Xanthomonas vignicola* sp. nov. pathogenic on cowpea and beans. *Phytopathology* 34: 430 - 32.

Gena, M., Shah, R., Jadon, K.S. *et al.* (2009). Host range of *Xantho-monas axonopodis* pv. *vignicola,* the incitant of bacterial blight in cowpea. *Indian Phytopathol.* 62: 539 – 40.

Gitaitis, R.D., Bell, D.K. and Smittle, D.A. (1986). Epidemiology and control of bacterial blight and canker of cowpea. *Plant Dis.* 70: 187 - 90.

Jindal, K.K. and Thind, B.S. (1990). Microflora of cowpea seeds and its significance in the biological control of seed borne infection of *Xanthomonas campestris* pv. *vignicola*. *Seed Sci. & Technol.* 18: 393 - 403.

Jindal, K.K. and Thind, B.S. (1990a). Control of bacterial blight of cowpea (*Vigna unguiculata* (L.) Walp.). *Trop. Pest Manag.* 36: 211 - 13.

Jindal, K.K. and Thind, B.S. (1993). Biological control of bacterial blight of cowpea (*Vigna unguiculata* (L.) Walp.). *Phytopathol. Medit.* 32: 193 - 200.

Jindal, K.K., Thind, B.S. and Soni, P.S. (1989). Physical and chemical agents for the control of *Xanthomonas campestris* pv. *vignicola* from cowpea seeds. *Seed Sci. & Technol.* 17: 371 - 82.

Khan, A.N.A. (1996). Incidence and perpetuation of bacterial blight of cowpea caused by *Xanthomonas campestris* pv. *vignicola*. *Proc. 9th Int. Conf. Plant Path. Bact.* Chennai, India, pp. 349 - 56.

Khatri-Chhetri, G.B., Wydra, K. and Rudolph, K. (2003). Metabolic diversity of *Xanthomonas axonopodis* pv. *vignicola*, causal agent of cowpea bacterial blight and pustule. *Eur. J. Plant Pathol.* 109: 851 - 60.

Okechukwu, R.U. and Ekpo, E.J.A. (2004). Sources of resistance to cowpea bacterial blight disease in Nigeria. *Phytopathol. Z.* 152: 345 - 51.

Patel, P.N., Jindal, J.K. and Shekhawat, G.S. (1972). Bacterial blight (canker) of cowpea. In: P.N. Patel (ed.). *Plant Bacteriology.* Vol. I, *Bacterial Diseases of Plants in India*. Indian Agricultural Research Institute, New Delhi, pp. 79 - 83.

Sikirou, R. and Wydra, K. (2004). Persistence of *Xanthomonas axonopodis* pv. *vignicola* in weeds and crop debris and identification of *Sphenostylis stenocarpa* as a potential new host. *Eur. J. Plant Pathol.* 110: 939 - 47.

Singh, D. and Patel, P.N. (1977). Studies on resistance in crops to bacterial diseases in India. VII. Inheritance of resistance to bacterial blight disease in cowpea. *Indian Phytopathol.* 30: 99 - 102.

Soni, P.S. (1982). Detection of Phytopathogenic Bacteria from Cowpea and Mung Seeds. Ph.D. Thesis. Punjab Agricultural University, Ludhiana, 180 pp.

Soni, P.S. and Thind, B.S. (1987). Detection of *Xanthomonas campestris* pv. *vignicola* (Burkh.) Dye from cowpea seeds and *X. campestris* pv. *vignaeradiatae* (Sabet *et al.*) Dye from mung bean seeds by growing on tests. *Phytopathol. Medit.* 26: 1 - 6.

Soni, P.S. and Thind, B.S. (1991). Detection of *Xanthomonas campestris* pv. *vignaeradiatae* (Sabet *et al.*) Dye from green gram seeds and *X. campestris* pv. *vignicola* (Burkh.) Dye from cowpea seeds with the help of bacteriophages. *Plant Dis. Res.* 6: 6 - 11.

Wydra, K., Khatri-Chhetri, G., Mavridis, A. *et al.* (2004). A diagnostic medium for the semi-selective isolation and enumeration of *Xanthomonas axonopodis* pv. *vignicola*. *Eur. J. Plant Pathol.* 110: 991 - 1001.

Zandjanakou-Tachin, M., Fanou, A., Le Gall, P. *et al.* (2007). Detection, survival and transmission of *Xanthomonas axonopodis* pv. *manihotis* and *X. axonopodis* pv. *vignicola*, causal agents of cassava and cowpea bacterial blight, respectively, in/by insect vectors. *Phytopathol. Z.*, 155: 159 - 69.

13.31. BACTERIAL LEAF SPOT OF GREEN GRAM

Green gram (*Vigna radiata* (L.) Wilczek) suffers from two bacterial diseases, namely, bacterial leaf spot and halo blight. Out of these two, the former is economically more important and widespread. Bacterial leaf spot of green gram or *mung* bean (local name of green gram in India) was first reported from China by Fang and associates in 1964. In India, its presence has been mentioned in the past but detailed information on certain aspects of the disease has been published for the first time by Patel and his associates (Patel and Jindal, 1972; Patel *et al.*, 1972). The disease is widespread in green gram growing areas of India. In addition to china and India, it also occurs in Pakistan (Iqbal *et al.*, 2003).

13.31.1. Symptoms

The symptoms on cotyledons of the seedlings, raised from infected seeds, start as pinhead, circular, water-soaked spots, which are lighter green in colour as compared to neighbouring healthy tissue. The spots increase in size, turn brown, become necrotic and raised. In case of severe infection, the spots may coalesce to cover the whole leaf. Longitudinal lesions, slits or cankers extending from soil level to the growing tip develop on the stems of seedlings grown from severely infected seeds.

On trifoliate leaves, brown, circular to irregular and distinctly raised spots appear. Generally, the spots are dry and necrotic from the beginning. Initially, they appear on either surface as superficial eruptions but gradually develop through the entire thickness of the leaflet and become corky and rough. After attaining some height, the eruptions begin to collapse due to disintegration of infected cells, resulting in the formation of rifts. Sometimes, the rifts are also caused due to the disintegration of leaf tissue by the action of wind. In older spots, the superficial eruptions collapse leaving behind saucer-shaped depressions. In a few cases, the shot-hole symptoms are also observed. Generally, the spots are isolated but in severe infections, they coalesce to form necrotic patches but never develop into a typical leaf blight phase as in common blight of bean (Fig. 13.31a & b). Lesions on petioles, petiolule or pulvinus are brown, flat or raised and may be elongated in the form of long streaks. Normally, the infected plants do not defoliate.

On stems, dark brown, elongated lesions are produced. In severely infected plants, dark brown, longitudinal slits or cankers develop on the stems, which often extend from soil level to the growing tip.

Lesions on pods are brown and raised but are rarely seen in nature. In artificially inoculated pods, brown and raised lesions appear. Sometimes, a thin layer of shining gummy bacterial ooze can also be seen on the outer surface. On the inner side of green pods, brownish lesions are formed just beneath the bacterial ooze. The seeds touching these lesions also develop brown spots. The seeds produced in the infected pods are generally small, shriveled and discoloured. Most of the discoloured seeds are brown and yellowish but occasionally black also. The normal or effective disease development occurs only on the monsoon season crop. In the summer crop, the symptoms develop on the cotyledonary and first trifoliate leaves, but subsequent spread of the disease does not occur.

As the pathogen is largely parenchymatous, infected plants are not completely killed. However, severe infections result in the necrosis of considerable leaf area leading to reduction in photosynthesis and general weakening of the plants.

13.31.2. Causal organism

Xanthomonas axonopodis pv. *vignaeradiatae* (Sabet *et al.* 1969) Vauterin *et al.* 1995

Syn. *Xa. campestris* pv. *vignaeradiatae*

The description of *Xa. axonopodis* Starr and Garces 1950 emend. Vauterin *et al.* 1995 is given below. Cells are straight rods, Gram-negative, nonsporing and measure 0.4 - 0.6 × 0.8 - 2.0 μm; occur mostly singly or in pairs, occasionally short chains and rarely in filaments. Motile by a single polar flagellum, obligately aerobic and metabolism is strictly respiratory type. Colonies are usually yellow, smooth and butyrous, viscid or mucoid. Yellow pigments, xanthomonadins are produced. The viscous or mucoid growth of the culture is due to the production of an extracellular heteropolysaccharide called xanthan.

Catalase positive, oxidase is negative or weak. Urease and indole are not produced and nitrate is not reduced to nitrite. Litmus milk is not acidified. The bacterium utilizes dextrin, cellobiose, gentiobiose, maltose, D-alanine, L-alanine and L-glutamic acid but does not utilize β-methyl-D-glucoside, L-rhamnose, D-glacturonic acid, D-glucuronic acid and *p*-hydroxyphenylacetate. Most of the strains utilize D-galactose, melibiose, acetic acid, DL-lactic acid, malonic acid, L-leucine, L-serine and glycerol.

The members of this species are slow growing and metabolically less active than those of *Xa. campestris*. Both these species can be clearly differentiated on the basis of phenotypic characters. *Xa. axonopodis* is the most heterogeneous species currently recognized in genus *Xanthomonas* and includes several pathovars. In general, these pathovars are not distinguishable by their phenotypic characters and their identification is reliable on the basis of their host range.

Xa. axonopodis pv. *vignaeradiatae* being a pathovar of *Xa. axonopodis,* has the same morphological, cultural and biochemical characters as those of the latter. In addition to above mentioned characteristics, *Xa. axonopodis* pv. *vignaeradiatae* produces H_2S, liquefies gelatin, hydrolyses starch and ferments glucose, sucrose, mannose, xylose, lactose and raffinose producing only acid (Patel and Jindal, 1972). Thakur *et al.* (1977) confirmed the existence of six races in the bacterium. They further reported that (1) resistance in a differential to a race is governed by a single gene, (2) gene governing resistance in a differential is dominant to its recessive allele governing susceptibility in another differential and (3) that a gene or allele governing resistance or susceptibility in the differentials to different races does not overlap.

Patro and Jindal (2000) have shown the existence of plasmids in the bacterium and their correlation with virulence. The highly virulent isolate contained the maximum number of plasmids. The plasmid-borne genes were shown to govern virulence, exopolysaccharide production, antibiotic resistance and pigmentation.

The bacterium produces exopolysaccharide, proteolytic and cellulolytic enzymes which play role in pathogenesis. Higher proteolytic and cellulolytic activities were observed in susceptible green gram cv. Pusa Baisakhi after 120 hr of inoculation with a virulent culture (Dutta and Jindal, 2002).

13.31.3. Disease cycle

Earlier, the bacterium was considered to be seed-borne but Soni (1982) established the seed-borne nature of the bacterium and showed that seed is an important source of survival for the bacterium. He further reported that the bacterium survived for one year in the seeds and inoculation of the bacterium only at inflorescence and at flat green pod stage led the bacterium to become seed-borne. Deformity of seeds like shriveling, discolouration and small size was only an indication but not a sure test of seed-borne infection.

Soni and Thind (1987) compared different growing on tests for the detection of *Xa. axonopodis* pv. *vignaeradiatae* from green gram seeds and found that traditional growing on test was more suitable in monsoon season when conditions were highly favourable for disease development while incubation growing on test proved more effective in summer crop season. They further reported that wounded seed growing on test was most effective for detecting the lowest level of seed-borne infection and it detected 0.1% infected seeds in a seed lot. Bacteriophages have also been found effective for the detection of bacterium from shriveled and discoloured green gram seeds (Soni and Thind, 1991).

The bacterium enters the host through stomata and wounds, and is largely confined to the parenchyma tissue. Soni (1982), on the basis of recovery of the bacterium only from roots, primary and trifoliate leaves, suggested the non-systemic nature of the bacterium. Secondary spread of the bacterium is mainly through wind-splashed rain. Dissemination to distant places and new areas is mainly through the infected seed.

Most of the commercially grown varieties in India including Pusa Baisakhi are susceptible to the bacterium while Jalgaon 781 is resistant.

13.31.4. Host range

Patel and Jindal (1972) reported that in addition to green gram the bacterium was moderately virulent on bean, lentil, and Indian bean (*Dolichos lablab*) but weakly virulent on lima bean (*Phaseolus lunatus* and *P. bracteatus*. The progress of infection on bean was not as fast and destructive as with the common and fuscous blight pathogens, the actual pathogens of this host. On bean, the spots were water-soaked but not raised as in green gram; they coalesced and developed large necrotic patches.

13.31.5. Management

1. As the bacterium perpetuates through infected seed, the use of bacteria-free seed and the eradication of seed-borne inoculum are the most effective and viable methods of disease control.
2. Seed treatment with hot water at 52°C for 30 min, solar heat (for details see bacterial blight of cowpea), Streptocycline (100 μg/ml) + captan (0.2%) or Streptocycline (100 μg/ml) + Agallol (0.2%) have been found very effective for the eradication of seed-borne inoculum (Jindal and Thind, 1987).
3. For controlling the secondary infection of the bacterium, 4 sprays, i.e., first two sprays of Streptocycline (100 μg/ml) + Bavistin (0.05%) followed by two sprays of Streptocycline (100 μg/ml) + Blitox (0.2%), at 10 days intervals, have been found effective by Jindal and Thind (1987). They also reported that the combinations of these sprays with hot water and solar heat treatments of seed proved most effective in controlling the disease and caused 69.9 and 72.5% reduction in disease intensity, respectively. Gour and Ahmed (1983) recommended 3 sprays of Agrimycin - 100 (0.025%) or Plantomycin (0.025%) + Bavistin (0.1%) for the control the disease. Borah *et al.* (2000) have advocated the use of a phylloplane bacteria, *Bacillus* sp. for the control of disease and recommended its preventive application.

REFERENCES

Borah, P.K., Jindal, J.K. and Verma, J.P. (2000). Biological management of bacterial leaf spot of mungbean caused by *Xanthomonas axonopodis* pv. *vignaeradiatae*. *Indian Phytopathol*. 53: 384 - 94.

Dutta, S. and Jindal, J.K. (2002). Hydrolytic enzymes and exopolysaccharide production of *Xanthomonas axonopodis* pv. *vignaeradiatae* inciting bacterial leaf spot of mungbean. *Annl. Plant Prot. Sci.* 10: 295 - 98.

our, R.B. and Ahmed, S.R. (1983). Studies on the chemical control of bacterial blight of moong incited by *Xanthomonas phaseoli* (Smith) Dowson. *Pesticides* 17: 22 - 23.

qbal, S.M., Zubair, M., Anwar, M. *et al.* (2003). Resistance in mungbean to bacterial leaf spot disease. *Mycopathologia* 1: 81 - 83.

indal K.K. and Thind, B.S. (1987). Control of Cercospora and bacterial leaf spot of green gram. *Indian J. Agric. Sci.* 57: 372 - 75.

atel, P.N. and Jindal, J.K. (1972). Bacterial leaf spot and halo blight diseases of mung bean (*Phaseolus aureus*) and other legumes in India. *Indian Phytopathol.* 25: 517 - 25.

atel, P.N., Jindal, J.K. and Singh, D. (1972). Studies on resistance in crops to bacterial diseases in India. IV. Resistance in mung bean (*Phaseolus aureus*) to *Xanthomonas phaseoli. Indian Phytopathol.* 25: 526 - 29.

atro, T.S.S.K. and Jindal, J.K. (2000). Expression of plasmid-borne characters in *Xanthomonas axonopodis* pv. *vignaeradiatae*. *Trop. Agr. Res.* 12: 86 - 94.

oni, P.S. (1982). Detection of Phytopathogenic Bacteria from Cowpea and Mung Seeds. Ph.D. Thesis. Punjab Agricultural University, Ludhiana, 180 pp.

oni, P.S. and Thind, B.S. (1987). Detection *of Xanthomonas campestris* pv. *vignicola* (Burkh.) Dye from cowpea seeds and *X. campestris* pv. *vignaeradiatae* (Sabet *et al.*) Dye from mung bean seeds by growing on tests. *Phytopathol. Medit.* 26: 1 - 6.

oni, P.S. and Thind, B.S. (1991). Detection of *Xanthomonas campestris* pv. *vignaeradiatae* (Sabet *et al.*) Dye from green gram seeds and *X. campestris* pv. *vignicola* (Burkh.) Dye from cowpea seeds with the help of bacteriophages. *Plant Dis. Res.* 6: 6 - 11.

hakur, R.P., Patel, P.N. and Verma, J.P. (1977). Studies on resistance in crops to bacterial diseases in India. XI. Genetic make-up of mungbean differentials of the races of bacterial leaf spot pathogen, *Xanthomonas phaseoli. Indian Phytopathol.* 30: 217 - 21.

3.32. HALO BLIGHT OF GREEN GRAM

Green gram or *mung* bean (vernacular name of green gram in India) uffers from two bacterial diseases, namely, bacterial leaf spot and halo light. Halo blight is comparatively of less economic significance than acterial leaf spot. Patel and Jindal (1970) observed halo blight for the rst time in 1969 at Indian Agricultural Research Institute, New Delhi on ome *mung* bean lines resistant to bacterial leaf spot. Subsequently, chmitthenner *et al.* (1971) reported a severe epiphytotic of the disease om Ohio in the USA.

13.32.1. Symptoms

On the seedlings developing from infected seeds, water-soaked spots appear first on cotyledonary leaves and then spread to the trifoliate leaves. Severe infection at this stage may result in the death of seedlings. On trifoliate leaves, the spots may be small or large, round or irregular, flat and water-soaked. The spots are surrounded by halo-like zones of greenish-yellow tissue, which result from the toxin produced by the bacteria multiplying in the water-soaked lesions. Young developing leaves on severely infected plants show systemic chlorosis due to the translocation of the toxin. The formation of halo zones and development of systemic chlorosis are favoured by low temperature (15 - 20°C) at night, as in halo blight of common bean. The spots on old leaves are sometimes angular and do not show distinct halo. Frequently, the infection spread along the veins and veinlets turning them brown and necrotic and causing distortion of the developing leaflets.

On the petioles, long streaks appear which are first water-soaked but later on turn red. Pulvinus and petiolules of the severely infected leaves appear brown and sunken. Generally, there is no defoliation of the plants. On the stem, first water-soaked spots appear which gradually enlarge and turn into reddish streaks extending longitudinally. Such stems often split. On pods, water-soaked lesions appear and turn brown with time.

13.32.2. Causal organism

Pseudomonas syringae pv. *phaseolicola* (Burkholder 1926) Young *et al.* 1978 *mung* bean strain

Syn. *Psm. phaseolicola mung* bean strain, *Psm. savastanoi* pv. *phaseolicola mung* bean strain

The first isolation of the *mung* bean halo blight pathogen appears to have been made by D.C. Graham of Scotland in 1966 from *mung* bean seeds. He named the bacterium as *Pseudomonas phaseolicola* but presumably did not work with it further. Schroth *et al.* (1971) compared this culture with *Psm. phaseolicola* from different hosts and found it to be distinct from bean isolates in utilization of several nutritional substrates and host range. Schmitthenner *et al.* (1971) found that the *mung* bean bacterium was similar to both *Psm. phaseolicola* and *Psm. glycinea* in cultural characteristics. As only the *mung* bean isolates were virulent to *mung* bean, they proposed that the *mung* bean bacterium be named as *mung* bean strain of *Pseudomonas phaseolicola*. Keeping in view the host range and overall similarity in bacteriological properties of the *mung* halo blight pathogen to that of bean halo blight pathogen, Patel and Jindal (1972) also identified it as *mung* strain of *Psm. phaseolicola*. The below given description of the *mung* bean halo blight bacterium is based on the studies carried out by Patel and Jindal (1972a).

The colonies on nutrient agar are circular, white, glistening, flat, entire and produce green fluorescent pigment on King's medium B. Cells are Gram-negative rods, motile with 1 - 4 polar flagella and strictly aerobic. The bacterium utilizes glucose oxidatively, utilizes asparagine and can grow in the presence of 4.0% sodium chloride. The hydrolysis of gelatin is slow and milk is peptonized without curd formation. The tests for production of ammonia, urease and catalase, and starch hydrolysis are positive while those for production of indole, acetoin and H_2S, and reduction of nitrate are negative. The acid is produced from glucose, mannose, sucrose, fructose, galactose, ribose, arabinose, melibiose, cellobiose, lactose, xylose, raffinose and mannitol but not from trehalose, rhamnose, sorbose, sorbitol, adonitol, dulcitol, inulin, salicin, inositol, dextrin, glycogen and *meso*-erythrit. The bacterium utilizes sodium propionate and succinate but not sodium oxalate and tartrate.

13.32.3. Disease cycle

The bacterium is seed-borne as reported by Schmitthenner *et al.* (1971) and its spread to different places occurs through seed. Not much information is available on this disease and it is presumed that the disease cycle is similar to that of bean halo blight. The disease assumed serious proportions in Ohio, USA and the crop raised from infected seed gave 60% less yield than the crop raised from healthy seed (Schmitthenner *et al.*, 1971). The disease has not appeared in a severe form in India. Hence, there is a need to investigate if the environmental or other factors are responsible for its low incidence in India, particularly in *Kharif* crop.

13.32.4. Host range

Schmitthenner *et al.* (1971) reported susceptible reaction on red kidney bean (*Phaseolus vulgaris*), lima bean (var. Fordhook 242) and soybean (var. Acme), in addition to *mung* bean. Schroth *et al.* (1971) also found that *mung* strain could infect few varieties of bean, lima bean and soybean. Patel and Jindal (1972a) found that the infection on *Phaseolus atropurpureus, P. bracteatus, Dolichos lablab* and *D. biflorus* was as severe as on *mung* bean. On *P. mungo, P. trilobus* and *Lathyrus sativus,* only few spots with distinct haloes were produced. Among the non-legumes, *Ficus religiosa* was susceptible. Only *P. bracteatus* and *P. atropurpureus* are susceptible to both the pathogens. Bean plants showed hypersensitive reaction within 24 hr when inoculated with the *mung* bean pathogen. Bean bacterium attacked 10 of the 15 *Phaseolus* spp. and 2 other legumes whereas *mung* bean pathogen infected only 5 of 13 *Phaseolus* spp., 4 other legumes and a non-legume (Patel and Jindal, 1972a). *Mung* bean strain of the bacterium has a distinct host range than those of bean strain and they do not cross infect (Patel and Walker, 1965).

13.32.5. Management

1. As the bacterium perpetuates only through infected seed, the use of pathogen-free seed is the best approach to manage the disease.
2. There is no information available on seed treatment. However, it is possible that seed treatments recommended for the eradication of *Xanthomonas axonopodis* pv. *vignaeradiatae* from *mung* bean seed may also work for this pathogen. Hence, these treatments should be tried to see their efficacy.
3. Restricted use or avoidance of overhead-irrigation will help in reducing the incidence of the disease.

REFERENCES

Patel, P.N. and Jindal, J.K. (1970). Bacterial diseases in seed legumes in 1968 and 1969. Paper presented at the Fourth Annual Workshop Conference on Pulse Crops held at Ludhiana (Punjab) in April, 1970.

Patel, P.N. and Jindal, J.K. (1972). Halo blight of mung (green gram). In: P.N. Patel (ed.). *Plant Bacteriology*. Vol. I, *Bacterial Diseases of Plants in India*. Indian Agricultural Research Institute, New Delhi, pp. 75 - 78.

Patel, P.N. and Jindal, J.K. (1972a). Bacterial leaf spot and halo blight diseases of mung bean (*Phaseolus aureus*) and other legumes in India. *Indian Phytopathol.* 25: 517 - 25.

Patel, P.N. and Walker, J.C. (1965). Resistance in *Phaseolus* to halo blight. *Phytopathology* 55: 889 - 94.

Schmitthenner, A.F., Hoitink, H.A.J. and Kroetz, M.E. (1971). Halo blight of mungbean incited by a new strain of *Pseudomonas phaseolicola* (Abstr.). *Phytopathology* 61: 909.

Schroth, M.N., Vilma Vitanza, B. and Hildebrand, D.C. (1971). Pathogenic and nutritional variation in the halo blight group of fluorescent pseudomonads of bean. *Phytopathology* 61: 852 - 57.

13.33. BACTERIAL LEAF SPOT OF BLACK GRAM

Black gram or mash (*Phaseolus mungo*), also called *urad* in India, suffers from a bacterial leaf spot. Natural occurrence of the disease was first observed during 1961 - 62 around Annamalainagar, Tamil Nadu by Rangaswami and Gowda (1963). Patel and Jindal (1970) observed it in varying intensities in experimental trials laid out in different parts of India, namely, Pantnagar (U.P.), Delhi, Ludhiana (Punjab), Jabalapur (M.P.) and Amravati (Maharashtra). The disease probably occurs also in other black gram growing areas of the country.

13.33.1. Symptoms

Circular or irregular, flat or raised spots appear on the affected leaves. The spots may be water-soaked or necrotic from the beginning of their formation but eventually turn necrotic. Generally, the spots coalesce leading to the development of a limited blight. Similar spots also develop on the stems. Extensive pod infection is rarely seen in the nature.

13.33.2. Causal organism

Xanthomonas phaseoli mash strain.

The bacterium is similar in bacteriological characters to *Xa. phaseoli* strains from hyacinth bean and *moth* bean but differs in host range from these bacteria. Hence, it is identified as *Xa. phaseoli* mash strain. The morphological, physiological and biochemical characters of *Xa. phaseoli moth* bean strain are given under the description of bacterial blight of *moth* bean.

13.33.3. Disease cycle

Very little information is available on the epidemiological aspects of the disease. However, the field observations suggest that the bacterium is transmitted through seed and the disease development is favoured by warm and wet climate prevailing in monsoon (July to September). The bacterium enters the host through stomata and wounds (Patel and Jidnal, 1972).

13.33.4. Management

1. Use of resistant varieties.
2. Use of pathogen-free seed and treatment of seed in case the seed is not free from the pathogen
3. The possibility of bactericidal chemical sprays for the management of disease should also be explored.

REFERENCES

Patel, P.N. and Jindal, J.K. (1970). Bacterial diseases in seed legumes in 1968 and 1969. *Proceedings of the Fourth Annual Workshop Conference on Pulse Crops*, I.C.A.R., New Delhi.

Patel, P.N. and Jindal, J.K. (1972). Bacterial leaf spot of urid (black gram). In: P.N. Patel (ed.). *Plant Bacteriology*. Vol. I, *Bacterial Diseases of Plants in India*. Indian Agricultural Research Institute, New Delhi, pp. 92 - 93.

Rangaswami, G. and Gowda, S.S. (1963). On some bacterial diseases of ornamentals and vegetables in Madras state. *Indian Phytopathol*. 16: 74 - 85.

13.34. BACTERIAL BLIGHT OF *MOTH* BEAN

Moth bean (*Phaseolus acontifolius* L.), also called *moth* or *matki* in India, suffers from a serious bacterial disease called bacterial blight. Patel and Jindal (1970) observed it for the first time at Amravati in Maharashtra, India during a survey carried out in 1968 - 1969. The disease also occurs in other states, namely, Rajasthan, Punjab, Haryana, Uttar Pradesh, Delhi, Gujarat etc.

13.34.1. Symptoms

Numerous, small or large, irregular, brown, necrotic spots appear on the leaves. The spots are more prominent on the upper than on the lower surface of leaves. The spots coalesce forming large necrotic blotches. The centres of the several spots and blotches torn out to give a ragged appearance to the leaves. Severely diseased leaves sometimes turn yellow but do not defoliate prematurely. Brown spots appear on the petioles, petiolules, stems and pods, which expand when the disease advances (Patel and Jindal, 1972a).

13.34.2. Causal organism

Xanthomonas phaseoli moth strain.

The bacterium shows morphological, physiological and biochemical properties similar to those of *Xa. phaseoli*, causal agent of common blight of bean. The below given description of the bacterium is taken from Patel and Jindal (1972).

Cells are Gram-negative, nonsporing, not acid-fast rods having a single polar flagellum. On nutrient agar, the colonies are circular, convex, glistening, yellow and slimy. On slants, the growth is bold and slimy up to a week but later it becomes flat and dry. The bacterium is strict aerobe, utilizes glucose oxidatively, hydrolyses starch, liquefies gelatin and produces ammonia, H_2S and catalase. It does not utilize asparagine as a sole source of carbon and nitrogen and the tests for production of indole, acetoin and urease, and nitrate reduction are also negative. Milk is turned alkaline and peptonized without coagulation and litmus is reduced but reaction for methyl red test is negative. The growth is retarted by 3.0% and suppressed by 4.0% NaCl. Acid but no gas is produced from glucose, sucrose, mannose, fructose, xylose, galactose, ribose, arabinose, cellobiose and dextrin while from lactose, sorbose, raffinose, rhamnose, melibiose, sorbitol, mannitol, adonitol, dulcitol, inositol, inulin, salicin and *meso*-erythrit even the acid is not produced. Sodium propionate and succinate are utilized but not oxalate and tartrate.

In addition to *moth* bean, the bacterium caused moderate infection on *P. vulgaris* (bean), *P. bracteatus* and *P. atropurpureus* (Patel and Jindal, 1972).

13.34.3. Disease cycle

The bacterium is seed-borne as shown by Patel and Jindal (1972). It enters the host through stomata. The spread of the pathogen and disease severity increase under warm and high rainfall conditions. Detailed studies are needed on the epidemiological aspects of the disease.

13.34.4. Management

1. As the bacterium is seed-borne, use of pathogen-free seed is recommended. In the absence of a pathogen-free seed, the seed treatments recommended for bacterial blight of cowpea or bacterial leaf spot of green gram can be tried.
2. The possibility of disease management with foliar sprays of bactericides appears to give positive results and it should be explored.

REFERENCES

Patel, P.N. and Jindal, J.K. (1970). Bacterial diseases in seed legumes in 1968 and 1969. *Proceedings of the Fourth Annual Workshop Conference on Pulse Crops*, ICAR, New Delhi.

Patel, P.N. and Jindal, J.K. (1972). Occurrence of bacterial blight disease on moth bean (*Phaseolus acontifolius*) in India. *Indian Phytopathol.* 25: 464 - 66.

Patel, P.N. and Jindal, J.K. (1972a). Bacterial blight of moth bean. In: P. N. Patel (ed.). *Plant Bacteriology*. Vol. I, *Bacterial Diseases of Plants in India*. Indian Agricultural Research Institute, New Delhi, pp. 90 - 91.

13.35. BACTERIAL BLIGHT OF CLUSTERBEAN

Bacterial blight of clusterbean (*Cyamopsis tetragonaloba* (L.) Taub.) is a widespread disease, which is particularly destructive to the rainy season crop. The loss is mainly due to blightening and mortality of the seedlings in the early stage. Severe cankering of the stem and death of grown up plants also results in heavy losses. In *Kharif* 1971, the disease caused 50 - 70% loss in yield amounting to rupees 3 - 5 lacs in 600 acres of cv. Pusa Naubahar grown for certified seed production by National Seed Corporation of India.

Patel *et al.* (1953) reported a bacterial leaf spot of *guar* (local name for clusterbean in India) from Patna and Bombay and named the pathogen as *Xanthomonas cyamopsidis*. Later on, Patel and Patel (1958) described a severe leaf blight of *guar* from Anand, Gujarat and suggested the resemblance of its symptoms with those of black rot caused by *Xa. campestris* in crucifers. They presumed this disease different from that described by Patel *et al.* (1953) and hence, named the bacterium as *Xa. cyamophagus*. Srivastava and Rao (1963) described a bacterial leaf and stem blight of *guar* and concluded that all the different types of symptoms are caused by the same pathogen. They also suggested the invalidation of *Xa. cyamophagus*.

13.35.1. Symptoms

Both leaf spot and blight phases of the disease appear simultaneously. Infection through stomata results in leaf spots, which are interveinal, round, water-soaked or oily in appearance and well-defined on the dorsal surface of the leaf. They may enlarge or coalesce, ultimately leading to blight phase. Then the pathogen rapidly invades vascular tissue and causes flaccidity of the affected portions. These flaccid portions eventually become necrotic and turn brown. Several flaccid areas formed on a leaf, including those on the margins, coalesce resulting in blightening of the entire lamina.

Infections resulting from hydathodal penetrations appear as V-shaped areas at the leaf margins and progress toward the midrib. From the leaf, infection moves systemically through the petiole, into the stem producing black longitudinal streaks on the stem. Infection of the stem results in wilting of the apical leaves and curvature of the stem. Finally the stem splits and breaks off.

On pods, black streaks appear, presumably resulting from local infection. In contrast to other systemic bacterial disease of legumes like common or fuscous blight of bean and cowpea blight, infection on suture is rarely seen. The seedlings, developing from infected seeds, show rapid infection of primary leaves from where the infection progresses to the tender stems, resulting in the death of seedlings.

13.35.2. Causal organism

Xanthomonas axonopodis pv. *cyamopsidis* (Patel *et al.* 1953) Vauterin *et al.* 1995

Syn. *Xa. campestris* pv. *cyamopsidis*, *Xa. cyamopsidis*

The below given description of the bacterium is taken primarily from the publications of Patel *et al.* (1953) and Patel and Patel (1958).

Cells are short rods (0.4 - 0.8 × 1.2 - 1.5 μm), occur singly or rarely in pairs, encapsulated, facultatively anaerobic, nonsporing, not acid-fast, Gram-negative and motile by a polar flagellum. Colonies on potato dextrose agar are yellow, round having entire margins, shining, butyrous, pulvinate and attain diameter of 1.1 cm after an incubation period of 4 days. Colonies on nutrient agar appear as mustard yellow, flat, smooth, glistening having entire margins and grow up to 1.1 cm in diam after 6 days of incubation.

The bacterium hydrolyses starch and casein, reduces litmus, produces hydrogen sulphide and ammonia but not indole and does not reduce nitrates. Acid is produced without gas from glucose, lactose, maltose, dextrose, xylose and mannitol. The optimum temperature for growth ranges from 25 - 31°C and thermal death point is 52°C. The bacterial growth in culture media is inhibited by aqueous leaf extracts of *Emblica officinalis* (aonla), *Eucalyptus citriodora* and *Tagetes* sp. (marigold). *E. officinalis* leaf extracts exhibited maximum antibacterial activity followed by *E. citriodora* (Kaur *et al.*, 2010).

Two pathogenic races, namely, race 0 and race 1 have been reported from USA. Race 1 infects cultivar Brooks while race 0 does not. In India, this cultivar has been found highly susceptible to the bacterium under natural conditions, indicating that Indian strains are more virulent.

13.35.3. Disease cycle

The bacterium is internally seed-borne. Seed is the most important source of perpetuation, particularly in areas where only one crop of *guar* is taken in a year. Volunteer plants are another source of perpetuation where *guar* is planted in the same fields every year. This source becomes even more important when the variety grown is of shattering type. The pathogen does not survive in infected plant debris under North Indian conditions where the gap between two crops is of 9 months (Srivastava and Rao, 1970).

The bacterium enters the host through stomata, hydathodes or wounds and its multiplication and movement in the host is largely vascular. The movement of the bacterium from the leaves, through the petiole, to the stem and its further advancement to the seed clearly suggests its systemic nature. The disease is favoured by high relative humidity and a warm temperature of 28 - 32°C, which usually prevail during monsoon in India. Therefore, the disease is usually more severe in July and August. Secondary spread of the bacterium occurs mostly through wind-splashed rain. Long distance spread or spread to the new areas occurs usually through infected seed. The disease is more severe in low-lying areas where water stagnates temporarily. The disease is also severe in furrow-sown fields where water contaminated with pathogen moves from one end to another end of the furrow in sloppy fields.

Kaur *et al.* (2004) evaluated 10 genotypes under field conditions and found that cv. HG 75 was highly resistant and cv. PNB most susceptible to the disease.

Shah *et al.* (1996) found that a temperature range of 20 - 28°C and relative humidity between 70 and 80% favoured rapid progress of the disease. On the basis of epidemiological studies carried out for five years, from 1991 - 1995, they also developed a model to forecast the disease based on these parameters of the preceding week. Yadav and Nath (2006) reported that a temperature range of 29 - 34°C and relative humidity above 68% favoured the disease. They further reported that the maximum disease incidence of 58.3% was observed in the crop sown in the last week of July.

Most of the Indian improved varieties like Pusa Naubahar and Pusa Sadabahar are highly susceptible to the disease. The branching-type varieties are also susceptible but they suffer less from the disease because the new shoots compensate to some extent for the dead shoots. In addition to *guar*, bean acts as a natural host of the bacterium (Patel and Jindal, 1972).

13.35.4. Management

1. Use of pathogen-free seed and eradication of seed-borne inoculum are the most effective and practical means of disease management. The seeds from surface irrigated summer grown crops are free from infection (Srivastava and Rao, 1963a, 1970).
2. Hot water treatment at 56°C for 10 min eradicates the seed-borne inoculum. Seed dip in Streptocycline solution (200 μg/ml) for 3 hr proved highly effective for eradicating the seed-borne inoculum (Rathore, 2000). Seed treatment with Streptocycline (500 μg/ml) followed by its foliar applications gave minimum disease incidence and highest grain yield (Yadav and Nath, 2006a).
3. Fresh leaf extracts of *Emblica officinalis* and *Eucalyptus* sp. have been demonstrated to be effective in reducing the disease under field conditions (Kaur *et al.*, 2005, 2006).
4. Use of resistant varieties.

REFERENCES

Kaur, K., Jalali, I. and Chhabra, M.L. (2005). Efficacy of botanicals for the management of bacterial blight of clusterbean under disease stress conditions. *Proc. Natl. Symp. Plant Disease Diagnosis, Epidemiology and Management*, Hisar. 47pp.

Kaur, K., Jalali, I. and Pal, V. (2006). Factors affecting the bioefficacy of phyto-extracts against *Xanthomonas axonopodis* pv. *cyamopsidis*. *Pl. Dis. Res.* 21: 164 - 67.

Kaur, K., Jalali, I. and Pal, V. (2010). Toxicity of botanicals against clusterbean bacterial blight pathogen. *J. Mycol. Pl. Pathol.* 40: 124 - 27.

Kaur, B., Purkayastha, S., Dilbaghi, N. *et al.* (2004) Evaluation of clusterbean genotypes for resistance to bacterial blight. *Ann. Agri. Bio. Res.* 9: 213 - 15.

Patel, M.K., Dhande, G.W. and Kulkarni, Y.S. (1953). Bacterial leaf-spot of *Cyamopsis tetragonaloba* (L.) Taub. *Curr. Sci.* 22: 183.

Patel, P.N. and Jindal, J.K. (1972). Bacterial blight of guar. In: P.N. Patel (ed.). *Plant Bacteriology*. Vol. I, *Bacterial Diseases of Plants in India*. Indian Agricultural Research Institute, New Delhi, pp. 66 - 69.

Patel, A.J. and Patel, M.K. (1958). A new bacterial blight of *Cyamopsis tetragonaloba* (L.) Taub. *Curr. Sci.* 27: 258 - 59.

Rathore, B.S. (2000). Management of bacterial blight of clusterbean through seed dressers. *Plant Dis. Res.* 15: 89 - 92.

Shah, R., Dashora, P.K., Bhatnagar, M.K. *et al.* (1996). Forecasting model for bacterial blight of clusterbean. *Proc. 9th Int. Conf. Plant Path. Bact.* Chennai, India, pp. 525 - 27.

Srivastava, D.N. and Rao, Y.P. (1963). Bacterial blight of guar. *Indian Phytopathol.* 16: 69 - 73.

Srivastava, D.N. and Rao, Y.P. (1963a). Seed transmission and control of bacterial blight of guar (*Cyamopsis tetragonaloba* (L.) Taub.). *Indian Phytopathol.* 16: 390 - 91.

Srivastava, D.N. and Rao, Y.P. (1970). Epidemiology and control of bacterial blight of guar (*Cyamopsis tetragonaloba* (L.) Taub.). Indian Phytopathol. Soc. Bull. No 6, 1 - 4.

Yadav, S. and Nath, R. (2006). Effect of weather parameters and different sowing times on the occurrence of clusterbean bacterial blight. *J. Mycopathol. Res.* 44: 267 - 69.

Yadav, S. and Nath, R. (2006a). Efficacy of chemicals against bacterial blight of clusterbean. *J. Mycopathol. Res.* 44: 271 - 273.

13.36. BACTERIAL LEAF SPOT OF HYACINTH BEAN

Lablab bean, hyacinth bean, Indian bean or Indian butter bean (*Lablab purpureus,* syn. *Dolichos lablab, Lablab niger*) is also called by another name in India, i.e., *sem*. The first natural occurrence of bacterial leaf spot on hyacinth bean was reported by Rangaswami and Gowda (1963) from Tamil Nadu, India. Subsequently, Patel and Jindal (1970) found it at Indian Agricultural Research Institute, New Delhi and Amravati in Maharashtra, India. The disease also occurs in Sudan (Sabet, 1959).

13.36.1. Symptoms

The disease is more severe during rainy season. The symptoms appear mainly on leaves, which are round or irregular, water-soaked spots. These spots eventually turn brown. Several such spots may coalesce. In a relatively dry weather, tissues surrounding closely situated spots may turn yellow and form large blotches but it never leads to an extensive blightening. Due to severe spotting, leaves turn yellow and fall off prematurely. On stems, oblong, reddish-brown lesions, sometimes having greyish centres, appear. On pods, the spots are first water-soaked but later on turn brown.

13.36.2. Causal organism

Xanthomonas phaseoli hyacinth bean strain.

13.36.3. Disease cycle

The bacterium is seed-borne and also survives on vines growing perennially in the kitchen gardens. Warm and wet weather prevailing during monsoon season favours the disease development. In hyacinth or Indian bean, the bacterium is largely confined to parenchymatous tissue while in common bean (*Phaseolus vulgaris*), it enters the vascular system. New growth on infected vines is generally free from infection under dry conditions. Considerable reduction in yield occurs when plants are badly infected or disease severity is high in monsoon season. Seedling mortality occurs very rarely (Patel and Jindal, 1972).

13.36.4. Host range

Rangaswami and Gowda (1963) reported that the bacterium infected *D. biflorus, Vigna catjang, P. vulgaris, Tophrosia purpurea and Cyamopsis psoralioides.* However, Patel and Jindal (1973) did not obtain infection on the above-mentioned hosts except *P. vulgaris.* The susceptibility of *D. lablab* to *Xa. fuscans* subsp. *fuscans* and *Xa. axonopodis* pv. *glycines* has been reported. Whether *D. lablab* serves as a natural host of these pathogens is not known.

Patel and Jindal (1973) evaluated 334 germplasm lines against the pathogen and found none of them to be immune or highly resistant. However, PLS 63, PLS 101, M.S. 9491 and D.L. 2434 showed moderate resistance as the lesions remained small, turned necrotic and the infected leaves did not defoliate.

13.36.5. Management

1. As the bacterium is seed-borne, use of pathogen-free seed or treatment of infected seed can prove useful. Seed treatment recommended for bacterial blight of cowpea can be tried.

2. As the bacterium is mainly confined to the parenchyma, foliar sprays of Streptocycline @ 100 μg/ml can also prove effective.

REFERENCES

Patel, P.N. and Jindal, J.K. (1970). Bacterial diseases in seed legumes in 1968 and 1969. Paper presented at the Fourth Annual Workshop Conference on Pulse Crops held at Ludhiana (Punjab) in April, 1970.

Patel, P.N. and Jindal, J.K. (1972). Bacterial leaf spot of *Dolichos lablab*. In: P.N. Patel (ed.). *Plant Bacteriology*. Vol. I, *Bacterial Diseases of Plants in India*. Indian Agricultural Research Institute, New Delhi, pp. 84 - 86.

Patel, P.N. and Jindal, J.K. (1973). Studies on resistance in crops to bacterial diseases in India. VI: Reaction of moth bean (*Phaseolus acontifolius*) and hyacinth bean (*Dolichos lablab*) lines to bacterial blight and leaf spot diseases incited by *Xanthomonas phaseoli* strains. *Indian Phytopathol.* 26: 607 - 08.

Rangaswami, G. and Gowda, S.S. (1963). On some bacterial diseases of ornamentals and vegetables in Madras state. *Indian Phytopathol.* 16: 74 - 85.

Sabet, K.A. (1959). Studies in the bacterial diseases of Sudan crops. III. On the occurrence, host range and taxonomy of the bacteria causing leaf blight diseases of certain leguminous plants. *Ann. Appl. Biol.* 47: 318 - 31.

13.37. BACTERIAL LEAF SPOT AND STEM CANKER OF PIGEON PEA

Bacterial leaf spot and stem canker of pigeon pea (*Cajanus cajan*), also called *arhar* in India, was first reported from Bombay state by Kulkarni *et al.* (1950). Later on, it was reported from other states, namely, Tamil Nadu, Punjab, Uttar Pradesh, Madhya Pradesh, Delhi and Haryana. The disease also occurs in China and Sudan.

13.37.1. Symptoms

The symptoms first appear as small, round, light brown, water-soaked spots on the upper surface of leaves, which later on turn dark brown, become angular in shape and are surrounded by haloes. The upper surface of the spots becomes raised due to exudation and drying of the bacterial ooze. The spots may coalesce to from large lesions of 2 mm in diameter. Due to excessive spotting, the leaves turn yellow and defoliate prematurely. In severe infection, the spots appear all over the leaf, infection extending to main and lateral veins, and petioles; resulting in yellowing of the leaves and ultimately shedding of the leaves. On the main stem and lateral branches, dark brown cankers measuring 12.5 mm long and 2.5 mm wide appear. When these cankers are numerous and close to each other, they cause peeling of the bark. Generally, the cankers on the stem are not very deep and do not cause girdling of the stems. A gummy bacterial exudate comes out from these cankers. An early severe infection may cause death of the seedlings (Patel and Jindal, 1972).

13.37.2. Causal organism

Xanthomonas axonopodis pv. *cajani* (Kulkarni *et al.* 1950) Vauterin *et al.* 1995

Syn. *Xa. campestris* pv. *cajani*

Kulkarni *et al.* (1950) described the bacterium for the first time and named it as *Xanthomonas cajani* sp. nov. Dye (1978) changed it to *Xa. campestris* pv. *cajani* and later on, Vauterin *et al.* (1995) reclassified it as *Xa. axonopodis* pv. *cajani*. The below given description of the bacterium is based primarily on Kulkarni *et al.* (1950).

Cells are short rods (0.9 - 1.4 × 1.3 - 2.2 μm). Gram-negative, encapsulated, not acid-fast, nonsporing and motile by a single polar flagellum. Colonies on potato dextrose agar are naphthaline yellow, shining, pulvinate, smooth having entire margins and 1.5 cm in diam after 7 days of incubation. Colonies on nutrient agar are round, slightly raised, yellow, shining and attain a diam of 0.7 cm after 4 days of incubation. The bacterium reduces litmus, peptonizes milk, liquefies gelatin, hydrolyses starch and casein, produces hydrogen sulphide and ammonia but nitrate reduction, M.R. and V.P. tests are negative. The bacterium produces acid but no gas from glucose, sucrose, dextrose, lactose and maltose and does not utilize salicin. Citrates are utilized. Optimum temperature for growth is 26 - 32°C and thermal death point at 51°C. Growth is aerobic.

13.37.3. Disease cycle

The bacterium is seed-borne. There is no information on its survival in the infected plant debris. The bacterium enters the host through stomata and wounds. The pathogen does not infect any other host (Kulkarni *et al.*, 1952)

The disease is generally more severe during monsoon (July - September) when the relative humidity is 80 - 90%, air temperature ranges from 24 - 31°C and the crop is succulent. New leaves formed after the rains are generally free from infection. When the pigeon pea is grown as a mixed crop with sorghum, the high humidity and low temperature resulting from shading by the sorghum plants also increase the disease severity.

13.37.4. Management

1. Use of resistant varieties is the most effective and economical method.

2. Use of pathogen-free seed will ensure the disease free crop because the infected seed is the only source of perpetuation of the bacterium. To eliminate the seed-borne infection, seed treatments used for managing bacterial blight of cowpea and bacterial blight of clusterbean can be tried.

REFERENCES

Dye, D.W. (1978). Genus IX *Xanthomonas* Dowson 1939. In: Young, J.M., Dye, D.W., Bradbury, J.F. *et al.* A proposed nomenclature and classification for plant pathogenic bacteria. *N. Z. J. Agric Res.* 21: 162 - 66.

Kulkarni, Y.S., Patel, M.K. and Abhyankar, S.G. (1950). A new bacterial leaf-spot and stem canker of pigeon pea. *Curr. Sci.* 19: 384.

Kulkarni, Y.S., Patel, M.K. and Abhyankar, S.G. (1952). A new bacterial leaf-spot and stem canker of pigeon pea. *Indian Phytopathol.* 5: 21 - 22.

Patel, P.N. and Jindal, J.K. (1972). Bacterial leaf spot and stem canker of pigeon pea. In: P.N. Patel (ed.). *Plant Bacteriology.* Vol. I, *Bacterial Diseases of Plants in India.* Indian Agricultural Research Institute, New Delhi, pp. 87 - 89.

Vauterin, L., Hoste, B., Kersters, K. *et al.* (1995). Reclassification of *Xanthomonas. Int. J. Syst. Bacteriol.* 45: 472 - 89.

13.38. BACTERIAL BLIGHT OF SESAME

Bacterial blight of sesame (*Sesamum indicum* L.) was first reported from Sudan by Boughey in 1942. Sabet and Dowson (1960) described the symptoms and identified the bacterium as *Xanthomonas sesami.* The disease occurs in most of the sesame growing regions of the world and causes considerable losses under favourable weather conditions. The complete loss of crop, particularly under rain fed conditions, has been reported from Sudan (Sabet and Dowson, 1960). Vijayat and Chakravarti (1977) reported 60% loss in the capsules under field conditions from Rajasthan, India. Approximately 20% loss in yield have been reported from Jabalpur area of Madhya Pradesh, India by Shukla *et al.* (1972). The disease has also been reported from Korea, Pakistan and Venezuela.

13.38.1. Symptoms

The plants are susceptible at all the growth stages. Small, water-soaked, dark green, marginal spots develop on the lower surface of cotyledons of the seedlings developing from infected seeds. In severe infection, the lesions spread rapidly covering the whole cotyledons, which eventually dry. However, in mild infection, small, scattered, brown, necrotic spots appear on the cotyledons. In highly susceptible varieties, the infection of the growing point may cause the death of seedlings. In surviving seedlings, dark brown, water-soaked spots appear on the true leaves. These spots turn dark olive-green, increase in size (2 - 3 mm) and turn dark reddish-brown to black. The spots are initially round but often attain angular shape. The spots may remain small or may coalesce to form large patches/blotches. Later on, the dead tissue of the spots dries out, becomes brittle and ultimately the leaves fall. The infection may also start from the margin of the leaves and proceeds towards the midrib forming light brown spots. The infected leaves ultimately show severe blightening and finally dry.

On stem, the infection starts as light brown discolouration and spreads along the stem forming slightly raised dark reddish-brown lesions. Honey dew-like bacterial ooze comes out from the infected areas of the stem. On capsules, oval, slightly raised, dark reddish-brown lesions are produced. The infection also occurs on petioles and flowers.

Bashir *et al.* (2007) found that the combined infection of *Pseudomonas syringae* pv. *sesami* and *Xa. campestris* pv. *sesami,* causal agents of bacterial leaf spot and bacterial blight of sesame, respectively, resulted in more disease severity than the individual infections of these pathogens. The percentages of stem and leaf infection, and defoliation were more in combined infection than the individual infections. Under such conditions, it becomes difficult to distinguish the symptoms of the individual diseases.

13.38.2. Causal organism

Xanthomonas campestris pv. *sesami* (Sabet & Dowson 1960) Dye 1978

Syn. *Xa. sesami*

The bacterium being a pathovar of *Xa. campestris* species, has the same morphological, cultural and biochemical characters as those of *Xa. campestris.* It can be distinguished from *Xa. campestris* and its other pathovars on the basis of host range. The description of *Xa. campestris* is given under black rot of crucifers.

13.38.3. Disease cycle

Infected seeds carry the bacterium and effectively transmit it to the seedlings. The bacterium survives in the infected seeds up to 16 months and in soil from 4 - 6 months (Habish and Hamad, 1971). As the bacterium does not survive in the soil up to the sowing of the next crop, infected seed is the only source of perpetuation of the bacterium. The bacterium enters the host mainly through stomata and quickly invades vascular tissue. Secondary spread is mainly through the wind-splashed rain. The spread to the distant areas occurs through the infected seed (Chand and Singh, 1972).

High temperature and high humidity favour the disease development. The disease mainly develops in the rainy season or when the relative humidity is high at night. Maximum disease intensity was observed in September 1993 and 1994 in Kanpur, U.P, India when the mean temperature ranged between 29.0 and 29.4°C, relative humidity was 88.0 to 90.5% and the rainfall was 8.90 to 9.97 mm. There was considerable reduction in disease severity by the end of September due to the decrease in temperature, relative humidity and rainfall (Srivastava *et al.*, 1997). The severity of disease is also related to soil moisture (30 - 40%) and relative humidity (75 - 85%).

Age of leaves influences the severity of disease. The disease index on older leaves is much less than that on fully expanded young leaves. Multiplication of the pathogen in the artificially inoculated leaves reached maximum on 3rd day in both young and old leaves but the number of bacterial cells in young leaves was 10 times more than those in old leaves (Chand *et al.*, 1970). Nitrogen and photoperiod influence the susceptibility of the varieties. Application of nitrogen under long photoperiod increased the susceptibility of variety Venezuela 51 while its susceptibility decreased under the short photoperiod. However, nitrogen under short photoperiod increased the susceptibility of variety Early Russian with little effect under long photoperiod. Recently, Firdous *et al.* (2007) reported that tomato gave the better hypersensitive reaction than potato to the bacterium, and high relative humidity and $1x10^9$ cfu/ml were required for the hypersensitive reaction.

13.38.4. Management

The following measures are helpful in reducing the disease incidence.

1. Use of resistant varieties like T 58.
2. Sow the crop early, i.e., immediately after the onset of monsoon.
3. Use of pathogen-free seed. In the absence of pathogen-free seed, the seed should be treated with hot water at 52°C for 10 minutes followed by soaking in streptomycin solution (100 μg/ml) for 6 - 8 hr. Chand *et al.* (1970) found that seed treatment with Ceresan Wet (0.1%), streptomycin (100 μg/ml) and their combination for 80 minutes reduced seedling infection but did not completely eliminate it, and the secondary spread was so rapid that there were no differences in disease incidence among the treated and untreated seed plots.
4. Foliar sprays of Streptocycline solution (100 μg/ml) are recommended for the control of the disease but the best control is achieved with the combination of seed treatment and foliar

sprays. Rai and Srivastava (2003) obtained an effective control o the disease along with an increase in yield by seed treatment with Streptocycline (100 μg/ml) followed by 3 sprays of Streptocycline (100 μg/ml) + Blue Copper 50 (copper oxychloride @ 500 μg/ml) at 10 days intervals.

REFERENCES

Bashir, S., Irfan ul-Haque, M., Mukhtar, T. *et al.* (2007). Pathogenic variation in *Pseudomonas syringae* and *Xanthomonas campestris* pv. *sesami* associated with blight of sesame. *Pak. J. Bot.* 39: 939 - 43.

Chand, J.N., Mandloi, S.C. and Kulkarni, S.N. (1970). Studies on bacterial blight of sesamum caused by *Xanthomonas sesami.* Symposium on approaches to control important bacterial diseases of plants in India. Indian Phytopathol. Soc. Bull. No. 6, 19 - 23.

Chand, J.N. and Singh, D. (1972). Bacterial blight of sesamum. In: P.N. Patel (ed.). *Plant Bacteriology*. Vol. I, *Bacterial Diseases of Plants in India.* Indian Agricultural Research Institute, New Delhi, pp. 169 - 72.

Firdous, S.S., Asghar, R., Haque, H.I. *et al.* (2007). Development of Hypersensitive response by *Xanthomonas campestris* pv. *sesami* on *Lycopersicon esculentum* L. and *Solanum tuberosum* L. leaves. *Pak. J. Bot.* 39: 2135 - 39.

Habish, H.A. and Hamad, A.H. (1971). Survival and chemical control of *Xanthomonas sesami. F.A.O. Plant Prot. Bull.* No. 19: 36 - 40.

Rai, M. and Srivastava, S.S.L. (2003). Field evaluation of some agro-chemicals for management of bacterial blight of sesame. *Farm Sci. J.* 12: 81 - 82.

Sabet, K.A. and Dowson, W.J. (1960). Bacterial leaf spot of sesame (*Sesamum orientale* L.). *Phytopathol. Z.* 37: 252 - 58.

Shukla, B.N., Chand, J.N. and Kulkarni, S.N. (1972). Changes in sugar contents of sesame leaves infected with *Xanthomonas sesami. Nature* (London) 213 (5078): 813.

Srivastava, S.S.L., Rai, M. and Singh, D.V (1997). Epidemiological study on bacterial blight of sesame. *Indian Phytopathol.* 50: 289 - 90.

Vijayat, R. and Chakravarti, B.P. (1977). Yield losses due to bacterial leaf spot of *Sesamum orientale* in Rajasthan. *Indian J. Plant Pathol.* 7: 97.

13.39. BACTERIAL LEAF SPOT OF SESAME

The disease was first observed by Malkoff in Bulgaria in 1903 and he named it as "Black rot". Later on, 1906, Malkoff attributed it to *Bacterium sesami* but the Koch's postulates of the bacterium were proved by Kovacevski in 1930. Since then, the disease has been reported from Sofia, the USA, Japan, Greece, Yugoslavia, Egypt and India. In India, the disease was first reported from Delhi by Durgapal and Rao (1967). The disease has now spread to most states of India through infected seed.

13.39.1. Symptoms

The symptoms develop on all the above ground plant parts as blackish spots varying in size. On leaves, the spots appear oil-soaked and have a star-like external border or they are delimited by the veins, giving them a distinct angular shape. From the spots, infection moves along the veins, turning these veins sooty in appearance. In severe infection, the spots may cover the entire leaf area. Finally, the infected leaves curl, die and defoliate. The infection on the stem appears in the form of lesions, which generally encircle the stem forming rings. The lesions appear on the whole length of the stem and frequently coalesce forming large elongated areas. The capsules, infected at an early stage, turn black and do not bear seeds. Infection on older capsules results in the formation of minute spots (about 2 mm diam). Severely infected plants may die.

13.39.2. Causal organism

Pseudomonas syringae pv. *sesami* (Malkoff 1906) Young *et al.* 1978

Syn. *Psm. sesami*

The major characteristics of the bacterium are given below.

Bacterial cells are Gram-negative, nonsporing, non-encapsulated rods measuring 0.6 - 0.8 by 1.2 - 3.8 μm and motile by 2 - 5 polar flagella. The cells may occur singly or in pairs. Colonies on beef agar are circular, flat, striate, smooth and white having entire margins. Green fluorescent pigment is produced in culture. The bacterium liquefies gelatin rapidly, and gives an alkaline reaction in litmus milk without coagulation. The tests for starch hydrolysis, nitrate reduction and production of indole and hydrogen suphide are negative. Acid but no gas is produced from glucose, however, from sucrose, lactose and glycerol, even acid is not produced. Temperature for optimum growth is 30°C. The bacterium is facultatively aerobic.

Pathogenic variability exists in the bacterium. Strains differing in virulence and races giving differential reactions on sesame varieties have been reported. Thomas and Orellana (1962) identified 2 races of the pathogen, namely, race 1 and race 2. Race 2 infects variety Margo while race 1 does not. Varieties Delco and Venezuela 51 are susceptible to both the races while variety Early Russian is resistant. Two races are similar morphologically and also react alike in several standard physiological tests, but differ in glucose-asparagine ratio requirements. At high sugar levels, the race 2 requires less asparagine than the race 1. Race 2 also liquefies gelatin and peptonizes milk faster than race 1.

Thomas and Orellana (1962, 1963) observed that the reactions of Margo, Early Russian and Venezuela 51 varieties to two races of the pathogen were related to differences in concentration of aspartic acid, leucine, histidine, cysteine and certain other amino acids and were associated closely with apparent differences in the ratio of reducing sugars to these amino acids. Delco and Venezuela 51, the most susceptible varieties, had the highest concentration of amino acids. Their findings substantiate that the two races differ in the glucose asparagine ratios at which they grow.

13.39.3. Disease cycle

The bacterium is transmitted through infected seed and this probably is the only source of perpetuation of the bacterium. Fawzi *et al.* (1991) showed that the bacterium is internally seed-borne. After germination, the bacteria from the infected seeds infect the cotyledonary leaves and from there the bacteria spread to all the above ground plant parts. The pathogen enters the host through stomata and wounds and the infection remains confined to the parenchymatous and vascular tissues. The infection of the capsules results in the infection of the seeds. Warm temperature and wind-splashed frequent rains are the most favourable factors for the development of epiphytotics (Chand and Singh, 1972).

Firdous *et al.* (2009) conducted histological studies to elucidate the mode of infection of *Psm. syringae* pv. *sesami* and reported that periodical examination of cleared discs taken from inoculated leaves showed the presence and multiplication of bacterial cells in depressions and around trichome bases for 24 hr before penetration through stomata and trichome basal cells. They further reported that appearance of pin-point-sized spots three days after inoculation coincided with chloroplast damage; and early appearance of small necrotic spots four days after inoculation revealed the presence of the bacterium in parenchymatous tissues and from there it moved to transverse vascular systems.

Sesame appears to be the only natural host although some other hosts are mildly infected artificially (Sutic and Dowson, 1962). However, these hosts do not play any role in the disease cycle.

Application of nitrogen and photoperiod considerably influence the reaction of sesame varieties to the pathogen. Thomas (1965) studied the effect of 12 and 16 hr photoperiod per day along with and without supplemental nitrogen on the reaction of resistant variety Early Russian and susceptible variety Venezuela 51. The most favourable combination for disease development was short light period coupled with supplemental nitrogen and with this combination both the varieties gave susceptible reaction. However, with either light period and without supplemental nitrogen, the original reactions of the varieties remained unchanged.

13.39.4. Management

1. Grow resistant varieties. White-seeded early varieties perform better than other varieties.
2. Seed treatments and foliar sprays recommended for bacterial blight of sesame are also effective against this disease.

REFERENCES

Chand, J.N. and Singh, D. (1972). Bacterial leaf spot of sesame. In: P.N. Patel (ed.). *Plant Bacteriology.* Vol. I, *Bacterial Diseases of Plants in India.* Indian Agricultural Research Institute, New Delhi, pp.173 - 77.

Durgapal, J.C. and Rao, Y.P. (1967). Bacterial leaf spot of sesamum (*Sesamum orientale* L.) in India. *Indian Phytopathol.* 20: 178 - 179.

Fawzi, F.G., El Wakil, A.A., El Deeb *et al.* (1991). Association of *Pseudomonas syringae* pv. *sesami* with some sesame seed lots in Egypt. *Egyptian J. Agric. Res.* 69: 685 - 93.

Firdous, S.S., Asghar, R., Ul-Haque, M.I. *et al.* (2009). Pathogenesis of *Pseudomonas syringae* pv. *sesami* associated with sesame (*Sesamum indicum* L.) bacterial leaf spot. *Pak. J. Bot.* 41: 927 – 34.

Sutic, D. and Dowson, W.J. (1962). Bacterial leaf spot of sesamum in Yugoslavia. *Phytopathol. Z.* 45: 57 - 65.

Thomas, C.A. (1965). Effect of photoperiod and nitrogen on reaction of sesame to *Pseudomonas sesami* and *Xanthomonas sesami. Plant Dis. Rptr* 49: 119 - 20.

Thomas, C.A. and Orellana, R.G. (1962). Resistance of sesame varieties and pathogenicity of strains of *Pseudomonas sesami* in relation to amino acids and reduced sugars (Abstr.). *Phytopathology* 52: 30.

Thomas, C.A. and Orellana, R.G. (1963). Amino compounds and reducing sugar in relation to resistance of sesame to *Pseudomonas sesami. Phytopathol. Z.* 46: 101 - 04.

13.40. CROWN GALL OF PLANTS

Most of the bacterial plant pathogens cause necrotic or hypoplastic effect on the host tissue. However, crown gall bacterium *(Agrobacterium tumefaciens)* has an opposite effect, i.e., it causes hyperplasia and hypertrophy of the affected tissue. The disease is worldwide in occurrence and the causal bacterium has the widest host range of any plant pathogen. It infects a great variety of dicotyledonous and some gymnospermous plants. However, it does not infect monocots (grasses). More than 643 plant species belonging to 331 genera and 93 families are infected by the bacterium. However, the disease is more severe on all stone fruits; apple, pear, blueberry, grapes and some ornamental plants such as rose.

Many hosts are only known through artificial inoculation. Tomato acts as a very good indicator host, but has never been found as a natural host. The detailed host range can be found in De Cleene and De Ley (1976) and Bradbury (1986) and also on http://www.cabicompendium.org/cpc/home.asp.

The disease is known to occur in the USA since 1870. In 1897, Cavara from Italy was the first to isolate the bacterium from stem galls of grapes and prove its pathogenicity. However, between 1892 and 1894, numerous investigators from Europe and America worked on the disease without isolating the bacterium. Smith and Townsend (1907) described the bacterium and named it as *Bacterium tumefaciens,* which later on, was changed to *Agrobacterium tumefaciens* by Conn in 1942.

In India, the occurrence of the disease was reported by Singh in 1943 but its authentic report was made by Durgapal (1971) who isolated the bacterium and proved its pathogenicity. The disease occurs on cherry, peach, plum and pecan nut (*Carya illinoensis*).

Two other diseases, having some similarities with crown gall are caused by two other *Agrobacterium* species. *Ag. rhizogenes* causes hairy root of apple and some other plants. *Ag. rubi* causes galls on canes of *Rubus* spp. (brambles). Hairy root of apple is described separately.

13.40.1. Symptoms

The galls or overgrowths occur commonly on the subterranean roots and crowns of trees and shrubs (Fig. 13.40). They may also be formed on stem, branches, leaves and leaf-veins. Aerial galls appear generally on grapevines, raspberry and black berry. Aerial galls are not common on fruit trees, although they can occur. Galls are usually formed in late spring or early summer.

Young galls are more or less spherical, white or flesh-coloured and soft. As they grow, their surface becomes convoluted. In advanced stages, outer tissue of the galls turns dark brown to black due to death and decay of peripheral cells. Breakdown of peripheral tumour tissue releases crown gall bacteria into the soil. As the galls age, they become hard, rough, woody and more cleft. Relatively soft galls are disintegrated while those consisting of a corky exterior and a hard woody interior persist on the host. The size of galls may vary from 0.5 – 10 cm in diameter but in extreme cases they may be 30 cm or more in diameter.

Occasionally, complex tumours called **teratomas** are formed. These tumours may give rise to roots and buds and the buds give rise to shoots carrying leaves. In some hosts, secondary tumours develop at points distant from the primary tumours. These tumours are sterile, i.e., they do not contain bacteria and their tissue can be grown on chemically defined media. When tissues of such tumours are transplanted into a healthy plant, sterile tumours are produced. Crown galls or tumours caused by *Ag. tumefaciens* differ from other hyperplastic diseases that the former result from uncontrolled growth or proliferation of host cells due to the transformation of infected cells, caused by the pathogen.

Aerial plant parts of mildly infected plants may appear normal. However, in case of severe infection, the plants are stunted, unthrifty and bear small and chlorotic leaves. In general, infected plants are more susceptible to adverse environmental conditions, particularly to winter injury. The yield of severely infected plants is reduced and most severely infected plants and vines may die.

Crown gall of grape is an important disease of wine grapes in cold climates. In grapes, the galls are formed mostly on the aerial parts of the vines. The galls on grape roots are not typical; however the bacteria can induce a localized necrosis of roots. Most grape cultivars are susceptible to the disease. The incidence of grape crown gall from year to year is generally related to the severity of the preceding winter and the age of the vines. For more details on the disease, the readers can refer to the review article by Burr and Otten (1999).

Cane gall: Small protuberances or elongate ridges appear on the fruiting canes of *Rubus* spp. somewhat before the middle of growing season. They increase in size and may completely cover sections of the surface, especially in the lower portion of the cane. The gall tissue is whitish at first but turns brown with age. As the season advances, severely infected canes split open, the leaves wither and the berries dry up.

The simultaneous occurrence of crown gall, hairy root and wound overgrowths is commonly found in nature. Wound overgrowths, non-parasitic in nature, are generally confused with crown gall since they may occur along with the latter. Such overgrowths often occur on apple trees at the margin of root and scion, particularly when an imperfect union has been made (Walker, 1969).

Besides crown gall, some plant pathogenic bacteria produce galls on twigs, limbs and trunks of plants. In the latter case, bacterial pockets are found in light green to light yellow soft parenchymatous tissues surrounded by white woody tissues. Hence, the section of a tumour or gall should be examined for identification/differentiation of the disease.

13.40.2. Causal organism

Agrobacterium tumefaciens (Smith & Townsend 1907) Conn 1942

Syn. *Bacterium tumefaciens, Rhizobium radiobacter.*

For detailed taxonomic discussion and reclassification of *Agrobacterium* species into *Rhizobium*, see Young *et al.* (2001). Garrity *et al.* (2005) have recognized both *Agrobacterium* and *Rhizobium* as separate genera in the family Rhizobiaceae. *Ag. tumefaciens* is used here as a valid name.

Bacterial cells are Gram-negative, nonsporing rods measuring 0.6 - 1.0 × 1.5 - 3.0 μm and motile by 1- 6 peritrichous flagella. Aerobic, chemooganotrophs, soil inhabitants and possess a respiratory type of metabolism with oxygen as the terminal electron acceptor. Colonies are generally circular, smooth, convex, and colourless to light beige. Growth on media containing carbohydrates is usually copious and slimy. The species is divided into three biovars, i.e., biovar 1, 2 and 3. The mol% of G + C of the DNA is 57 - 63 (T_m). *Ag. tumefaciens* strongly absorbs Congo red and aniline blue in contrast to little or no absorption by *Ag. rhizogenes*. *Ag. tumefaciens* produces an abundant growth on sodium selenate agar and calcium glycerophosphate medium with mannitol in contrast to no or very little growth by *Ag. rhizogenes*.

The genome of *Ag. tumefaciens* str. C 58 consists of a circular chromosome (2.8 Mb), a linear chromosome (2.0 Mb) and two plasmids, namely, tumour-inducing plasmid called Ti- plasmid (206.479 kbp) and At-plasmid called the cryptic plasmid. The pathogenic ability, i.e., the ability to cause galls/tumours on the host plants is governed by the Ti-plasmid. The strains, which lack Ti-plasmid are non-pathogenic. The plants infected with *Ag. tumefaciens* str. C 58 produce nopaline type of opine. If the bacterium is grown near its maximum temperature of 30ºC, then the bacterium loses its Ti-plasmid as well as its pathogenicity. Plasmid At of str. C 58 has been shown to be involved in the metabolism of opines and conjugation. The genome of str. C 58 was simultaneously sequenced by Goodner *et al.* (2001) and Wood *et al.* (2001).

Al-Karablieh and Khlaif (2002) found that the detection of *Ag. tumefaciens* from infected and symptomless plants by detecting *tmr* gene was more efficient than the routine isolation of the bacterium because the former method was more sensitive.

Ag. tumefaciens is an excellent genetic engineering tool. The DNA transformation capabilities of the bacterium have been extensively exploited in biotechnology as a means of inserting foreign genes into plants. Nester (2008) has rightly called it a **natural genetic engineer** and also a **household name** in plant genetic engineering and plant molecular genetics.

Ag. vitis Ophel & Kerr 1990 causes crown gall of grape. Earlier, the strains of agrobacteria, causing crown gall of grapes, were classified as biovar 3 of *Ag. tumefaciens*. Taxonomic evaluation of biovar 3 strains by Ophel and Kerr in 1990 resulted in the naming of a new species, *Ag. vitis*. The bacterium is insensitive to agrocin 84 produced by *Ag. radiobacter* strain K 84. In contrast to *Ag. tumefaciens*, it has a limited host range. According to Kawaguchi (2009), *Rhizobium vitis* (syn. *Ag. vitis*) strains are genetically heterogeneous and can be separated into several genetic groups. Although *Ag. vitis* is mainly responsible for causing crown gall of grapes, *Ag. rhizogenes* and *Ag. tumefaciens* also infect grapes. Out of 56 tumorigenic Spanish grapevine strains of *Agrobacterium* spp., three strains were identified as *Ag. rhizogenes* (biovar 2), three as *Ag. tumefaciens* (biovar 1) and the remaining 50 as *Ag. vitis* (biovar 3) (Palacio-Bielsa *et al.*, 2009). The strains of these three species were clearly distinguished on the basis 16S rRNA gene sequencing. However, no universal primer pair was found for the detection of these three species.

13.40.3. Disease cycle

The bacterium is soil inhabitant and remains viable in soil for several years. It can also survive in infected plant debris, on and around root surfaces utilizing nutrients leaked and in the outer portion of primary galls. The bacterium is widely distributed on the living planting stock. The bacterium penetrates the host through wounds, fresh and deep wounds are more conducive for infection. The wounding of the plants results from pruning, cultural operations, growth cracks, grafting and chewing insects. The dissemination of the bacterium occurs through wind-splashed rain and chewing insects. Abundant bacterial cells present on the surface of galls are washed down by rainwater and carried away by the flowing water. Long distance spread of the pathogen occurs through the infected nursery and planting material, bacterial cells present on the surface or inside the host tissue. The tumour formation occurs between 10 and 33°C. In some hosts like chrysanthemum, the bacterium also causes systemic infection.

Infection process and the host pathogen interaction in this disease have been elucidated thoroughly. Nester (2008) has given a comprehensive account of host-parasite interaction leading to the development of tumours. A single bacterial cell can cause infection under favorable conditions. The wounding of host cells results in the release of cell sap containing phenolic and sugar compounds. The bacterial cells are attracted by these chemicals such as acetosyringone, swim through soil water with the help of flagella and attach to the wounded plant cells. Attachment of bacteria to plant cells is accomplished through two steps. After an initial weak and reversible attachment, the bacteria synthesize cellulose fibrils that anchor them to the wounded plant cells. Four main genes involved in this process are *chvA, chvB, pscA* and *att*. The products of the first three genes appear to be involved in the actual synthesis of the cellulose fibrils. These fibres also anchor the bacteria to each other, resulting in the formation of a microcolony. After production of cellulose fibrils, a Ca^{2+} dependent outer membrane protein called rhicadhesin is produced that also helps in sticking the bacteria to the cell wall.

A large fragment of Ti-plasmid, called transfer DNA or T-DNA is excised from the plasmid, which after insertion into the host cell gets incorporated at a semi-random location into the plant genome. For transferring T-DNA, *Ag. tumefaciens* uses a Type IV secretion system, involving the production of T-pilus. Two-component sensor system of Vir A/Vir G is able to detect phenolic signals, particularly acetosyringone and hydroxyacetosyringone released by wounded plant cells. This results in a signal transduction event, activating the expression of 11 genes within the Vir B operon, which are responsible for the formation of T-pilus. First of all, the Vir B propilin is formed, which is a polypeptide of 121 amino acids and is processed by the removal of 47 residues to form a T-pilus subunit. The subunit becomes circular by the formation of a peptide bond between two ends of the polypeptide. Products of the other Vir B genes transfer the subunits across the plasma membrane. Evidence provided by various studies shows that all the genes from Vir B6 to Vir B10 might encode components of the transporter. An ATP-ase is also required for the active transport of the subunits (Wikipedia, 2008).

Before transferring T-DNA, it is removed from the Ti-plasmid. A Vir D1/D2 complex cuts the DNA at the left and right border sequences. The Vir D2 protein is covalently attached to the 5' end. Vir D2 contains a motif that leads to the nucleoprotein complex being targeted to the type IV secretion system. In the cytoplasm of the host, T-DNA complex gets coated with Vir E2 proteins, which are exported through the type IV secretion system independently from the T-DNA complex. Nuclear localization signals located on the Vir E2 and Vir D2 are recognized by the importin alpha protein, which then associates with importin beta and the nuclear pore complex to transfer the T-DNA into the nucleus. Another protein, VIP 1 also plays an important role in the process, possibly by acting as an adapter to bring the Vir E2 to the importin. After entering the nucleus, VIP 2 may target the T-DNA to the areas of chromatin that are being actively transcribed so that the T-DNA can integrate into the host genome (Wikipedia, 2008). Abraham *et al.* (2005) put forth the hypothesis that *Arabidopsis thaliana* Vir E2 interacting protein 1 (AtVIP 1), which associates with the T-complexes by binding to their Vir E2 component, acts as a molecular link between the T-complex and the histone constituents of the host chromatin. The hypothesis was based on their observations of interaction of AtVIP 1 with the histones in *in vitro* and in planta. Gelvin (2010), in his review article, has highlighted the role of plant proteins involved in *Agrobacterium*-mediated genetic transformation. Integration of T-DNA into a plant genome establishes a permanent transformation event, permitting stable expression of T-DNA-encoded transgenes. The transformation process is complex and requires participation of numerous plant proteins. The review deals with current information on plant proteins that contribute to *Agrobacterium*-mediated transformation, the roles of these proteins in the transformation process and the modern technologies that have been used to elucidate the cell biology of transformation.

T-DNA region contains genes, which cause gall formation. Among them, genes *iaam* and *laah* encode for the synthesis of indole acetic acid via the IAM pathway. As this biosynthetic pathway is not used in many plants for the production of IAA, it indicates that the plant has no molecular means of regulating it and the IAA will be produced constitutively. The *ipt* and other genes encode the sythensis of cytokinins. The production of IAA and cytokinins stimulates cell division and gall formation and explains the hormonal independence of crown gall tissue. T-DNA region also contains genes which encode the synthesis of opines. Opines, namely, nopaline, octopine, agropine and agrocinopine are present in the crown gall tissue and not in the healthy plant tissue. The infecting bacterium utilizes these opines for its nourishment as a source of energy, carbon and usually nitrogen. Strains of *Ag. tumefaciens* are frequently classified according to the opines synthesized. Although the presence of the bacterium is necessary for the induction of crown gall, once the gall is initiated, the bacterium can be withdrawn without affecting the growth of galls.

13.40.4. Management

As there is no cure of infected plants, prevention of infection is essential. Therefore, the main emphasis should be laid on the production of clean nursery stock.

1. Raise the disease-free nursery by adopting the following measures.
   (a) Nursery should be raised in the soil, free from the pathogen and chewing insects.
   (b) Destroy or dispose of the infected stock.
   (c) Well-fitted grafts wrapped with antiseptic tape should be used. Budding should be preferred over grafting.
   (d) Inspection and certification of the stock should be mandatory.
   (e) Avoid or minimize the wounding of nursery stock.
   (f) Budding and grafting tools should be properly disinfected.
2. Application of antibiotics and other bactericides has generally not been successful, partially due to their short-lived activity in soil and partially due to presence of the bacterium in the host tissue.
3. Soil fumigation tried in nurseries has also met with poor success mainly due to the occurrence of bacterium in the plant tissue.

However, in some cases, the incidence of crown gall has increased after fumigation resulting from the elimination of microorganisms which compete with the crown gall bacterium (Burr, 2001).

4. *Ag. radiobacter* strain K 84 is very effective in controlling crown gall. The seeds, cuttings and roots of young seedlings are dipped in cell suspension of strain K 84 before sowing. Since its first release in 1973 in Australia, it has been adopted in Europe, Africa, North America and South America and is used on commercial scale for the control of crown gall on stone fruits and roses. It was the first case of a microorganism to be used commercially for the control of a plant disease. However, in 1973, some reports from Greece indicated its non-effectiveness for controlling crown gall disease (Kerr, 1980).

Ag. radiobacter str. K 84 produces a bacteriocin called agrocin 84, which is inhibitory to many pathogenic strains of *Agrobacterium*. The biosynthesis of agrocin 84 is encoded by a 48 kb plasmid, called pAgK84. It is a conjugative plasmid and can be transferred to the agrobacteria including pathogens. The transfer of this plasmid makes the recipient bacterium resistant to agrosin 84, thereby resulting in non-control of the recipient bacterium with str. K 84. This might has been one of the reasons for the failure of control of crown gall in Greece. Very recently, a failure of biological control in one peach nursery in Italy, due to transfer of pAgK84 from *Ag. rhizogenes* K 84 to natural pathogenic agrobacteria has also occurred (Raio *et al.*, 2009)

The transfer of pAgK84 to plant pathogenic agrobacteria posed a potential threat to the continued success of the biological control of crown gall. Prof. Allen Kerr of Waite Agricultural Research Institute, Adelaide, Australia in collaboration with Dr. Farrand of University of Illinois identified the region of DNA that mediates the transfer of the plasmid pAgK84 to other bacteria. Prof. Kerr's team deleted this region and constructed a deletion mutant (transfer deficient, Tra) and named it as strain K 1026. In June 1987, strain K 1026 was released for experimental field trials in Waite Agricultural Research Institute, Adelaide, Australia. It was only the third recombinant DNA organism to be field-tested anywhere in the world. In 1988, Department of Agriculture, New South Wales, released strain K 1026 for commercial use as a pesticide, under the trade name **'No Gall'**. Strain K 1026 was the first recombinant DNA organism to be released for commercial use anywhere in the world. Strain K 1026 has not been approved for release in any other country, except Australia, where strain K 84 is still used for control. Strain K 1026 is not intended to be used on fruit bearing crops. Out of two biovars of *Ag. tumefaciens,* biovar 2 is more sensitive to biocontrol by strain K 84 or K 1026 of *Ag. radiobacter*.

The following precautions should be taken to ensure effective control with biocontrol agents.

1. Do not mix the biocontrol agent with chlorinated water.
2. Do not expose the control agent to direct sunlight.
3. Do not expose the control agent to either very high or freezing temperatures.

Strains K 84 and K 1026 are only effective in controlling crown gall caused by nopaline type strains of *Ag. tumefaciens* and *Ag. rhizogenes* and not effective for crown gall of grapes caused by *A. vitis*. Dipping of basal ends of grape cuttings in cell suspension of *Rahnella aquatilis* HX2 reduced the incidence of grape crown gall to 30.8% compared to 93.5% in control in a three-year field trial (Chen *et al*., 2007). F 2/5, a non-tumorigenic strain of *A. vitis* was also found effective for the control of crown gall of grapes. F 2/5 produces an antibiotic that inhibits growth of many tumorigenic *A. vitis* strains *in vitro*, but antibiotic minus mutants of F 2/5 were found to be as effective as the wild-type strain for controlling crown gall (Burr *et al*., 1997). F 2/5 gives better disease control when applied 24 hr prior to pathogen inoculation. Kawaguchi (2009) obtained an effective control of grapevine crown gall with a nonpathogenic *Rhiz. vitis* (syn. *Ag. vitis*) strain VAR03 - 1. The soaking of roots of grapevine seedlings in a cell suspension of strain VAR03 - 1 before planting in infested soil caused significant reduction in tumour formation in a greenhouse test and also gave an effective control in the field. The established population of VAR03 - 1 strain was10^6 cfu/g in the grapevine rhizosphere soil and it persisted for 2 years.

The submersion of dormant grape cuttings in water bath at 55°C for 30 min eradicates the inoculum from the cuttings except the survival of low levels of inoculum in tissues near the galls. The most effective way of producing *A. vitis* free grapevines is by propagating vines from shoot tips *in vitro* because the bacterium does not systemically invade grape shoots. The establishment of such a planting requires the selection of an area where soil is free from *A. vitis*.

REFERENCES

Abraham, L., Joseph. R., Zakai, N. *et al.* (2005). The plant VirE2 interacting protein 1. A molecular link between the *Agrobacterium* T-complex and the host cell chromatin. *Plant Physiology* 138: 1318 - 21.

Al-Karablieh, N. and Khlaif, H. (2002). Occurrence and distribution of crown gall disease in Jordan. *Phytopathol. Medit.* 41: 226 - 34.

Bradbury, J.F. (1986). *Guide to the Plant Pathogenic Bacteria*. CAB International, Wallingford, UK.

Burr, T. (2001). Crown gall. In: O.C. Maloy and T.D. Murray (eds). *Encyclopedia of Plant Pathology*. Vol. 1, John Wiley and Sons, Inc., New York, pp. 263 - 65.

Burr, T.J. and Otten, L. (1999). Crown gall of grape: Biology and disease management. *Annu. Rev. Phytopathol*. 37: 53 - 80.

Burr, T.J., Reid, C.L., Tagliati, E. *et al.* (1997). Biological control of grape crown gall by strain F2/5 is not associated with agrocin production or competition for attachment sites on grape cells. *Phytopathology* 87: 705 – 11.

Chen, F., Guo, Y.B., Wang. J.H. *et al.* (2007). Biological control of grape crown gall by *Rahnella aquatilis* HX2. *Plant Dis*. 91: 957 - 63.

De Cleene, M. and De Ley, J. (1976). The host range of crown gall. *Bot. Rev.* 42: 389 – 466.

Durgapal, J.C. (1971). A preliminary note on bacterial diseases of temperate plants in India. *Indian Phytopathol.* 24: 379 - 82.

Garrity, G.M., Bell, J.A. and Lilburn, T. (2005). The revised road map to the manual. In: D.J. Brenner, N.R. Krieg, J.T. Staley and G.M. Garrity (eds). *Bergey's Manual of Systematic Bacteriology*. 2nd Ed. Vol. 2, The *Proteobacteria*, Part A, *Introductory Essays*. Springer-Verlag, New York, pp 159 - 200.

Gelvin, S.B. (2010). Plant proteins involved in *Agrobacterium*-mediated genetic transformation. *Annu. Rev. Phytopathol*. 48: 45 – 68.

Goodner, B., Hinkle, G., Gattung, S. *et al.* (2001) Genome sequence of the plant pathogen and biotechnology agent *Agrobacterium tumefaciens* C58. *Science* 294: 2323 - 28.

Kawaguchi, A. (2009). Studies on the diagnosis and biological control of grapevine crown gall and phylogenetic analysis of tumorigenic *Rhizobium vitis*. *J. Gen. Plant Pathol.* 75: 462 – 63.

Kerr, A. (1980). Biological control of crown gall through production of agrocin 84. *Plant Dis*. 64: 25 - 30.

Nester, E. (2008). *Agrobacterium*: The natural genetic engineer: 100 years later. APSnet Feature Story, Online [http://www.apsnet.org/online/feature/Agrobacterium/].

Palacio-Bielsa, A., Gonzalez-Abolafio, R., Alvarez, B. *et al.* (2009). Chromosomal and Ti plasmid characterization of tumorigenic strains of three *Agrobacterium* species isolated from grapevine tumours. *Plant Pathol*. 58: 584 – 93.

Raio, A., Peluso, R., Puopolo, G. *et al.* (2009). Evidence of pAgK84 transfer from *Agrobacterium rhizogenes* K84 to natural pathogenic *Agrobacterium* spp. in an Italian peach nursery. *Plant Pathol*. 58: 745 – 53.

Smith, E.F. and Townsend, C.O. (1907). A plant tumour of bacterial origin. *Science*, N.S. 26: 671 - 73.

Walker, J.C. (1969). *Plant Pathology*. 3rd Ed. McGraw-Hill Book Co. New York, 819 pp.

Wikipedia (2008). *Agrobacterium tumefaciens*- Wikipedia, the free encyclopedia. Online [http://en.wikipedia.org/wiki/*Agrobacterium_tumefaciens*].

Wood, D.W., Setubal, J.C., Kaul, R. *et al.* (2001). The genome of the natural genetic engineer *Agrobacterium tumefaciens* C58. *Science* 294: 2317 - 23.

Young, J.M., Kuykendall, L.D., Martinez-Romero, E. *et al.* (2001). A revision of *Rhizobium* Frank 1889, with an emended description of the genus, and the inclusion of all species of *Agrobacterium* Conn 1942 and *Allorhizobium undicola* de Lajudie *et al.* 1998 as new combinations: *Rhizobium radiobacter, R. rhizogenes, R. rubi, R. undicola* and *R. vitis*. *Int. J. Syst. Evol. Microbiol.* 51: 89 - 103.

13.41. HAIRY ROOT OF APPLE

Hairy root is another important disease of apple though it is less serious than crown gall. It is also known by other names such as **'woolly root'** and **'woolly knot'** and was long considered merely a form of crown gall. The pathogen also infects Maluspumila, *Rosa* spp., *Rubus* spp; rose, pear, raspberry, cotoneaster, hollyhock, honey locust, honey suckle, mulberry, pea vine, peach, quince, Russian olive and spirea. Artificially inoculated host range extends to more than 37 plant species belonging to 30 genera and 15 families (De Cleene and De Ley, 1981). The disease is widely distributed in apple nurseries and orchards in the USA. It is also found on rose in California and Texas states. The disease also occurs in Japan, Bulgaria and France.

13.41.1. Symptoms

The disease commonly occurs on grafted apple nursery trees, which are usually 1, 2 or 3 years old. The production of large number of small roots either directly from stems or from roots or from localized hard swellings, that often occur at graft unions of root and scion, is the characteristic feature of the disease (Walker, 1969). The production of roots may be as profuse as that of witches' broom. It can appear simultaneously with crown gall on the same plant and the symptoms can be confused in the early stage. On rose, the hairy roots originate at the ends of cuttings or at disbud scars. These malformations may be fleshy at first but eventually become fibrous. Infected apple trees become stunted.

13.41.2. Causal organism

Agrobacterium rhizogenes (Riker *et al.* 1930) Conn 1942 emend. Sawada *et al.* 1993

Syn. *Ag. radiobacter* biovar *rhizogenes, Ag. radiobacter* pv. *rhizogenes*

The below given description of the bacterium is taken from Sawada *et al.* (1993) and Hayward and Waterston (1998). Cells are Gram-negative, non-sporing rods measuring 0.6 - 1.0 × 1.5 - 3.0 μm, motile by means 1 - 4 peritrichous flagella. Aerobic, possessing a strictly respiratory type of metabolism. Colonies are circular, convex, smooth and non-pigmented to light beige. Growth on carbohydrate containing media is usually accompanied by copious, extracellular polysaccharide slime. Optimum temperature for growth is between 25 and 28ºC and no growth occurs at 35ºC. Strains require growth factors like biotin and do not grow in simple mineral salts medium. Citrate and L-tyrosine are utilized. Acid is produced from dulcitol and α-methyl-D-glucoside in the Ayers-Rupp-Johnson's synthetic medium containing 0.02% yeast extract. 3-Ketolactose is not produced from lactose and the arginine dihydrolase test is also negative. A large tumour-inducing plasmid (Ti) or a hairy root-inducing plasmid (Ri) is detected in many strains. The mol% G + C of the DNA is 59.0 - 63.0 (Tm). Like *Ag. tumefaciens*, it is also used as a tool in genetic engineering, though less frequently.

13.41.3. Disease cycle

The epidemiology and other aspects of the disease are comparatively much less worked out than that of crown gall. Like *Ag. tumefaciens, Ag. rhizogenes* is also an exclusively wound pathogen. It is readily disseminated on nursery stock, which may become infected from soil or by tools contaminated with the pathogen when cuttings and graftings are made. On rose, the bacteria may penetrate roots through injuries caused by the root-lesion nematode, *Pratylenchus vulnus*. The bacterium survives in steamed or untreated soil for more than one year. The virulence of the bacterium has been maintained in the culture for over 19 years. The other fundamental aspects of the disease cycle are similar to those of crown gall.

13.41.4. Management

Management approaches are the same as those for crown gall.

REFERENCES

De Cleene, M. and De Ley, J. (1981). The host range of infectious hairy-root. *Bot. Rev.* 47: 147 - 94.

Hayward, A.C. and Waterston, J.M. (1998). *Agrobacterium rhizogenes. IMI Description of Fungi and Bacteria*. Sheet 41, CAB International, Wallingford, UK.

Sawada, H., Ieki, H., Oyaizu, H. *et al.* (1993). Proposal for rejection of *Agrobacterium tumefaciens* and revised descriptions for the genus *Agrobacterium* and for *Agrobacterium radiobacter* and *Agrobacterium rhizogenes. Int. J. Syst. Bacteriol*. 43: 694 - 702.

Walker, J.C. (1969). *Plant Pathology*. 3rd Ed. McGraw-Hill Book Co. New York, 819 pp.

13.42. FIRE BLIGHT OF POME FRUIT TREES

Fire blight is probably the most devastating bacterial disease of some pome fruits and also the most feared by growers (Thomson, 2001). It is most destructive on pears but certain apple and quince varieties are also highly susceptible. Crabapple (*Malus* spp.) firethorn (*Pyracantha* spp.) and hawthorn (*Crataegus* spp.) are also frequently attacked. The bacterium also occasionally attacks rock spray (*Cotoneaster* spp.) The bacterium infects more than 180 plant species including fruit trees, wild and ornamental plants, most of them belonging to family Rosaceae. Kim *et al.* (2001) have reported a similar disease on Asian pear, *Pyrus pyrifolia* from Korea and Japan caused by *Erwinia pyrifoliae*.

Fire blight is of historical importance because this was the first plant disease shown to be caused by a bacterium. Professor T.J. Burrill from the University of Illinois, USA, through a series of experiments starting from 1878 to 1884 proved that the disease is caused by a bacterium. It is considered that fire blight originated from America and was known to be there for many years before it was discovered in the Hudson Valley of New York in 1794. The disease spread to most areas of the US by the 1900s. Since then, it has spread to the whole of Europe, except Latvia and Ukraine, Iran, Israel, Jordan, Lebanon, North America, New Zealand and Turkey. The disease is known to occur in more than 40 countries. So far there is no confirmed report of its occurrence in India.

A survey in California in 1976 estimated a loss of $ 4.7 million due to fire blight while in Southwest Michigan alone the loss in 1991 was $ 3.8 million. Again in 2000, a severe outbreak of fire blight of apples occurred in Michigan, which resulted into a loss of $ 42 million due to the removal of 350000 - 450000 apple trees covering approximately 1550 - 2300 acres (Longstroth, 2001). In 1998, apple and pear growers in Washington and northern Oregon suffered a loss of $ 68 million due to fire blight damage (Douglas, 2006).

13.42.1. Symptoms

The symptoms are found on all the above ground plant parts as well as on roots. The infected blossoms, leaves, twigs and branches give a burnt or scorched appearance, hence the name fire blight. The symptoms appear in the spring when the trees start the new growth. The first sign is the exudation of watery, light tan bacterial ooze from the twig, branch or trunk cankers (holdover cankers) formed in the previous season(s). The ooze turns dark after exposure to air, spreads to form a thin glistening layer or dries down to form hard beads or films. The following different types or phases of symptoms appear. However, not all the types or phases appear in all the orchards or years or areas.

13.42.1.1. Blossom blight: It is the most devastating phase of th disease. It is usually the first symptom to appear within one to two week after bloom. Infected blossoms first appear water-soaked and then the begin to shrivel, wilt and eventually turn brown to black. The infecte blossoms may fall but often remain hanging on the trees. Infection of single flower in the cluster usually results in the death of whole spu During the warm and humid weather, bacteria come out on the surface infected blossoms in the form of ooze and are disseminated by pollinatin insects (Fig. 13.42a). This phase is critical for the outbreak of the fi blight epiphytotics because the bacteria can be readily disseminate throughout the orchard by pollinating insects.

13.42.1.2. Fruit blight: Usually immature and unripened fruits a infected. Infection occurs through infected spur or directly throug wounds or lenticels of the fruits. The infected fruits become water-soake turn brown, shrivel and finally turn black. The infected fruits usuall remain clinging to the trees for several months. Bacterial ooze comes ou on the surface of infected fruits in warm and humid weather (Fig.13.42b)

13.42.1.3. Shoot or twig blight: Also called **"blight strikes"**, thes symptoms develop on actively growing vegetative shoots. Twigs or shoo may be infected directly or infection from blossoms may move into then Infected shoots wilt from the tip downward and form a diagnost **"shepherd's crook"** (hooked) at the end. The bark of infected twig turns brownish-black and is soft in the beginning but shrinks an hardens later on. The leaves on the infected twigs turn black and remai clinging to the twig (Fig. 13.42c). This phase is more destructive to youn and vigorously growing trees.

13.42.1.4. Leaf blight: In addition to blightening of leaves c infected twigs, leaf blight occurs from independent infection of leave through wounds or through stomata and hydathodes. The necrotic, brow to black lesions appear along the midrib or main veins, between the vei and along the margins. As the blackening spreads, the leaves curl an shrivel, hang downwards, and generally remain attached to the hooke and blighted twigs.

13.42.1.5. Trauma blight: The symptoms are similar to shoot bligh but are usually more random and widespread throughout the orchar The symptoms result on account of damage caused to the host b traumatic events like hail, frost or storm. The wounds caused by thes traumatic events allow the bacteria to penetrate the host, bypassing th normal defenses even in the resistant hosts.

13.42.1.6. Canker blight: It is also called by other names lik **"limb"**, **"trunk"** or **"body blight"**, depending on which part the infectio occurs. Cankers are formed as a result of bacteria traveling systemical into the wood tissue of the tree. From infected twigs and fruit spur infection moves down to the branches, scaffold limbs and trunks, formin cankers. In the beginning, the bark where canker is formed appears water-soaked and later on, becoming discoloured, sunken and dry. The cankers formed on large limbs and trunk appear slightly to deeply depressed and occasionally develop cracks at the margins. When a canker enlarges and girdles the branch, the portion of the branch above the cankers dies. Many small cankers, less than 2.5 centimeters in diameter, usually not sunken and rarely cracked, often appear on small twigs. These usually go unnoticed due to their small size but are a major source of inoculum for infection in the next spring. One active, overwintering canker contains sufficient bacteria to severely contaminate flower blossoms of trees present in ¼ to ½ acre (Smith, 2010). Cankers without cracks are termed indefinite and are more viable source of inoculum.

***13.42.1.7. Collar and root blight*:** When the cankers develop at the base of the trunk, it is called the **collar blight**. The formation of cankers at collar or ground level results in an early death of plants.

The infection of the root by the bacterium results in **root blight** or **rootstock blight**. The primary route of entry of the bacterium into the rootstock is internally, often through apparently healthy limbs and trunks of trees that had only a few infected blossoms or shoots of the scion. After reaching the susceptible rootstock, the bacteria initiate the formation of new cankers that completely girdle and kill the tree in one to a few months. Some of the rootstocks, namely, M - 26 and M - 9, generally preferred in tree size controlling orchards, are highly susceptible to fire blight pathogen.

13.42.2. Causal organism

Erwinia amylovora (Burrill 1882) Winslow *et al.* 1920

Syn. *Bacterium amylovorum*

The bacterium is rod-shaped, Gram-negative, motile with peritrichous flagella and requires nicotinic acid as a growth factor. The colonies on nutrient agar or King's agar B are smooth, creamy-white and round with a glistening shine. Miller and Schroth's selective (MSS) medium, which utilizes sorbitol as the carbon source, is suitable for the isolation of the bacterium due to elimination of common bacterial species. The modification of MSS medium by Ishimaru and Klos (1984) is helpful for isolation of the bacterium in the presence of high population of *Pantoea agglomerans*. The optimum temperature for growth is around 27°C. Gelatin is liquefied, levan is produced from sucrose but nitrate reduction and H_2S production are negative. Lactate, citrate, formate and trehalose are utilized as a sole source of carbon. The G + C content of the DNA is 53.6 – 54.1 mol%.

The bacterium contains a 40 kb segment of DNA that contains two large regions termed *hrp* and *dsp*. Genes present in these regions code for harpin and probably for other proteins that are responsible for pathog enicity, hypersensitivity and perhaps host range. An extracellular polys accharide (EPS) called amylovoran and type III secretion system are tw major but separate virulence factors of the bacterium. The other virulenc factors include lipopolysaccharides like levan, glucan and levansucrase.

Most strains of the bacterium contain a nonconjugative plasmi called the ubiquitous plasmid pEA29. The loss of this plasmid results i the reduction of virulence of the bacterium and the strains cured of thi plasmid show thiamine auxotrophy. The presence of pEA29 influences th expression of chromosomal amylovoran biosynthetic operon durin infection. The amylovoran is a pathogenicity determinant of the fir blight pathogen and thiamine biosynthesis is required for the high leve expression of the amylovoran biosynthetic operon during infectior Among the other plasmids found in *Er. amylovora*, plasmid pEL60 is (special interest because it is an IncL/M plasmid with a high degree (similarity to the backbone of pCTX-M3, a multiple antibiotic resistanc plasmid identified in the opportunistic human pathogen *Citrobacte freundii*. pEL60 is a cryptic plasmid, encoding only genes involved i replication, maintenance and conjugative transfer, while pCTX-M contains two additional insertions totaling 29.6 kb that encode a numbe of antibiotic resistance genes. Koczan *et al.* (2009) also stressed th importance of amylovoran and levan in pathogenesis of *Er. amylovorc* They showed that biofilm formation played a critical role in th pathogenesis; amylovoran was required for biofilm formation and leva contributed to biofilm formation. Scanning electron microscopic analysi revealed an extensive biofilm formation in xylem vessels after inoculatio with *Er. amylovora* wild-type cells. On the other hand, Δams mutant cell were nonpathogenic and died rapidly following inoculation. Lee *et a.* (2010) reported that virulence factor expression levels influenced th disease severity caused by wild isolates of *Er. amylovora* on apple trees On apple seedlings, three virulence characteristics, namely, amylovora production, biofilm formation and growth in immature apple frui accounted for >70% of the variation in disease severity while i greenhouse-grown trees, amylovoran production, biofilm formatior growth in immature apple fruit, hypersensitive cell death elicitation an sorbitol utilization accounted for >75% of the variation in disease severity

13.42.3. Disease cycle

The bacteria present in the dormant or holdover cankers, formed o the twigs, branches and stems in the previous year, are the main sourc of perpetuation. The bacteria present at the margins of the cankers ar more active. The bacteria present in buds and apparently healthy woo also contribute to the survival but to a lesser extent.

In the next spring, when weather warms up, the bacteria in the cankers begin to multiply and invade the healthy adjoining bark. During humid or wet weather, bacterial cells ooze out through the cracks and lenticels and this coincides with the opening of pear blossoms. Due to sweet and sticky nature of bacterial ooze, various insects like bees, flies and ants are attracted and get smeared with the ooze when they feed on it and spread it to all the flowers they visit subsequently. Splashing rain also spreads the bacteria from oozing cankers to flowers and other plant parts. Under arid conditions, the bacterial ooze dries to form long, narrow aerial strands, which are spread by wind currents.

The bacteria present on the stigmas of flowers do not infect immediately but grow epiphytically. Stigmas have a moist, nutrient-rich surface, which provides a favourable habitat for the multiplication of bacteria increasing the number of cells up to $10^{6\text{-}7}$ per stigma. These bacteria are again spread to other flowers by rain or honeybees or other insects before the onset of environmental conditions favourable for infection. A mean temperature of 15.5°C is required for the epiphytic growth of the bacteria on the stigmas. With the help of rain or dew, the bacteria move from stigmas to the hypanthium or floral cup of the flower and enter the flowers through secretory cells (nectarthodes). After penetration into flowers, the bacteria multiply rapidly, producing necrotic symptoms.

Flowers can become infected within minutes after a rain or heavy dew when the average daily temperature is 16°C or higher.

From the infected flower, infection spreads down the pedicel and then into the fruit spur. Infection of the spur causes the death of all the flowers, fruits and leaves borne on it. Shoot blight usually develops after blossom blight. The bacteria from infected flowers are disseminated by splashing rain and they enter leaves, twigs and branches. The penetration of leaves occurs mostly through wounds caused by insects, hailstorms, strong winds but less frequently through stomata and hydathodes. From the leaf, infection spreads to the petiole and then to the twig or branch. Twigs and branches are penetrated directly through the lenticels and wounds and indirectly via flower or leaf infections. Once inside the host tissue, the bacteria move intercellularly through the parenchyma cells and in some cases through the xylem and phloem leading to systemic infection. In systemic infection, surprisingly a single flower infection can cause the death of a tree in a few weeks. Unlike other wilt bacteria, *Er. amylovora* moves rapidly from the vessels to the other tissues and causes the death of cells of these tissues. The blight and canker symptoms result from the death of these tissues.

Ideal conditions for infection and disease development include wet or humid weather coupled with daytime temperature ranging from 24 to 29°C and night temperature above 13°C. High soil fertility including excessive nitrogen which promotes succulent growth also favours disease development. Vigorously growing shoots are most severely infected. The infection progresses very rapidly in the succulent tissue and may advance 10 to 30 cm in a day (Thomson, 1992).

Pirc *et al.* (2009) developed specific, sensitive and reliable TaqMan real-time PCR assays for the detection of *Er. amylovora* targeting its chromosomal DNA (*ams*C gene and ITS region). The real-time PCR assays in combination with an automated DNA-extraction method based on magnetic beads (QuickPick TM) reliably detected at least 10^3 cells ml^{-1} (c. four cells per reaction) of the pathogen from blighted woody plant material. The assays were also able to detect the bacterium from symptomless plants including cases where subsequent isolation of the bacterium failed, giving an absolute quantification of *Er. amylovora* before and after enrichment in liquid media, which provided proof for viability and the ability of the bacterium to multiply.

Certain orchard management practices, adopted in the US during the last decade, have increased the frequency of more devastating outbreaks of fire blight. These are briefly summarized as below:

1. **High tree density:** Instead of 100 - 200 apple trees per acre, now 500 - 1000 trees per acre are planted.
2. **Tree size control:** In order to accomplish high tree densities, tree size-controlling rootstocks, namely, M - 9 and M - 26 are widely used. Both these rootstocks are highly susceptible to the disease (Steiner, 1998).
3. **Susceptible varieties:** Apple varieties, namely, Gala, Fuji, Gingergold, Jonagold and Braeburn grown due to their demand in the fresh market, are highly susceptible to the disease. The combination of these varieties on highly susceptible rootstocks like M - 9 and M - 26 further aggravates the problem.
4. **Training system:** New tree training systems adopted to make high-density plantings more productive, push the trees into early bearing. In this process, some natural physiological mechanisms that resist the progress of infection are reduced.

13.42.4. Management

No single control measure is effective. Therefore, an integrated and aggressive fire blight management programme should be followed. The following measures will help in minimizing the disease severity.

1. Sanitation

(a) During the winter, all blighted twigs, branches, cankers and even entire trees, if needed, should be cut and burnt. The infected twigs and branches should be cut at least 20 - 25 centimetres below the visible injury or canker.

(b) The orchard should be inspected in the early season to locate the cankers, which were not removed during the previous pruning because these become more prominent once the bacteria start oozing from them. These cankers should be removed by cutting the twigs and branches at least 30 centimetres below the point of visible infection. The cutting tools should be disinfected after every cut with 10% commercial sodium hypochlorite solution (Wilcox, 2001).

2. Horticultural practices

(a) To reduce excessive succulency, trees should be grown in soil having balanced nutrition and normal dose of nitrogen. The trees should also receive limited pruning.

(b) A proper insect control should be followed in the post-bloom period to control the insects responsible for the spread of bacteria.

(c) Moderately resistant and resistant varieties, if available, should be grown in areas where disease is destructive.

(d) Growth-regulating acylcyclohexanedione compounds, namely, prohexadione-calcium (trade names Apogee® and Regalis®) and trinexapac-ethyl (trade names Moddus® and Primo®) offer a new, promising and safe method for induction of resistance in pome fruit trees against fire blight and other pathogens and also for maintaining vegetative and reproductive equilibrium. The application of these compounds activates plant defence reactions leading to formation and storage of a biocidal compound, luteoforol in the treated tissue. The release of this compound as a result of pathogen infection causes the death of the pathogen and the adjacent plant cells. Spinelli *et al.* (2004) reported that the combination of *Pa. agglomerans* and acylcyclohexanediones significantly reduced both blossom blight incidence and the percentage of shoot infection. The enhanced protection achieved with the combination resulted from the reduction in blossom blight incidence caused by the former and prevention of shoot infection due to the activation of plant defenses by the latter. Recently, Spinelli *et al.* (2007) found that both prohexadione-calcium and trinexapac-ethyl caused significant reduction in blossom blight incidence leading to significant reduction in shoot blight incidence. Both the compounds also caused 60 - 66% reduction in the migration of the bacterium in the plant tissue.

3. Chemicals

Various chemicals, namely, copper oxychloride, Bordeaux mixture (0.5%) and streptomycin (100 μg/ml) have been used for the control of the disease. Bordeaux mixture and copper oxychloride, as dormant sprays, have offered some, but not much, protection from the disease. Moreover, copper fungicides may cause a russeting or scarring of the fruit surface (Teviotdale, 2003). Streptomycin is highly effective against blossom blight and also against trauma blight but the timings of its sprays are critical for effective control because this antibiotic is active only for three days after its application. For maximum benefit, streptomycin must be applied on the day when an infection event occurs or a day after it. A delay in spraying for 24 hr will result in 10 - 20% reduction in control. For the prediction of infection event accurately, fire blight forecasting models, namely, **MARYBLYT** and **cougarblight** can be effectively used.

The continuous use of streptomycin in many areas has resulted into the development of bacterial strains resistant to this antibiotic, hence making its use ineffective in these areas. Pathogen strains resistant to streptomycin were first observed in California pear orchards in 1971 and in Washington and Oregon in 1972 and then in the western United States as well as in Michigan and Missouri. Very recently, Russo *et al.* (2008) reported the occurrence of these resistant strains from New York. Streptomycin resistance have also been reported from British Columbia (Canada), Israel, Lebanon and New Zealand. In such areas, another broad-spectrum antibiotic, oxytetracycline can be effectively used. For best results, it should be applied within 24 hr before infection and also within 12 - 24 hr after the infection.

Messenger, a biochemical pesticide has shown some promise in controlling blossom blight phase when used in combination with streptomycin sprays. It is not bactericidal but acts by activating the systemic acquired resistance genes of the host, for which about 5 - 7 days are needed. Therefore, its application should be made sufficiently in advance of blossom blight infection.

4. Biological control

Johnson and Stockwell (1998) have highlighted the use of biocontrol agents for the management of fire blight. Blight-Ban A 506 (a product of *Pseudomonas fluorescens* strain A 506) is now available commercially for its control. Two applications of Blight-Ban caused 40 - 60% reduction in blossom blight. In another trial, its single application caused 50% reduction in fire blight but when this application was followed by weakly applications of streptomycin, the reduction in disease was 70%. The reduction in frost injury was also more in the combination treatment. Svircev *et al.* (2006) obtained an effective control of the disease, comparable to streptomycin, using *Pa. agglomerans* (a biocontrol agent of the pathogen) and phages specific to *Er. amylovora.* In this case, *Pa. agglomerans* was also used to deliver and sustain phages on the blossom surface.

5. Transgenic plants

Transgenic apple lines, over-expressing the apple gene *MpNPR 1,* have shown increased resistance to fire blight pathogen (Malnoy *et al.*, 2006).

REFERENCES

Douglas, S.M. (2006). Fire blight. The Connecticut Agri. Exptl. Sta. Online [http://www.caes.state.ct.us/FactSheetFiles/PlantPathology/fspp029f.htm].

Ishimaru, C. and Klos, E.J. (1984). New medium for detecting *Erwinia amylovora* and its use in epidemiological studies. *Phytopathology* 74: 1342 - 45.

Johnson, K.B. and Stockwell, V.O. (1998). Management of fire blight: A case study in microbial ecology. *Annu. Rev. Phytopathol.* 36: 227 - 48.

Kim, W.S., Hildebrand, M., Jock, S. *et al.* (2001). Molecular comparison of pathogenic bacteria from pear trees in Japan and the fire blight pathogen *Erwinia amylovora. Microbiology* 147: 2951 – 59.

Koczan, J.M., McGrath, M.J., Zhao, Y. *et al.* (2009). Contribution of *Erwinia amylovora* exopolysaccharides amylovoran and levan to biofilm formation: Implications in pathogenicity. *Phytopathology* 99: 237 – 44.

Lee, S.A., Ngugi, H.K. and Halbrendt, N.O. (2010). Virulence characteristics accounting for fire blight disease severity in apple trees and seedlings. *Phytopathology* 100: 539 – 50.

Longstroth, M. (2001). The 2000 fire blight epidemic in Southwest Michigan apple orchards. *The Compact Fruit Tree.* 34: 16 – 19.

Malony, M. *et al.* (2006). Transgenic apple trees over-expressing the apple gene MpNPR 1 have increased resistance to fire blight. *Acta Hortic.* 704: 521 - 26.

Pirc, M., Ravnikar, M. Tomlinson, J. *et al.* (2009). Improved fire blight diagnostics using quantitative real-time PCR detection of *Erwinia amylovora* chromosomal DNA. *Plant Pathol.* 58: 872 – 81.

Russo, N.L., Burr, T.J., Breth, D.I. *et al.* (2008). Isolation of streptomycin-resistant isolates of *Erwinia amylovora* in New York. *Plant Dis.* 92: 714 - 18.

Smith, T.J. (2010). Fire blight management in the Pacific Northwest USA. Online [http://county.wsu.edu/chelan-douglas/agriculture/treefruit/Pages/Fire_Blight.aspx].

Spinelli, F., Vanneste, J.L., Ciampolini, F. *et al.* (2007). Potential and limits of acylcyclohexanediones for the control of blossom blight in apple and pear caused by *Erwinia amylovora. Plant Pathol.* 56: 702 - 10.

Spinelli, F., Vanneste, J.L., Cornish, D. *et al.* (2004). Growth-regulating acylcyclohexanediones, trinexapac-ethyl and prohexadione-calcium, decrease blossom blight incidence in pome fruits. *Acta Hort.* 704: 245 - 48.

Steiner, P.W. (1998). Problems in managing fire blight in high density orchards on M-9 and M-26 rootstocks. Presented in the Annual Meeting of the Va./W.Va. State Horticultural Societies, Roanoke, Va.

Svircev, A.M., Lehman, S.M., Kim, W. *et al.* (2006). Control of the fire blight pathogen with bacteriophages. *Proc. Int. Symp. Biol. Control Bact. Plant Dis., 1st, Seeheim/Darmstadt, Germany,* 259, Berlin: Dtsch. Bibl., CIP-Einh.aufn.

Teviotdale, B.L. (2003). Fire blight. IPM Education and Publications, Univ. Calif. Agric. Nat. Res. Publ. 7414.

Thomson, S.V. (1992). Fire blight of apple and pear. In: J. Kumar, H.S. Chaube, U.S. Singh and A.N. Mukhopadhyay (eds). *Plant Diseases of International Importance.* Vol. III, *Diseases of Fruit Crops.* Prentice Hall, Englewood Cliffs, New Jersey, pp.32 - 65.

Thomson, S.V. (2001). Fire blight. In: O.C. Maloy and T.D. Murray (eds). *Encyclopedia of Plant Pathology.* John Wiley and Sons, Inc., New York. pp. 459 - 61.

Wilcox, W. (2001). Fire Blight. Tree fruit and berry pathology. Online [http://www.nysaes.cornell.edu/pp/extension/tfabp/firepm.shtml].

13.43. CITRUS CANKER

Citrus canker is one of the most devastating diseases of citrus (*Citrus* spp.) trees. It is also one of the most feared diseases in the United States and some other countries. The losses occur due to the reduction in yield and market value of the infected fruits and also due to the premature fruit drop. Sometimes badly infected fruits may become unsalable. Fruits produced in canker-endemic regions suffer from a restricted market due to the enforcement of international plant quarantines.

The disease was first observed in 1913 in Florida. Hasse identified the causal bacterium as *Pseudomonas citri* in 1915. The presence of the disease on citrus specimens collected from India (*C. medica* from Dehra Dun, Uttarakhand, India in 1827 - 1831) and Indonesia (*C. aurantifolia* in 1842 - 1844) and deposited in Kew Herbarium suggests that the disease probably originated in India or Indonesia. The reference to its Chinese origin is also found in the literature (Goto, 1992). The disease is endemic in Southeast Asia and Japan from where it has spread to many citrus-growing countries. The disease occurs in severe form in seasons and/or regions characterized by warm and humid weather conditions. According to Goto (1992), the disease is particularly epidemic and damaging on seedlings and young trees, especially after storms (hurricanes) accompanied with rain under warm conditions. Full-grown trees show much less disease severity and damage.

Citrus canker has a long history in Florida where it was repeatedly introduced and eradicated. It was introduced in Florida in 1912 along with infected nursery plants from Japan and from there it spread to all the Gulf States of the US and beyond. It took about 20 years to eradicate the disease from Florida by destroying and burning of more than three million nursery and a quarter million fruit bearing trees (Agrios, 2004). It took 17 more years (until 1949) to eliminate the disease from the entire United States. The appearance of a bacterial leaf spot (Xanthomonas leaf spot or citrus bacterial spot) resembling to citrus canker in symptoms in 1984 and the appearance of actual citrus canker in 1986 in Manatee county of Florida forced the government to start a new series of eradicating measures, which continued till 1992 and resulted in the destruction of more than 20 million nursery and orchard trees. Florida was declared free of citrus canker in early 1994 but unfortunately the disease appeared again in urban Miami in October 1995, in commercial citrus in Manatee county in June 1997 and subsequently in many other counties of the state. In spite of removal and destruction of 1.56 million commercial trees and 600,000 dooryard citrus trees statewide, the infected area increased to 1701 Km^2 till March 2002.

Citrus canker is also present in Oceania, Argentina, Brazil, Congo, Democratic Republic, Madgascar, Mauritius, Gabon, the Netherlands, Paraguay and Uruguay. The eradication programmes have succeeded in eradicating the disease from Australia, South Africa, New Zealand and Mozambique. The latest outbreak and eradication of citrus canker in Australia occurred in 1991. In Brazil, the disease was noticed in 1957 from where it subsequently spread to Uruguay, Paraguay and Argentina. The disease has become established in these countries in spite of the attempts to eradicate it. However, the eradication efforts in Brazil have kept the disease under good control (incidence below 0.22%) in Sao Paulo, the major citrus producing state of the country. It is one of the most important diseases of the international quarantine and all citrus-producing countries free of citrus canker strictly prohibit the import of citrus plants and fruits from disease affected countries.

13.43.1. Symptoms

The symptoms appear on all the above ground plant parts. Young citrus tissues during the last half of the expansion phase of the growth are at maximum susceptibility. Water-soaked, translucent, tiny spots, usually dark green with raised convex surface first appear on lower surface and then on upper surface of the leaves. Initially, the spots are corky having oily appearance but later on turn brown and are surrounded by glossy, dark brown margins. As the disease advances, the spots turn grayish or whitish, centre of the spot ruptures exposing a light brown spongy mass and forming a crater-like depression. Yellow haloes are usually formed around such mature lesions. The centres of the spots may fall down giving a shot-hole effect. The lesions are round to irregular in shape and measure 5 - 10 mm in diameter. As the disease severity increases, the spots coalesce giving a scabby appearance (Fig. 13.43a). The formation of lesions on midrib or petioles of leaves results in more defoliation than the scattered lesions only on the leaf blade. The feeding activities of Asian citrus leafminer, *Phyllocnistis citrella,* increase the disease severity on leaves by exposing the leaf mesophyll tissues to splashed inoculum and thereby increasing the leaf infection.

As the disease progresses, the symptoms also appear on twigs, branches, fruits and thorns. The lesions on twigs are observed generally on more susceptible plant species like grapefruit, lime, sweet orange and trifoliate orange etc. The lesions on twigs and branches are raised, corky and surrounded by an oily or water-soaked margin but no chlorotic haloes. The lesions may vary in size from 1 to several centimeters in length. Severely infected twigs start drying from tip to backward and severe defoliation occurs.

The symptoms on fruits are similar to those found on leaves except that yellow haloes are not always present. The lesions are rough and corky, having more prominent crater-like depressions and remain confined to the rind. Severely infected fruits become highly blemished and fetch very low price or may become unmarketable (Fig. 13.43b).

The vigorously growing citrus trees are highly susceptible to infection as each growth flush of the tree during a growing season is accompanied by a period of increased susceptibility. When leaves, twigs and fruits reach maturity, they become more resistant to infection due to the thickening of the cuticle and hardening of the tissue.

Till now, five types of citrus cankers or related forms have been reported affecting *Citrus* species. **Asiatic** or **true citrus canker** or **canker/ cancrosis A**, originally found in Asia, is by far the most widespread and severe form of the disease (Gottwald *et al.,* 2002). It also affects large number of *Citrus* species. The symptoms described above are of typical true citrus canker.

False canker or **cancrosis B**, first found in South America, occurs mainly on lemons and to a lesser extent on lime, sour orange, sweet orange and citron but not on grapefruit. It is found in Argentina, Paraguay and Uruguay.

Cancrosis/canker C or **Mexican lime canker,** originally found in Sao Paulo state of Brazil, primarily affects Mexican lime and to a lesser extent lemon. In Mexican lime, it is most severe on Galego lemon but causes only slight infection on Tahiti lemon.

Rodoriguez *et al.* (1985) reported a disease causing pustules on leaves and young shoots of Mexican lime but not infecting fruits, and on other itrus species along the Pacific coast of Mexico in 1981. It is called **anker D**.

In 1984, another bacterial disease of citrus called nursery-form anker was found in a nursery in Florida and subsequently in many other urseries of the state (Schoulties *et al.*, 1987). The symptoms appeared on eaves and shoots of nursery plants but not on trees. The lesions produced vere flat but occasionally sunken, necrotic having an extremely water-oaked appearance and chlorosis but never looked like erupted canker esions. The host range of the bacterium included grapefruit, sweet and rifoliate oranges, tangor, tangerine, tangalo, mandarin, citrumelo and itrange. The disease was called as Xanthomonas leaf spot or citrus bacte-ial spot and the pathogen was named as *Xa. campestris* pv. *citrumelo.* Now the disease is called as **canker E**. The strains of pathogen causing ankers B, C and D were designated as *Xa. campestris* pv. *aurantifolia.*

Citrus canker can be confused with other diseases like scab, anthrac-ose, leprosis and melanose but the following characteristic symptoms of itrus canker will distinguish it from other diseases (Patel and Desai, 972).

1. The lesions are raised on both the surfaces of the leaves.
2. The lesions have a distinguished margin.
3. The lesions have crater-like depressions in the centre, which can be seen more clearly with the help of a hand lens.

.3.43.2. Causal organisms

Xanthomonas citri subsp. *citri* (ex Hasse 1915) Gabriel *et al.* 1989, subsp. nov. (syn. *Xa. citri, Xa. campestris* pv. *citri, Xa. axonopodis* pv. *citri*) causes Asiatic or true citrus canker or canker A.

Xa. fuscans subsp. *aurantifolii* Schaad *et al.* 2007, subsp. nov. (syn. *Xa. campestris* pv. *aurantifolii*) causes cankers B, C and D.

Xa. alfalfae subsp. *citrumelonis* Schaad *et al.* 2007, subsp. nov. (syn. *Xa. campestris* pv. *citrumelo*) causes canker E.

The bacterium causing Asiatic or true citrus canker was first dentified as *Pseudomonas citri* by Hasse in 1915. After the establishment of genus *Xanthomonas,* it was transferred to this genus as *Xa. citri.* With he introduction of pathovar system, it was changed to *Xa. campestris* pv. *citri* and in 1995, Vauterin *et al.* reclassified it as *Xa. axonopodis* pv. *citri.* The important morphological, physiological and biochemical characters of his bacterium described by Goto (1992) are given below.

Cells are Gram-negative, nonsporing, straight rods and measure 0.5 -).75 × 1.5 - 2.0 μm. They are predominantly single, aerobic, nonencapsu-ated, having a polar flagellum and chemoorganotrophic. The cells do not accumulate poly-β-hydroxybutyrate granules and the metabolism of glucose is oxidative. The colonies on solid media are circular, conve having an entire margin. The colonies are creamy-yellow and stra yellow on potato glucose agar and on nutrient agar, respectively due t the production of xanthomonadins. Catalase test is positive while oxidas is negative or weak. Methionine or cysteine is essential for growth bu serine inhibits the growth. Asparagine is not used as a sole source c carbon and nitrogen. The bacterium hydrolyses starch, esculin, casein and Tween 80, liquefies gelatin and pectate gel and produces tyrosinase The tests for nitrate reduction, arginine dihydrolase, urease and amin acid dicarboxylases are negative. Glucose, sucrose, fructose, maltos lactose, galactose, xylose, mannose, glycerol, dextrin, starch and citrat are utilized as the sole sources of carbon. Optimum temperature fo growth is 28ºC.

Till now, five different types of citrus cankers or related forms hav been reported and these have been named as cankers A, B, C, D and E Schaad *et al.* (2005) studied 44 strains of bacteria representing causa agents of all these five citrus canker types and on the basis of DNA - DN relatedness assays, sequence of 16S – 23S intergenic spacer regions an AFLP analysis grouped them into three distinct genotypes: taxon included all "A" strains, taxon II contained all "B", "C" and "D" strain and taxon III contained all "E" strains. They named taxon I strains a *Xanthomonas smithii* subsp. *citri* (ex Hasse 1915) sp. nov. nom. rev. comb nov., taxon II strains as *Xanthomonas fuscans* subsp. *aurantifolii* (e Gabriel *et al.* 1989) sp. nov. nom. rev. comb. nov. and taxon III strains a *Xanthomonas alfalfae* subsp. *citrumelo* (ex Riker and Jones) Gabriel *et al* 1989 sp. nov. nom. rev. comb. nov.

Schaad *et al.* (2006) emended the classification of xanthomona pathogens on citrus as their earlier publication (Schaad *et al.*, 2005 contained nomenclature errors. *Xa. smithii* subsp. *citri* and *Xa. alfalfa* subsp. *citrumelo* were reclassified as *Xa. citri* subsp. *citri* and *Xa. alfalfa* subsp. *citrumelonis*, respectively. *Xa. fuscans* subsp. *aurantifolii* and *Xa alfalfae* subsp. *citrumelonis* share many common characteristics. Th below given description of these three pathogens is taken from Schaad *e al.* (2006).

13.43.2.1. Xa. citri subsp. citri: The colonies of *Xa. citri* subsp. *citr* appear on YDC and FS agars after 40 - 44 and 56 - 60 hr of incubation respectively at 28 - 30°C. The bacterium utilizes arabinose, maltose lactose, cellobiose, mannitol and aspartic acid; hydrolyzes pectate liquefies gelatin, and causes an alkaline hydrolysis of litmus milk. *Xa citri* subsp. *citri* produces large amounts of extracellular polysaccharide (EPS) both in culture media and in host tissue. The cells of the bacteriun are rapidly killed when diluted in water, i.e., < 10^6 cells/ml. Therefore, th EPS protects the bacterial cells in water and from desiccation in air (i the bacterial ooze). Strains of *Xa. citri* subsp. *citri* are susceptible t bacteriophages CP1 and CP2 whereas those of *Xa. fuscans* subsp. *aurantifolii* are not. The bacterium may be distinguished from *Xa. campestris* and most other *Xanthomonas* species, subspecies and pathovars by DNA - DNA reassociation assays, ITS sequencing, rep-PCR profiles and phenotypic traits.

13.43.2.2. Xa. fuscans subsp. aurantifolii: The colonies of *Xa. fuscans* subsp. *aurantifolii* appear on YDC and FS agars after 50 – 60 and 70 – 76 hr of incubation, respectively at 28 - 30°C. The strains of the bacterium produce a water-soluble brown pigment on several common agar media including YDC. The bacterium utilizes lactose, cellobiose and mannitol, and precipitates litmus milk. *Xa. fuscans* subsp. *aurantifolii* is distinguished from *Xa. citri* subsp. *citri* and *Xa. alfalfae* subsp. *citrumelonis* as it precipitates litmus milk and hydrolyzes gelatin. *Xa. fuscans* subsp. *aurantifolii* does not utilize maltose or hydrolyze pectate whereas *Xa. citri* subsp. *citri* does. *Xa. fuscans* subsp. *aurantifolii* is differentiated from most other *Xanthomonas* species, subspecies and pathovars by DNA - DNA reassociation assays, rep-PCR profiles, ITS sequences and phenotypic traits.

13.43.2.3. Xa. alfalfae subsp. citrumelonis: The colonies of the bacterium on YDC and FS agars appear after 30 - 34 and 40 - 44 hr of incubation, respectively at 28 - 30°C. *Xa. alfalfae* subsp. *citrumelonis* utilizes raffinose whereas *Xa. citri* subsp. *citri* does not. *Xa. alfalfae* subsp. *citrumelonis* can be differentiated from *Xa. fuscans* subsp. *aurantifolii* due to its more rapid growth on agar media, liquefaction of gelatin and utilization of maltose. It is distinguished from *Xa. citri* subsp. *citri* by its ability to utilize raffinose, to produce acid from cellobiose and mannitol and faster growth on YDC and FS agars. *Xa. alfalfae* subsp. *citrumelonis* is distinguished from *Xa. campestris* pv. *campestris* and other *Xanthomonas* pathovars, subspecies and species by DNA - DNA reassociation assays, rep-PCR profiles, ITS sequences and phenotypic traits. These three subspecies can be distinguished from each other by serological assays.

13.43.3. Disease cycle

The bacteria multiply in lesions formed on leaves, twigs and fruits. During wet weather, the bacteria ooze out from these lesions and are dispersed to new growth and other plants. Wind-driven rain is the main natural dispersal agent and the wind speed >18 mph helps in the stomatal penetration of bacteria. The rainstorms also increase the water congestion of leaf tissue and citrus foliage can hold 7 microlitres of water per centimeter square of leaf area. Even one cell forced through stomatal openings in water-congested tissues can lead to successful infection. Pria *et al.* (2006) found that temperature had a greater effect than leaf wetness duration on the incidence of Asiatic citrus canker in Brazil. As

100% disease incidence occurred during 4 hr leaf wetness period, they suggested that minimum leaf wetness period is less than 4 hr. Wind-driven rain is the main agent for short distance or local spread of the bacteria but in some cases it may also cover a considerable distance. In Argentina, the wind-blown inoculum was detected up to 32 meters from the infected trees. Greater wind speeds consistently disperse more bacteria. In Florida, severe rainstorms and tropical storms carried the inoculum up to a distance of 7 miles. Long distance spread occurs generally through infected nursery plants, rainwater or floods. Strong winds and rainstorms increase the disease severity leading to disease outbreaks by causing injuries to the host tissue, increasing humidity and by aiding in the dispersal of the inoculum. However, the size of injuries does not affect the disease severity because a single bacterium can cause the infection under most favourable conditions (Goto, 1992). Prolonged humidity for about 12 hr after artificial inoculation significantly increases the infection rate whereas for stomatal infection it is essential only for several minutes. The healing of wounds affects the chances of infection and the healing is much faster at higher temperatures around 25°C.

The survival of the bacterium from one growing season to next occurs mainly in the lesions formed on branches and stems. The bacteria in stem lesions can remain viable for a long period of 5 - 7 years. The bacterium is frequently found in soil, on citrus roots or weeds growing in orchards. Its population on weed roots does not decline in winter. The bacterium can survive for 2 - 3 months in leaf lesions, if the infected leaves are kept under dry conditions. The soil heavily infected with *Xa. citri* subsp. *citri* supports the inoculum level sufficient for infection only for a few weeks during the citrus growing season. However, till now, there is no direct evidence that the bacteria surviving in low numbers on weed hosts or in soil can lead to the development of epiphytotics.

The penetration of the bacterium into the host tissue occurs through stomata and more frequently through wounds caused by pruning, thorns and insects. The penetration through stomata generally occurs in the young tissue. The number and size of stomata does not affect the disease severity because the number of lesions does not differ significantly on leaves having more and wider stomata than on those having less and narrower stomata. Instead, the size of the lesion increases in susceptible varieties compared to resistant ones. The wounds caused by citrus leaf-miners also serve as avenues for the entry of the bacterium and even the penetration of the resistant hosts also occurs due to the overcoming of the natural barriers of the host. The contaminated larvae of leafminer also transport the bacteria through feeding galleries. The leafminer wounds differ from natural wounds that they take more time for healing, i.e., 10 - 14 days in comparison to 1 day for natural wounds, and also require lower inoculum dose (1/1000 of the dose required for infection through natural openings) for infection. On some trees, citrus canker infections are confined entirely to leafminer wounds. A study conducted in South Broward county of Florida in 2002 showed that approximately 60% of the diseased citrus canker leaves were injured by leafminers and the number of canker lesions increased significantly on the leafminer infested leaves in comparison to leaves infected through stomata. Christiano *et al.* (2007) showed that wounds caused by Asian citrus leafminer (*P. citrella*) on leaves of Tahiti lime led to more disease severity than mechanical wounds and intact leaves. The maximum disease severity of 87% on leaves having pupal stage injury and of 63% on those having third instar stage injury, which was five times higher than that on intact leaves, was due to greater amount of damage that exposes mesophyll cells to infection.

The optimum temperature for infection of host lies between 20 and 30°C. Excessive nitrogen fertilization leads to more canker development due to flushing of more shoots and succulent host tissue. The disease is most severe on young trees that flush many angular shoots than on mature or old trees. The angular shoots that develop from summer to autumn usually have a prolonged disease cycle due to the presence of young and susceptible shoots for a long time. The fruits also show infection for several months because their continuous growth provides suitable stomata for infection.

13.43.4. Host range

Xa. citri subsp. *citri* infects a large number of citrus species. Among the most susceptible hosts are grapefruit, Mexican (Key) lime and lemons, trifoliate oranges, some sweet oranges and limes. In India, *kagzi* lime is the most susceptible host. These hosts suffer badly from the disease in moist tropical and subtropical climates. Mandarin oranges like satsuma, cleopatra and ponkan are resistant. The other natural hosts include *Swinglea glutinosa, Fortunella margarita, F. japonica* and *Severinia buxifola*.

13.43.5. Management

1. The enforcement of strict quarantine is the best method of control in disease-free areas. *Xa. citri* subsp. *citri* is EPPO A 1 quarantine pest. Every effort should be made to eradicate the bacterium from such areas when it is introduced or found in these areas.
2. In areas where the disease is endemic, the following integrated management measures should be adopted to contain it.
   (a) Use of disease-free nursery and production of pathogen-free budwood are the foundations of an integrated management programme.

(b) Use of resistant or field resistant cultivars is the most important feature of the integrated programme. Eighty-five percent of citrus production in Japan is from the cultivation of a moderately resistant cultivar, Unshu (satsuma) orange.

(c) Cultural practices like use of windbreaks and pruning of diseased summer and autumn shoots are very effective. Pruning should be done during the dry season to prevent the spread of the pathogen. Windbreaks are most effective for the control of disease on susceptible varieties.

(d) Periodical sprays of insecticides should be given to control the citrus leafminer.

(e) Goto (1992) has given the criteria on the basis of which an epiphytotic can be predicted 1 - 2 months in advance. Based on that, the citrus growers can be advised to take timely steps for an effective chemical control.

(f) Preventive sprays of copper-based bactericides are used throughout the world to manage citrus canker. Copper sprays alone may control the disease on resistant or moderately resistant cultivars while the control on susceptible or highly susceptible cultivars will require the integration of several control measures. The timing and number of copper sprays will vary with weather conditions, age of trees and reaction of host cultivars but in general, 3 – 5 copper sprays are sufficient on cultivars having intermediate level of resistance. A NASBA (nucleic acid sequence based amplification) protocol using primers for *gum*D mRNA has been developed to assess the viability of this pathogen to a bactericidal treatment. The method is rapid, specific and sensitive, and is able to detect viable bacterial cells using a hybridization device, which allows the visualization of the results in only 30 minutes (Scuderi *et al.,* 2010).

The continuous use of copper bactericides has resulted in the appearance of copper tolerant strains of the bacteria in Argentina. In such areas, copper-mancozeb, a mixture of copper and mancozeb, should be used instead of copper alone. On susceptible or highly susceptible cultivars and in areas prune to disease epiphytotics, prevention of primary infection on spring shoots is very essential for the management of the disease.

In India, Patel and Desai (1970) found that the disease was effectively controlled within three years by removing the infected twigs every year and spraying the trees with 1% Bordeaux mixture four times during the year. The first spray should be given in November-December, second in February-March, third before monsoon or first fortnight of June and fourth in July-August or when there is break in rains.

(g) Recently, Balogh *et al.* (2008) have obtained an effective and consistent control of the disease with a bacteriophage. The phage application caused 59% reduction in disease severity. However, phage treatment was ineffective, when applied with skim milk.

REFERENCES

Agrios, G.N. (2004). *Plant Pathology.* 5th Ed. Elsevier Academic Press, San Diego, 922 pp.

Balogh, B., Canteros, B.I., Stall, R.E. *et al.* (2008). Control of citrus canker and citrus bacterial spot with bacteriophages. *Plant Dis.* 92: 1048 - 52.

Christiano, R.S.C., Pria, M.D., Jesus, W.C. Jr. *et al.* (2007). Effect of citrus leaf-miner damage, mechanical damage and inoculum concentration on severity of symptoms of Asiatic citrus canker in Tahiti lime. *Crop Prot.* 26: 59 - 65.

Goto, M. (1992). Citrus canker. In: J. Kumar, H.S. Chaube, U.S. Singh and A.N. Mukhopadhyay (eds). *Plant Diseases of International Importance.* Vol. III, *Diseases of Fruit Crops.* Prentice Hall, Englewood Cliffs, New Jersey, pp. 170 - 208.

Gottwald, T.R., Graham, J.H. and Schubert, T.S. (2002). Citrus canker: The pathogen and its impact. APSnet Feature Story, [http://www.plantmanagementnetwork.org/php] review/citrus canker.

Patel, R.S. and Desai, M.V. (1970). Control of citrus canker. *Indian J. Hort.* 27: 93 - 98.

Patel, R.S. and Desai, M.V. (1972). Citrus canker. In: P.N. Patel (ed.). *Plant Bacteriology.* Vol. I, *Bacterial Diseases of Plants in India.* Indian Agricultural Research Institute, New Delhi, pp. 184 - 91.

Pria, M.D., Christiano, R.C.S., Furtado, E.L. *et al.* (2006). Effect of temperature and leaf wetness duration on infection of sweet oranges by Asiatic citrus canker. *Plant Pathol.* 55: 657 - 63.

Rodoriguez, S.G., Stapleton, J.J. and Civerolo, E.L. (1985). *Xanthomonas campestris* involved in Mexican lime bacteriosis in Colima, Mexico. *Proc. 6th Int. Conf. Plant Path. Bact.*, Maryland, pp. 658 - 62.

Schaad, N.W., Postnikova, E., Lacy, G. *et al.* (2005). Reclassification of *Xanthomonas campestris* pv. *citri* (ex Hasse 1915) Dye 1978 forms A, B, C, D and E as *X. smithii* subsp. *citri* (ex Hasse) sp. nov. nom. rev. comb. nov.; *X. fuscans* subsp. *aurantifolii* (ex Gabriel 1989) sp. nov. nom. rev. comb. nov., and *X. alfalfae* subsp. *citrumelo* (ex Riker and Jones) Gabriel *et al.*, 1989 sp. nov. nom. rev. comb. nov.; *X. campestris* pv. *malvacearum* (ex Smith 1901) Dye 1978 as *X. smithii* subsp. *smithii* sp. nov. comb. nov nom. nov.; *X. campestris* pv. *alfalfae* (ex Riker and Jones, 1935) Dye 1978 as *X. alfalfae* subsp. *alfalfae* (ex Riker *et. al.*, 1935) sp. nov. nom. rev.; and "var. *fuscans*" of

X. campestris pv. *phaseoli* (ex Smith, 1897) Dye 1978 as *X. fuscans* subsp. *fuscans* sp. nov. *Syst. Appl. Microbiol.* 28: 494 - 518.

Schaad, N.W., Postnikova, E., Lacy, G. *et al.* (2007). Validation of the publication of new names and new combinations previously effectively, but not validly published outside the IJSEM. 57: 893 – 97. Effective Publication: Schaad N.W., Postnikova, E., Lacy, G. *et al.* (2006). Emended classification of xanthomonad pathogens on citrus. *Syst. Appl. Microbiol.* 29: 690 – 95.

Schoulties, C.L., Civerolo, E.L., Miller, J.W. *et al.* (1987). Citrus canker in Florida. *Plant Dis.* 71: 388 - 95.

Scuderi, G., Golmohammadi, M., Cubero, J. *et al.* (2010). Development of a simplified NASBA protocol for detecting viable cells of the citrus pathogen *Xanthomonas citri* subsp. *citri* under different treatments. *Plant Pathol.* 59: 764 – 72.

13.44. CITRUS HUANGLONGBING

It is the most serious and devastating disease of citrus. The disease is also called by various other names such as **citrus greening disease, yellow dragon disease** or **yellow shoot disease**. Yellow dragon disease and yellow shoot disease are the English translations of Chinese word huanglongbing. International Organization of Citrus Virologists, in its congress held in China in 1995, decided to adopt the original Chinese name "huanglongbing" (HLB) as the official name for the disease.

Southern china has long been considered as the area of origin for the Asian form of the disease. However, Gottwald *et al.* (2007) are of the view that earlier descriptions of citrus die-back in India suggest that the disease was present there earlier. Capoor (1963) attributed citrus die-back to citrus tristeza virus after biological indexing studies but Raychaudhuri *et al.* (1969, 1974) later showed that huanglongbing was a major component of the problem. Beattie *et al.* (2006) have hypothesized that the disease might have originated in Africa, possibly in a symptomless host such as *Verpris lanceolata*.

The disease is widespread in Asia (eastern, central and southern), Africa, Arabian Peninsula and Brazil. The disease was detected in Florida State on 24 August 2005 in a pummelo tree in a commercial tropical fruit nursery in Florida City. Since then, it has spread to most of the southern Florida. In Brazil, the disease was first reported in 2004 from Sao Paulo state and both Asian and American forms of the disease occur there. According to Pietersen *et al.* (2010), till todate only '*Candidatus* Liberibacter africanus' is associated with the disease in commercial citrus in South Africa in addition to '*Ca.* Li. africanus subsp. capensis', which has been found only in South Africa infecting an indigenous rutaceous species, *Calodendrum capense* (cape chestnut). The disease has not been reported from Australia, Mexico, Central America and the Mediterranean citrus-growing areas. The citrus-growing countries, where the disease is absent, are actively guarding against the entry of the pathogen.

Bove (2006) rightly stated that HLB is the most important, serious, severe, destructive and devastating disease of citrus in the world. There are no known sources of resistance and management of the disease is difficult and expensive. The reduction in yield can reach to the extent of 30 to 100% depending on proportion of affected canopy. The affected orchards become economically unviable in 7 - 10 years after planting. About 100 million trees have been affected and destroyed in many countries of South and Southeast Asia, Indonesia, the Philippines, India, Arabian Peninsula and South Africa. Since 2004, more than 500 thousand trees have been officially destroyed in Brazil due to HLB and around 300 - 400 thousand trees unofficially destroyed by commercial citrus growers (Gottwald *et al.,* 2007). Catastrophic losses due to the disease from India during 1960s were reported by Fraser *et al.* (1966). During the past two decades, seven reviews have been published on the disease (Bove, 2006; da Graca, 1991; da Graca and Korsten, 2004; Garnier and Bove, 1993; Halbert and Manjunath, 2004; Gottwald *et al.,* 2007; Gottwald, 2010). The review published by Gottwald *et al.* (2007) deals with epidemiology of HLB and the implications of temporal increase and spatial spread of the disease on prospects for disease control and mitigation. The review also contains the concise information on the main features and symptoms of the disease, the causal organisms and their vectors. The most recent review published by Gottwald (2010) deals with the current epidemiological information of the disease.

13.44.1. Symptoms

The symptoms vary with host species/cultivar, age of host, time of infection and biotic and abiotic factors that affect the growth of trees. The most characteristic foliage symptoms are the blotchy mottling and yellowing of leaves (Fig. 13.44a). In blotchy mottling, the leaves develop a pattern of yellow and dark green areas lacking clear limits between the colours. The patterns of blotchy mottling are asymmetrical on two halves of the leaf. The small yellow leaves appear on a single shoot/branch or on a section of the tree canopy and yellowing gradually spreads to the entire tree (Fig. 13.44b). The appearance of a yellow shoot or shoots has given the name "yellow shoot disease". The newest leaves may show symptoms, similar to zinc deficiency with yellow venation, while the older leaves have the characteristic greening mottle. The other symptoms are reduced foliage, severe die-back of twigs or limbs, poor flowering and stunting of trees.

The fruits on diseased trees are small, poorly coloured and often misshaped. Seeds in symptomatic fruits are often small, dark and aborted and the fruits remain green on stylar end even after maturity (Fig. 13.44c), hence the name greening. The diseased fruits have a sour bitter taste and drop prematurely. The following symptoms of huanglongbing help in differentiating it from chronic citrus decline caused by citrus tristeza virus, blight and citrus variegated chlorosis.

1. Yellowing of leaf midrib and veins and slight swelling of veins in early stage of the disease.
2. Seeds are aborted in small and misshaped fruits. The fruits, especially small, have a bitter taste compared to sweet taste found in other diseases.

The above given symptoms are sufficient to diagnose the disease under field conditions but final confirmation by other methods is required for regulatory measures. Before the development of molecular techniques, the diagnosis of the disease was done by biological indexing, thin layer chromatography for determining the presence of gentisoyl-β-glucoside, although not all HLB-infected citrus species produce it, use of monoclonal antibodies and use of indicator plants like sweet orange, Duncan grapefruit and sweet tangor seedlings. A staining technique employing gentisic acid has been found very useful in the survey work. Recently developed techniques such as DNA hybridization and PCR are rapid, sensitive and specific for the diagnosis of HLB in the laboratory. PCR is now the main confirmatory test and is routinely used in many laboratories including those in Florida and Brazil. Das *et al.* (2007) confirmed the presence of the disease in Northeastern region of India by PCR technique. Recently, Manjunath *et al.* (2008) have developed a TaqMan-based real-time quantitative PCR method to detect 'Ca. Li. asiaticus' from nymphs and adults of *Diaphorina citri* collected from apparently healthy and diseased citrus trees. This finding is highly significant in a way that the presence of the bacterium in an area can be detected one to several years before the development of disease symptoms and hence the strategies for disease management can be initiated in an advance. Li *et al.* (2008) using TaqMan real-time PCR established a universal standard curve (based on DNA extracted from naturally infected plants) for quantification of the pathogen in various tissues of different citrus species planted in different geographical regions for studying the host pathogen interactions and epidemiology, and for devising disease management strategies.

There are three types of HLB, namely, Asian, African and American. The symptoms of Asian and African forms are generally the same but the Asian form is considered more severe because it causes more extensive and severe die-back eventually killing the trees. The bacterium causing Asian form is also more heat tolerant. The symptoms of African form are suppressed by long exposure to temperature above 30°C. Hence, it is usually found in citrus at elevations above 700 m, while Asian form occurs in lower lying, hotter areas. American form appears to resemble closely the Asian form in symptom expression and severity but its causal agent is also intolerant to heat like that of African type.

Chen *et al.* (2009) have shown the presence of a phytoplasma related to '*Ca.* Phytoplasma asteris' in citrus infected with '*Ca.* Li. asiaticus'. Of the 141 samples collected from Guangdong Province, P. R. China, 69 (48.9%) samples were positive for both '*Ca.* Phy. asteris' and '*Ca.* Li. asiaticus'. After the transmission from symptomatic citrus to periwinkle (*Catharanthus roseus*) via dodder (*Cuscuta campestris*), both phytoplasma and '*Ca.* Li. asiaticus' were detected from the infected periwinkle. In addition to yellowing/mottling, the infected periwinkle plants showed symptoms of virescence and phyllody which are commonly associated with phytoplasma diseases.

13.44.2. Causal organisms

'*Candidatus* Liberibacter asiaticus' Jagoueix-Eveillard *et al.* 1994 causes Asian form of HLB.

'*Candidatus* Liberibacter africanus' Jagoueix-Eveillard *et al.* 1994 causes African form of HLB.

'*Candidatus* Liberibacter africanus subsp. capensis' Garnier *et al.* 2000 causes HLB of cape chestnut (*Calodendrum capense*).

'*Candidatus* Liberibacter americanus' Teixeira *et al.* 2005 causes American form of HLB.

The above-mentioned four pathogens are fastidious, phloem-limited, Gram-negative bacteria. '*Ca.* Li. asiaticus' requires high temperature than '*Ca.* Li. africanus' for optimum expression of the disease.

For many years, HLB was considered to be caused by mineral deficiencies/toxicities or water logging. Later on, the transmission of the casual agent through grafting and insects led to the conclusion that it was a virus. The electronic microscopic studies revealed the presence of mycoplasma-like organisms (MLOs) in the phloem sieve elements of infected plants but not in healthy plants. However, Garnier and Bove in 1978, after close examination of these organisms, found that they have thicker envelops than MLOs and hence they are true bacteria

Sechler *et al.* (2009) designed a medium called Liber A to culture all the three '*Ca.* Liberibacter spp.' that cause HLB. The medium containing citrus vein extract and a growth factor supported the growth of '*Ca.* Liberibacter spp.' for four or five single colony transfers before the decline of viability. The colonies of '*Ca.* Li. asiaticus' were irregularly-shaped, convex and 0.1 – 0.3 mm after 3 – 4 days of incubation. Two strains of '*Ca.*

Li. asiaticus' and one of '*Ca.* Li. americanus' grown on this medium proved pathogenic to citrus and were isolated from noninoculated tissues of inoculated trees and seedlings 9 and 2 months later, respectively. The identity of the bacteria was confirmed by RT-PCR and 16S rDNA sequencing. This is the first report of the culturing and pathogenicity of '*Ca.* Li. asiaticus' and '*Ca.* Li. americanus' associated with HLB.

Most recently, Duan *et al.* (2009) have reported the complete genome sequence of '*Ca.* Li. asiaticus', one of the causal agents of HLB. A complete circular genome of this bacterium was obtained by metagenomics, using the DNA extracted from a single '*Ca.* Li. asiaticus'-infected psyllid. The 1.23 mb genome has an average G + C content of 36.5%. A high percentage of genes involved in both cell motility (4.5%) and active transport in general (8.0%) may be contributing to its virulence. The bacterium appears to have a limited ability for aerobic respiration and is likely auxotrophic for at least five amino acids. As the bacterium is intracellular in nature, it lacks type III and type IV secretion systems as well as typical free-living or plant-colonizing extracellular degradative enzymes. However, it seems that it has all type I secretion system genes needed for both multidrug efflux and toxin effector secretion. This is the first genome sequence of a fastidious alphaproteobacteria that is both an intracellular plant pathogen and insect symbiont. Wulff *et al.* (2009) have reported rRNA operons and genome size of another liberibacter, '*Ca.* Li. americanus'. The sequences of the entire 23S rRNA and 5S rRNA genes from '*Ca.* Li. americanus' (Lam) have been obtained, using a consensus primer designed on known $tRNA^{Met}$ sequences of rhizobia. The size of the Lam genome was determined by PFGE, using Lam-infected periwinkle plants for bacterial enrichment, and was found to be close to 1.31 mbp. In order to determine the number of ribosomal operons on the Lam genome, probes designed to detect the 16S rRNA gene and the 3' end of the 23S rRNA gene were developed and used for Southern hybridization with I-*Ceu*I-treated genomic DNA. These results suggest that there are three ribosomal operons in a circular genome. Lam is the first liberibacter species for which such data are available.

Tomimura *et al.* (2009) investigated the genetic diversity and relationship among '*Ca* Li. asiaticus' isolates from different hosts and distinct geographical areas in Southeast Asia. Genetic diversity was estimated by sequencing four well-characterized DNA fragments, namely, the 16S ribosomal DNA (rDNA) and 16S/23S intergenic spacer regions, the outer membrane protein (*omp*) gene region, the *trm*U-*tuf*B-*sec*E-*nus*G-*rp*/KAJL-*rpo*B region (gene cluster region) and the bacteriophage type DNA polymerase region. The sequences of the 16S rDNA and 16S/23S intergenic spacer regions were identical in all the isolates but nucleotide substitutions were observed in both the *omp* gene and the gene cluster regions. The extended bacteriophage-type DNA polymerase sequences acquired by thermal asymmetric interlaced polymerase chain reaction provided the most sequence diversity among the isolates. Phylogenetic analysis of the bacteriophage-type DNA polymerase sequences revealed three clusters in the Southeast Asian population of the pathogen and all the Indonesian isolates were clustered in one group. Remaining two clusters were not correlated with geographic distribution and the differences in genetic sequences also did not reflect differences in the hosts, i.e., mandarin or pummelo from which they were isolated. The bacteriophage-type DNA polymerase region would be useful for molecular differentiation among Southeast Asian isolates of '*Ca*. Li. asiaticus'.

13.44.3. Disease cycle

As the host trees infected by the bacteria are perennial in nature, the bacteria survive in the infected hosts. The pathogen is spread by infected nursery plants, by grafting diseased bud wood, dodder and two species of psyllid insects, namely, Asian citrus psylla, *Diaphornia citri* and African citrus psylla, *Trioza erytreae*. Natural transmission of Asian and American HLB is by *D. citri* while that of African HLB is by *T. erytreae*. The temperature tolerance and sensitivities of these two species match those of HLB type they are naturally associated with; i.e., *D. citri* is heat tolerant, whereas *T. erytreae* is sensitive to high temperature. However, under experimental conditions, both the insect species can transmit both the pathogens causing Asian and African HLB. Although, the pathogens are bacteria, they do not spread by casual contamination of personnel and tools or by wind and rain. Hartung *et al*. (2010) could not prove the seed transmission of '*Ca*. Li. asiaticus' and were also unable to confirm an earlier report of transmission of the bacterium through seed.

Adults of both the insect species can acquire and transmit the bacteria but the acquisition periods reported by different workers vary greatly. An acquisition period of 15 - 30 min and 5 hr minimum for *D. citri* and 24 hr for *T. erytreae* has been reported. Nymphs of both the species acquire the bacteria, which are transmitted by the adults in a persistent manner. There is also evidence for transovarial transmission by *T. erytreae*. Other species of psyllid have been found infecting citrus but none of them have been found to act as a vector.

Most recently, Kim *et al*. (2009) have given the microscopic and microarray analyses of response of sweet orange (*Citrus sinensis*) to '*Ca*. Li. asiaticus' infection. The microarray analysis indicated that the pathogen infection affected expression of 624 genes, and the proteins encoded by some of these genes were associated with sugar metabolism, plant defence, phytohormone and cell wall metabolism. The anatomical analyses indicated that HLB bacterium infection caused phloem disruption, sucrose accumulation and plugged sieve pores. The phloem blockade occurred due to the plugging of sieve pores rather than the formation of bacterial cell aggregates because '*Ca.* Li. asiaticus' does not form aggregates in citrus. The up-regulation of three key starch genes including ADP-glucose pyrophosphorylase, starch synthase, glucose-bound starch synthase and starch debranching enzyme likely contributed to accumulation of starch in HLB-affected leaves. The up-regulation of *pp2* gene is related to callose deposition to plug sieve pores in the infected plants.

13.44.4. Host range

The bacteria infect most citrus species, hybrids and many citrus relatives. Most severely infected hosts are sweet oranges, mandarins, mandarin hybrids and tangelo (Halbert and Manjunath, 2004). HLB in India and Africa is more severe on mandarins and sweet oranges while limes, lemons, pummelos and grapefruit show tolerant reaction. However, Asian forms of HLB found in Taiwan and Florida and the American form reported from Brazil affect all commercial citrus species severely. It is possible that strains of HLB pathogen adapt to citrus species over a period of time, which is evident from the fact that grapefruit and pummelo, once considered resistance to the pathogen in Taiwan are now quite susceptible.

Some citrus relatives, namely, *Atalantia missionis, Microcitrus australasica, Swinglea glutinosa, Clausena indica, Balsamocitrus dawei* and *Severinia buxifolia* are also severely infected. The other citrus relatives infected are *Limonia acidissima, Verpris lanceolata* and *Murraya paniculata*. There is disagreement in the literature about the status of *M. paniculata* as a host of HLB pathogens. It is a host in Brazil but not in Taiwan. *Catharanthus roseus* (pink periwinkle) and *Nicotiana xanthii* (tobacco) are the non-rutaceous reported hosts. Most recently, tomato has also been shown the host of '*Ca.* Li. asiaticus' and the bacterium was transmitted through dodder (Duan *et al.,* 2008).

13.44.5. Management

1. Enforcement of quarantines to prevent the entry of the pathogen into disease-free areas.
2. Use of disease-free propagating plant material.
3. Removal and destruction of infected trees as soon as they are noticed. In South Africa, a management strategy consisting of production of disease-free bud wood and nursery trees, psyllid control and removal of infected trees was implemented. Le Roux *et al.,* (2006) found that as a result of this strategy, the disease incidence of 38% found in 1960s dropped to less than 1% in 2006. When HLB was reported in Brazil, citrus industry in Sao Paulo state of Brazil had already moved to bud wood isolation, a

nursery disease certification programme, nursery industry isolation and requiring all nursery production to be in secure insect-proof screen houses. The Florida Department of Agriculture and Consumer Services (FDACS) has strengthened its existing bud wood nursery certification programme by expanding testing for HLB and movement of certified bud wood sources to an isolated greenhouse location in northern Florida outside the commercial production area. The FDACS has desired all the field-grown nursery production to cease and that no field-grown citrus nursery trees be sold after January 2008. Further, all subsequent nursery production must be in insect-proof greenhouses or screen houses with a requisite 1-mile buffer where no HLB-infected commercial or residential citrus has been detected (Gottwald *et al.,* 2007).

4. Tetracycline injections or sprays reduced foliar and fruit incidence ranging from 10 - 97% (da Graca, 1991). However, Buitendag and von Broembsen (2007) have discouraged the use of antibiotics for the control of the disease because only partial success was achieved. Most recently, Zhang *et al.* (2010) devised an optimized system using '*Ca*. Li. asiaticus'-infected periwinkle cuttings to screen chemicals effective for the control of this bacterium. The vermiculite was used as a growth medium and a fertilization routine consisted of half-strength Murashige and Tucker medium supplemented with naphthalene acetic acid (4 μg/ml) and indole-3-butyric acid (4 μg/ml). The presence of '*Ca*. Li. asiaticus' in the infected cuttings was detected using both nested PCR and quantitative real-time PCR. This system allowed a plant regeneration rate of 60.6% for '*Ca*. Li. asiaticus'-infected cuttings in contrast to <1% regeneration rate with water alone. When treated with penicillin G sodium at 50 μg/ml, all plants regenerated from '*Ca*. Li. asiaticus'-infected cuttings were free from infection of '*Ca*. Li. asiaticus'. Treatment with 2, 2-dibromo-3-nitrilopropionamide also significantly reduced the percentage of infected plants and the titer of the bacterium at 200 μl/litre.

5. Thermotherapy of bud wood at 47°C for 4 hr also gave 100% control (da Graca, 1991). However, this method is considered impracticable for a large-scale use.

6. The control of psyllid vectors through chemicals is probably the easiest strategy to control HLB. Biological control of psyllid vectors have also been tried and found successful in some cases. Two hymenopterous ectoparasites, namely, *Tamarixia dryl* and *T. radiatus* parasitize nymphs of both the psyllid vectors. On Reunion Island, these parasites significantly reduced the psyllid

populations and lessened the damage of HLB. However, the biological control with these parasites in Taiwan was not effective due to the presence of indigenous populations of hyperparasites, which attack these hymenopterous biocontrol agents. As these hyperparasites are not present on Reunion Island, the biocontrol is more effective. In Florida, the introduction of *T. radiatus* caused 4 - 70% reduction in psyllid populations. Another internal parasitoid, *Diaphorencyrtus aligarhensis*, has also been found to attack *D. citri.* Unfortunately, the use of insecticides to control psyllid vectors of HLB also reduces the population of bicontrol insects (Gottwald *et al.,* 2007).

7. In an international workshop on HLB held in Ishigaki, Japan, a novel strategy to control HLB was approved. The strategy is to interplant citrus with guava, which apparently inhibits the psyllid vector. The inhibition can be by attracting the vector to guava and killing it when it feeds, presumably by some toxin or by repelling or confusing the vector through the release of volatile compounds. The exact effect is not known but preliminary data obtained are quite encouraging (Beattie *et al.,* 2006).

REFERENCES

Beattie, G.A.C., Holford, P., Mabberley, D.J. *et al.* (2006). Aspects and insights of Australia-Asia collaborative research on Huanglongbing. *Proc. Int. Workshop for the Prevention of Citrus Greening Disease in Severely Infected Areas.* Int. Res. Div., Agric. Forestry Fisheries Res. Counc. Secretariat, Ministry of Agric., Forestry and Fisheries, Tokyo, Japan, pp. 47 - 64.

Bove, J.M. (2006). Huanglongbing: A destructive, newly-emerging, century-old disease of citrus. *J. Plant Pathol.* 88: 7 - 37.

Buitendag, C.H. and von Broembsen, L.A. (2003). Living with citrus greening disease. In: P. Moreno, J.V. da Graca and L.W. Timmer (eds). *Proc. 12th Conf. Int. Organ. Citrus Virol.* IOCV, Riverside, CA, pp. 269 - 71.

Capoor, S.P. (1963). Decline of citrus trees in India. Bull. Nat. Inst. Sci. India. 24: 48 - 64.

Chen, J., Pu, X., Deng, X. *et al.* (2009). A phytoplasma related to '*Candidatus* Phytoplasma asteris' detected in citrus showing huanglongbing (yellow shoot disease) symptoms in Guangdong, P. R. China. *Phytopathology* 99: 236 – 42.

da Graca, J.V. (1991). Citrus greening disease. *Annu. Rev. Phytopathol.* 29: 109 - 36.

da Graca, J.V. and Korsten, L. (2004). Citrus huanglongbing: Review, present status and future strategies. In: S.A.M.H. Naqvi (ed.). *Diseases of Fruits and Vegetables.* Vol. I. Kluwer Academic Press, Dordrecht, pp. 229 - 245.

Das, A.K., Rao, C.N. and Singh Shyam (2007). Presence of citrus greening (huanglongbing) disease and its psyllid vector in North-eastern region of India confirmed by PCR technique. *Curr. Sci.* 92: 1759 - 63.

Duan, Y.P., Gottwald, T., Zhou, L.J. *et al.* (2008). First report of dodder transmission of '*Candidatus* Liberibacter asiaticus' to tomato (*Lycopersicon esculentum*). *Plant Dis.* 92: 831.

Duan, Y., Zhou, L., Hall, D.G. *et al.* (2009). Complete genome sequence of citrus huanglongbing bacterium, '*Candidatus* Liberibacter asiaticus' obtained through metagenomics. *Mol. Plant-Microbe Interact.* 22: 1011 – 20.

Fraser, L.R., Singh, D., Capoor, S.P. *et al.* (1966). Greening virus, the likely cause of citrus dieback in India. *FAO Plant Prot. Bull.* 14: 127 - 30.

Garnier, M. and Bove, J.M. (1993). Citrus greening disease and the greening bacterium. In: P. Moreno, J.V. da Graca and L.W. Timmer (eds). *Proc. 12th Conf. Int. Organ. Citrus* Virol. IOCV, Riverside, CA, pp. 212 - 19.

Gottwald, T.R. (2010). Current epidemiological understanding of citrus huanglongbing. *Annu. Rev. Phytopathol.* 48: 119 – 39.

Gottwald, T.R., da Graca, J.V. and Bassanezi, R.B. (2007). Citrus huanglongbing: The pathogen and its impact. APSnet Feature Story, Online [http://www.apsnet.org/huanglongbing].

Halbert, S.E. and Manjunath, K.L. (2004). Asian citrus psyllids (Sternorrhyncha: Psyllidae) and greening disease in citrus: A literature review and assessment of risk in Florida. *Florida Entomologist* 87: 330 - 54.

Hartung, J.S., Halbert, S.E., Pelz-Stelinski, K. *et al.* (2010). Lack of evidence for transmission of '*Candidatus* Liberibacter asiaticus' through citrus seed taken from affected fruit. *Plant Dis.* 94: 1200 – 05.

Kim, J.S., Sagaram, U.S., Burns, J.K. *et al.* (2009). Response of sweet orange (*Citrus sinensis*) to '*Candidatus* Liberibacter asiaticus' infection: Microscopy and microarray analyses. *Phytopathology* 99: 50 - 57.

Le Roux, H.F., van Vuuren, S.P. and Manicom, B.Q. (2006). Huanglongbing in South Africa. *Proc. Huanglongbing-Greening Int. Workshop*, Ribeirao Preto, Brazil, pp. 5 - 9.

Li, W., Li, D., Twieg, E. *et al.* (2008). Optimized quantification of unculturable '*Candidatus* Liberibacter spp.' in host plants using real-time PCR. *Plant Dis.* 92: 854 - 61.

Manjunath, K.L., Halbert, S.E., Ramadugu, C. *et al.* (2008). Detection of '*Candidatus* Liberibacter asaticus' in *Diaphorina citri* and its importance in the management of citrus huanglongbing in Florida. *Phytopatholohy* 98: 387 - 96.

Pietersen, G., Arrebola, E., Breytenbach, J.H.J. *et al.* (2010). A surrey for '*Candidatus* Liberibacter' species in South Africa confirms the presence of only '*Ca*. L. africanus' in commercial citrus. *Plant Dis.* 94: 244 - 49.

Raychaudhuri, S.P., Nariani, T.K., Gosh, S.K. *et al.* (1974). Recent studies on citrus greening in India. In: L.G. Weathers and M. Cohen (eds). *Proc. 6th Conf. Int. Organ. Citrus Virol.* Univ. California, Div. Agric. Sci. pp. 53 - 57.

Raychaudhuri, S.P., Nariani, T.K. and Lele, V.C. (1969). Citrus die-back problem in India. *Proc. 1st Int. Citrus Symp.* 3: 1433 - 37.

Sechler, A., Schuenzel, E.L., Cooke, P. *et al.* (2009). Cultivation of '*Candidatu* Liberibacter asiaticus', '*Ca.* L. africanus' and '*Ca.* L. americanus' associate with huanglongbing. *Phytopathology* 99: 480 – 86.

Tomimura, K., Miyata, S., Furuya, N. *et al.* (2009). Evaluation of genetic diversit among '*Candidatus* Liberibacter asiaticus' isolates collected in Southeas Asia. *Phytopathology* 99: 1062 - 69.

Wulff, N.A., Eveillard, S., Foissac, X. *et al.* (2009). rRNA operons and genome siz of '*Candidatus* Liberibacter americanus', a bacterium associated with citru huanglongbing in Brazil. *Int. J. Syst. Evol. Microbiol.* 59: 1984 - 91.

Zhang, M., Duan, Y., Zhou, L. *et al.* (2010). Screening molecules for control o citrus huanglongbing using an optimized regeneration system fo '*Candidatus* Liberibacter asiaticus'-infected periwinkle (*Catharanthu roseus*) cuttings. *Phytopathology* 100: 239 - 245.

13.45. CITRUS STUBBORN DISEASE

The symptoms of the disease reflect the name of the disease as th diseased trees remain stunted and fail to grow despite the best efforts o the orchardists. It is a serious disease of certain citrus species and occur in hot dry areas. It is considered as a greatest threat to the production o grapefruit and sweet oranges in California and some Mediterranea countries. Due to slow progress of the symptoms, the spread of the diseas often goes unnoticed. The loss occurs due to reduction in yield and lov market value of the produce.

The disease was first reported from California in 1915 on nave oranges. It affected most commercial citrus varieties in California bu grapefruit, tangelos and navel and Valencia oranges were most severel infected. During the mid 1950s and early 1960s, the disease appeared o more than 2 million trees, incidence ranging from 50 - 100% in som orchards in California. This large-scale infestation of trees caused a alarm for the research scientists and the investigations for th management of the disease were intensified. Since its first appearance the disease has spread to many countries including western Unite States, eastern Mediterranean Basin, Middle East, Brazil and norther Africa. The disease is not present in Florida and strict quarantines ar enforced to regulate the entry of citrus material into the state. Th presence of the disease in Australia is doubtful.

Etiological studies during 1930s showed that the causal agent coul be transmitted by grafting. For many years, the disease was thought to b caused by a virus due to its graft transmissibility and non-isolation of a pathogen. The suppression of the disease by tetracycline and not by peni cillin demonstrated by Igwegbe and Calavan (1970) and later on electro microscopic observations indicating the presence of mycoplasma-like organisms in the phloem element of diseased but not healthy citrus trees indicated that the causal agent is a member of the class Mollicutes. Fudl-Allah *et al.* (1972) cultured an organism from infected citrus trees, followed by its characterization and naming (named as *Spiroplasma citri)* by Saglio *et al.* (1973).

13.45.1. Symptoms

The symptoms of the disease are most pronounced under warm conditions. In cool weather, even highly susceptible cultivars may remain symptomless. The symptoms vary greatly and often some of the symptoms are expressed at one time. The leaves on infected trees are small, cupped, abnormally upright, often mottled and chlorotic.

Excessive defoliation of affected trees occurs in the winter. Mottled appearance and veinal chlorosis of leaves can be confused with symptoms caused by citrus tristeza virus and/or the cachexia viroid (Brown, 1992). Twigs and branches on the affected trees generally show upright growth. Shortening of internodes and production of more shoots may result into witches' broom appearance. Some of the affected twigs die-back and the bark may be abnormally thickened. Wilting of affected branches is a characteristic symptom in the advanced stage of the disease. The disease is rarely lethal but the trees may show severe stunting. The affected trees may bloom prolifically more than once during the season, many fruits drop prematurely leading to reduction in yield.

Affected trees usually produce smaller and fewer fruits, which are generally lopsided and resemble acorns. The rind of such fruits is abnormally thick and green at the base and thin and coloured at the tip. Fruits from affected trees are usually sour, bitter or insipid. The seeds of such fruits are poorly developed and frequently aborted (Agrios, 2004). The symptoms on mature trees are generally less conspicuous. Symptoms of the disease are persistent, remaining evident even when the tree is top-worked with healthy bud wood, hence the name 'stubborn'. Some hosts do not show any symptoms, while others show only mild stunting and deformation of the leaves. Non-citrus hosts may show symptoms of stunting and other deformities. The titre of *Sp. citri* but not the bacterial genotype is, at least in part, responsible for the severity of the disease in affected citrus trees. Quantitative polymerase chain reaction (PCR) showed that the spiroplasma titre was more than 6000 times higher in fruits from severely symptomatic than from mildly symptomatic trees (Mello *et al.*, 2010)

As the symptoms of citrus stubborn are readily confused with other diseases/maladies, the proper identification of disease and its confirmation by greenhouse tests is essential for proper diagnosis of the disease. The diagnosis of the disease in field should be carried out in the warm weather, when the day temperature continuously exceeds 27°C, because the symptoms are most prominent at this temperature. Its confirmation should be done by grafting on indicator plants and by culturing the causal organism. Duncan or Marsh grapefruit and Madam Vinous sweet orange are very good indicator hosts and the grafting should be done on the seedlings of these hosts in a warm greenhouse (day temperature 27 - 35°C and night temperature 21 - 24°C). Three growth flushes should be observed for leaf symptoms (misshapen and mottling) and stunting of seedlings. The confirmation of the disease can also be done by identification of the pathogen on the basis of helical morphology and motility by dark field microscopy and/or serological techniques like Elisa. Nowadays, DNA probes and polymerase chain reaction techniques are routinely used in many laboratories.

The detection of *Sp. citri* by PCR has been done using primers based on gene sequences for spiralin, the most abundant membrane protein of the bacterium. However, these primers lack the sensitivity needed for consistent pathogen detection from field citrus trees. Because the spiralin gene is present as a single copy in the chromosome of all *Sp. citri* strains examined, PCR detection based on multiple-copy membrane protein genes like P89 or P58 will be certainly more efficient. Yokomi *et al.* (2008) developed a PCR technique for the detection of *Sp. citri* using primers based on sequences of the P89 putative adhesion gene and P58 putative adhesion multigene of the bacterium. PCR results matched with those of pathogen culturing 85 - 100% of the time depending on the primers used. Moreover, the method also detected the bacterium from 5 - 10% of trees giving negative results by culturing of the pathogen, suggesting that PCR performed as well or better than pathogen culturing. However, Real-time PCR developed with detection limits estimated to be between 10^{-4} and 10^{-6} ng by serial dilution of a recombinant *Sp. citri* plasmid into DNA extracts from healthy sweet orange proved to be the best method for detection.

13.45.2. Causal organism

Spiroplasma citri Saglio *et al.* 1973.

The bacterium is motile, helical in shape, filamentous and lacks a true cell wall. Helical filaments are usually 100 - 200 nm in diameter and 2 - 4 um in length. Colonies on solid media containing 20% horse serum and 0.8% Noble agar (Difco) are umbonate; 60 - 150 μm in diam. Moderate turbidity is produced in liquid cultures. Acid is produced from glucose and fructose but delayed from mannose. Arginine hydrolysis and phosphatase activity are positive and tetrazolium chloride is anaerobically reduced. Optimum temperature for growth is 32°C. The mol% G + C of the DNA is 25 - 27 (Tm, BD). DNA hybridization tests between *Sp. citri* and other established species of the genus show no significant genetic relatedness (<1%). Its viability is increased by passage through insect vectors (leafhoppers).

13.45.3. Disease cycle

The pathogen survives in infected hosts. It is found in the phloem of infected hosts. The pathogen is spread naturally in citrus orchards by several leafhoppers, like *Scaphytopius nitridus*, *Circulifer tenellus* (beet leafhopper) and *Neoaliturus haemoceps*. The transmission (spread) of the pathogen also occurs through budding and grafting but neither through seed nor mechanically. The natural spread of the pathogen is most rapid in young citrus orchards.

Among the vectors, *C. tenellus* is the most efficient vector in the nature. It has a wide host range including numerous weeds of family Brassicaceae. *Sp. citri* multiplies in *C. tenellus* before its transmission to the host. After ingestion by the leafhopper, the pathogen traverses the gut wall, enters the haemocoel and multiplies during the latent period ranging from days to weeks. Then the cells move to salivary glands where they continue to multiply in the cell cytoplasm. Finally, the pathogen cells are deposited with salvia into the phloem cells of host plants. In some insects, symptoms such as shortened life span, reduction in egg numbers and weakening of flight muscles and other organs are produced. It indicates that the insects also act as hosts for the pathogen. However, there is no evidence of transovarial transmission (Fletcher, 2001).

Colonization of phloem sieve tube cells by the pathogen results in the occlusion and necrosis of these cells. However, the sieve cells are not uniformly invaded; some sieve tubes have no or a few cells while others are completely occluded. Occlusion of tubes interferes with the movement of nutrients leading to manifestation of some of the disease symptoms. Loss of chlorophyll may result due to the breakdown of anthocyanin pigments. Stunting can occur either due to the reduction in the availability of nutrients or variation in the normal balance of plant growth regulators. *Sp. citri* has been reported to cause changes in the levels of both auxins and cytokinins.

13.45.4. Host range

Citrus is the main economic host of *Sp. citri*. Most citrus species and cultivars are infected, some of them symptomlessly. Sweet oranges, tangelos, grapefruit, mandarins and their hybrids show maximum disease severity. Trifoliate oranges and their hybrids, acid limes and lemons show mild symptoms on artificial inoculation. Many dicots and some monocots including weeds (several species of family Brassicaceae) are also the hosts of *Sp. citri* but many of them do not show any symptom. Non-citrus host species include pink periwinkle (*Catharanthus roseus*), London rocket (*Sisymbrium irio*), horseradish (*Armoracia rusticana*), wild radish

(*Raphanus raphanistrum*) turnip, Bermuda grass, Johnson grass, foxglove, rice, peach, cherry, pear, sesame, zinnia, aster and marigold. The bacterium also causes brittle root of cultivated horseradish (Anonymous, 2004).

13.45.5. Management

The following measures should be taken to manage the disease.

1. Enforcement of strict quarantines to prevent the entry of the pathogen in disease-free areas like Australia and Florida state of USA.
2. Use of pathogen-free bud wood and rootstocks (Anonymous, 2003).
3. Early detection, removal and destruction of diseased trees.
4. Destroy all the Brassica weed hosts in field borders and within groves to limit the spread of the pathogen to healthy trees. Potential insect vectors on the weeds should be killed before their removal.
5. As tetracycline treatment of young citrus trees has shown encouraging results, it should be extended to commercial scale to test its viability.

REFERENCES

Agrios, G.N. (2004). *Plant Pathology*. 5th Ed. Elsevier Academic Press, San Diego, 922 pp.

Anonymous (2003). Citrus stubborn disease. University of California, IPM online [http://www.ipm.ucdavis.edu/PMG/r107101211.html].

Anonymous (2004). Exotic threats of citrus: Stubborn disease of citrus. Fact sheet, Plant Health Australia. Online [www.planthealthaustralia.com.au].

Brown, L.G. (1992). Stubborn disease of citrus. Plant Pathology Circular No. 351. Fla. Dept. Agric. and Consumer Services, Division of Plant Industry, Gainesville, FL.

Fletcher, J. (2001). Citrus stubborn disease. In: O.C. Maloy and T.D. Murray (eds). *Encyclopedia of Plant Pathology*. Vol. 1, John Wiley and Sons, Inc., New York. pp. 222 - 24.

Fudl-Allah, A.E.-S.A., Calavan, E.C. and Igwegbe, E.C.K. (1972). Culture of a mycoplasmalike organism associated with stubborn disease of citrus. *Phytopathology* 62: 729 - 31.

Igwegbe, E.C.K. and Calavan, E.C. (1970). Occurrence of mycoplasmalike bodies in phloem of stubborn-infected citrus seedlings. *Phytopathology* 60: 1525 - 26.

Mello, A.F.S., Yokomi, R.K., Melcher, U. *et al.* (2010). Citrus stubborn severity is associated with *Spiroplasma citri* titre but not with bacterial genotype. *Plant Dis.* 94: 75 - 82.

Saglio, P., Lhospital, M., Lafleche, D. *et al.* (1973). *Spiroplasma citri* gen. and sp. n.: A mycoplasma-like organism associated with "stubborn" disease of citrus. *Int. J. Syst. Bacteriol.* 23: 191 - 204.

Yokomi, R.K., Mello, A.F.S., Saponari, M. *et al.* (2008). Polymerase chain reaction-based detection of *Spiroplasma citri* associated with citrus stubborn disease. *Plant Dis.* 92: 253 - 60.

13.46. BACTERIAL CANKER AND GUMMOSIS OF STONE FRUIT TREES

Bacterial canker and gummosis affects mainly stone fruit trees. The disease is most severe on sweet cherry (*Prunus avium*); particularly on young trees and the trees under stress are much more susceptible than healthy trees. The disease is also a major limitation for timber production from wild cherry. Pome fruits like apple and pear, rootstocks and some ornamental trees are also infected but to a lesser extent. Out of *Pseudomonas syringae* pv. *syringae* and *Psm. syringae* pv. *morsprunorum*, the causal agents of the disease, the latter has a limited host range and is confined mainly to tart cherry and plum. The disease also occurs on sour cherry (*P. cerasus*), apricot (*P. armeniaca*), almond (*P. amygdalus*) and peach (*P. persica*). *Psm. syringae* pv. *syringae* has a wide host range and is more virulent on sweet cherry and other stone fruits. The disease is also known by other names like **bud blast, blossom blast, spur and twig blight, sour sap** and **die-back**.

Disease outbreaks are usually associated with prolonged cold wet weather accompanied with late spring frost. *Psm. syringae* pv. *morsprunorum* is widespread in Europe and is present in eastern North America and South Africa. In South Africa, bacterial canker caused by *Psm. syringae* pv. *syringae* is one of the most important diseases of stone fruit trees, and annual losses probably exceed 10 million US dollars (Hattingh *et al.*, 1989).

Third pathogen, namely, *Psm. syringae* pv. *persicae* causing leaf spots, cankers and gummosis of fruits on peach trees has been reported from France. It killed more than one million trees in France. One or more of these pathogens have also been reported from Australia, Canada, India, Japan, Lebanon, UK and USA.

13.46.1. Symptoms

The symptoms appear on all the above ground plant parts. The most characteristic symptom is the formation of cankers accompanied by the exudation of gum. The exudation of gum occurs in most cankers early in the growing season. The gum comes out through injuries or cracks in the bark and runs down on the surface of limbs. New cankers are generally formed in late winter and early spring and these generally develop on trunk, limbs and twigs. Cankers usually develop at the bud union, in pruning wounds or when the pathogen invades woody tissue following infection of blossoms or dormant buds. The spread of the cankers is mostly upward and slightly downwards and to the sides. The cankered areas are slightly sunken and often darker than the surrounding healthy bark. On cherry, the cankers are typically elongated in shape. Cankers expand rapidly in spring and may girdle trunks, branches and twigs. Leaves on the terminal portions of cankered limbs and branches become curled, droop, first turn light green, then yellow and finally die. Within a few weeks, the branch or the plant parts above the cankers also die. The removal of cankered bark exposes the underlying bright orange to brown tissue, often emitting a strong sour smell. The cankers on the shoots also develop from infection occurring through leaf scars. Infected young shoots wilt and die-back from the tip.

On leaves, water-soaked spots appear, which are small, circular to angular in shape and often surrounded by chlorotic haloes in the early stage. Later on, spots become brown, dry and brittle, and eventually fall out forming shot-holes or giving a tattered appearance to the leaves. The spots may also coalesce, especially at the margins of the leaves, to from large patches of dead tissue. In sweet cherry, leaf spots typically occur on the margins of the leaves. However, symptoms on leaves are apparently limited to some fruit cultivars growing in regions where summers are wet and humid (Hattingh and Roos, 1992).

Infected dormant buds turn brown and fail to open or break in spring and this phase is referred to as **"dormant bud blast"** or **"dead bud condition"**. Small cankers often develop at the base of these buds. Cherry, apricot and pear generally suffer heavily from dormant bud blast in certain areas. Some infected dormant buds open in spring but they collapse in early summer exhibiting wilted leaves and dried-up fruits. Both leaf and flower buds are equally damaged. In severe cases, 90% or more buds on a tree may be killed.

The infection of the blossoms occurs in spring and the infected blossoms turn brown to black and die. The blossom infection usually spreads into the fruit-bearing spur causing the collapse of the whole blossom cluster. During full bloom, the poor bloom of the infected trees is quite evident.

On green cherry fruits, the lesions are chocolate-brown, surrounded by water-soaked tissue. As the lesions and fruit age, the affected tissues collapse forming deep black depressions in the flesh with margins turning yellow to red. Elliptical and brown lesions having water-soaked margins appear on fruit stalks. Fruit infections occur sporadically but they can be severe when prolonged wet and cold weather prevails during or shortly after bloom (Jones, 2001).

The symptoms of bacterial canker can be confused with those of Cytospora canker but these can be distinguished on the basis of differences in the canker margin. There is a distinct margin at the top and bottom of Cytospora cankers, while the margins of bacterial cankers are irregular and streaks of discoloured strands frequently extend into healthy tissue.

13.46.2. Causal organisms

1. *Pseudomonas syringae* pv. *syringae* van Hall 1902

Syn. *Psm. syringae* pv. *japonica*

2. *Psm. syringae* pv. *morsprunorum* (Wormald 1931) Young *et al.* 1978

3. *Psm. syringae* pv. *persicae* (Prunier *et al.* 1970) Young *et al.* 1978

Syn. *Psm. morsprunorum* f. sp. *persicae*

The cells are straight rods, measure 0.7 - 1.2 × 1.5 μm, Gram-negative, aerobic, motile having polar flagella and do not accumulate poly-β-hydroxybutyrate as a carbon reserve. The bacteria give negative reaction for oxidase and arginine dihydrolase. Colonies on King's medium B are circular, smooth and glistening with entire or lobate margins, and produce light green fluorescent pigment. Both the pathovars can be differentiated from each other on the basis of GATTa (gelatin liquefaction, esculin hydrolysis, tyrosinase activity and tartrate utilization) scheme. *Psm. syringae* pv. *syringae* is positive for the first two tests and negative for the last two tests while reverse is true for *Psm. syringae* pv. *morsprunorum.* LOPAT scheme can also be used for the differentiation of these pathovars. Both the pathovars elict hypersensitive reaction on tobacco leaves.

Most strains of *Psm. syringae* pv. *syringae* produce the phytotoxin called syringomycin, which plays a role in the virulence of the pathogen. Two genes, namely, *syrA* and *syrB* involved in the production of this toxin have been identified and analysed in strain B 301D of the bacterium (Xu and Gross, 1988). The study further revealed that *syrA* is a regulatory gene and is required for both syringomycin production and pathogenicity. Both the genes are required for the formation of proteins Sr4 and Sr5, which are believed to be components of the syringomycin synthetase complex. Probes and PCR primers for genes required for syringomycin toxin are highly reliable for detecting pathogenic strains of this pathovar.

Many strains of *Psm. syringae* pv. *syringae* act as ice-nuclei. *Psm. syringae* pv. *syringae* contains plasmid-mediated genes for copper resistance. Many strains of *Psm. syringae* pv. *morsprunorum* contain genes for coronatine production. The complete genome sequencing of *Psm. syringae* pv. *syringae* strain B 728a by Feil *et al.* (2005) revealed that the genome (6.1 mb) contains a circular chromosome and no plasmid. Among the 976 protein-encoding genes, there are genes that encode for syringopeptin, syringomycin, indole acetic acid biosynthesis, arginine degradation and production of ice-nuclei. Certain unique genes such as, for ectoine synthase, DNA repair and antibiotic production may contribute to the epiphytic fitness and stress tolerance of the bacterium.

Third pathogen, *Psm. syringae* pv. *persicae* is more closely related to *Psm. syringae* pv. *morsprunorum.* It does not produce fluorescent pigment on King's medium B but it does on casamino acid sucrose agar.

13.46.3. Disease cycle

The bacteria perpetuate in the margins of active cankers, infected buds and systemically in the vascular systems of certain hosts. The bacteria also survive epiphytically on blossoms, buds and leaves of healthy and infected plants and also on nonhost plants including weeds. The bacterial population on blossoms can build up to $10^{4\text{-}6}$ cfu/blossom (Kennelly *et al.,* 2007). In fact, the concept of a plant pathogen surviving as an epiphyte on healthy plant tissue was first proposed in studies of *Psm. syringae* pv. *morsprunorum.* Epiphytic populations of these bacteria also establish on the surface of leaves. The bacteria present in the buds colonize the new leaves as they emerge in spring. Environmental conditions influence the multiplication of bacteria causing fluctuations in population levels. During adverse atmospheric conditions like warm, dry spells, epiphytic bacteria enter through stomata and colonize the intercellular spaces of spongy parenchyma. On reaching the uninvaded stomatal chambers, they multiply profusely and the cell masses extrude through stomata. It appears that epiphytic populations are constantly replenished in this way (Hattingh and Roos, 1992). In plum, *Psm. syringae* pv. *syringae* spreads from colonized blossoms, which are not killed, to the developing seeds and survives in the seed.

The dissemination of the bacteria occurs through wind-splashed rain, water splashes, pruning tools, and infected budwoods and nursery plants. The penetration of the bacteria occurs through wounds, stomata and injuries caused by frost and severe storms on blossoms, buds and leaves. Most recently, Spotts *et al.* (2010) showed that all the seven types of injuries and wounds, whether natural or operational, facilitated the infection of sweet cherry by *Psm. syringae* pv. *syringae.* Infection of inoculated wounds made in spring and summer (heading cuts when trees were planted, scoring cuts and summer pruning) resulted in maximum canker incidence and severity. Inoculation of heading cuts resulted in the highest tree mortality of 86%. Bacterial cankers did not develop in summer or winter when active cankers were cut and the same pruning tool was used immediately to make heading cuts on healthy trees. Heading cuts became resistant to infection after 1 week in summer and 3 weeks in winter.

The infection and progress of disease is greatly influenced by environmental factors. Free water on leaf surface and high relative humidity are essential for at least 24 hr for significant leaf infection to occur. Symptoms appear after about 5 days at 21 - 27°C. The disease is usually severe after cold winters and prolonged spring rains. A warning system based on temperature and free moisture conditions for the prediction of blossom blight infection in the field has been developed (Latorre *et al.*, 2002). The development of the cankers is governed largely by the weather and the growth of the host. The cankers expand in the fall and winter but the maximum development occurs in the period between the end of cold weather and the beginning of the rapid tree growth in the spring. The active growth of the host and high temperature in the spring result in the formation of callus tissue around the cankers, which checks the further advance of the cankers. Some cankers are inactivated permanently while others restart the growth when favourable conditions for the growth of bacteria return. Fresh infections during the active growing season hardly cause any significant loss as their development is quickly arrested by the callus tissue.

Frost injury predisposes the host tissue to infection but the infection is governed by the extent of wet weather during the thawing period. Many strains of *Psm. syringae* pv. *syringae* act as ice-nuclei and hence cause frost injury to plants by forming ice at relatively higher freezing temperatures. These bacteria also produce bacteriocins, which inhibit the growth of non-ice nucleation-active bacteria, thereby making their own growth less competitive. Frost injury predisposes the buds, flowers and young leaves to infection. Infection of buds starts at the base of outer bud scales and spreads killing the bud tissue leading to their death. The infection may spread downwards killing the stem tissues around the base of bud. The bacteria penetrate flowers through natural openings and wounds, and spread quickly through floral parts under humid conditions. The infection may advance into the spur and further to the twig. Young and succulent leaves are frequently infected during cool and wet springs and the bacteria penetrate through stomata. Bacteria spread intercellularly and cause collapse and death of cells leading to formation of necrotic and angular spots. The infections on mature leaves are rare. Under humid conditions, the bacteria ooze out from infected plant parts through stomata, lenticels and wounds formed in the infected tissue and are spread to other plant parts by various agencies.

The limbs generally get infected during the fall and early winter. The entry of bacteria occurs through natural and pruning wounds, leaf scars and through the bases of infected buds and spurs. After entry into the host, bacteria multiply and move intercellularly into the bark and then further into the ray parenchyma of the phloem and xylem. In advanced stages of infection, bacteria breakdown parenchyma cells forming cavities full of bacterial cells. Occasionally, the bacteria invade the xylem vessels but they do not move for long distances in these vessels (Agrios, 2004).

13.46.4. Management

Currently, the effective management of the disease is not possible due to the lack of effective chemicals and biocontrol agents, lack of host resistance and the endophytic nature of the pathogen during some phases of the diseases cycle. However, the following cultural practices and protective measures will certainly reduce the incidence and severity of the disease.

1. Most if not all, factor that reduce the vigour of trees probably favour disease development. Hence, avoiding or reducing the effect of these stress factors will certainly minimise the severity of the disease.
2. Plant only the healthy nursery trees in the orchards.
3. Use only the healthy budwood for propagation.
4. Susceptible varieties should be propagated on rootstocks resistant to the disease and grafting should be as high as possible. Mazzard seedlings, particularly the F12-1, are the most resistant rootstocks.
5. Prune the cankered limbs at least 15 cm below the visible cankers. Pruning should not be done in early spring and fall, as the bacteria are active during this period. Pruning tools should be disinfected frequently with 10% clorox (household bleach) or shellack thinner (70% ethyl alcohol). Cankers on trunks and large branches can be removed by cauterization with a hand held propane burner. The flame is directed particularly at the margins of the canker till the under lying tissue crackles and chars. It has been used with some success in New Zealand.
6. Destroy badly infected young plantings.
7. Protective sprays of fixed copper compounds (copper oxychloride @ 8 - 12 lb/A) or Bordeaux mixture (12: 12: 100) has resulted in partial control of the disease in nursery and orchards. However, pathogen strains resistant to copper are common in orchards in Pacific Northwest and other areas with a history of copper use. Moreover, copper is phytotoxic to most stone fruits, particularly

to sweet cherry and its repeated applications aggravate it further. Cherry nursery plants should be sprayed before digging for planting, and in the orchard; sprays should be given in the fall and in the early spring before blossoming. Airblast sprayer should be preferred over other sprayers as it would blow off all except terminal leaves from the trees and thereby reduce the chances of leaf infection through leaf scars.

8. Acibenzolar-S-methyl (ASM, sold under the trade name, Actigard in the United States and Bion in Europe) is a compound that enhances the natural defences of plants by inducing systemic acquired resistance. Monthly applications of ASM moderately reduced bacterial decline of hazelnut caused by *Psm. syringae* pv. *syringae* in Italy.
9. Avoid ring nematodes infested soils for planting orchards, as the high population of these nematodes is generally associated with high disease incidence.

REFERENCES

Agrios, G.N. (2004). *Plant Pathology*. 5th Ed. Elsevier Academic Press, San Diego, 922 pp.

Feil, H., Feil, W.S., Chain, P. *et al.* (2005). Comparison of the complete genome sequences of *Pseudomonas syringae* pv. *syringae* B728a and pv. *tomato* DC 3000. *Proc. Natl. Acad. Sci.* USA, 102: 11064 - 69.

Hattingh, M.J. and, Roos, I.M.M. (1992). Bacterial canker of stone fruits. In: J. Kumar, H.S. Chaube, U.S. Singh and A.N. Mukhopadhyay (eds). *Plant Diseases of International Importance.* Vol. III, *Diseases of Fruit Crops.* Prentice Hall, Englewood Cliffs, New Jersey, pp. 394 - 404.

Hattingh, M.J., Roos, I.M.M. and Mansvelt, E.L. (1989). Infection and systemic invasion of deciduous fruit trees by *Pseudomonas syringae* in South Africa. *Plant Dis.* 73: 784 - 89.

Jones, A.L. (2001). Bacterial canker of stone fruit trees. In: O.C. Maloy and T.D. Murray (eds). *Encyclopedia of Plant Pathology*. Vol. 1, John Wiley and Sons, Inc., New York, pp. 74 - 76.

Kennelly, M.M., Cazorla, F.M., de Vicente, A. *et al.* (2007). *Pseudomonas syringae* diseases of fruit trees: Progress toward understanding and control. *Plant Dis.* 91: 4 - 17.

Latorre, B.A., Rioja, M.E. and Lillo, C. (2002). The effect of temperature on infection and a warning system for pear blossom blast caused by *Pseudomonas syringae* pv. *syringae. Crop. Prot.* 21: 33 - 39.

Spotts, R.A., Wallis, K.M., Serdani, M. *et al.* (2010). Bacterial canker of sweet cherry in Oregon– infection of horticultural and natural wounds, and resistance of cultivar and rootstock combinations. *Plant Dis.* 94: 345 - 50.

Xu, G.W. and Gross, D.C. (1988). Physical and functional analyses of the *syrA* and *syrB* genes involved in syringomycin production by *Pseudomonas syringae* pv. *syringae*. *J. Bacteriol.* 170: 5680 - 88.

13.47. BACTERIAL CANKER OF MANGO

The disease is also called **bacterial black spot of mango** (*Mangifera indica* L.). Doidge in 1909 reported the disease for the first time from South Africa. Since then, it has been reported from Australia, the Comoro Islands, Kenya, Japan, Malaysia, Mauritius, New California, Pakistan, the Philippines, Reunion Island, Taiwan, Thailand and the United Arab Emirates. The disease probably occurs throughout southern and eastern Africa and in most parts of Asia. The disease also occurs in Brazil (Gagnevin and Pruvost, 2001). In India, it was first reported by Patel *et al.* (1948) as bacterial leaf spot and is prevalent in Uttar Pradesh, Bihar, Maharashtra, Tamil Nadu, Delhi, Haryana and Punjab.

The disease is a limiting factor in mango production because it is very difficult to control the disease, particularly on susceptible varieties. It can be very destructive in areas where high temperature and rainfall occur concomitantly. Most of the commercial varieties are highly susceptible to the pathogen and the infection results in drastic yield losses. These losses result from premature fruit drop, reduction in fruit quality, severe defoliation of plants and fruit rot in storage. Insect and fungal invasions through fruit cankers further aggravate the loss. The fruit infections ranging from 50 - 80% are common on highly susceptible cultivars. In 1966 and 1967, severe epiphytotics of the disease caused almost 100% fruit loss on most susceptible cultivars in most of the mango growing areas of South Africa. The fruit loss was mainly due to the fruit drop and non-marketability of the fruit. The estimated loss alone in 1966 was approximately US $ 1 million (Gagnevin and Pruvost, 2001).

In India, the pathogen has been observed to cause 20 - 50% fruit infection in cultivar Langra and 10 - 85% in seedling mangoes under natural conditions. In hybrids and cv. Bangalora, 37 - 40% fruits showed infection, of which 20 - 25% dropped. In many hybrids, even 100% fruit infection was observed. The infected fruits, including those invaded by insects and fungi rot in the orchard during ripening (Shekhawat and Patel, 1972). Kishun (1982) observed 20 - 80, 10 - 70, 10 - 40 and 10 - 55% fruit drop in Alphanso, Pairi, Totapuri and local cultivars, respectively and the maximum fruit drop in all the cultivars occurred 30 - 60 days after fruit set. He further reported that the fruit rot in storage was 5 - 80, 11 - 67, 10 - 100 and 5 - 80% in Alphanso, Pairi, Totapuri and local varieties, respectively.

13.47.1. Symptoms

The symptoms appear on the aerial parts of the mango trees. Leaf and fruit symptoms are most common but symptoms on twigs and branches occur on highly susceptible cultivars or when the infection is very severe. On leaves, the symptoms appear as mintute, water-soaked, irregular spots, which enlarge, turn brown or dark brown, become angular due to delineation by veins and finally cankerous and raised. Individual spots are usually smaller than 0.5 cm^2. Most of the lesions are surrounded by chlorotic haloes, which are more pronounced on young leaves and may extend 5 mm beyond the necrotic areas. On old leaves, the haloes are narrow and can be best seen when the leaf is held against light. Several lesions may coalesce to form larger necrotic cankerous patches. In old lesions, the upper epidermis becomes loose, greyish white and sometimes peels-off. Severe leaf infection may result in yellowing of the leaves and defoliation (Fig. 13.47).

On fruits, the symptoms appear as small, water-soaked spots on lenticels. These spots later become star-shaped, develop into cankers, which are much more raised and darker than those on leaves, and often burst open releasing a highly contagious gummy substance containing bacterial cells. Insect larvae and fungi enter fruits through the openings of these lesions and cause rotting and fruit drop. Often, a **"tear stain"** infection pattern is observed on the fruit. Many times fruits in bunches look healthy but their close examination reveals the presence of lesions at the points where they touch each other.

On twigs and branches, the lesions are raised having longitudinal rifts, predominantly formed on the under surface but without any gummy ooze. Several lesions may run together forming longitudinal scars. The disease is not systemic. Bud necrosis and inflorescence cankers are rarely observed and only on highly susceptible cultivars.

Susceptibility of the plant organs varies with time for a given cultivar. Young leaves are resistant, likely because they do not have functional stomata. However, they become very susceptible when they enlarge, with lesions appearing just after leaf hardening. Leaf susceptibility then decreases with time. Fruit susceptibility increases over time and is highest during the month preceding harvest. This correlates with maximum receptivity of the lenticels.

13.47.2. Causal organism

Xanthomonas citri pv. *mangiferaeindicae* (Patel *et al.* 1948) Ah-You *et al.* 2009

Syn. *Xa. campestris* pv. *mangiferaeindicae, Xa. axonopodis* pv. *mangiferaeindicae*

Patel *et al.* (1948) first described the bacterium and named it *Pseudomonas mangiferae-indicae* sp. nov. Robbs *et al.* (1974) reclassified the bacterium as *Xa. campestris* subsp. *mangiferae-indicae* sp. nov. With the introduction of pathovar system for the classification of plant pathogenic bacteria, it was changed to *Xa. campestris* pv. *mangiferaeindicae*. Later on, Ah-You *et al.* (2007) changed it to *Xa. axonopodis* pv. *mangiferaeindicae* and finally to *Xa. citri* pv. *mangiferaeindicae* (Ah-You *et al.*, 2009).

The bacterium has all the characteristics of genus *Xanthomonas* except the production of yellow pigment. It produces non-pigmented (creamy-white) colonies on agar media. However, a few yellow-pigmented strains have been isolated from mango in Brazil, Florida, South Africa and Reunion Island. Atypical strains, including yellow-pigmented strains collected from various countries and the non-pigmented strains isolated from mango in Brazil and from ambarella (*Spondias cytherea* Sonn.) in the French West Indies are phenotypically and genetically distinct from *Xa. citri* pv. *mangiferaeindicae.* The below given description of the bacterium is taken primarily from Manicom and Wallis (1984).

Bacterial cells are Gram-negative, aerobic, nonsporulating rods (0.4 - 0.5 × 1.0 - 1.5 μm) and motile by means of a single polar flagellum. Colonies on nutrient agar or King's medium B are round, shallow, convex having entire margins. The colony colour is initially smoke grey but soon changes to creamy-white. On potato dextrose agar, the colonies are circular, smooth, glistening, pulvinate having entire margins and measure 1.0 - 1.5 cm in diameter after 7 days of incubation. The bacterium is strictly oxidative. The optimum temperature for growth is 27°C but the growth can occur up to 37°C. The thermal death point is around 55°C.

The tests are positive for esculin hydrolysis, caseinase, catalase, cellulase, gelatinase, hydrogen sulphide, levan and lipase production and starch hydrolysis but negative for acetoin production, arginine dihydrolase activity, indole and ammonia production, 2-ketogluconate production, nitrate reduction, cytochrome oxidase, poly-β-hdroxybutyrate synthesis, urease production, growth on asparagine as sole source of C and N, soft rot of potatoes and hypersensitive reaction on tobacco. The bacterium produces acid from glucose, sucrose, xylose, galactose, mannose, D-ribose, trehalose, cellobiose, melibiose, melezitose, esculin, dextrin, glycerol and sorbitol when growth occurs utilizing these compounds. SXTPA, a semi-selective medium was developed by Chand and Kishun (1995) for the isolation of the bacterium. Two semi-selective media, namely, KC and NCMT3 having plating efficiencies ranging from 76 - 104% and from 78 - 132%, respectively have also been developed for the isolation of the bacterium (Pruvost *et al.*, 2005). In the latter case, the repeated isolations of the pathogen were possible even from asymptomatic leaves.

Most of the strains produce heat stable toxin (Kishun and Shukla, 2000). Gagnevin *et al.* (1997) assessed the genetic diversity among 138 strains of the bacterium isolated from mango, ambarella and pepper tree in 14 countries and found that 11 strains did not belong to *Xa. campestris* pv. *mangiferaeindicae*. They further revealed that the remaining 127 strains formed four groups, of which the group with the greatest diversity consisted of strains from Southeast Asia where mango originated. Another group contained only strains isolated from pepper trees in Reunion, indicating that pepper tree may not be a collateral host for bacterial leaf spot pathogen.

Ah-You *et al.* (2007), on the basis of pathogenic variation and host range, divided and reclassified *Xa. campestris* pv. *mangiferaeindicae* into three pathovars, namely, *Xa. axonopodis* pv. *mangiferaeindicae* pathogenic on mango and cashew, *Xa. axonopodis* pv. *anacardii* pathogenic on cashew and *Xa. axonopodis* pv. *spondiae* pathogenic on ambarella and mombin. These three pathovars also correspond to three genetically divergent groups of *Xa. axonopodis* identified by Rademaker on the basis of AFLP data. Most recently, Ah-You *et al.* (2009), using amplified fragment length polymorphism, multilocus sequence analysis and DNA - DNA hybridization for genotypic classification of *Xanthomonas* pathovars associated with the plant family Ancardiaceae, identified causal agents of mango bacterial canker and cashew bacterial spot as *Xa. citri* pv. *mangiferaeindicae* and *Xa. citri* pv. *anacardii*, respectively.

13.47.3. Disease cycle

The bacterium survives in leaf lesions and in cankers formed on the twigs and branches but the survival is more efficient in leaf lesions. Three-month-old leaf lesions on a susceptible cultivar may contain 10^7 cfu per lesion while the population in 18-month-old lesions was 10^5 cfu per lesion. The population of the bacterium in leaf lesions decreases relatively slowly under controlled conditions. The role of twig cankers is also important in the survival of the bacterium as these persist on the trees while the infected leaves prematurely defoliate.

The population of the bacterium on mango tree surfaces can average as high as 10^6 cfu per leaf or fruit. Surface population varies with climatic conditions but is correlated with the number of lesions per leaf or fruit. Humid conditions favour the epiphytic growth. The presence of free water allows the release and redistribution of bacteria from the ruptured epidermis. The population quickly decreases to undectable levels during dry conditions. Populations of $10^{4\text{-}6}$ cfu per fruit can be found on young fruits under natural conditions and probably these constitute inoculum associated with latent infection, which will be responsible for symptoms when the fruit receptivity increases. Kishun and Chand (1993) have reported the epiphytic survival of the bacterium on 18 weeds found in and around the mango orchards.

Long distance dissemination or dissemination to the pathogen-free areas occurs through contaminated or infected propagating plant material. Seed-plant transmission of the bacterium has not been proved. Wind-driven rains and overhead-irrigation are important for local or short distance dissemination of the bacterium. In India, Kishun (1986) reported that *Myllocerus discolor* var. *variegata* (ashy weevil), *Orthaga evadrusalis* (leaf webber) and *Cantheconidia furcellata* (bug) transmitted the pathogen from the diseased to healthy plants. Later on, Kishun and Chand (1989) extended the list of such insects to 22 including the first two insects mentioned above.

The bacterium enters the host through stomata, lenticels and wounds. Under North Indian conditions, the infection on fruits starts by the end of April and the disease progresses considerably even during summer months. On leaves and twigs, fresh infections develop only after the rains and the severity of the disease is maximum in July-August. During winter, the disease remains practically dormant.

13.47.4. Host range

Xa. citri pv. *mangiferaeindicae* have been isolated from cashew (*Anacardium occidentale* L.) in India but these strains were lost. Non-pigmented strains were also isolated from ambarella in the French West Indies and from Brazilian pepper (*Schinus terebinthifolius* Raddi) in Reunion. The ambarella strains are not only taxonomically distinct from mango isolates but are also much less aggressive on mango. In contrast, isolates from Brazilian pepper cannot be distinguished from mango isolates by most phenotypic and genetic analyses but they are also less aggressive on mango compared to mango isolates. Hence, both the ambarella and Brazilian pepper strains do not pose any serious threat to mango.

13.47.5. Management

1. Use of disease-free nursery is the most effective approach for the management of the disease. Both the rootstocks and scion wood used should be free from the pathogen. However, it is very difficult to determine whether the planting material is free from the pathogen or not due to latent infection of the pathogen or presence of its epiphytic population.
2. Good nursery management practices including disinfection of budwood using bactericides, protection of plants from wind-driven rains by the use of windbreaks, destruction of source of inoculum and reduction of epiphytic bacterial populations by copper sprays should be followed.

3. Chemical control alone is generally ineffective but better control of the disease has been achieved by following an integrated approach that includes destroying inoculum sources by pruning, use of windbreaks and drip irrigation and regularly spraying copper fungicides. In South Africa, more than 20 annual sprays of copper are given on susceptible cultivars. There is every likely hood that it will result in the development of pathogen population resistant to the copper. Three sprays of Streptocycline (100 μg/ml) at 15 days intervals, starting when fruits are of pea seed size, have been tried in India and found effective in some cases. Fruit stones used for seedlings should be dipped in Streptocycline solution (100 μg/ml) for four hr before sowing (personal communication from Ram Kishun).
4. Attempts have been made to manage the disease by the use of antagonistic bacteria (Kishun *et al.*, 2006), where the single application of these bacteria reduced the disease.
5. Use of resistant varieties is another very effective method of disease management but unfortunately there is hardly any cultivar with complete resistance to the pathogen. Heidi and Sensation cultivars are partially resistant to the pathogen. In India, Bombay Green has been found resistant to the pathogen under both natural and artificial inoculation conditions.

REFERENCES

Ah-You, N., Gagnevin, L., Chiroleu, F. *et al.* (2007). Pathological variations within *Xanthomonas campestris* pv. *mangiferaeindicae* support its separation into three distinct pathovars that can be distinguished by amplified fragment length polymorphism. *Phytopathology* 97: 1568 - 77.

Ah-You, N., Gagnevin, L., Grimont, P.A.D. *et al.* (2009). Polyphasic characterization of xanthomonads pathogenic to members of the Anacardiaceae and their relatedness to species of *Xanthomonas. Int. J. Syst. Evol. Microbiol.* 59: 306 – 18.

Chand, R. and Kishun, R. (1995). A semi selective medium for the isolation of bacterial pathogen *Xanthomonas campestris* pv. *mangiferaeindicae. African Plant Prot.* 1(1): 45 – 48.

Gagnevin. L., Leach, J.E. and Pruvost, O. (1997). Genomic variability of the *Xanthomonas* pathovar *mangiferaeindicae*, agent of mango bacterial black spot. *Appl. Environ. Microbiol.* 63: 246 – 53.

Gagnevin, L. and Pruvost, O. (2001). Epidemiology and control of mango bacterial black spot. *Plant Dis.* 85: 928 - 35.

Kishun, R. (1982). Loss in mango fruit due to bacterial canker *Xanthomonas mangiferaeindicae. Proc. 5th Int. Conf. Plant Path. Bact.* Cali., Colombia, pp. 181 - 84.

Kishun, R. (1986). Role of insects in transmission and survival of *Xanthomona campestris* pv. *mangiferaeindicae. Indian Phytopathol.* 39: 509 - 11.

Kishun, R. and Chand, R. (1989). Mechanical transmission of *Xanthomonas campestris* pv. *mangiferaeindicae* through insects. *Indian J. Plant Pathol.* 7: 11 - 14.

Kishun, R. and Chand, R. (1993). Epiphytic survival of *Xanthomonas campestri* pv. *mangiferaeindicae* on weeds and its role in MBCD. *Plant Dis. Res.* 9: 35 - 40.

Kishun, R., Mishra, D., Ram, R.A. *et al.* (2006). Management of mango bacteria canker through antagonists. *J. Eco-friendly Agric.* 1(1): 54 - 56.

Kishun, R. and Shukla, R.S. (2000). Evaluation of *Xanthomonas campestris* pv *mangiferaeindicae* for toxins and their phytotoxicity. *Proc. Int. Conf. Integrated Plant Dis. Mang. for Sustainable Agriculture*, pp. 743 - 44.

Manicom, B.Q. and Wallis, F.M. (1984). Further characterization of *Xanthomonas campestris* pv. *mangiferaeindicae. Int. J. Syst. Bacteriol.* 34: 77 - 79.

Patel, M.K., Moniz, L. and Kulkarni, Y.S. (1948). A new bacterial disease of *Mangifera indica* L. *Curr. Sci.* 17: 189 - 90.

Pruvost, O., Roumagnac, P., Gaube, C. *et al.* (2005). New media for the semiselective isolation and enumeration of *Xanthomonas campestris* pv. *mangiferaeindicae,* the causal agent of mango bacterial black spot. *J. Appl. Microbiol.* 99: 803 - 15.

Robbs, C.F., Ribeiro, R. de L.D. and Kimura, O. (1974). Sobre el posicao taxonomica de *Pseudomonas mangiferaeindicae* Patel *et al.* 1948, agente causal da "Mancha bacteriana" das folhas da Mangira (*Mangifera indica* L.) Arq. Univ. Fed. Rural Rio de Janeiro 4: 11 - 14.

Shekhawat, G.S. and Patel, P.N. (1972). Bacterial canker of mango. In: P.N. Patel (ed.). *Plant Bacteriology*. Vol. I, *Bacterial Diseases of Plants in India.* Indian Agricultural Research Institute, New Delhi, pp. 192 - 95.

13.48. PIERCE'S DISEASE OF GRAPEVINE

Pierce's disease (hereafter referred to as PD), named after N.B. Pierce, is a lethal disease of grapevines. It was first observed in 1884 around Anaheim when a mysterious epiphytotic affected vines in southern California. Earlier the disease was believed to be caused by a virus. In 1973, a xylem-limited rickettsia-like bacterium was implicated with the disease on the basis of an electron-microscopic examination of the infected plant tissue. Later on in 1978, Davis and associates isolated the bacterium and proved its pathogenicity.

The disease occurs in the United States, Mexico, Argentina, Brazil, Paraguay, Costa Rica, Slovenia and Venezuela. In North America, PD is mainly confined to the areas having mild winters. The disease is less prevalent where winter temperatures are cold. It is present in southern United States of Florida, Texas and California, and in the eastern it extends up to Virginia. In Florida and other Southeastern states, the disease has prohibited the commercial cultivation of European grape varieties. With the introduction and establishment of glassy-winged sharpshooter in California after 1990, status of the disease changed from minor to severe with enormous increase in losses (Anonymous, 2000).

13.48.1. Symptoms

The symptoms of PD observed in summer and fall are more reliable for identification than spring symptoms. The vines infected in spring show first symptom of water stress in midsummer. The margins of leaves become slightly yellow or red, in white and red varieties, respectively. As the infection progresses, leaf margins dry and turn brown in concentric zones (Fig. 13.48). Dried leaf laminas fall, leaving behind dried petioles having burnt-appearing tips, called **matchsticks**, attached to the vines, even after normal leaf fall. The wood of infected canes matures irregularly, forming patches of immature green bark surrounded by mature brown bark and vice versa. Grape clusters on infected vines stop growing, shrivel and dry up. Leaf symptoms may vary with different grape cultivars. In Chenin Blanc and Thompson Seedless, discolouration and scorching of leaves may appear in sectors instead of along the margins (Anonymous, 2006).

Infection gradually spreads from the point of infection to the upper side and more slowly towards the base of the vines. The die-back of shoots and roots occurs. Climatic differences of different regions can influence the severity and timing of the symptoms but not the type. In hot climates, the disease progresses rapidly even under adequate soil moisture conditions due to higher moisture stress. The shoots of infected vines grow slowly and vines deteriorate rapidly.

Infected plants show a delayed and stunted growth in the next spring season. Some canes and spurs may fail to bud out. Newly developed leaves show chlorotic areas between the veins. Late in the season, leaves become scorched and the scorching occurs first on the oldest leaves. Yellow to brown streaks are present in longitudinal and cross sections of the current season wood of the infected canes. The gum formation in vessels and other cells, and formation of tyloses also occur in this wood (Agrios, 2004).

Young vines are more susceptible to disease due to fast movement of the bacteria than the old vines. The length of time for which an infected vine can survive depends on climate, age and variety of vines. One-year-old wine of a susceptible variety like Pinot Noir can die the same year after infection while chronically infected 10-year-old vines of Chenin Blanc or Ruby Cabernet can live for more than 5 years. Krell *et al.* (2006)

reported that diagnosis of the disease based on foliar symptoms was the only consistent indicator of infection in *Xylella fastidiosa*-positive grapevines and samples from the basal portion of a cane increased the probability of its detection. Schaad *et al.* (2002) has highlighted the use of a real-time polymerase chain reaction requiring only one hour for the on site diagnosis of PD from asymptomatic vines.

13.48.2. Causal organism

Xylella fastidiosa subsp. *fastidiosa* Wells *et al.* 1987

Syn. *Xyl. fastidiosa*

In 1973, the pathogen reported to be associated with this disease was referred to as "rickettsia-like" because of its similarities in morphology and ultra structure with members of the Rickettsiaceae. Wells *et al.* (1987) proposed a new genus *Xylella* and included all the strains of fastidious, Gram-negative, xylem-limited bacteria under a single species called *Xyl. fastidiosa.* The following description of the bacterium is taken mainly from Wells *et al.* (1987).

The species epithet *fastidiosa* refers to nutritional fastidiousness, particularly for primary isolation of the bacterium. It does not grow on most common bacteriological media. However, it can be cultured on specialized media like buffered cysteine-yeast extract (BCYE) agar containing charcoal. The PW medium has gained the widest acceptance, and supports the growth of all known strains of *Xyl. fastidiosa.* Almeida *et al.* (2004) has proposed a simpler solid medium for supporting the growth of PD strains of *Xyl. fastidiosa.* Simpson *et al.* (2000) determined the complete genome sequence of clone 9a5C, the causal agent of citrus variegated chlorosis. It was the first plant pathogenic bacterium for which the complete genome was sequenced. The mechanisms associated with pathogenicity and virulence involve toxins, antibiotics, ion sequestration and bacterium-host interactions mediated by a range of proteins.

Cells are Gram-negative, non-flagellate, straight rods measuring 0.25 - 0.35 × 0.9 - 3.5 μm. Cells are predominantly single but sometimes form filamentous strands under certain conditions. It is an obligate aerobe, inhibited by 2.5% carbon dioxide. It is non-pigmented, non-halophilic, Kovac's oxidase negative and catalase positive. The bacterium hydrolyzes gelatin, utilizes hippurate and most strains produce beta-lactamase. It does not ferment D-glucose, does not produce indole and H_2S, and is also negative for beta-galactosidase, lipase, amylase, coagulase and phosphatase tests. Optimum temperature for growth is 26 - 28°C and optimum pH is 6.5 - 6.9. The G + C content of the DNA is 51.0 - 52.4 mol%.

Although Wells *et al.* (1987) recognised only one species to include all the strains, there are definitely different pathotypes, most of which have not been characterized and compared with each other (Hopkins and

Purcell, 2002). Schaad *et al.* (2004) proposed three new subspecies, namely, *Xyl. fastidiosa* subsp. *piercei, Xyl. fastidiosa* subsp. *multiplex* and *Xyl. fastidiosa* subsp. *pauca*. *Xyl. fastidiosa* subsp. *pauca* has not been validated due to the lack of deposition of a culture of the type strain. As the type strain designated for subsp. *piercei* is the type strain of the species, the name *Xyl. fastidiosa* subsp. *fastidiosa* takes priority over *Xyl. fastidiosa* subsp. *piercei*. The study of isolates on the basis of which Schaad *et al.* (2004) proposed three new subspecies of *Xyl. fastidiosa*; strains from grape, alfalfa, almond and maple were classified as *Xyl. fastidiosa* subsp. *piercei,* those from peach, plum, elm, pigeon grape, sycamore and almond as *Xyl. fastidiosa* subsp. *multiplex* and those only from citrus as *Xyl. fastidiosa* subsp. *pauca*.

13.48.3. Disease cycle

After penetration into grapevines, the bacteria multiply for several months in water-conducting system of plants and spread through the xylem system. The multiplication of the bacteria causes blocking of xylem vessels and this blockade is responsible for most of the external symptoms of the disease. Recently, Roper *et al.* (2007) hypothesized on the basis of presence of genes in the genome of *Xyl. fastidiosa* that several cell wall degrading enzymes like beta-1,4 endoglucanases, xylanases, xylosidases and polygalacturonase (PG) produced by *Xyl. fastidiosa* breakdown pit membranes. They further demonstrated that among them, PG encoded by *pglA* gene is a critical virulence factor and is required by *Xyl. fastidiosa* for successful infection of grapevines because a mutant in *pglA* lost the pathogenicity.

The multiplication and movement of the bacterium varies in different plant species. In V*itis vinifera* and wild grapes, the bacterium multiplies and reaches a concentration of 10 million to 1 billion bacterial cells per gram of host tissue and it spreads systemically inside the host. In blackberry, the infection is systemic but the multiplication (1/100 to 1/1000 times less) and movement of the bacterium is slower than in grapevines. In mugwort, the bacterium multiplies without causing any systemic infection.

The bacterium is not seed-borne. The pathogen is transmitted by grafting and insects. The sharpshooters and spittlebugs belonging to subfamily Cicadellinae and family Cercopidae, respectively are the insect vectors of the pathogen. These insects acquire the bacteria from the xylem vessels of the infected vines and transmit them to the healthy plants. Four species of sharpshooters, namely, blue-green (*Graphocephala atropunctata*), green (*Draeculacephala minerva*), red-headed (*Carneocephala fulgida*) and glassy-winged sharpshooter (*Homalodisca coagulata*) are the main vectors of *Xyl. fastidiosa* subsp. *fastidiosa* in North America. In Southeast Europe, *Cicadella viridis and Philaenus spumarius* (meadow spittlebug) are the main vectors. As glassy-winged sharpshooter moves faster and flies longer distances in grapevine yards than the other species of sharpshooters, it is a serious threat to vineyards. It was found in high numbers in 1998-99 in Kern county vineyards of California and in 2001, hundreds of vines were found infected with Pierce's disease.

Some vectors are highly efficient in transmitting the Pierce's disease bacterium as they require less than 100 bacteria per insect for transmission and transmit it immediately after acquiring it from the host. An infectious blue-green sharpshooter has more than 90% chances of transmission. An infectious adult is capable of transmitting the bacterium throughout its life. However, the infectious nymphs transmit the bacterium until they molt and they have to reacquire the bacteria after every molt. *Xyl. fastidiosa* is transmitted from the foregut and foregut lining is shed during molting. Xylem feeding insects require succulent plant tissue for feeding. As grapes are pruned every year and produce a succulent new growth, they serve as good feeding host (Varela, L.G.). Grape leafhoppers are not vectors as they feed on mesophyll cells and phloem tissue.

Some vines may recover from the disease in the first winter following infection but the recovery depends on the variety and time of infection. The vines in which the bacteria have survived the first winter, i.e., remained infected for over a year, their chances of recovery are remote. Infections that occur till June have the highest chances to continue in the next year. However, late infections, i.e., after June by blue-green, green and red-headed sharpshooters are less likely to persist in the following season. However, the late infections caused by glassy-winged sharpshooters survive better because the insect feeds on leaves near the base of cane and also on 2-year-old dormant wood and these infected parts are not removed by pruning.

Recently, Francis *et al.* (2008) reported *Nicotiana tabacum* cv. SR1 (Petite Havana) as an excellent bioassay host to detect the bacterium from infected plants and to investigate the host pathogen interactions. It was a better bioassay host than *N. tabacum* cvs. Havana, RP1 and TNN reported previously.

Rootstock species and hybrids vary greatly in susceptibility to the pathogen. Many rootstock species are resistant to the pathogen but a resistant rootstock does not impart resistance to a susceptible grape variety grafted on it.

13.48.4. Host range

Strains of *Xyl. fastidiosa* subsp. *fastidiosa* causing PD have a very wide host range and it has been further expanded by including the hosts showing positive insect transmission of the bacterium. Hopkins (1989) stated that many grasses, legumes and perennial trees belonging to at least 28 families are infected by the bacterium but most of them do not show any symptom. Hill and Purcell (1997) found that 94 out of 155 tested species were the host of the bacterium. Wistrom and Purcell (2005), on the basis of greenhouse studies, showed that out of 29 plant species tested, 27 were the host of the bacterium. Common sunflower, cocklebur, annual bur-sage, morning glory, horseweed, sacred datura, poison hemlock and faba bean were the most frequently infected hosts. Purcell (2006) has compiled a comprehensive list of host species of PD strains.

There are other strains of the bacterium than PD strain. These strains cause different diseases in different hosts in different regions. These diseases include phony peach disease, alfalfa dwarf disease, citrus variegated chlorosis, plum leaf scald and leaf scorch diseases of coffee, oleander, oak, elm, maple, mulberry, plum and sycamore. Some of these strains also cross infect hosts.

13.48.5. Management

The management of the disease is difficult due to the systemic nature of the pathogen and wide habitats of the vectoring insects. However, the following measures will help in reducing the disease incidence.

1. Insecticidal control of insect vectors in habitats adjacent to the vineyards helps in reducing the disease incidence but it is not viable in highly susceptible cultivars or in vineyards having less than 3-year-old vines.
2. Plant, less susceptible or tolerant cultivars, in vineyards that are near a known or suspected hot spot of the disease.
3. Use systemic insecticides to control insect vectors on vines and also on a 50 - 100 feet wide strip of natural vegetation along the vineyard edge. Tubajika *et al.* (2007) obtained 88%, 78% and 78% disease control with the application of kaolin, harpin and imidacloprid insecticides, respectively in Kern county of California.
4. Insect vectors breeding on irrigated grasses should be controlled by weed control rather than insecticide control.
5. Remove and destroy chronically infected vines, i.e., vines showing disease symptoms for more than one year.
6. Also remove vines showing extensive foliar symptoms and severe die-back even if it is their first year of infection.
7. Early disease detection and vine removal is recommended in vineyards that experience influxes of glassy-winged sharpshooters.
8. Hot water treatment of cuttings and propagation buds at 45°C for 3 hr.

9. *Xyl. fastidiosa* subsp. *fastidiosa* is a quarantined pest (EPPO A1) in many countries. Enforcement of quarantine regulations will prevent the entry of the pathogen in disease-free areas.
10. Tetracycline is effective in reducing the disease incidence but its efficacy and viability needs to be tested at the field level.
11. The cross protection studies carried out by Donald Hopkins at the University of Florida, USA, using a benign strain of PD bacterium, reduced the death of Cabernet Sauvignon vines to less than 20%. Although 20% vine death may not sound that encouraging, it is significant in the sense that the cultivar like Cabernet Sauvignon cannot normally be grown in Florida, the hottest PD spot in the US.
12. Fletcher and Wayadande (2002) have suggested a novel approach of paratransgenesis to control plant diseases caused by *Xyl. fastidiosa.* In this process, non-*Xylella* enteric bacteria, i.e., normally gut bacteria of leafhoppers will be transformed to secrete toxin(s) lethal to *Xyl. fastidiosa* subsp. *fastidiosa.* These transformed bacteria will pass from insect to insect via normal copraphagus activity (ingestion of faecal matter) and these paratransgenic insects will harm the *Xyl. fastidiosa* subsp. *fastidiosa* cells acquired by them.

REFERENCES

Agrios, G.N. (2004). *Plant Pathology.* 5th Ed. Elsevier Academic Press, San Diego, 922 pp.

Almeida, R.P.P., Mann, R. and Purcell, A.H. (2004). *Xylella fastidiosa* cultivation on a minimal solid defined medium. *Curr. Microbiol.* 48: 368 - 72.

Anonymous (2000). Pierce's disease. Hortguard threat data sheet. Department of Agriculture, Government of Western Australia, pp. 10.

Anonymous (2006). Grape Pierce's disease. IPM Programme, University of California. Online [http://www.ipm.ucdavis.edu/PMG/r302101211.html].

Fletcher, J. and Wayadande, A. (2002). Fastidious vascular-colonizing bacteria. APS Education Center, [http://www.apsnet.org/education/introplantpath/pathogengroups/fastidious].

Francis, M., Civerolo, E.L. and Bruening, G. (2008). Improved bioassay of *Xylella fastidiosa* using *Nicotiana tabacum* cultivar SR1. *Plant Dis.* 92: 14 - 20.

Hill, B.L. and Purcell, A.H. (1997). Populations of *Xylella fastidiosa* in plants required for transmission by an efficient vector. *Phytopathology* 87: 1197-1201.

Hopkins, D.L. (1989). *Xylella fastidiosa*: Xylem-limited bacterial pathogen of plants. *Annu. Rev. Phytopathol.* 27: 271 - 90.

Hopkins, D.L. and Purcell, A.H. (2002). *Xylella fastidiosa*: Cause of Pierce's disease of grapevine and other emergent diseases. *Plant Dis.* 86: 1056 - 66.

Krell, R.K., Perring, T.M., Farrar, C.A. *et al.* (2006). Intraplant sampling of grapevines for pierce's disease diagnosis. *Plant Dis.* 90: 351 - 57.

Purcell, A.H. (2006). *Xylella fastidiosa.* A scientific and community internet resource on plant diseases caused by the bacterium *Xylella fastidiosa.* Online [http://nature.berkeley.edu/xylella/central.hosts.htm].

Roper, M.C., Greve, L.C., Warren, J.G. *et al.* (2007). *Xylella fastidiosa* requires polygalacturonase for colonization and pathogenicity in *Vitis vinifera* grapevines. *Mol. Plant-Microbe Interact.* 20: 411 - 19.

Schaad, N.W., Opgenorth, D. and Gaush, P. (2002). Real-time polymerase chain reaction for one-hr on-site diagnosis of Pierce's disease of grape in early season asymptomatic vines. *Phytopathology* 92: 721 - 28.

Schaad, N.W., Postnikova, E., Lacy, G. *et al.* (2004). *Xylella fastidiosa* subspecies: *X. fastidiosa* subsp. *fastidiosa*, subsp. nov., *X. fastidiosa* subsp. *multiplex* subsp. nov. and *X. fastidiosa* subsp. *pauca* subsp. nov. *Syst. Appl. Microbiol.* 27: 290 - 300.

Simpson, A.J.G., Reinach, F.C., Arruda, P. *et al.* (2000). The genome sequence of the plant pathogen, *Xylella fastidiosa. Nature* 406: 151 - 57.

Tubajika K.M., Civerolo, E.L., Puterka, G.J. *et al.* (2007). The effects of kaolin, harpin and imidacloprid on development of Pierce's disease in grape. *Crop Prot.* 26: 92 - 99.

Varela, L.G. Pierce's disease in the North coast. University of California. Cooperative Extension and Statewide IPM Project. Online [http://www.cnr.berkeley.edu/xylella/pd97.html].

Wells, J.M., Raju, B.C., Hung, H.Y. *et al.* (1987). *Xylella fastidiosa* gen. nov., sp. nov: Gram-negative, xylem-limited, fastidious plant bacteria related to *Xanthomonas* spp. *Int. J. Syst. Bacteriol.* 37: 136 - 43.

Wistrom, C. and Purcell, A.H. (2005). The fate of *Xylella fastidiosa* in vineyard weeds and other alternate hosts in California. *Plant Dis.* 89: 994 - 99.

13.49. BACTERIAL CANKER OF GRAPEVINE

Grapevine (*Vitis vinifera* L.) suffers from several bacterial diseases, namely, bacterial blight (*Xylophilus ampelinus*), bacterial necrosis (*Pseudomonas syringae* pv. *syringae*), Pierce's disease (*Xylella fastidiosa* subsp. *fastidiosa*), crown gall (*Agrobacterium vitis*) and bacterial canker (*Xanthomonas campestris* pv. *viticola*). Pierce's disease and crown gall are described elsewhere in the book and bacterial canker is described here. It is also called by another name, i.e., **bacterial leaf spot of grapevine**.

The disease was first observed by Nayudu (1972) in Andhra Pradesh, India. Subsequently, the disease has been reported from other states of India, namely, Punjab, Karnataka, Maharashtra and it also occurs in other grape growing areas of the country. In Punjab, the disease is severe on commercial grown cultivars like Perlette, Anabe-e-shahi and Thompson seedless. In certain areas of Punjab, the disease incidence as high as 93% and disease intensity up to 29% has been reported. The disease also occurs in Brazil, where it was first observed in 1998 in vineyards in the Sumedio of the Sao Francisco river basin area of Pernambuco state. The disease incidence was as high as 100% and in some cases; the yield losses were nearly total (Lima *et al.*, 1999). The disease also occurs in other states of Brazil, namely, Piaui, Bahia, Ceara and Roraima. In Brazil, it is a quarantined disease, officially restricted to these states including Pernambuco (Halfeld-Vieira and Nechet, 2006).

13.49.1. Symptoms

The symptoms are found on all the above ground plant parts. The symptoms first appear on young leaves as small, water-soaked lesions on the lower surface surrounded by yellow haloes. Lesions enlarge; become angular, cankerous and dark brown. Lesions then also appear on the upper leaf surface. Sometimes, the lesions coalesce and cause death of extensive leaf areas. Spots are rough to touch due to drying of gummy bacterial ooze. Cankers on the veins are also common. Vein infection may range from 1 mm long to entire length of the vein. On petioles, the lesions are brown, elongated and cankerous, and measure 1 - 5 mm in length (Fig. 13.49).

On stems, brown to black, elongated, raised cankers appear. In advanced stage, cankers split longitudinally in the centre exposing the vascular discolouration. Severe infection on young twigs and branches results in drying and killing of twigs. Brown to black, irregular to circular, raised, skin-deep cankers appear on the berries. The cankers are mainly formed on the inner berries of the bunch. The symptoms on berries appear near maturity.

In Punjab, India, the disease appears in 3rd or 4th week of March and the progress of the disease continues on the new plant growth till middle of May. From mid May to mid July, the fresh infections do not occur on young plant parts. With the onset of rains, the fresh infections take place from mid July and the disease continues to progress up to September.

13.49.2. Causal organism

Xanthomonas campestris pv. *viticola* (Nayudu 1972) Dye 1978.

Syn. *Pseudomonas viticola*

The colonies are white, convex, mucoid and no fluorescent pigment is produced in culture media. Cells are Gram-negative rods, nonsporing and grow aerobically. The bacterium hydrolyses esculin and starch, and produces acid from trehalose, glucose, cellobiose, melibiose and arabinose;

weakly positive for glycerol. Acid is not produced from dulcitol and the tests for oxidase and urease are also negative. Asparagine is not utilized as a sole source of carbon and nitrogen. The bacterium produces hypersensitive reaction on tomato and tobacco in 48 hours.

On the basis of combined analysis of PCR patterns obtained with primers REP, ERIC and BOX, Trindade *et al.* (2005) found a high degree of similarity between Brazilian strains and Indian type strain NCPPB 2475. Trindade *et al.* (2007) designed primers Xcv1F/Xcv3R and RST2/Xcv3R based on the partial sequence of *hrpB* gene for the detection and identification of the bacterium. The detection of primer RST2/Xcv3R was 104 cfu/ml and the sensitivity of Xcv1F/Xcv3R and RST2/Xcv3R was 10 pg and 1 pg of purified bacterial DNA, respectively. The presence of the bacterium in tissues of grapevine petioles, previously inoculated with it, could not be detected by PCR using macerated extract added directly in the reaction. However, amplification was positive with the introduction of an agar-plating step prior to PCR.

13.49.3. Disease cycle

The bacterium survives mainly in the stem cankers. The survival also occurs in the infected crop residue. The local spread of the bacterium occurs through wind-splashed rain, harvesting and pruning tools. The long distance dissemination of the pathogen occurs through infected propagating material.

13.49.4. Management

1. Use of disease-free propagating material. Hot water treatment of cuttings at 52°C for 30 min reduces the infection.
2. Rouging and drastic pruning of the infected vines. The time of pruning also affects the disease management. Chand *et al.* (1991) found that disease intensity was 76.8% at three locations in Karnataka, India when vines were pruned between 1st and 15th September while it was 5.6% in vineyards pruned after 10th October. Pruning the vines from the second week of October and onwards gave the best results in the grape producing areas of Maharashtra and North Karnataka.
3. Disinfection of harvesting and pruning tools.
4. Use of windbreaks to reduce pathogen dissemination.
5. Give protective sprays of copper and thiocarbamate fungicides (Nascimento and Mariano, 2004). Under Punjab (India) conditions, 4 sprays at 15 days intervals starting from 15th July gave an effective control of the disease. Bordeaux mixture (0.8%) proved most effective followed by Streptocycline (200 μg/ml).

REFERENCES

Chand, R., Patil, B.P. and Kishun, R. (1991). Management of bacterial canker disease (*Xanthomonas campestris* pv. *viticola*) of grapevine (*Vitis vinifera*) by pruning. *Indian J. Agri. Sci.* 61: 220 - 22.

Halfeld-Vieira, B.A. and Nechet, K. de Lima (2006). Bacterial canker of grapevine in Roraima, Brazil. *Fitopatologia Brasileira* 31: 604.

Lima, M.F., Ferreira, M.A.S.V., Moreira, W.A. *et al.* (1999). Bacterial canker of grapevine in Brazil. *Fitopatologia Brasileira* 24: 440 - 43.

Nascimento, A.R.P. and Mariano, R. de L.R. (2004). Bacterial canker of grapevine: Etiology, epidemiology and control strategies. *Ciencia Rural* 34: 301 - 07.

Nayudu, M.V. (1972). *Pseudomonas viticola* sp. nov., incitant of a new bacterial disease of grapevine. *Phytopathol. Z.* 73: 183.

Trindade, L.C., Lima, M.F. and Ferreira, M.A.S.V. (2005). Molecular characterization of Brazilian strains of *Xanthomonas campestris* pv. *viticola* by rep-PCR fingerprinting. *Fitopatologia Brasileira* 30: 46 - 54.

Trindade, L.C. da, Marques, E., Lopes, D.B. *et al.* (2007). Development of a molecular method for detection and identification of *Xanthomonas campestris* pv. *viticola*. *Summa Phytopathologica* 33: 16 - 23.

13.50. BACTERIAL CANKER OF KIWI

Bacterial canker of kiwi (staminate kiwi, *Actinidia deliciosa*) was first reported from California, USA in 1980. Later on in 1983, the causal bacterium was identified as *Pseudomonas syringae*. Serizawa *et al.* (1989) reported the disease on Chinese gooseberry (*A. chinensis*) from Japan in early 1980s, where its severity is one of the limiting factors in kiwifruit production. According to Goto (1992), the disease found in Japan is more destructive and such form of disease has not been found in other regions of the world. In the EPPO region, the disease was first noticed in northern Italy in 1992 where it remained sporadic and with a low incidence for 15 years. However, in 2007/2008 economic losses started occurring, particularly in the Lazio region, and the possible spread of the disease to other kiwifruit producing regions in Italy began to raise concerns. Due to the emerging problem of *P. syringae* pv. *actinidiae* in the Mediterranean region, the EPPO Secretariat decided to add it to the EPPO Alert List. Ferrante and Scortichini (2010) reported a severe epiphytotic of the disease during 2008-09 from central Italy on *A. chinensis* (yellow kiwifruit). In Italy, it is estimated that the economic losses, including impact on trade, due to *P. syringae* pv. *actinidiae* have reached two million euros (Balestra *et al.,* 2009). The disease has also been reported from China, New Zealand and Korea. The disease also poses a serious problem in Korea, where it was first noticed in 1988.

13.50.1. Symptoms

The symptoms appear on almost all the above ground plant parts. The symptoms appear on trunks and limbs from late winter to early spring, after the break in dormancy of vines. The most conspicuous symptom is the red-rusty exudation that appears on buds, joints of branches and leaf, and pruning scars. The bark of affected parts deeply shrivels and dries often, leading to the formation of cracks.

On leaves, the infection is generally limited to young expanding leaves. Initially, small, water-soaked, light green, scattered spots appear on leaf lamina. The spots increase in size (2 - 3 mm in diam), turn brown to dark brown and become angular in shape due to the restrictions imposed by veins. The spots are surrounded by bright yellow haloes. Bacterial ooze is generally formed on the lower surface of leaves. In severe infection, the entire leaf may be blighted and shriveled (Fig. 13.50a).

Water-soaked, dark green, elongated lesions appear on the infected canes. As the infection advances, longitudinal cracks often appear leading to wilting and blightening of the whole shoot and exudation of bacterial ooze (Fig. 13.50b). Removal of the bark usually reveals a brown discolouration of the external vascular tissues and reddening of the tissues beneath lenticels. Infected flowers turn brown and often wither before opening. Infected flowers that open, generally have under-developed petals. Necrotic lesions are also formed on the sepals (Goto, 1992).

Another disease of kiwi, called **bacterial blossom blight** or **blossom rot** has some similarities with bacterial canker. Blossom blight occurs only on young flowers and is characterized by browning of petals and shedding of premature flowers. The severity of the disease increases with rainfall during the flowering period. *Psm. syringae*, *Psm. viridiflava* and *Psm. syringae* pv. *actinidiae* form the causal complex, which causes blossom blight.

13.50.2. Causal organisms

1. *Pseudomonas syringae* pv. *actinidiae* Takikawa *et al.* 1989

Syn. *Psm. syringae* pv. *avellanae*

2. *Psm. syringae* pv. *syringae* van Hall 1902

Syn. *Psm. syringae* pv. *japonica*

13.50.2.1. Psm. syringae pv. actinidiae: It was first identified by Takikawa *et al.* (1989) as a new pathovar of *Psm. syringae*. Ushiyama *et al.* (1992) are of the view that *Psm. syringae* pv. *actinidiae* might have originated from *Actinidia* wild species, namely, *A. arguta* and *A. kolomikta*, which are distributed in northern areas of Japan and show natural infection.

Bacterial cells are Gram-negative, nonsporing rods, motile by polar flagella. It does not produce fluorescent pigment on King's medium B. Neither β-galactosidase and urease are produced, nor gelatin is liquefied. The bacterium does not utilize DL-lactate, caprate, erythritol, xylose, L-histidine and trigonelline (Goto, 1992). Some strains of the bacterium produce phaseolotoxin, as does *Psm. syringae* pv. *phaseolicola*, the causal agent of bean (*Phaseolus vulgaris*) halo blight. However, the Korean strains of the bacterium were found to produce coronatine but not phaseolotoxin (Han *et al.*, 2003). Most recently, Ferrante and Scortichini (2010) reported that none of the strains isolated from central Italy during 2008 – 9 epiphytotic possessed genes coding for phaseolotoxin or coronatine but all had an effector protein, namely, *hopA1*, differentiating them from the strains causing past outbreaks in Japan and Italy. They further found that strains, which caused outbreak in central Italy appears to be distinct from those found previously in Japan, South Korea and Italy.

***13.50.2.2. Psm. syringae pv. syringae*:** In contrast, *Psm. syringae* pv. *syringae*, the bacterium first isolated from kiwi canker in 1983, differs from *Psm. syringae* pv. *actinidiae* in liquefying gelatin and producing fluorescent pigment. Strains of both the pathogens resistant to streptomycin occur in Korea and Japan (Han *et al.*, 2004). Strains of *Psm. syringae* pv. *actinidiae* resistant to copper sulphate occur in Japan and the resistance to copper sulphate operates through copper-trapping system (Masami *et al.*, 2004). Strains of *Psm. syringae* resistant to copper have also been reported from New Zealand (Vanneste and Voyl, 2003).

Rees-George *et al.* (2010) found that several polymerase chain reaction (PCR) primers designed earlier to identify *Psm. syringae* pv. *actinidiae* were not specific. Therefore, they designed two new sets of PCR primers, PsaF1/R2 and PsaF3/R4 to be complementary to a portion of the 16S - 23S rDNA intertranscribed spacer regions. These primers amplified a DNA fragment from strains of *Psm. syringae* pv. *actinidiae* and a strain of the tea pathogen, *Psm. syringae* pv. *theae*, but not from 56 strains of bacteria belonging to six genera and 17 species. It was not possible to distinguish *Psm. syringae* pv. *actinidiae* from the phylogenetically similar *Psm. syringae* pv. *theae* using the ITS, *gyrB, acnB, rpoD, pgi* or *cts* gene regions to design PCR primers. As *Psm. syringae* pv. *theae* is unlikely to be present in kiwifruit, primers PsaF1/R2 and PsaF3/R4 can be used to detect *Psm. syringae* pv. *actinidiae* from kiwifruit tissue.

13.50.3. Disease cycle

The bacteria survive in cankers formed on the infected vines. In Japan, infected *Actinidia* wild spp. (*A. arguta*, *A. kolomikta*) also help in the perpetuation of the bacterium and act as a source of infection for cultivated vines. The bacterial cells ooze out from cankers formed on trunks, limbs and canes, and are distributed by wind-splashed rain and overhead-irrigation. Long distance spread occurs through infected planting material. The bacteria enter the host through stomata, lenticels and wounds caused by pruning and cultural operations. Mature leaves and canes are not infected in summer and even the penetration through wounds does not result in infection. The disease is severe when the mean temperature in early spring remains below 15°C and is coupled with frequent rains. Winter cankers cause more yield loss as they reduce the size of productive vines by damaging the main vine structure. Spring cankers have less direct effect on the yield but are important for the formation of winter cankers in the next year. Four species of *Actinidia*, namely, *A. deliciosa, A. chinensis, A. arguta* and *A. kolomikta* are susceptible to the pathogen. There is no data on the susceptibility of other *Actinidia* species. Observations made in Italy suggest that the damage is more severe on yellow-fleshed kiwifruit, namely, *A. chinensis* cvs. Hort 16A and Jin Tao than on the more widely grown green-fleshed cv. Hayward of *A. deliciosa.*

Li *et al.* (2005) reported that the resistance of cultivars was related to the contents of lignin and soluble sugars in annual twigs and foliage. The contents of both the compounds were higher in resistant cultivars than those of susceptible cultivars. After infection, soluble sugar decreased in both the cultivars but the decrease was much more in susceptible cultivar compared to resistant cultivar. In contrast, the lignin content increased in both the cultivars and the increase in resistant cultivar was much higher in comparison to susceptible cultivar. Balestra *et al.* (2001) reported that staminate kiwi cv. GTH, highly resistant to *Psm. syringae* pv. *syringae*, acquired susceptibility when transformed with *rolABC* of *Agrobacterium rhizogenes*. The susceptibility was inherited by the T1 offspring containing *rolABC* genes, derived by crossing pollen of transgenic *rolABC* GTH with the pistilate cv. Hayward (highly susceptible). The susceptibility was probably correlated to high nitrogen content in the leaves of *rolABC* plants. The high resistance of the staminate GTH was inherited by all the offsprings lacking *rolABC* genes.

13.50.4. Host range

Psm. syringae pv. *actinidiae* naturally infects only kiwi plants, both cultivated and wild. It can infect peach and Mume on artificial inoculation. *Psm. syringae* pv. *syringae* is pathogenic to cowpea and carrots in addition to kiwi.

13.50.5. Management

The following measures are helpful in minimizing the disease incidence.

1. Use of disease-free nursery plants
2. Use of windbreaks to prevent injuries to the vines from strong winds.
3. During pruning, infected vines show bacterial ooze at the cut surface. In such cases, these should be cut back to the healthy portion.
4. During pruning, the scissors should be disinfected regularly with disinfectants like sodium hypochlorite solution.
5. Avoidance or minimum use of overhead-irrigation
6. Streptomycin and copper sulphate are generally used to control the disease in orchards. However, their continuous use has resulted in the development of pathogen strains resistant to both the chemicals and this has created a serious problem in the management of the disease. The strains of *Psm. syringae* pv. *actinidiae* resistant to streptomycin and copper sulphate occur in Japan (Goto *et al.*, 1994; Lee *et al.*, 2005) and those of *Psm. syringae* in New Zealand (Vanneste and Voyl, 2003). Han *et al.* (2004) also showed that Japanese and Korean isolates of *Psm. syringae* pv. *actinidiae* and *Psm syringae* pv. *syringae* harboured *strA-strB* streptomycin resistance genes for a streptomycin-resistance determinant.

REFERENCES

Balestra, G.M., Mazaglia, A., Quattrucci, A. *et al.* (2009). Occurrence of *Pseudomonas syringae* pv. *actinidiae* in Jin Tao kiwi plants in Italy. *Phytopathol. Medit.* 48: 299 – 301.

Balestra, G.M., Rugni and Varvaro, L. (2001). Increased susceptibility to *Pseudomonas syringae* pv. *syringae* and *Pseudomonas viridiflava* of kiwi plants having transgenic *rolABC* genes and its inheritance in the T1 offspring. *Phytopathol. Z.* 149: 189 - 94.

Ferrante, P. and Scortichini, M. (2010). Molecular and phenotypic features of *Pseudomonas syringae* pv. *actinidiae* isolated during recent epidemics of bacterial canker on yellow kiwifruit (*Actinidia chinensis*) in central Italy. *Plant Pathol.* 59: 954 – 62.

Goto, M. (1992). *Fundamentals of Bacterial Plant Pathology*. Academic Press, Inc, San Diego, 342 pp.

Goto, M., Hikota, T., Nakajima, M. *et al.* (1994). Occurrence and properties of copper-resistance in plant pathogenic bacteria. *Ann. Phytopathol. Soc. Japan* 60: 147 - 53.

Han, H.S., Koh, Y.J., Hur, J.-S. *et al.* (2003). Identification and characterization of coronatine-producing *Pseudomonas syringae* pv. *actinidiae. J. Microbiol. Biotechnol.* 13: 110 - 18.

Han, H.S., Koh, Y.J., Hur, J.-S. *et al.* (2004). Occurrence of the *strA-strB* streptomycin resistance genes in *Pseudomonas* species isolated from kiwifruit plants. *J. Microbiol.* 42: 365 - 68.

Lee, J.H., Kim, J.H., Kim, G.H. *et al.* (2005). Comparative analysis of Korean and Japanese strains of *Pseudomonas syringae* pv. *actinidiae* causing bacterial canker of kiwifruit. *Plant Pathol. J.* 21: 119 - 26.

Li, M., Tan, G., Li, Y. *et al.* (2005). Relationship between contents of lignin and soluble sugar in plants of kiwifruit cultivars and their resistance to kiwifruit bacterial canker infected by *Pseudomonas syringae* pv. *actinidiae. Acta Phytophylacica Sinica* 32: 138 - 42.

Masami, N., Masao, G., Katsumi, A. *et al.* (2004). Nucleotide sequence and organization of copper resistance genes from *Pseudomonas syringae* pv. *actinidiae. Eur. J. Plant Pathol.* 110: 223 - 26.

Rees-George, J., Vanneste, J.L., Cornish, D.A. *et al.* (2010). Detection of *Pseudomonas syringae* pv. *actinidiae* using polymerase chain reaction (PCR) primers based on the 16S-23S rDNA intertranscribed spacer region and comparison with PCR primers based on other gene regions. *Plant Pathol.* 59: 453 – 64.

Serizawa, S., Ichikawa, T., Takikawa, Y. *et al.* (1989). Occurrence of bacterial canker of kiwi-fruit in Japan: Description of symptoms, isolation of the pathogen and screening of bactericides. *Ann. Phytopathol. Soc. Japan* 55: 427 - 36.

Takikawa, Y., Serizawa, S., Ichikawa, T. *et al.* (1989). *Pseudomonas syringae* pv. *actinidiae* pv. nov.: The causal bacterium of canker of kiwifruit in Japan. *Ann. Phytopathol. Soc. Japan* 55: 437 - 44.

Ushiyama, K., Kita, N., Aono, N. *et al.* (1992). Bacterial canker disease of wild actinidia plants as the infection source outbreak of bacterial canker of kiwifruit caused by *Pseudomonas syringae* pv. *actinidae. Ann. Phytopathol. Soc. Japan* 58: 426 - 30.

Vanneste, J.L. and Voyl, M.D. (2003). Genetic basis of copper resistance in New Zealand strains of *Pseudomonas syringae. New Zealand Plant Prot.* 56: 109 – 12.

13.51. BACTERIAL BLIGHT OF MULBERRY

Bacterial blight of mulberry (*Morus alba* L.) was first reported from the USA in 1983. Prior to it, it was incompletely described by Boyer and Lambert from France in 1893. It is the earliest record of a disease caused by a *Pseudomonas* species. The occurrence of the disease in Japan was confirmed in 1901. The disease has also been reported from eastern Anatolia region of Turkey where its incidence was nearly 100% in spring 1999 (Sahin *et al.,* 1999). The disease also occurs in Pakistan, Korea DPR, Republic of Korea, China, Brazil, Bulgaria, Canada, Australia, Germany, Hungary, New Zealand, Romania, South Africa, UK, Uganda, Iran, Italy and India. In India, the disease is widespread in Karnataka state where mulberry is widely grown due to the presence of sericulture industry. The disease incidence was 45.9% and the percentage of infested gardens ranged from 80 - 85 in Kollegal area of Karnataka state in rainy season of 1998 (Sharma *et al.,* 2000).

Initially, the symptoms of the disease appear in cool, wet weather of spring season but the severity of the disease increases in rainy season during June to July. The severity of the disease decreases after the rainy season but it appears again in autumn in a mild form.

13.51.1. Symptoms

The symptoms appear on leaves, twigs and buds. On leaves, small, water-soaked, irregular, dark green, sunken spots appear. The spots turn brown to black with yellow borders and translucent edges as they age. The spots enlarge, become angular in shape and often fall out. Elongated, dark brown and slightly sunken spots appear on large veins, midribs and petioles of the leaves. The occurrence of these spots causes curling and distortion of the leaves due to the unbalanced growth of lamina. Diseased leaves at the tips of twigs wilt and later on die. Infected buds may become disfigured as they swell.

Dark brown to black, elongated, more or less sunken streaks appear on twigs. In humid weather, cream-yellowish bacterial ooze exudes from lenticels of the affected parts. The streaks may coalesce to girdle the shoots resulting in the die-back symptoms. Severely infected young trees often appear stunted.

13.51.2. Causal organism

Pseudomonas syringae pv. *mori* (Boyer & Lambert 1893) Young *et al.,* 1978

Syn. *Psm. mori*

Cells are Gram-negative, aerobic rods measuring 0.9 - 1.3 by 1.8 - 4.5 μm and motile with a polar flagellum. Green fluorescent pigment is produced on King's medium B. Colonies on agar media are white, slow growing, smooth, flat with entire margins and becoming undulate. The tests for gelatin liquefaction, nitrate reduction, hydrogen sulphide and oxidase production are negative while those for catalase and levan production are positive. Milk becomes clear with alkaline reaction. Arbutin is hydrolysed and D-gluconate, 2-ketoglutarate, inositol, mannitol, sorbitol and trigonelline are used as a sole source of carbon. DL-lactate, quinate, L-tartrate, erythritol and DL-homoserine are not used as a sole source of carbon. On the basis of fatty acid methyl ester analysis by gas chromatography, the bacterium confirms to *Psm. syringae* identification, with a similarity index ranging from 0.82 - 0.94 (Sahin *et al.,* 1999).

Sharma *et al.* (2000) reported that both *Psm. syringae* pv. *mori* and *Xanthomonas campestris* pv. *mori* were responsible for causing bacterial blight of mulberry and they also reported the synergistic effect of both the bacteria on disease development. There is a need to study the diseases caused by these bacteria individually and to distinguish these diseases on the basis of symptoms. Another report by Nagaraj *et al.* (2001), reporting *Xa. campestris* pv. *moricola* as the cause of bacterial blight of mulberry from Karnataka, India has further added to the confusion regarding the causal agent(s) of the disease. *Xa. campestris* pvs. *mori* and *moricola* are not included in the validly published names of plant pathogenic bacteria enlisted by Young *et al.* (2004). However, Bull *et al.* (2010) have included *Xa. campestris* pv. *mori* in the comprehensive list of names of plant pathogenic bacteria published during 1980 to 2007.

13.51.3. Disease cycle

The bacterium survives in the cankers formed on the twigs. Pruned shoots and fallen leaves on ground, if not removed and destroyed, also serve as a source of inoculum. The bacterium is not transmitted through seed. Bacterial cells oozing from twig lesions are disseminated by wind-splashed rain. Secondary spread of the bacterium occurs through rain from freshly formed lesions.

The increased severity of the disease during rainy season is due to the high humidity and dissemination of the bacterium caused by wind-splashed rain. Gangwar and Thangavelu (1995) found a highly significant positive correlation between disease severity and rainfall and relative humidity but a non-significant correlation with minimum and maximum temperatures. High doses of nitrogenous fertilizers, close spacing of trees, poor field sanitation and wounding of trees also increase the disease severity.

13.51.4. Host range

Mulberry is the only naturally infected host. In addition to *M. alba*, *M. alba* var. *taratarica*, *M. bombycis* (Japanese mulberry), *M. kagayamae*, *M. latifolia*, *M. nigra* (black mulberry) and *M. ruba* (red mulberry) have been reported natural hosts of *Psm. syringae* pv. *mori* (Janse, 2006). On artificial inoculation, the bacterium infects kidney beans (Goto, 1992).

Among the indigenous mulberry varieties, grown in India, S 3, S 13, S 14 and S 1531 are highly resistant and K 2, MR 2, S 54, S 25, S 34, S 1,

CSRS 1, Sujanpur and SLM 1 are resistant to the pathogen. Some of the exotic germplasms, namely, English Black, and *M. australis* are also resistant (Gangwar and Thangavelu, 1995).

13.51.5. Management

1. Use of resistant varieties and healthy planting material are the most effective management practices.
2. Minimize wounds to the limbs and new shoots.
3. Prune the affected shoots and branches in the late dormant season and destroy them. Pruning should be done back to healthy wood and the pruning tools should be disinfected with 10% bleach or 70% alcohol between cuts.
4. Do not apply excessive doses of nitrogenous fertilizers.
5. Plant the trees at proper spacing to provide good air circulation.
6. Spray Kocide 2000 T/N/O @ 0.75 - 3 lb/A as a dormant spray and @ 0.75 - 2 lb/A when new growth starts. The application of Kocide is not recommended on trees grown for edible fruit.

REFERENCES

Bull, C.T., De Boer, S.H., Denny, T.P. *et al.* (2010). Comprehensive list of names of plant pathogenic bacteria, 1980 - 2007. *J. Plant Pathol.* 92: 551 - 92.

Gangwar, S.K. and Thangavelu, K. (1995). Varietal and seasonal incidence of bacterial leaf blight disease of mulberry caused by *Pseudomonas mori* (Boyer *et* Lambert) Stevens. *Indian J. Sericulture* 34: 110 - 13.

Goto, M. (1992). *Fundamentals of Bacterial Plant Pathology*. Academic Press, Inc. San Diego, 342 pp.

Janse, J.D. (2006). *Phytobacteriology: Principles and Practice.* CABI Publishing, Wallingford, UK, 360 pp.

Nagaraj, M.S., Srinivasachary and Khan, A.N.A. (2001). Studies on the management of *Xanthomonas campestris* pv. *moricola* of mulberry. *J. Mycopathol. Res.* 39: 87 - 90.

Sahin, F., Kotan, R. and Donmez, M.F. (1999). First report of bacterial blight of mulberries caused by *Pseudomonas syringae* pv. *mori* in the eastern Anatolia region of Turkey. *Plant Dis.* 83: 1176.

Sharma, D.D., Baqual, M.F., Gupta, V.P. *et al.* (2000). A survey on the occurrence of bacterial blight disease complex in mulberry. *Indian J. Sericulture*, 39: 113 - 16.

Young, J.M., Bull, C.T., De Boer, S.H. *et al.* (2004). Names of plant pathogenic bacteria, 1864 - 2004. Online [http://www.isppweb.org/about_tppb.asp].

13.52. ANGULAR LEAF SPOT OF *PEEPAL*

Peepal (*Ficus religiosa* L.) is a tree, which is generally grown for shade in India. It is also a religious and sacred tree and is worshiped in India. An angular leaf spot of bacterial nature was first observed on this tree in 1968 at the Indian Agricultural Research Institute, New Delhi, India. Later on, it was also observed in Uttar Pradesh, India (Jindal and Patel, 1972) and also occurs in other states such as Punjab, Haryana etc.

13.52.1. Symptoms

Numerous water-soaked, translucent, angular spots appear on the infected leaves. Later on, the spots turn brown, necrotic and are surrounded by chlorotic haloes. Several spots coalesce covering large areas on the leaf lamina and may even cover the entire leaf (Fig. 13.52). Such leaves soon become dry and fall down. In severely infected trees, all the leaves on a tree are generally infected. Occasionally, the infection spreads to leaf veins and petioles. Light brown to dark brown, small lesions may also appear on the twigs.

13.52.2. Causal organism

Xanthomonas campestris pv. *fici* (Cavara 1905) Dye 1978

Syn. *Xa. fici*

Jindal and Patel (1972) reported the bacterium for the first time from *F. religiosa* and named it as *Xanthomonas fici* sp. nov. However, from the recent name of the bacterium given above, it seems that Cavara (1905) also described or named this bacterium earlier. The description of the bacterium given below is taken from Jindal and Patel (1972).

Cells are Gram-negative rods having rounded ends; usually occur singly but occasionally in pairs, non-encapsulated, nonsporing, not acid-fast and motile with a single polar flagellum. The bacterium is strictly aerobic.

The colonies on nutrient agar are circular, raised, glistening and yellow. The bacterium utilizes glucose oxidatively, hydrolyses starch, liquefies gelatin, produces H_2S and ammonia but not indole, acetoin and urease. Asparagine is not utilized as a sole source of carbon and nitrogen. Milk is turned alkaline and peptonized without coagulation, litmus is reduced and methyl red test is negative. The growth is retarted by 2.0% and suppressed by 3.0% sodium chloride. Catalase is produced. Acid but no gas is produced from mannose, sucrose, glucose, galactose, arabinose, cellobiose, lactose, trehalose, fructose, xylose, raffinose, melibiose, glycogen and dextrin. Neither acid nor gas is produced from *meso*-erythrit, ribose, sorbose, rhamnose, sorbitol, adonitol, dulcitol, inositol, mannitol, inulin and salicin. The bacterium utilizes sodium propionate and succinate but not oxalate and tartrate. Unlike other *Xanthomonas* spp., the bacterium reduces nitrate to nitrite.

13.52.3. Disease cycle

As all the leaves on *peepal* trees are shed off in the winter, the bacterium probably survives in lesions formed on branches, growing buds or leaf scars.

Generally, the disease does not appear on new leaves developing in summer months (April to June) until the onset of the monsoon. It indicates that high humidity and spattering rains are required for the multiplication and dissemination of the pathogen. The bacterium enters through stomata and remains confined largely to the parenchymatous tissue (Jindal and Patel, 1972a).

13.52.4. Host range

Jindal and Patel (1972) found that on artificial inoculation, *F. infectoria* and *F. glomerata* were as susceptible as the natural host, *F. religiosa*, whereas *F. carica, F. bengalensis* and *Morus alba* showed small, water-soaked spots, which quickly turned necrotic and did not expand. Hence, they are not considered susceptible hosts.

13.52.5. Management

No information is available on the management of the disease. As the disease is largely parenchymatous, it can be effectively controlled by 2 - 3 sprays of Streptocycline @ 100 μg/ml.

REFERENCES

Jindal, J.K. and Patel, P.N. (1972). Angular leaf spot of pipal (*Ficus religiosa*) due to *Xanthomonas fici* sp. nov. *Indian Phytopathol.* 25: 164 - 66.

Jindal, J.K. and Patel, P.N. (1972a). Angular leaf spot of pipal. In: P.N. Patel (ed.). *Plant Bacteriology*. Vol. I, *Bacterial Diseases of Plants in India*. Indian Agricultural Research Institute, New Delhi, pp. 245 - 46.

INDEX

C

D

S

Y

Figure 13.1a Bacterial blight of rice: Leaf blight phase

Figure 13.1b Bacterial blight of rice: Kresek phase

Figure 13.2 Bacterial leaf streak of rice
(source: www.agribusinessweek.com)

Figure 13.4a Bacterial stripe of barley: Infected plant

Figure 13.4b Bacterial stripe of barley: Stripes present on leaves

Figure 13.5a Bacterial stalk rot of maize: Stalks appear as boiled in water and on the right side, a toppled plant

Figure 13.5b Bacterial stalk rot of maize: Longitudinally splitted stem shows discolouration of the pith (source: www.cropdisease.cropsci.illinois.edu)

Figure 13.7a Stewart's wilt of corn: Wilted plant (courtesy: Robert Wick; www.umassvegetable.org)

Figure 13.7b Stewart's wilt of corn: Leaf blight phase (courtesy: Robert Wick; www.umassvegetable.org)

Figure 13.8a Red stripe of sugarcane: Red stripe symptoms on leaf (source: www.apsnet.org)

Figure 13.8b Red stripe of sugarcane: Top-rot phase (source: www.apsnet.org)

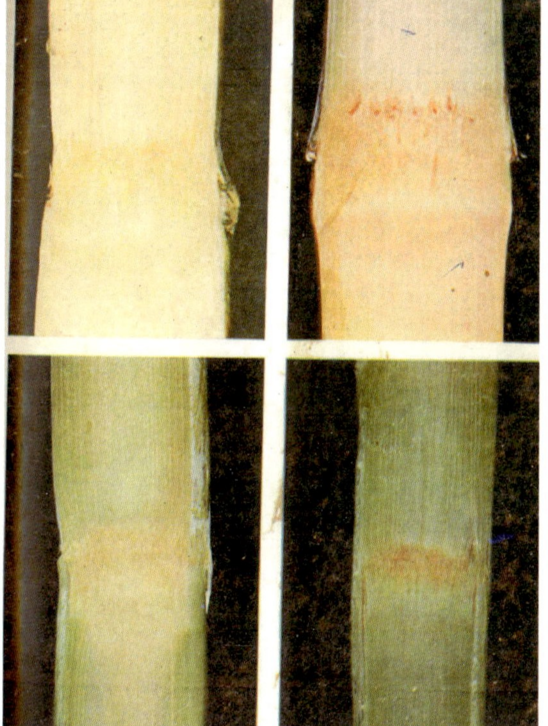

Figure 13.9 Ratoon stunting disease of sugarcane: Internal stalk symptoms.
Upper: Healthy (left) and diseased (right) mature canes
Lower: Healthy (left) and diseased (right) young canes

Figure 13.10 Gummosis of sugarcane

Figure 13.11 Leaf scald of sugarcane

Figure 13.12a Brown rot of potato: Cross section of an infected potato tuber shows greyish discolouration of vascular tissue and bacterial ooze (courtesy: Patrice Champoiseau_UF_Guatemala)

Figure 13.12b Bacterial wilt of tomato: Wilted plants
(courtesy: AVRDC – The World Vegetable Center)

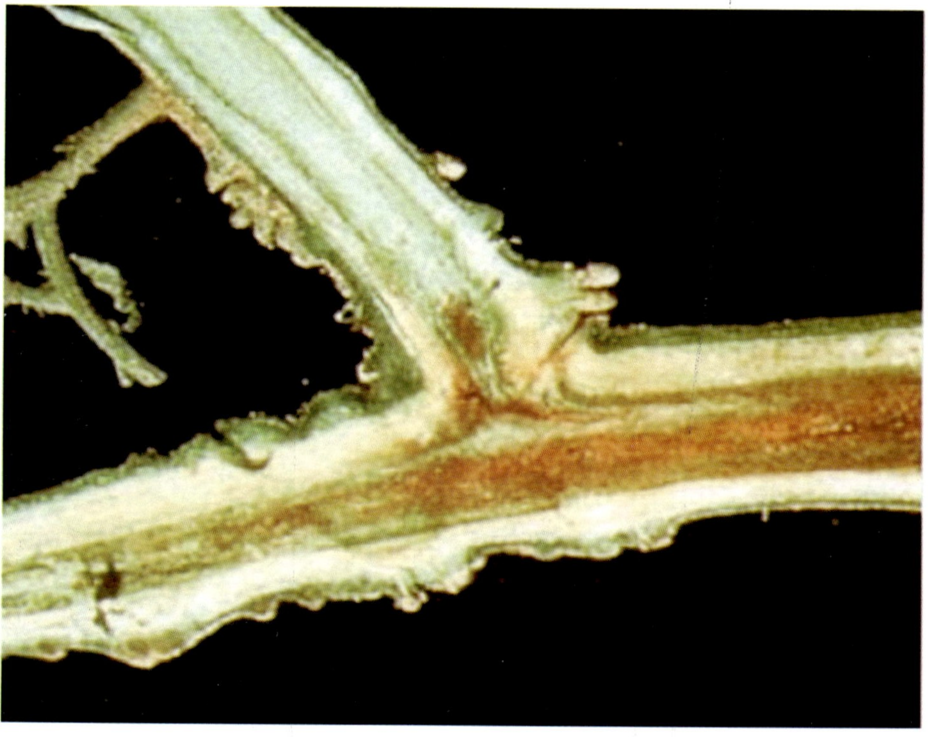

Figure 13.12c Bacterial wilt of plants:
Longitudinally splitted tomato stem shows reddish brown discolouration
(courtesy: AVRDC – The World Vegetable Center)

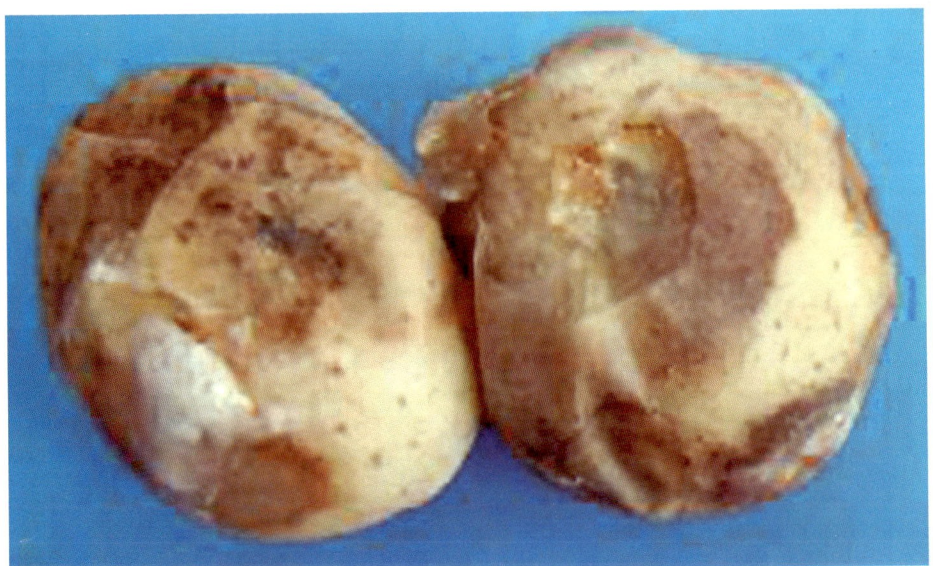

Figure 13.13a Bacterial soft rot: Advanced soft rot symptoms on potato tubers (courtesy: The Queen's Printer for Ontario; www.omafra.gov.on.ca)

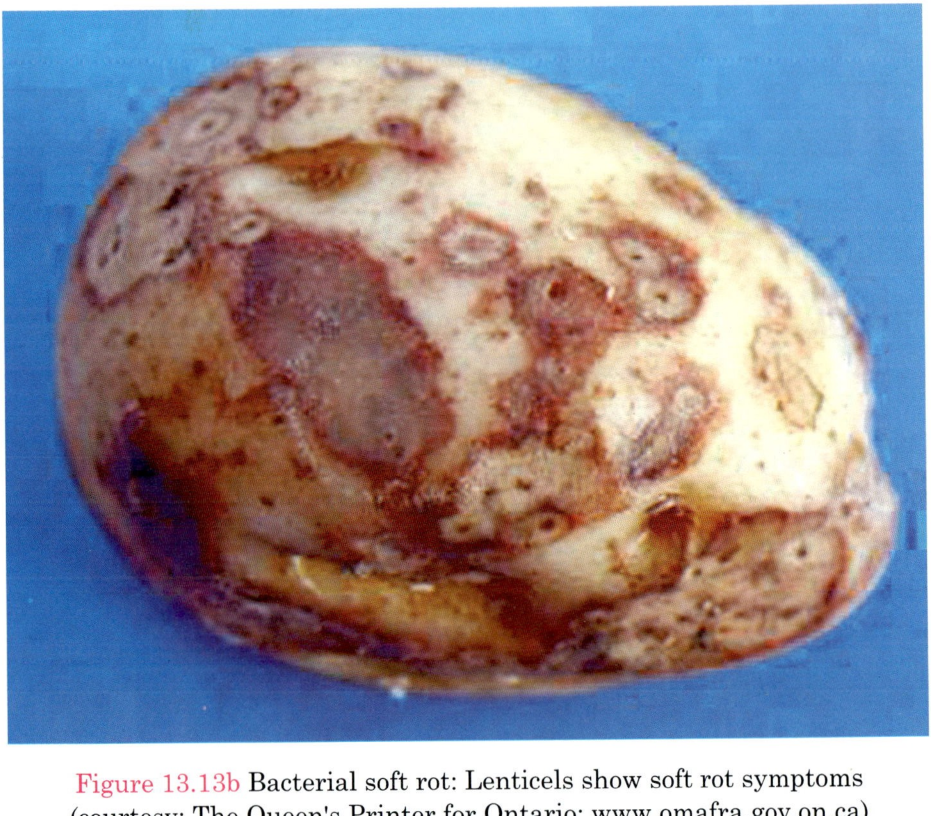

Figure 13.13b Bacterial soft rot: Lenticels show soft rot symptoms (courtesy: The Queen's Printer for Ontario; www.omafra.gov.on.ca)

Figure 13.14 Black leg symptoms on potato plant (right) and tubers (courtesy: Olszowy Walt; www.samdennigan.ieagronomy)

Figure 13.15 Scab lesions on potato tubers

Figure 13.16a Ring rot of potato: Symptoms on aerial plant parts (courtesy: Solke H. De Boer; Agriculture and Agri-Food, Canada)

Figure 13.16b Ring rot of potato: Cross section of a potato tuber showing vascular discolouration (courtesy: Solke H. De Boer; Agriculture and Agri-Food, Canada)

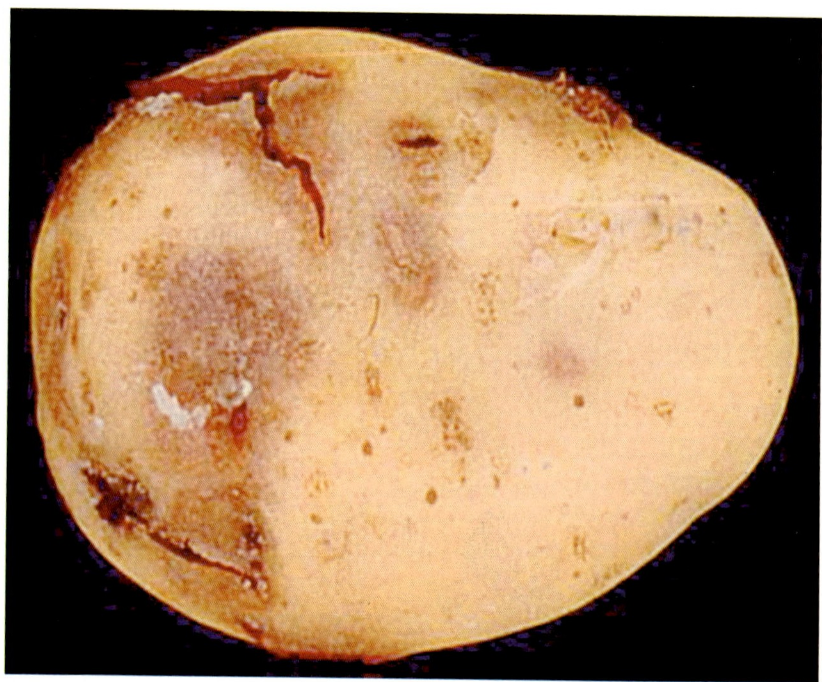

Figure 13.16c Ring rot of potato: Discolouration and cracking of potato tuber skin (courtesy: Solke H. De Boer; Agriculture and Agri-Food, Canada)

Figure 13.17a Black rot of cabbage: V-shaped lesions on field plants (courtesy: Meg McGrath; www.vegetablemdonline.com)

Figure 13.17b Black rot of cauliflower: V-shaped lesions on transplants (courtesy: Sally Miller; www.ohioonline.osu.edu)

Figure 13.18a Bacterial spot of tomato: Symptoms on foliage (source: Illinois Fruit and Vegetable News, Vol. 13, No. 8)

Figure 13.18b Bacterial spot of tomato: Symptoms on fruits (courtesy: M. Babadoost; Illinois Fruit and Vegetable News, Vol. 13, No. 8)

Figure 13.19a Bacterial canker of tomato: Wilted plant showing necrosis and death of leaves (courtesy: Diane Cupples; Agriculture and Agri-food, Canada)

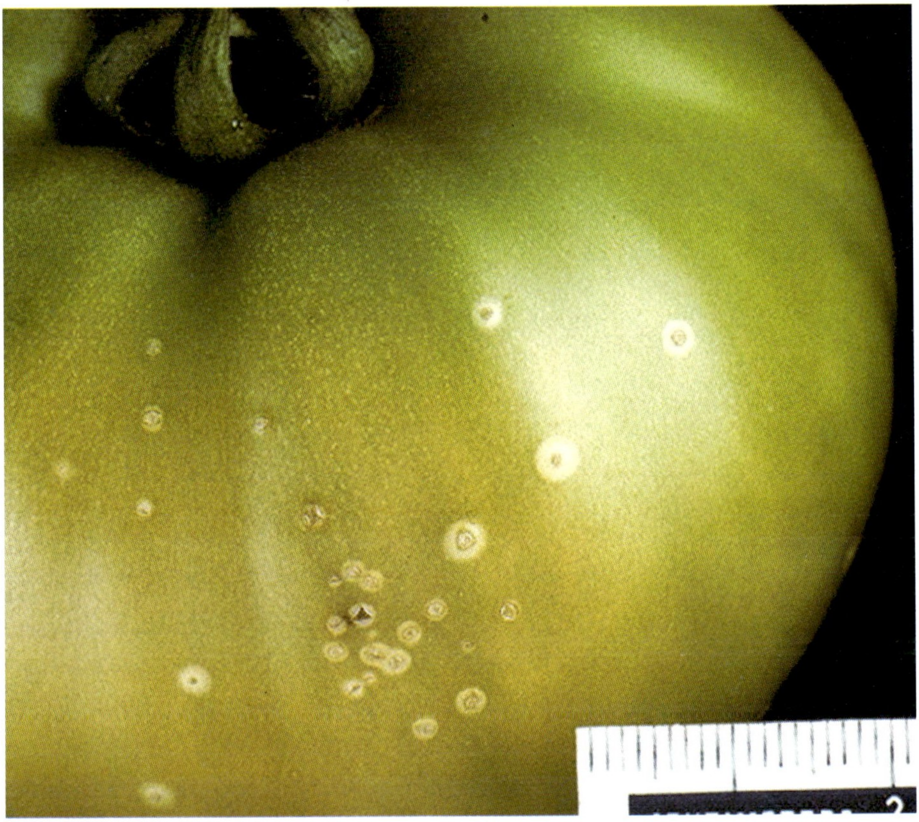

Figure 13.19b Bacterial canker of tomato: Bird's-eye like spots on fruits (courtesy: Diane Cuppels; Agriculture and Agri-Food, Canada)

Figure 13.20a Bacterial speck of tomato: Symptoms on leaf (source: www.aggie-horticulture.tamu.edu)

Figure 13.20b Bacterial speck of tomato: Symptoms on fruits (courtesy: M. Babadoost; Illinois Fruit and Vegetable News, Vol. 13, No. 8)

Figure 13.21a Bacterial blight of pea: Vegetable pea (courtesy: Julien Lanson; www.inra.frhyp3)

Figure 13.21b Bacterial blight of pea: Field pea
(courtesy: Julie Pont; Dept. of Primary Industries, NSW, Australia)

Figure 13.22a Bacterial wilt of cucurbits: Wilted plants
(courtesy: Missouri Botanical Garden)

Figure 13.22b Bacterial wilt of cucurbits: Strands of viscous bacterial ooze drawn from cut stem ends (courtesy: M.P. Hoffmann; www.apsnet.org)

Figure 13.23a Angular leaf spot of cucurbits: Angular spots on leaf (source: www.gov.mb.ca)

Figure 13.23b Angular leaf spot of cucurbits: Symptoms on fruit (source: www.gov.mb.ca)

Figure 13.24a Bacterial blight of cotton: Initial symptoms

Figure 13.24b Bacterial blight of cotton: Advanced stage of symptoms (courtesy: Stephen Allen; www.padil.gov.au)

Figure 13.24c Bacterial blight of cotton: Boll rot
(courtesy: Stephen Allen; www.csiro.au)

Figure 13.25a Common blight of bean: Symptoms on aerial plant parts
(courtesy: R. M. Harveson and H. F. Schwartz;
www.plantmanagement.network.org)

Figure 13.25b Common blight of bean: Symptoms on pods
(courtesy: R. M. Harveson and H. F. Schwartz; www.plantmanagement.network.org)

Figure 13.26a Halo blight of bean: Symptoms on aerial plant parts
(courtesy: R. M. Harveson and H. F. Schwartz; www.plantmanagement.network.org)

Figure 13.26b Halo blight of bean: Symptoms on pod
(courtesy: R. M. Harveson and H. F. Schwartz; www.plantmanagement.network.org)

Figure 13.27a Bacterial brown spot of bean: Symptoms on aerial plant parts
(courtesy: R.M. Harveson and H. F. Schwartz; www.plantmanagement.network.org)

Figure 13.27b Bacterial brown spot of bean: Symptoms on pod
(courtesy: R.M. Harveson and H.F. Schwartz; www.plantmanagement.network.org)

Figure 13.28 Bacterial pustule of soybean: Pustules on upper leaf surface (A), pustules on lower leaf surface (B) and magnified (10X) pustules on lower leaf surface (C) (courtesy: Loren J. Giesler; www.ianrpubs.unl.edu)

Figure 13.29 Bacterial blight of soybean (courtesy: M. Hunziker; www.Bugwood.org)

Figure 13.30 Bacterial blight of cowpea
(courtesy: Oliver Jeffrey; www.infonet-biovision.orgpustule)

Figure 13.31a Bacterial leaf spot of green gram: Symptoms on upper leaf surface

Figure 13.31b Bacterial leaf spot of green gram: Symptoms on lower leaf surface

Figure 13.40 Crown gall of plants
(courtesy: Carol Sheppard; RHS Garden, Wisley)

Figure 13.42a Fire blight of apple: Blossom blight (source: Illinois Fruit and Vegetable News, Vol. 14, No. 7)

Figure 13.42b Fire blight of pear: Symptoms on Bartlett pear fruit (courtesy: Sugar David; www.plant-disease.ippc.orst.edu)

Figure 13.42c Fire blight of apple: Shoot blight (courtesy: Patricia McManus; www.plantpath.wisc.edu)

Figure 13.43a Citrus canker: Leaf spots

Figure 13.43b Citrus canker: Twig canker

Figure 13.43c Citrus canker: Cankers on fruit

Figure 13.44a Citrus huanglongbing: Leaf symptoms

Figure 13.44b Citrus huanglongbing: Yellow shoot
(courtesy: Tim Gottwald; www.apsnet.org)

Figure 13.44c Citrus huanglongbing: Greening on stylar end of fruits (courtesy: Tim Gottwald; www.apsnet.org)

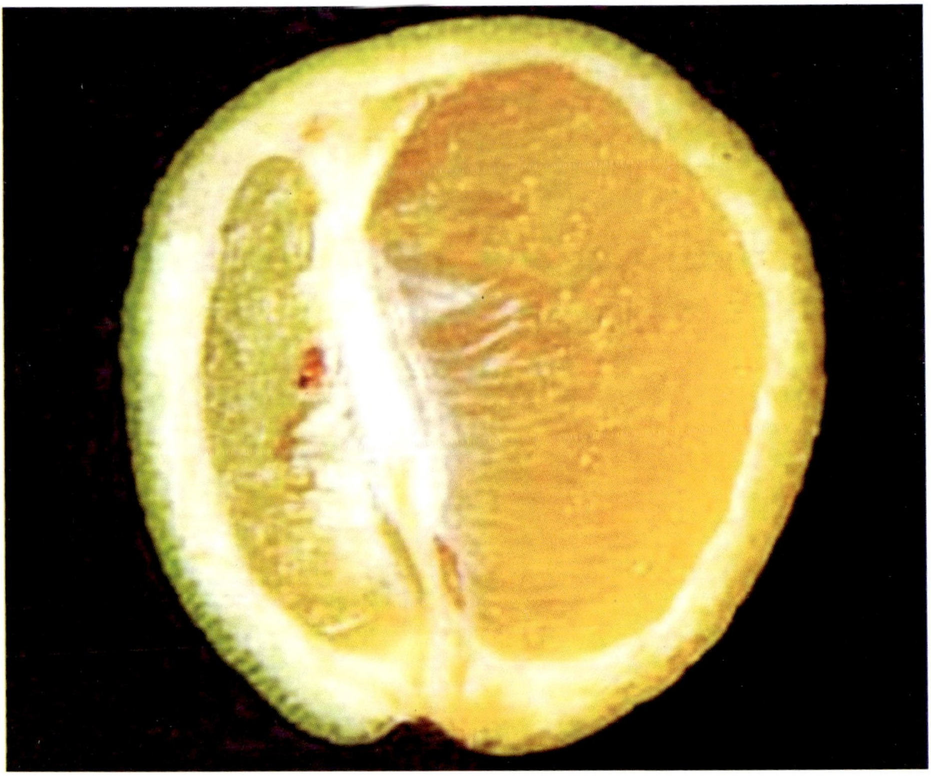

Figure 13.44d Citrus huanglongbing: Asymetrical lopsided fruit development

Figure 13.47 Bacterial canker of mango

Figure 13.48 Pierce's disease of grapevine
(courtesy: Calf. Dept. of Food and Agriculture, PD Control Programme; www.pdgwss.net)

Figure 13.49 Bacterial canker of grapevine

Figure 13.50a Bacterial canker of kiwi: Symptoms on leaf (courtesy: Plant Protection Service of Emilia-Romagna region, Italy)

Figure 13.50b Bacterial canker of kiwi: Stem canker (courtesy: Plant Protection Service of Emilia-Romagna region, Italy)

Figure 13.52 Angular leaf spot of *peepal:* Symptoms on upper leaf surface (left) and on lower leaf surface (right)